FIFTH INTERNATIONAL SYMPOSIUM

ON

CHROMATOGRAPHY

AND

ELECTROPHORESIS

Exclusively published in the United States by
ANN ARBOR — HUMPHREY SCIENCE PUBLISHERS, Inc.
Drawer No. 1425
Ann Arbor, Michigan 48106
—
1969
Ann Arbor and London

UNDER THE AUSPICES OF
THE MINISTRY OF NATIONAL EDUCATION AND CULTURE
(BELGIUM)

Copyright Ann Arbor — Humphrey Science Publishers, Inc. 1969
Drawer No. 1425, Ann Arbor, Michigan 48106

Ann Arbor — Humphrey Science Publishers, Ltd.
5 Great Russell Street, London, W.C.
Sole distributors in the Western Hemisphere
and the British Commonwealth of Nations.

Library of Congress Catalog Card No. 71-85202
SBN 87591 010 6

BELGIAN SOCIETY OF PHARMACEUTICAL SCIENCES
Société Belge des Sciences Pharmaceutiques
Belgisch Genootschap voor Pharmaceutische Wetenschappen

11. rue Archimède
BRUSSELS 4

FIFTH INTERNATIONAL SYMPOSIUM

ON

CHROMATOGRAPHY

AND

ELECTROPHORESIS

Brussels, 16-18 September 1968

ORGANIZING COMMITTEE :

Prof. A. DEFALQUE, Prof. P. DE MOERLOOSE, Mrs. J. DONY, Mrs. C. DORLET, Mrs. J. HANS-BERTEAU, Prof. M. HANS, Prof. A. HEYNDRICKX, Prof. G. LAGRANGE, Prof. Ch. LAPIÈRE, Mrs. J. OSLET, Miss L. RENOZ, Ph. C. A. STANDAERT, Prof. J. THOMAS.

Chairman : Prof. P. DE MOERLOOSE.
Secretaries : Miss L. RENOZ.
         A. STANDAERT.
Treasurers : Mrs. J. HANS-BERTEAU.
         Mrs. J. OSLET.

La Société Belge des Sciences Pharmaceutiques a organisé son Ve Symposium International et le présent volume réunit les conférences et les communications scientifiques.

L'intérêt suscité depuis le premier symposium n'a cessé de croître ; ce qui n'est pas surprenant si l'on considère le nombre de problèmes posés par la recherche biochimique et analytique, et résolus grâce aux techniques chromatographiques et électrophorétiques nouvelles.

Le Comité d'Organisation remercie Messieurs M. Lederer, J. Daussant, J. Meijer et G. Pataki, qui ont répondu spontanément à son invitation. Leurs conférences traitant de recherches personnelles entreprises dans le domaine de la chromatographie et de l'électrophorèse nous ont fourni des informations très importantes. Nous aimerions également féliciter les auteurs dont les communications sur leurs récents travaux sont éditées dans ce volume.

Notre gratitude s'adresse aux autorités académiques de l'Université Libre de Bruxelles dont les locaux furent mis à notre entière disposition, ainsi qu'aux Organismes et Industries pharmaceutiques pour leur appui financier lors de l'organisation de cette réunion internationale.

Nous espérons que tous les participants ont été intéressés par ce Ve Symposium et qu'ils auront gardé un souvenir agréable de leur séjour à Bruxelles.

Merci également à ceux dont la collaboration a hautement contribué au succès de ce Ve Symposium International sur la Chromatographie et l'Electrophorèse.

Le Comité d'Organisation.

Het is reeds de 5ᵉ maal dat het Belgisch Genootschap voor Farmaceutische Wetenschappen dit Internationaal Symposium inricht en de uitgave van de voordrachten en de wetenschappelijke mededelingen verzorgt.

Sedert het ontstaan van dit symposium is er steeds een groeiende belangstelling geweest. Dit hoeft ons niet te verwonderen als men nagaat welke nieuwe aspecten in het biochemisch en analytisch onderzoek tot uiting kwamen dank zij de aanwending van menigvuldige chromatografische en elektroforetische technieken. Dit wordt trouwens bevestigd door de mededelingen die in dit boek voorkomen.

Het inrichtend comité dankt de heren M. Lederer, J. Daussant, J. Meijer en G. Pataki, die onze uitnodiging om een voordracht te houden hebben aanvaard en ons zeer belangrijke gegevens hebben verstrekt over hun persoonlijke onderzoekingen in het gebied van de chromatografie en de elektroforese. Onze dank gaat ook naar de zoekers die een mededeling hebben gehouden en ons belangrijke nieuwigheden hebben bijgebracht. Dank zij hun medewerking is het ons mogelijk geweest dit boekdeel te laten verschijnen.

We danken de Akademische overheid van de Vrije Universiteit te Brussel, die ons de lokalen ter beschikking heeft gesteld ; eveneens gaat onze dank naar de Farmaceutische Bedrijven en Organismen die door hun financiële steun hebben bijgedragen tot het inrichten van dit symposium.

We hopen dat alle deelnemers van dit Vᵉ Symposium een aangenaam en interessant verblijf te Brussel hebben gehad.

Tenslotte gaat onze erkentelijkheid naar allen die medegewerkt hebben aan het inrichten en het welslagen van dit Vᵉ Internationaal Symposium voor Chromatografie en Elektroforese.

Het Inrichtend Comité.

The Belgian Pharmaceutical Science Society has organized its Vth International Symposium and has the pleasure to present herewith the proceedings of the lectures and scientific papers given on this occasion.

The interest created by the first symposium has been increasing ever since. This is not surprising, if we consider the number of problems connected with biochemical and analytical research which have been solved thanks to modern methods applied in the field of chromatography and electrophoresis.

The Organizing Committee would like to thank Messrs M. Lederer, J. Daussant, J. Meijer and G. Pataki, who kindly accepted to attend the symposium. Their lectures have yielded highly important information concerning their personal work in the field of chromatography and electrophoresis. We should also like to convey our gratitude to all scientists who have presented a paper at the conference thereby providing us with fresh information on their latest findings. Due to their collaboration, it has been possible to issue this volume.

We should also like to address our thanks to the academic Authorities of the Free University of Brussels for the opportunity to use their conference rooms and premises, as well as to the pharmaceutical organizations and industries for their financial support in organizing the international symposium.

We hope that all participants at the Vth International Symposium had an interesting and pleasant stay in Brussels.

Finally, we wish to convey our gratefulness to all those who have worked with us to help organize successfully this Vth International Symposium on Chromatography and Electrophoresis.

The Organizing Committee.

Nach der Veranstaltung ihres V. Symposiums hat die Belgische Gesellschaft für pharmazeutische Wissenschaften, die verschiedenen Vorträge und wissenschaftlichen Beiträge in dem vorliegenden Bande zusammengefasst.

Seit der Veranstaltung des 1. Symposiums ist diese Art von Tagung auf immer grösseres Interesse gestossen. Dies ist nicht verwunderlich angesichts der Vielzahl von Problemen, die sich auf den Gebieten der biochemischen und analytischen Forschung stellen und die dank der neuen Techniken der Chromatographie und der Elektrophorese gelöst werden konnten.

Das Organisationskommitee dankt den Herren M. Lederer, J. Daussant, J. Meijer und G. Pataki, dass sie der Einladung Folge geleistet haben. Ihre Vorträge über persönliche Forschungsarbeiten auf den Gebieten der Chromatographie und der Elektrophorese waren höchst aufschlussreich. Wir danken auch den Forschern, die einen Beitrag geliefert und uns von den Ergebnissen ihrer neuesten Arbeiten unterrichtet haben. Auf Grund dieser Zusammenarbeit war es uns möglich, den vorliegenden Band zu veröffentlichen.

Unser Dank gilt ebenfalls den Universitätsbehörden der Freien Universität Brüssel, die uns ihre Räumlichkeiten zur Verfügung gestellt haben. Auch den Pharmazeutischen Verbänden und den Industriebetrieben, die uns bei Veranstaltung dieser internationalen Tagung ihre finanzielle Unterstützung zuteil werden liessen, möchten wir unseren Dank aussprechen.

Wir geben der Hoffnung Ausdruck, dass die Teilnahme am V. Symposium in Brüssel für alle Beteiligten sowohl lehrreich als auch angenehm war.

Schliesslich möchten wir nicht versäumen, bei dieser Gelegenheit allen jenen, die zum guten Gelingen des V. Internationalen Symposiums über Chromatographie und Elektrophorese beigetragen haben, unseren Dank zu sagen.

Das Organisationskommitee.

MEMBRES D'HONNEUR
ERELEDEN

European Research Associates S.A. (Union Carbide).

Bayer-Pharma S.A.

Organon Belge S.A.

Ciba.

Brocades-Belga S.A.

Hoechst Belgium S.A.

Algemene Pharmaceutische Bond — Association Pharmaceutique Belge.

U.C.B. (Union Chimique — Chemische Bedrijven) S.A. — Dipha.

A. Christiaens S.A.

S.A. Upjohn N.V.

Roche S.A.

Janssen Pharmaceutica N.V.

Service de Controle des Médicaments — Dienst voor Geneesmiddelenonderzoek.

Pfizer S.A.

Ets. E. Baudrihaye.

R.I.T.

Phabelgo S.A.

Nestlé S.A.

Laboratoires Abbott.

Ets. A. de Bournonville & Fils.

R. Coles S.A.

Electa.

Federa S.C.

Flandria N.V.

N.V. Lab. Perfecta S.A.

Sandoz S.A.

Specia.

Squibb.

Lab. Tuypens N.V.

Lab. Dumas.

S.M.B. S.A.

Lab. Wolfs p.v.b.a.

CONFERENCES

VOORDRACHTEN

Some recent results in inorganic paper chromatography and electrophoresis

by

M. LEDERER

Laboratorio di Cromatografia del C.N.R.
Istituto di Chimica, Roma, Italy.

In this lecture we would like to survey the present state of inorganic chromatography and electrophoresis.

Actual analytical separations for qualitative and quantitative purposes have been worked out for many groups of substances and for so many special problems that on the whole on may consider this chapter more or less closed. There will be published no doubt improvements of techniques etc. improving or accelerating existing methods for many years to come but the overall picture will not change essentially.

The directions in which improved separations were sought may be summarised as follows :

1) *Partition chromatography on paper.* Most solvents are based either on a ketone (acetone or methylethylketone) or an alcohol (methanol to amyl alcohol) with various additions of acids, complexing agents, water, etc.

2) *Ion exchange papers.* Either cellulose ion exchange papers or papers containing ion exchange resins were used both with aqueous systems and organic solvents.

3) *Papers impregnated with liquid ion exchangers.* The results are very similar to those obtained with ion exchange papers but permit a larger range of capacities and ion exchange groups.

4) *Papers holding inorganic ion exchangers.* These were relative late-comers and have numerous analytical possibilities especially for specific separations. They seem however not to be widely used so far.

5) *Thin layer systems using mainly partition, ion exchange or liquid ion exchangers.* In the hundred odd papers on this topic there are few which describe real improvements in separation. The separation of all rare earths by Holzapfel et al. (1) should receive mention as such.

6) *Paper electrophoresis.* Improved separations and improved reproducibility was obtained with the modern high voltage equipment involving cooling to a constant temperature throughout the experimental run.

The use of cellulose acetate membrane led to the separation of all rare earths in very short times [Buchtela et al. (2)]. The new « free zone electrophoresis » apparatus of Hjertèn (3) is able to produce sharp separations in a matter of minutes. It seems to be the electrophoretic analogue to capillary gas chromatography.

On the other hand the « chromatographic investigation » of a reaction, a procedure which is very usual in biochemistry is now being employed more and more also in inorganic chemistry and its possibilities and scope will be discussed here.

In inorganic investigations we can divide the types of studies which are possible into completely reversible systems and slow reactions, unfortunately chromatography is only of limited use in the intermediate range i.e. in relatively fast reactions.

Reversible reactions such as complexing reactions can be studied by ion exchange paper chromatography, paper electrophoresis etc. and the order of the instability constants estimated roughly in very little time. Such studies reveal quickly the existence of the complex and offer comparative values. Accurate results must then be obtained by equilibrium methods. Numerous slow reactions have been studied by chromatographic and electrophoresis methods and often the intermediate were thus isolated or established for the first time. We shall list some few major fields such as condensed phosphates and arsenates, the halogen complexes of rhodium and ruthenium, the chemistry of thionates.

The lecturer feels that the future of inorganic chromatography and electrophoresis lies in this field and like in many fields of organic chemistry much progress can still be expected by the introduction of suitable separation methods.

To illustrate the scope of chromatographic investigations two examples from the lecturer's laboratory will be discussed ; one dealing with slow reactions and the other with completely reversible systems.

1) *The solution chemistry of technetium.*

Our interest in the solution chemistry of technetium started with the observation that when pertechnetate is dissolved in conc. HCl or conc. HBr, the solution developed on cellulose paper with dilute HCl (or HBr) yields a slow moving spot in addition of the colourless fast moving pertechnetate and the yellow (resp. red for bromide) tetravalent hexachloro (or bromo) technetate (IV).

By running chromatograms at various time intervals after dissolving TcO_4^- in the acid (Fig. 1), it could be established that the slow-moving from must be a valency intermediate between 7 and 4 i.e. 6 or 5 and other evidence has since confirmed our assumption that it is pentavalent.

When increasing the concentration of acid used to elute the Tc (V), the spot increases its Rf value in HCl and HBr but not in $HClO_4$ nor does it separate into various spots (Fig. 2) indicating that both chloro and bromo complexes are formed and that these are reversible.

Careful studies by Shukla (4) of the reduction of TcO_4^- in HCl and HBr at various temperatures showed that it is possible to produce a solution

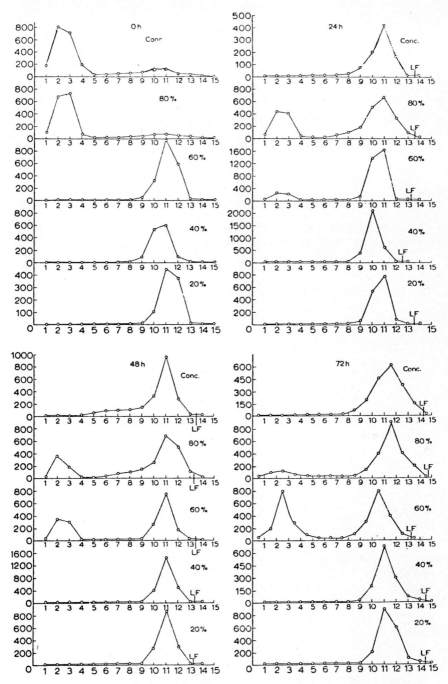

Fig. 1. — The reaction of TcO_4^- with HBr at room temperature. Chromatograms on Whatman N° 3MM paper developed with 0,9 N HBr as solvent. From top to bottom : conc. HBr, 80 % HBr, 60 % HBr 40 % HBr and 20 % HBr. On cellulose paper Tc(V) is separated from all Tc(VII) species. LF = liquid front.

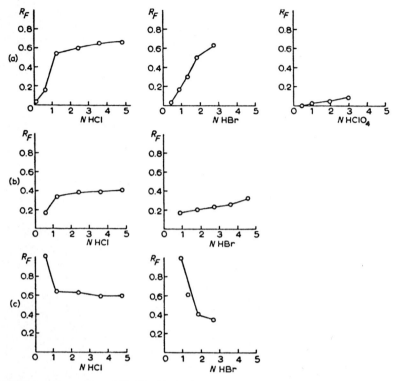

Fig. 2. — R_F values of Tc(V). (a) On Whatman N° 3MM paper with HCl, HBr and HClO$_4$ as eluants ; (b) on Macherey Nagel strong anion cellulose paper with HCl and HBr as eluants ; (c) R_F values on the anion exchange paper « corrected » for adsorption on the cellulose, *i.e.* R_F anion exchange paper/R_F cellulose paper.

of Tc(V) essentially free of Tc(7) and Tc(4) by simply storing it at a temperature near 0° C. Incidentally he also showed that only at low temperatures are strictly quantitative separations obtained on the paper chromatogram (Fig. 3).

While studying the reduction of TcO$_4^-$ in HCl and HBr at various dilutions, good separations were obtained on Macherey, Nagel quarternary ammonium cellulose papers. However in some solutions the Tc(IV) separated into two well-separated spots. By preparing pure TcCl$_6^{--}$ and TcBr$_6^{--}$ and ageing these in e.g. 1.2 NHCl (for the chloride) we could show that several spots are formed with time, all yellow which could thus only be ascribed to hydrolysis products of Tc(IV) i.e. TcCl$_5$H$_2$O$^-$, TcCl$_4$(H$_2$O)$_2^{--}$ etc. (see Fig. 4).

To sum up what we could deduce from paper chromatograms : Pertechnetate reduces in HCl or HBr via an intermediate valency which is very stable in HCl and less so in HBr to form mixtures of Tc(IV) halo-aquo complexes, the bromo-complexes being less stable than the corresponding chloro-complexes. In a given solution thus three valencies and several

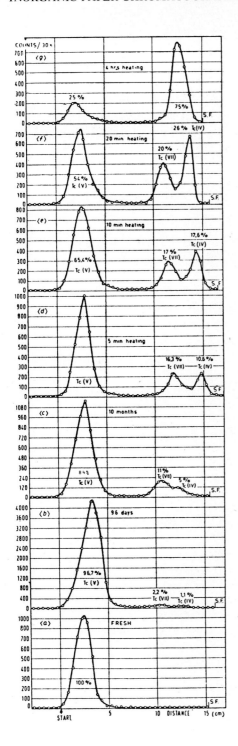

Fig. 3.

Radiochromatograms on HCl-washed Whatman 3MM (pure cellulose) paper at 8° C in 0.6 N HCl: *(a)* Fresh solution of ammonium pertechnetate in conc. HCl at 0° C; *(b)* Solution *(a)* aged for 96 days at 0° C; *(c)* Solution *(a)* aged for 10 months at 0° C; *(d)* Solution *(c)* heated on water bath for 5 minutes; *(e)* Solution *(c)* heated on water bath 10 minutes; *(f)* Solution *(c)* heated on water bath 20 minutes; *(g)* Solution *(c)* heated on water bath for 4 hours.

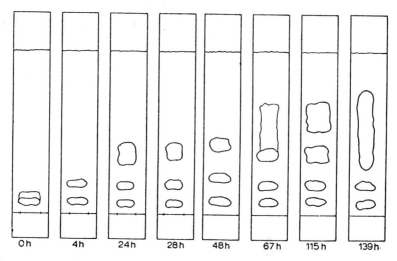

Fig. 4. — Hydrolysis of $TcCl_6^{2-}$. Schematic drawings of chromatograms on Macherey, Nagel strong anion exchange cellulose paper developed with 1.2 N HCl. From left to right : $TcCl_6^{2-}$ immediately after dissolution in 1.2 N HCl and after 4 h., 24 h., 48 h., 67 h., 115 h. and 139 h.

species of Tc(IV) may be present together and can be identified by paper chromatograms.

The stability of TcO_4^- in H_2SO_4 and HNO_3 was then studied and no change detected chromatographically at room temperature for 120 days (5). Tc(IV) and Tc(V) are converted back to TcO_4^- in sulphuric acid medium or other non-complexing acids.

A synthesis for TcF_6^{--} mentioned in the literature was examined. At the best only traces of TcF_6^{--} form in a fusion of $TcCl_6^{--}$ in NH_4HF_2 the rest being reoxidised to TcO_4^-.

The paper electrophoresis of outer-sphere complexes.

Complexes with a complete coordination sphere such as $Co(NH_3)_6^{+++}$ and $Fe^{III}(CN)_6$ may be still form complexes with further ligands (usually ions of opposite charge). This has first been observed by Werner in 1913. To date only two reviews (6, 7) survey the work and the techniques used on these outorsphere complexes. The techniques employed are rate of dialysis, polarography, conductivity measurements, liquid-liquid extraction, ion exchange, pH measurements, nuclear magnetic resonance, spectrophotometry and kinetic studies.

Beck (7) lists in a table the stability constants of outer-sphere complexes from the literature. This table is remarkable by the fact that it contains practically only complexes formed by $Co(NH_3)_6^{+++}$, $Co(en)_3^{+++}$, $Co(pn)_3^{+++}$ and a number of $Co(NH_3)_5X^{++}$ complexes where X is Cl,F,H_2O,N_3,NO_2 etc., some corresponding Cr complexes, only 3 Pt(IV) complexes and only one Fe complex [$Fe(phen)_3^{2+}$].

Our studies began when we were asked to check the purity of a series of ortho-phenanthroline complexes. Paper chromatography with acid-butanol mixtures produced reduction comets with some of the complexes and we decided to employ high voltage paper electrophoresis.

Here the spots moved without « tailing », however we thought that the differences in mobilities of the complexes should be bigger. In order to investigate the type of interaction between the electrolyte used and the complex ion we tried a large range of monovalent. divalent and trivalent anions in the electrolyte. Typical results are shown in figure 5. We see

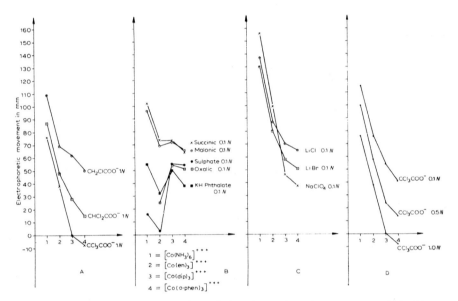

Fig. 5. — Graphical representation of the electrophoretic movement of Co(II) complexes in various electrolytes. (A) Comparison of mono-, di- and trichloro-acetate (1 N). (B) Comparison of several divalent anions (all 0.1 N). (C) Comparison of lithium chloride, lithium bromide and sodium perchlorate (0.1 N). (D) Comparison of 0.1, 0,5 and 1 N trichloroacetate. Th order of the complexes is : (1) Hexamminecobalt(III) ; (2) tris(ethylenediammine)cobalt(III) ; (3) tris(dipy-ridyl)cobalt(III) ; (4) tris(o-phenanthroline)cobalt(III).

here (Fig. 5A) that the speeds of the dipyridyl and the orthophenanthroline complexes slow down relative to $Co(NH_3)_6^{+++}$ as the size of the anion increases. Figure 5B shows that in some divalent anions the sequence is completely reversed $Co(NH_3)_6^{+++}$ and $Co(en)_3^{+++}$ complexing much more than the heavier complexes. Figure 5C shows that with inorganic anions the same rule holds as with the chloracetates ; perchlorate the least hydrated anion complexes much more than chloride and bromide. The fact that complexing is much accentuated by increasing concentration is illustrated in figure 5D and this holds for all electrolytes. Actual anionic movement was only observed with the orthophenanthroline complex in trichloracetate and with $Co(en)_3^{+++}$ in sulphate.

Separations of a range of cis-trans complexes of the type $Co(en)_2X_2$ were possible in sulphate where the mobility ratios were about 1 : 0.6 (cis always slower) while in acetate the mobility ratios were of the order of 1 : 0.9 requiring long distances for poor separations.

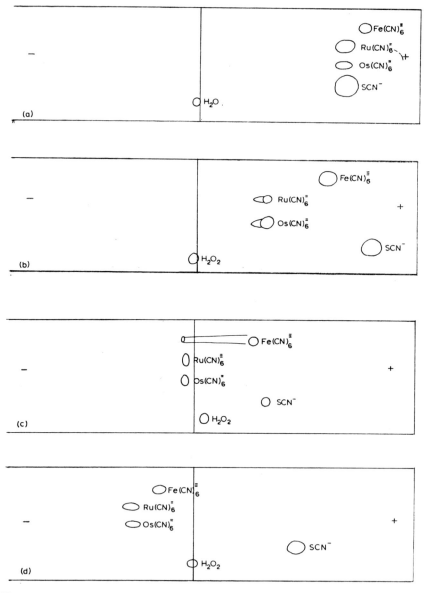

Fig. 6. — Some electropherograms of CNS^-, $Os(CN)_6^{2-}$, $Ru(CN)_6^{2-}$ and $Fe(CN)_6^{3-}$ in various electrolytes ($^3/_{10}$ × actual size) : (a) in 0.1 N LiCl for 30 min. with 1500 V ; (b) in 0.1 N $BaCl_2$ for 30 min. with 1500 V ; (c) in 0.5 N $AlCl_3$ for 30 min. with 1000 V ; (d) in 0.1 N $ZrOCl_2$ for 30 min. with 1500 V. Hydrogen peroxide is used as indicator for electroosmotic flow.

Rather spectacular results are obtained with anions of the type $Fe^{III}(CN)_6^{---}$, $Ru(CN)_6^{--}$, $Os(CN)_6^{--}$, as shown in figure 6. Cationic movement of these complexes occurs in 0.1 zirconyl chloride. It should be emphasised that these ions form « ion-pairs » very strongly even with monovalent cations as is also shown in the movement relative to SCN^- in 1 N LiCl.

A survey of quarternary ammonium salts in inorganic electrolytes showed that ion-pair or outer-sphere complex formation occurs although differences between electrolytes were noticed mainly in concentrations of more than 0.5 N. It was possible to change the sequence by the use of the correct electrolyte in separations of synthetic curarising compounds and some improved separations were obtained.

Finally a few words on what sort of results one can expect from a « chromatographic investigation ».

Coming back to the work on the solution chemistry of technetium, two spots were found for Tc(IV) in the less concentrated regions of HCl and HBr. Our deduction was that the hexachlorotechnetate hydrolysed to yield a mono-aquo-pentachlorotechnetate (IV). This opinion could be strengthened by studying the solution of pure $Tc(IV)Cl_6^{--}$ in dilute HCl over some length of time where the same second spot and others were found. Although the position of the spot on the ion exchange paper chromatogram also gives an indication of its charge, we have of course no chromatographic method which could identify completely this second spot. For its identification classical methods such as analysis of the isolated compound must be resorted to.

What we would like to emphasise however is that the chromatographic investigation usually shows up the existence of the « second spot ». It will also indicate whether a complexing reaction is reversible or forms slowly reacting compounds. In studying new reactions we can thus gain an overall picture (which often suffices anyhow) or establish how and what should be studied by classical methods.

In the case of the outer-sphere complexes (which are reversible) we were able to examine a much wider range of complexes than were studied previously and this was done very quickly. We were also able to point out some of the principles of outer-sphere complex formation e.g. that large poorly hydrated ions like trichloracetate and perchlorate complex more strongly than small hydrated ions and that large ligand groups like o-phenan-throline have a greater tendency to outer-sphere complexation than smaller groups.

References

1. H. HOLZAPFEL, LE VIET LAN and G. WERNER. *J. Chromatog.,* **24** : 153, 1966.
2. K. AITZETMÜLLER, K. BUCHTELA and F. GRASS. *Anal. Chim. Acta,* **38** : 249, 1967.
3. S. HJERTÉN. *Chromatog. Revs.,* **9** : 122, 1967.
4. S.K. SHUKLA. *Ric. Sci.,* **36** : 725, 1966.
5. L. OSSICINI and G. BAGLIANO. *Ric. Sci.,* **36** : 348, 1966.
6. F. BASOLO and R.G. PEARSON. *Mechanism of Inorganic Reactions.* Wiley. New York, p. 376, 1958.
7. M.T. BECK. *Coordin. Chem. Rev.,* **3** : 91, 1968.

L'analyse immuno-électrophorétique et ses domaines d'application

par

Jean DAUSSANT et Pierre GRABAR

Introduction.

Depuis longtemps, la précipitation spécifique antigène-anticorps est utilisée dans l'étude des protéines. C'est ainsi que Wells et Osborne (1912) choisissant les protéines végétales, parce que pouvant être à l'époque obtenues à un état mieux défini que les protéines animales, ont comparé plusieurs albumines d'origines végétales différentes en employant, entre autres méthodes immunologiques, la précipitation spécifique. Mais les méthodes utilisées alors ne pouvaient rendre compte que de l'ensemble des réactions d'un mélange d'antigènes avec les anticorps correspondants.

C'est le mérite des méthodes de précipitation spécifique en gel d'avoir permis de rendre mieux perceptibles les réactions individuelles des antigènes présents dans un mélange avec les anticorps correspondants. La première en date de ces méthodes, la diffusion simple, fût décrite par Oudin (1946) qui étudia différents facteurs intervenant dans la précipitation spécifique en gel. Deux autres types de méthodes, la double diffusion selon Ouchterlony (1949) qui permet la comparaison de plusieurs systèmes antigéniques et l'analyse immuno-électrophorétique suivant Grabar et Williams (1953) qui présente le même avantage et qui possède de plus un pouvoir de définition plus élevé, firent apparaître la précipitation spécifique en gel comme un moyen puissant dans l'étude des macromolécules, en particulier des protéines.

L'analyse immuno-électrophorétique (A.I.E.) étant décrite depuis de nombreuses années, nous n'insisterons pas sur les détails d'application, nous référant pour cela aux revues et traités déjà existants. Nous exposerons simplement les principes et les modalités générales de mise en œuvre de ces méthodes en insistant sur certains de ses développements récents. Nous donnerons ensuite quelques exemples choisis dans différents domaines de l'application de l'A.I.E.

L'ANALYSE IMMUNO-ELECTROPHORETIQUE, METHODE

1. Généralités et mise en œuvre.

L'analyse immuno-électrophorétique (A.I.E.) est applicable à l'étude des antigènes et haptènes qui forment avec les anticorps des complexes précipitants, ce qui est le cas de nombreuses macromolécules, en particulier des protéines.

Cette méthode est fondée sur deux caractéristiques indépendantes des macromolécules : la mobilité électrophorétique, qui est liée à l'existence de groupes polaires sur les molécules et la spécificité conférée à celles-ci par l'existence de sites antigéniques. Ces sites antigéniques, dans le cas des protéines, sont déterminés par une disposition particulière dans l'espace, d'un nombre limité d'acides aminés (Lapresle et Webb, 1964). La spécificité antigénique d'une protéine résulte donc de la structure de la molécule.

La méthode est d'une application simple : la solution contenant les antigènes est déposée dans un réservoir creusé dans un gel transparent qui constitue le support pour l'électrophorèse puis pour la diffusion ; après électrophorèse, l'immunsérum est placé dans un canal découpé dans ce même gel parallèlement à l'axe de migration des antigènes. Après un temps suffisant de diffusion, lorsque antigènes et anticorps se rencontrent en proportions convenables, ils forment dans le gel des précipités visibles. Les gels peuvent ensuite être lavés, afin d'éliminer les antigènes et les anticorps qui ne constituent pas les précipités spécifiques, puis séchés. Les arcs de précipitation peuvent être alors colorés par des colorants des protéines et aussi des lipoprotéines, des nucléoprotéines, des glucides ou d'autres groupements prosthétiques éventuellement présents dans les précipités spécifiques.

Les détails de mise en œuvre de la méthode ont été depuis longtemps décrits (Grabar et Burtin, 1960 ; Grabar, 1963). Des gels d'agar, d'ionagar ou d'agarose sont souvent utilisés. Le gel d'agarose présente l'avantage d'être constitué de polysaccharides neutres, les risques d'interaction avec les substances ionisables soumises à l'électrophorèse sont donc réduits (Uriel, Avrameas et Grabar, 1964). Des indications ont été données dans l'emploi de gels mixtes agarose poly-acrylamide (Uriel, 1966). Des feuilles d'acétate de cellulose ont également été employées comme support dans l'A.I.E. A titre d'indication, dans les gels du genre agar préparés en tampon véronal 0,025 M pH 8,2, la tension appliquée généralement est de l'ordre de 4 à 8 volts par centimètre, la durée de l'électrophorèse variant de 1 à 4 heures suivant les mélanges protéiques à séparer et la dimension des gels utilisés. Dans les micro-méthodes de l'A.I.E., la séparation électrophorétique se fait sur une longueur de 5 cm environ au lieu de 12 à 18 cm. Le diagramme immunoélectrophorétique résultant de la microméthode est de lecture plus difficile que celui donné par la méthode ordinaire, surtout lorsque le nombre de constituants révélés est élevé. La microméthode présente par contre plusieurs avantages ; elle nécessite de moins grandes quantités de réactifs et les résultats sont obtenus plus rapidement. De plus, dans la microméthode, il n'est pas nécessaire d'effectuer un mélange de la solution d'antigène avec le gel chauffé pour emplir le réservoir de départ, ce qui évite une dénaturation éventuelle par la chaleur. Ceci rend de plus possible plusieurs applications de la solution dans le réservoir de départ avant l'électrophorèse, ce qui permet d'éviter parfois une concentration préalable de la solution d'antigènes.

L'A.I.E. fournit donc une définition des constituants d'un mélange par au moins deux caractéristiques indépendantes des molécules. Ainsi, des constituants de mobilités très voisines ont pu être discernés les uns des autres, ce qui ne pouvait être possible par électrophorèse seule. Cet avantage comporte cependant une difficulté qui apparaît lors de l'interprétation du tracé immuno-électrophorétique, surtout si le nombre d'antigènes décelés est élevé. En effet, la position d'un arc de précipité entre l'axe de migration électrophorétique et le canal d'où diffusent les anticorps dépend en particulier de la concentration de l'antigène. Si les rapports des concentrations d'antigènes de mobilités voisines peuvent varier dans les solutions à analyser, il sera très difficile de reconnaître avec certitude des constituants d'une analyse à l'autre.

Cette difficulté peut être tournée en certains cas en employant l'une ou l'autre des dispositions particulières de l'A.I.E. qui permettent d'identifier un constituant obtenu pur à l'un des constituants décelés sur un diagramme immuno-électrophorétique (Grabar et Burtin, 1960 ; Heremans, 1960). Certaines de ces dispositions permettent aussi d'identifier l'un des constituants d'un diagramme immuno-électrophorétique à son correspondant s'il existe, d'un autre tracé immuno-électrophorétique. Ces dispositions utilisent l'association de la double diffusion à l'A.I.E. ou réalisent par exemple l'A.I.E. simultanée des deux mélanges antigéniques déposés avant électrophorèse en différentes places d'un même gel sur le même axe de migration électrophorétique.

Une autre façon d'identifier un constituant pur à l'un des constituants décelés par l'A.I.E. consiste à absorber l'immunsérum spécifique du mélange antigénique par l'antigène pur et à observer quelle ligne de précipitation est éliminée du tracé immuno-électrophorétique donné avec l'immunsérum quand on utilise celui-ci après épuisement.

Une autre méthode particulièrement puissante pour l'identification d'un constituant dans un mélange consiste en l'utilisation de sérum mono-spécifique dans l'A.I.E. du mélange. Ceci permettra de révéler sans ambiguité parmi les constituants d'un mélange celui qui correspond à la protéine considérée. Cette particularité conférée à l'A.I.E. et dans une certaine mesure à la double diffusion, par l'utilisation de sérums monospécifiques est intéressante puisqu'elle permet une étude de certaines caractéristiques de l'antigène à partir de la solution mère en évitant les risques de dénaturation pouvant intervenir au cours de la purification. Cette propriété de l'A.I.E. constitue un moyen particulièrement puissant pour reconnaître un antigène dans un milieu biologique en évolution, même si cet antigène a été modifié dans certaines de ses propriétés physico-chimiques. Elle permet de préciser la nature de certaines modifications ; changement de mobilité électrophorétique, perte d'une partie de la spécificité antigénique. En association avec l'ultracentrifugation différentielle ou la chromatographie d'exclusion, elle peut indiquer à partir du mélange, si l'antigène a été modifié dans sa taille moléculaire. L'obtention d'antigènes purs et la préparation d'immunsérums monospécifiques restent encore difficiles. Il semble néanmoins que l'on tende vers l'utilisation de plus en plus fréquente des sérums monospécifiques.

Des moyens supplémentaires de définition des précipités spécifiques en gels sont offerts par une particularité de ces précipités, celle de laisser accessible à leur caractérisation des sites actifs de l'antigène inclus dans le précipité. Ainsi des méthodes de caractérisation enzymatique sur les arcs de précipités ont été largement développées, des méthodes de caractérisation d'anticorps spécifiques de certaines protéines ont été mises au point.

II. Caractérisations enzymatiques.

Il apparaît étonnant que des caractérisations enzymatiques aient pu être effectuées sur les arcs de précipités spécifiques en gel. En effet, l'activité enzymatique est fréquemment partiellement ou totalement inhibée lorsque, en tube à essai, le substrat est ajouté à l'enzyme en présence de ses anticorps en excès (Cinader, 1966). Les causes de cette inhibition ont été attribuées à l'empêchement stérique qui fait que les anticorps complexés aux enzymes en masquent les sites actifs. Il apparaît peu probable que sites antigéniques et sites enzymatiques coïncident ; en effet, les enzymes d'une même spécificité mais issus d'espèces différentes semblent devoir présenter des sites actifs semblables ou même identiques ; une tolérance naturelle envers ces sites devrait donc exister chez les différentes espèces (Cinader, 1963).

En gels, les phénomènes semblent différents puisque de nombreuses activités enzymatiques ont pu être caractérisées sur les arcs de précipitation. La cause de la disparité de ces résultats a pu être attribuée au fait que, dans les conditions expérimentales en gels, seul est présent le complexe antigène-anticorps insoluble, en particulier il n'y a pas d'anticorps en excès. Dans ces conditions, l'importance de l'empêchement stérique par les anticorps peut être réduite. Il a été également indiqué que le gel aussi peut contribuer à la limitation de l'empêchement stérique par les anticorps (Uriel, 1966).

De nombreuses réactions de caractérisation enzymatiques sur les lignes de précipitation en gel ont été décrites (Uriel, in Grabar et Burtin, 1960 ; Uriel, 1963 ; Uriel, en préparation). Les principes de ces caractérisations ont été développés récemment (Uriel, Avrameas, Grabar, 1964 ; Uriel, 1966 ; Avrameas, 1967). Ils sont fondés sur les différences de propriétés physicochimiques ou chimiques apparaissant entre le substrat et les produits de la réaction enzymatique.

On a utilisé des substrats dont la solution est peu ou pas colorée, mais dont les produits de dégradation par l'enzyme sont insolubles et colorés ou bien rendus tels par un réactif présent dans la solution. Par exemple, les estérases carboxyliques sont décelées en utilisant l'acétate de β-naphtol comme substrat. Le β-naphtol libéré par la réaction est couplé avec un sel de diazonium ajouté au substrat et donne un formazan insoluble. Sur le tracé immuno-électrophorétique, la ou les bandes constituées par des estérases carboxyliques apparaissent colorées en bleu.

On a aussi utilisé des substrats qui sont colorables, mais qui ne le sont plus après l'action de l'enzyme. Un exemple de ce type est donné par les amylases. Les gels sont incubés dans une solution d'amidon soluble puis dans une solution d'iode-iodure de potassium ; l'ensemble du gel est coloré en bleu, la présence d'amylase est révélée par une absence de coloration le long de l'arc de précipitation constitué par l'enzyme et ses anticorps due à la dégradation de l'amidon en dextrines et sucres incolorables par l'iode.

L'utilisation parallèle d'inhibiteurs permet de préciser la nature de l'enzyme. Dans l'exemple précédent, la réaction de caractérisation décèle la présence à la fois de l'α- et de la β-amylases. La distinction entre l'une et l'autre de ces enzymes a été rendu possible par l'utilisation du chlorure mercurique à faible concentration comme inhibiteur de la β-amylase (Daussant, 1966).

Enfin, lorsque le substrat et les produits de dégradation ne peuvent être colorés, on adjoint au substrat un autre système indirectement colorable sous l'effet de la réaction enzymatique. La ribonucléase est ainsi caractérisée dans un milieu non tamponné qui contient en plus du substrat un indicateur de pH. Le lieu de l'activité enzymatique est révélé par un changement de couleur de l'indicateur coloré dû à la variation de concentration des ions hydrogènes provoquée par l'hydrolyse enzymatique (Uriel, Avrameas, 1964).

III. Marquage des antigènes.

L'introduction d'éléments radioactifs (Yagi et al., 1962, 1963) ou la fixation d'enzymes (Avrameas et Uriel, 1966) sur les antigènes a permis une autre catégorie de caractérisation, celle de l'activité anticorps envers un antigène donné. Le procédé expérimental est semblable avec les deux types d'antigènes marqués (l'utilisation de ces antigènes marqués augmente la sensibilité de la méthode). Par exemple, l'activité anticorps contre l'insuline bovine a pu être décelée par radio-immuno-électrophorèse sur plusieurs types d'immuno-globulines de cobayes et d'hommes (Yagi et al., 1963).

C'est en utilisant le marquage de l'albumine par la peroxydase que les anticorps anti-albumine humaine d'immunsérums de lapins ont pu être localisés dans les γ_1 et γ_2-globulines. Dans ce cas, la méthode est conduite de la façon suivante : l'immunsérum de lapin a été soumis à l'A.I.E. en utilisant un immunsérum de cheval anti-sérum de lapin. Après lavage du gel, la solution d'albumine marquée à la peroxydase de raifort est introduite dans le canal où initialement avait été placé l'immunsérum de cheval. La caractérisation de l'activité peroxydasique qui décèle la présence d'albumine sur les arcs de précipitation contenant ses anticorps est effectuée après un second lavage des gels (Avrameas et Uriel, 1966).

IV. Caractéristiques générales de la méthode.

L'A.I.E. comporte un inconvénient majeur qui d'ailleurs est commun à toutes les méthodes immunochimiques : il réside dans l'emploi de réactifs biologiques, les immunsérums, qui sont variables d'un animal à l'autre. Cette variabilité peut, dans une certaine mesure, être limitée par la constitution de mélanges de plusieurs immunsérums. Par ailleurs, l'emploi des micro-méthodes est évidement indiqué pour permettre la réalisation d'un nombre suffisant d'expériences avec un même réactif. De toute façon, le nombre de constituants révélés dans un mélange protéique par l'A.I.E. ne peut qu'indiquer un nombre minimum de constituants présents dans le mélange.

L'étude des constituants antigéniques par l'A.I.E. n'est réalisable que si ceux-ci sont solubles. Ainsi, une catégorie d'antigènes échappe en principe à cette méthode d'analyse. Des exemples ont cependant été donnés montrant que cet inconvénient pouvait être tourné. L'A.I.E. des protéines insolubles du blé, gliadine et gluténine, a pu être effectuée en adjoignant au gel de l'urée pour la durée de l'électrophorèse. La présence de l'urée dans le gel empêchant les précipités antigène-anticorps d'apparaître les gels sont lavés pour en éliminer l'urée avant l'application de l'immunsérum.

Mais, à côté de ces inconvénients, l'A.I.E. présente l'avantage d'être très sensible ; elle peut déceler la présence de quelques microgrammes de substance antigénique et se caractérise par une haute spécificité dûe aux réactions antigène-anticorps. Elle présente de plus la possibilité particulièrement intéressante de permettre d'établir une définition multiple des différents antigènes décelés dans un mélange sur un même tracé immuno-électrophorétique.

DOMAINES D'APPLICATION DE L'ANALYSE IMMUNO-ELECTROPHORETIQUE

L'analyse immuno-électrophorétique est actuellement utilisée dans de nombreux domaines de la recherche sur les macromolécules et les liquides biologiques. Plusieurs revues existent qui font le point sur ces applications en général (Grabar et Burtin, 1960 ; Grabar, 1964 a ; Grabar, 1968 a). Les applications de l'A.I.E. à des domaines plus particuliers, comme l'étude des antigènes tissulaires (Grabar, 1964 b, c ; 1968 b) et celle des enzymes (Uriel, 1966 et 1967) ont également fait l'objet de revues. Comme les exemples cités dans les revues précédemment mentionnées sont tirés le plus souvent du règne animal, nous reprendrons ici l'exposé résumé des différents domaines d'application de l'A.I.E. en mentionnant certains des exemples classiques et en citant des exemples moins connus que nous prendrons le plus souvent de la biochimie végétale. Nous avons classé les principaux domaines d'application en trois groupes :

I. Dénombrement, définition et identification de constituants de fluides biologiques et d'extraits ;

II. Comparaison des spécificités antigéniques de protéines ;

III. Etude des modifications de protéines en milieux biologiques.

1. Dénombrement, définition et identification de constituants de fluides biologiques et d'extraits.

L'exemple le plus classique dans ce domaine est sans doute celui donné par l'A.I.E. du sérum humain qui a permis d'y déceler l'existence de constituants inconnus jusqu'à l'utilisation de la méthode. Cette analyse a servi à établir une carte de référence des constituants du sérum dont certains ont été caractérisés par leur activité enzymatique, d'autres définis par leur rôle physiologique. Un autre exemple connu est constitué par l'A.I.E. du pancréas de bœuf dans lequel 17 constituants antigéniques ont été décelés parmi lesquels 15 ont été identifiés par une activité enzymatique. Enfin, l'A.I.E. des extraits d'organes ou de tissus, en conjonction avec l'absorption d'immunsérums a permis de discerner les antigènes spécifiques de ces organes ou tissus de ceux existant aussi dans le sérum, dans d'autres organes ou tissus.

Dans le domaine végétal, les protéines de l'orge sont parmi celles qui ont été le plus étudiées, sans doute à cause de l'importance qu'elles tiennent dans les industries de la malterie et de la brasserie. Il a semblé opportun à un groupe de l'European Brewery Convention de proposer une nomenclature des protéines solubles de l'orge afin qu'on y puisse ultérieurement rassembler pour références l'ensemble des résultats obtenus au cours de leur étude. Cette nomenclature devait permettre d'identifier les protéines par le plus grand nombre possible de critères fondés sur des propriétés indépendantes des molécules et devait de plus être suffisamment souple pour permettre l'identification de constituants qui auraient subi des modifications au cours de phénomènes apparaissant au cours de la germination ou du brassage. Le système de définition proposé est fondé sur l'A.I.E. dont le diagramme peut être obtenu avec un immunsérum pris comme référence, qui permet de déceler 17 arcs de précipitation, parmi lesquels 3 qui ont été définis par une activité enzymatique, peuvent servir de repères (Grabar, Lontie et Djurtoft, 1967).

L'A.I.E., en même temps que d'autres méthodes immunochimiques, a été employée pour évaluer les degrés de parenté entre semences de variétés d'espèces et de genres différents. La comparaison des tracés immuno-électro-phorétiques donnés par les extraits des différentes semences en employant les immunsérums homologues et hétérologues ainsi que l'observation des diagrammes immuno-électrophorétiques obtenus avec des sérums épuisés par des extraits hétérologues ont permis de préciser certaines hypothèses concernant la classification de certaines espèces à l'intérieur d'un genre, et con-tribué à la détermination de lignes ancestrales d'hybrides. Ce genre d'études et la critique des résultats obtenus ont fait l'objet de mises au point (Moritz, 1964 ; Fair Brothers, 1966).

II. Comparaison des spécificités antigéniques de protéines.

La spécificité antigénique des protéines a permis, par l'A.I.E., à partir parfois de mélanges protéiques, d'apporter un élément de comparaison portant sur la structure de certaines protéines.

C'est ainsi que l'A.I.E. peut servir de fondement exérimental à l'étude des isoenzymes (Uriel, 1967); le problème est posé quand une hétérogénéité est décelée (par électrophorèse ou chromatographie) dans la population d'en-zymes d'une même origine catalysant une même réaction. S'agit-il d'une variabilité autour d'une même espèce moléculaire, ou bien cette hétéro-généité résulte-t-elle de l'expression de différents gènes ? Dans le premier cas, une relation structurale entre les différents constituants doit pouvoir être reflétée par des réactions immunochimiques communes. Si, par contre, cette hétérogénéité est due à l'existence de protéines d'espèces moléculaires différentes, celles-ci n'auront vraisemblablement pas d'identité de structure et par conséquent apparaîtront distinctes dans leurs spécificités antigéniques.

C'est ainsi que 3 carboxypeptidases A, décelées dans un extrait pan-créatique de porc, présentent une même spécificité antigénique. Elles repré-sentent des formes moléculaires intermédiaires apparaissant au cours de l'ac-tivation de la pro-carboxypeptidase comme l'a montré l'A.I.E. du zymogène à différents stades de l'activation par la trypsine (Uriel, 1966).

C'est grâce à l'A.I.E. que 2 trypsines de mobilités distinctes décelées dans le pancréas activé de rat ont été reconnues correspondre à des protéines de structure différentes. De plus, l'A.I.E. de l'extrait non activé de pancréas de rat, grâce à l'activation des pro-enzymes sur le gel après A.I.E., a pu montrer que les pro-enzymes eux-mêmes étaient de spécificités antigéniques différentes (Pascale et coll., 1966).

L'A.I.E. de la lacto-dehydrogénase, en précisant la parenté antigénique existant entre les isoenzymes révélés par électrophorèse, a permis de confirmer l'existence de 2 subunités de structures distinctes associées en tétramères de composition variable (Avrameas et Rajewsky, 1964).

Dans la semence d'orge mure, la β-amylase est constituée d'une population de constituants différant par la taille moléculaire. L'A.I.E. a montré que ces différents constituants avaient une même spécificité antigénique. Il pourrait donc s'agir de formes multiples d'une même espèce moléculaire (Nummi, 1967). La β-amylase du blé est formée d'une population de molécules hétérogènes du point de vue mobilité électrophorétique. Ces différents enzymes présentent dans ce cas aussi une spécificité antigénique commune (Abbott et Daussant, en préparation).

En biochimie végétale des protéines non actives ont été comparées du point de vue de leur spécificité antigénique. L'A.I.E. de la gliadine et de la gluténine effectuée en présence d'urée pendant la phase électrophorétique a permis de reconnaître qu'elles renfermaient toutes deux un même constituant antigénique. Ce résultat indique qu'il existe une parenté de structure entre ces deux types de protéines (Benhamou et al., 1965).

III. Etude des modifications de protéines en milieux biologiques.

C'est dans ce domaine que l'A.I.E. a, sans doute, fait l'objet du plus grand nombre d'applications. Cette méthode, grâce à la spécificité des réactions immunochimiques est particulièrement apte à déceler des modifications sur un antigène déterminé ou dans la composition d'un mélange d'antigènes (changements de propriétés physicochimiques, disparition, apparition de constituants).

Ainsi, en pathologie, l'A.I.E. est utilisée comme méthode courante pour déceler des anomalies dans le tracé immuno-électrophorétique du sérum humain par exemple (Grabar et Burtin, 1960). Cette méthode a été souvent employée dans l'étude du développement de tumeurs. C'est ainsi que l'A.I.E., en utilisant les immunsérums correspondants, a permis de déceler dans certains sérums d'adultes des fœto protéines qui, normalement ne sont plus synthétisées après la naissance. Il semble que la résurgence de l'une d'entre elles, la β-fœtoprotéine, accompagne diverses maladies hépatiques, tandis que la réapparition de l'α_1-fœtoprotéine semble spécifiquement associée aux cancers primitifs du foie (Uriel et al., 1968).

Dans l'étude de la synthèse de globulines in vitro à partir d'explants de rate de poulet au cours des réponses immunitaires primaires et secondaires, la radio immunoélectrophorèse a permis de confirmer des résultats obtenus par d'autres méthodes (Corvazier, 1966). Des fragments de rate provenant de poulets non immunisés et de poulets à différents stades d'immunisation étaient placés en milieu de culture contenant des acides aminés marqués. La radio-immuno-électrophorèse des surnageants des milieux de cultures prélevés après un ou deux jours d'incubation était effectuée en utilisant un immunsérum anti-sérum de poulet. Les résultats obtenus confirment que la réponse primaire s'accompgane d'une production particulière des immunoglobulines M (19 S) tandis qu'une réponse secondaire est caractérisée par une stimulation de la synthèse des immunoglobulines G.

Ajoutons que des exemples ont été cités (Wolff, 1964) de l'application de l'A.I.E. dans l'étude de l'apparition et de la disparition de constituants durant la vie embryonnaire.

En biochimie végétale, l'A.I.E. a été utilisée dans l'étude des protéines actives et des protéines de réserve des semences au cours de la maturation, de la germination et de la croissance.

Pendant la maturation du haricot, les synthèses des différentes protéines s'échelonnent dans le temps comme l'a montré l'A.I.E. des extraits de ces semences prises à différents moments après la pollénisation et en utilisant des immunsérums préparés avec des extraits de semences mures et certaines de leurs fractions protéiques ; quelques protéines qui sont décelées très tôt après pollénisation ont été identifiées à des constituants du testa ; elles semblent donc d'origine uniquement maternelle. Une étape particulière de la synthèse protéique apparaît au moment de la deshydratation. A ce moment, plusieurs arcs de précipitation sont mis en évidence sur les tracés immuno-électrophorétiques. Cette formation soudaine de plusieurs constituants a été attribuée à une association rapide de sub unités qui auraient été déjà synthétisées et qui s'assembleraient à ce stade de la maturation. Au cours de la germination, la protéine de réserve, la phaséoline, se distingue des autres constituants et apparaît avec une mobilité modifiée (Kloz et al., 1966). Ceci peut constituer un argument en faveur de la théorie suivant laquelle, durant la germination, les protéines de réserve passeraient par un stade intermédiaire, dit d'activation, avant d'être hydrolysées par les enzymes protéolytiques (Ghetie, 1966).

Dans l'orge, les enzymes amylolytiques sont soumis à des modifications profondes au cours de la germination. La β-amylase, qui forme une population de molécules nétérogène du point de vue taille moléculaire, apparaît plus homogène selon ce critère dans l'orge germée (Nummi, 1967). L'A.I.E. a permis de démontrer que les enzymes de l'orge mure et celle de l'orge germée présentaient une même spécificité antigénique, mais différaient par leurs mobilités électrophorétiques. Si donc, l'identité antigénique laisse supposer que nous pouvons avoir à faire à des enzymes de même espèce moléculaire, la différence de mobilité implique que les enzymes de l'orge mure sont modifiées au cours de la germination, non seulement dans leur taille moléculaire mais aussi dans leur charge. L'A.I.E., associée à l'absorption d'immunsérums, a démontré que dans l'orge non germée, il n'existait pas de protéine présentant quelque analogie de spécificité antigénique avec l'α-amylase décelée dans la semence germée. La semence non germée ne semble donc pas devoir contenir de précurseur de l'enzyme ; le fort accroissement de l'activité enzymatique décelé pendant la germination doit donc être attribué à une synthèse plutôt qu'à une activation. Ces transformations sont décelables dès le début de la germination (Daussant, 1966).

La germination introduit des modifications importantes dans le caractère antigénique de la semence de haricot, en particulier par la diminution du nombre de ses constituants. Par contre, certains antigènes persistent et semblent être communs à toutes les cellules de la plantule aux différentes phases de son développement (Ghetie et Buzila, 1963 a). L'A.I.E. effectuée avec un immunsérum anti-antigènes des feuilles de fève a permis de suivre l'apparition d'antigènes organo spécifiques au cours de la différenciation de la feuille et montre que les feuilles exposées à la lumière présentent plus de constituants que celles gardées à l'obscurité (Ghetie et Buzila, 1963 b).

Ces exemples ne sont que quelques-uns choisis parmi beaucoup d'autres, et nous n'avons pas traité des applications plus pratiques de l'A.I.E., comme par exemple, sa contribution à l'évaluation du degré de purification d'un constituant. Nous avons seulement cherché à illustrer les domaines d'application pour lesquels nous pensons que l'A.I.E. peut apporter des réponses rapides à certains problèmes biochimiques ou donner des indications assez précises sur certaines orientations de recherche.

Bibliographie

ABBOTT, D. and DAUSSANT, J. A classification system for soluble proteins of wheat based on immuno-electrophoretic analysis (en préparation).

AVRAMEAS, S. Identification d'enzymes par des méthodes d'immunodiffusion en gels. *Symp. Series immunol. Standard, vol.* **4** : 25. Basel/New York, Karger, 1967.

AVRAMEAS, S. and RAJEWSKY, K. Immunological characteristics of some lactic dehydrogenases, *Nature,* **201** : 405, 1964.

AVRAMEAS, S. et URIEL, J. Méthode de marquage d'antigènes et d'anticorps avec des enzymes et son application en immunodiffusion. *C.R. Ac. Sci. Paris,* **262** : 2543, 1966.

BENHAMOU-GLYNN, N., ESCRIBANO, M.J. et GRABAR, P. Etude des protéines du gluten à l'aide de méthodes immunochimiques. *Bull. Soc. Chim. Biol.,* **47** : 1.141, 1965.

CINADER, B. Immunochemistry of enzymes. *New York Ac. Sci. Ann.,* **103** : 495, 1963.

CINADER, B. and LEPOW, I.H. The neutralization of biologically active molecules. *In.* Antibodies to biologically active molecules. *Europ. Biochem. Soc.,* Vienna. 1965 l. Oxford and New York, Pergamon Press, 1966.

CORVAZIER, P. et CHRISTOL, G. 1966. Proportions relatives des globulines non agglutinantes et des agglutinines formées *in vitro* au cours des réactions immunitaires classiques. *Bull. Soc. Chim. Biol.* Coll. Immunochimie, Déc. 1967.

DAUSSANT, J. Etude des protéines solubles de l'orge et du malt par des méthodes immunochimiques. Paris, Thèse, 1966. Biotechnique, 1966, vol. 4, 5, 6.

FAIRBROTHERS, E. Comparative serological studies in plant systematics. *Serol. Mus. Bul.,* p. 2, May 1966.

GHETIE, V. The mechanisms of reserve protein hydrolysis during seed germination. *Rev. Roum. Biochim.,* **3**(4) : 353, 1966.

GHETIE, V. and BUZILA, L. Immunochemical investigations on the germination of bean. *Acad. Rep. Populaire Romine, Studii Cercetari Biochim.,* **6** : 51, 1963 a.

GHETIE, V. and BUZILA, L. Electrophoretic and immunoelectrophoretic studies of cytoplasmic proteins during the development of Vicia faba leaves. *Acad. Rep. Populaire Romine Studii Cercetari Biochim.,* **6** : 49, 1963 b.

GRABAR, P. Analyse immunoélectrophorétique *in* Techniques de Laboratoire, Loiseleur , tome I, 1390 p. Paris, Masson (édit.), 1963.

GRABAR, P. Utilisation des spécificités antigéniques dans l'étude des produits naturels. *Bull. Soc. Chim. Biol.,* XLVI, n° 12, p. 1727, 1964 a.

GRABAR, P. Immunoelectrophoretic analyse von Zell und Gewebe Komponenten. 15. Colloquium der Gesell. f. Physiol. Chem. 22-25 April 1964 in Mosbach/Baden, p. 36. Berlin/Heidelberg/New York, Springer-Verlag, 1964 b.

GRABAR, P. Etude immunochimique des constituants des cellules et des tissus, in Méthodes Nouvelles en Embryologie, p. 19, Paris, Hermann, 1964 c.

GRABAR, P. New aspects of immunoelectrophoretic analysis. Clin. Prot. Chem., 1 : 91, Basel/New York, Karger, 1968 a.

GRABAR, P. Immunochemical investigations on tissue antigens, 1968 b (sous presse).

GRABAR, P. et BURTIN, P. L'analyse immuno-électrophorétique ; ses applications aux liquides biologiques. 1 vol. Paris, Masson (édit.), 1968 b.

GRABAR, P., LONTIE, E., DJURTOFT, R. The E.B.C. system of reference for barley proteins. J. of the Inst. Brew., 73 : 381, 1967.

GRABAR, P. et WILLIAMS, C.A. Méthode permettant l'étude conjuguée des propriétés électrophorétiques et immunochimiques d'un mélange de protéines ; application au sérum sanguin. Biochim. Biophys. Acta, 10 : 193, 1953.

HEREMANS, J. Les globulines sériques du système gamma. Bruxelles, Masson et Cie Arscia S.A., 1960.

KLOZ, J., TURKOVA, V., KLOZOVA, E. Proteins found during maturation and germination of phaseolus vulgaris L. Biologia Plantarum (Prague), 8 : (2) 164, 1966.

LAPRESLE, C. et WEBB, T. Données actuelles sur les bases chimiques de la spécificité immunologique des protéines. Bull. Soc. Chim. Biol., 46 : 1701, 1964.

MORITZ, O. Some special features of serobotanical work, in Taxonomic Biochemistry. New York, C.A. Leone Ronald Press Co, p. 275, 1964.

NUMMI, M. Studies on the heterogeneity of soluble barley proteins with particular reference to β-amylase. Helsinki, Thèse, 1967.

OUCHTERLONY, O. Antigen-antibody reactions in gels. Acta pathol. Microbiol. Scand., 26 : 507, 1949.

OUDIN, J. Méthode d'analyse immunochimique par précipitation spécifique en milieu gélifié. C. R. Acad. Sci., 222 : 115, 1946.

PASCALE, J., AVRAMEAS, S. and URIEL, J. The characterization of rat pancreatic zymogens and their active forms by gel diffusion techniques. J. Biol. Chem., 241 : 2023, 1966.

URIEL, J. Les réactions de caractérisation des constituants protéiques après électrophorèse ou immuno-électrophorèse en gélose. In : analyse immuno-électro-phorétique ; ses applications aux liquides biologiques. GRABAR, P. et BURTIN, P. 1 vol., p. 33. Paris, Masson (édit.), 1960.

URIEL, J. Characterization of enzymes in specific immune-precipitates. Ann. New York Acad. Sci., 103 : 956, 1963.

URIEL, J. Méthode d'électrophorèse dans des gels d'acrylamide-agarose. Bull. Soc. Chim. Biol., 48 : 969, 1966 a.

URIEL, J. Immunoelectrophoresis of enzymes in Antibodies to biologically active molecules. Europ. Biochem. Soc., Vienna, vol. 1. Oxford and New York, Pergamon Press, 1966 b.

URIEL, J. Les isoenzymes. Annales de la Nutrition et de l'Alimentation, 21 : n° 3, B 67, 1967.

URIEL, J. Color reactions for the identification of antigen-antibody precipitates in gel diffusion media (en préparation), 1968.

URIEL, J. and AVRAMEAS, S. Study of enzyme-antibodies precipitates using non chromogenic substrates. *Anal. Biochem.,* **9** : 180, 1964.

URIEL, J., AVRAMEAS, S. et GRABAR, P. Caractérisation d'enzymes après électrophorèse et analyse immuno--électrophorétique en agarose. *Prot. of Biol. Fluids,* **11** : 355. Amsterdam, Elsevier Co., 1964.

URIEL, J., NECHAUD, B. de, STANISLAWSKI-BIRENCWAJG, M., MASSEYEFF. R., LEBLANC, L., QUENUM, L., LOISILLIER, F. et GRABAR, P. Le diagnostic du cancer primaire du foie par des méthodes immunologiques. *La Presse Méd.,* p. 1415. Juin 1968.

WOLFF, E. Méthodes nouvelles en embryologie, détection de molécules par l'immunochimie. Paris, Hermann, 1964.

YAGI, Y., MAIER, P.,and PRESSMANN, D. Immunoelectrophoretic identification of guinea pig anti-insulin antibodies. *J. Immunol.,* **89** : 736, 1962.

YAGI, Y., MAIER, P., PRESSMANN, D., ARBESMAN, G.E., REISMAN, R.E. and LENZNER, A.R. Multiplicity of insulin binding antibodies in human sera. *J. Immunol.,* **90** : 760, 1963.

Recent advances in thin-layer chromatography

by

G. PATAKI

Research Department, Robapharm. Ltd., Basle, Switzerland.

Thin-layer chromatography was developed because of a specific need for a rapid method which separates small amounts of compounds. At the beginning the technique has been used for *screening* in column chromatography; *reaction control* and for carrying out certain reactions (such as oxidation, reduction) directly on the chromatogram. Subsequently, chromatography on thin-layers has grown into a indispensable laboratory method. Many publications in nearly all fields have shown, that this technique is superior to paper chromatography. As regards the fields of application, the earlier preoccupation with lipids has recently been offset by considerable advances in TLC of hydrophilic substances, e.g. *amino acids, peptides, nucleic acid derivatives* and *carbohydrates* (1-6). Publications in TLC in the field of *amino acid* and *peptide* chemistry have increased exponentially (2). The same is true also for *nucleic acid derivatives* (6). Today, there are over 5000 papers on TLC. Some of the recent developments of this technique should be discussed in this paper. First of all, it should be mentioned that great progress has been made in the basic techniques, that means preparation of the layers, developments of the chromatograms. Moreover, the *combined application* of *TLC* with other techniques, such as *gas-chromatography* or different *spectroscopic methods, has* been described by many auhors. Finally, the quantitative evaluation of thin-layer chromatograms, especially by means of *in situ* techniques, has been investigated in different laboratories. It has been shown that it is often useful to carry out the chromatography on layers with a *discontinuous* or *continuous* change of their composition. These are the so-called *gradient-layer* techniques *. On the other hand, *continuous*- or *discontinuous*-elution column chromatography in separation of complex mixtures has been used for many years. In the contrary, solvent gradients in TLC are rarely used, and complex mixtures are often not completely resolved with the one-dimensional technique. Many components of the solute migrate either with the solvent front, or they remain near the origin. The use of a *gradient,* either in the *mobile* or in the *stationary phase*, promises a major improvement. First, we shall discuss gradients in the mobile phase : We

---

* An excellent review on gradient-techniques is given by Niederwieser (7).

start with the so-called *Polyzonal-TLC*. If solvents consist of several components, they are subject to a partial *demixing* in penetrating a dry adsorbent. This effect has also been observed in paper chromatography and is the basis of frontal analysis as described by *Tiselius*. In *polyzonal-TLC*, according to Niederwieser and Brenner (8) substances are separated with the intentional use of this phenomenon. Of course, a trough chamber with saturated gas phase is completely *unsuitable* for this purpose. The demixing effect of a multicomponent solvent can be maintained only by working in a cooled sandwich-chamber, especially in the horizontal position, for instance in the BN-chamber. Let us discuss now the so-called *Duozonal-TLC*. that means the chromatography with a two-component solvent. Demixing takes place if the solvents are of quite different polarity. The polar component is adsorbed preferentially by the adsorbent during chromatography, and a zone — α-*zone* — containing *only* the less polar component, is formed behind the solvent front, which we call α-*front*. If the sample solution is spotted several times at increasing distances from the immersion line, the substances to be separated obtain the opportunity to migrate at different times in solvents which have different compositions than the mobile phase in the solvent tank. In this manner, by the way, practically the same amount of information is obtained in *one chromatogram* as if the sample were chromatographed on *different plates in different solvents*. During chromatography with a two-component solvent, three *extreme* cases are observed : *all substances* migrate in the α-*zone,* they migrate in the β-*zone,* or they migrate in *both zones.* In praxi,, of course, a *superposition* of these three cases can be often observed. If several substances migrate with the *demixing line*, it is of advantage to select a solvent with components that do not differ as much in polarity. Frequently, a separation is achieved by changing to the *Polyzonal* technique, it means to chromatography with a *multi-component solvent.* An exchange of the *polar component* in a two-component system (e.g. substitution of methanol for ethanol) gives a *different position* of the β-*front*. The elution power of the mobile phase can be influenced by a replacement of the polar component with a number of solvents of *similar polarity*. In this case the *demixing line* is fragmented : a solvent consisting of *n-components* will give n-*zones,* separated by *(n - 1)-fronts*. Substances which have migrated with the *demixing-line* of a *two-component* solvent without separation may be separated in a *polyzonal-system* (8). *Polyzonal-TLC* represents the simplest technique of *gradient-elution*. The gradient is, of course, *step-like* and determined by the solvent composition and by the adsorbent. Despite some limitations, which should not be discussed here, *Polyzonal-TLC* is generally superior to « *normal* »-TLC. It should be mentioned that this technique is rather suited for survey of *different substance classes* present in a mixture, but is not satisfactory for the separation of isomers. Polyzonal-TLC is often more sensitive, regarding the detection of substances, than the conventional method, since the spots are more compact and smaller. Therefore, the *Polyzonal-technique* is suitable for *micropreparative work*.

Let us discuss shortly the so-called *solvent-change-TLC* : This technique consists of *discontinuous* alteration of solvent during chromatography without any interruption of the separation process. Solvent change during chromatography is a common practice in column work. Elution is usual started with a *nonpolar* (or less polar) solvent, and after regular intervals changes take place to increasingly *more polar solvents*. The simplest possibility is putting the plate in several solutions, each after the other, without intermediate drying. This method, for instance, has been used by *Randerath* (5) for the separation of *nucleotides*.

We shall discuss now *Continuous Gradients in the mobile phase*. These techniques have been very intensively studied by Niederwieser (7). Since the space is limited, I would like to show you only one possibility : The arrangement for *Elution-Gradient-TLC* with capillary feed of solvent. Other possibilities were for instance, two or multi-chamber mixing systems. *Figure 1*

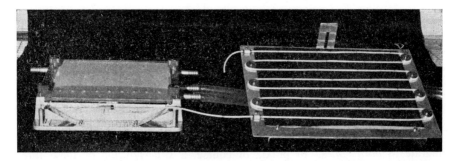

Fig. 1. — Apparatus for Elution Gradient TLC according to Niederwieser (7).

shows Niederwieser's apparatus for Elution-Gradient TLC : A Teflon tubing is strained, wave-like, around posts on a sleigh-like table, as one can see on the *right hand side* of the picture. The tube length depends on the solvent volume needed for a chromatogram. The slope of the table can be adjusted. The Teflon tubing is filled with the solvent (generally one starts with the less polar one). On the *left hand side* of the picture we can see the plate which is put out the cooling block of the *BN-chamber*. The plate is covered with the cover plate and a stainless steel solvent distributor, which is very easy to construct, connected with the filled capillary tubing solvent reservoir is pushed carefully between thin--layer plate and cover plats until it just touches the layer. Raising the end of the reservoir tube forms a capillary film between the two glass plates and enters the thin-layer. The slope of the reservoir tube is now adjusted. The optimal slope is generally valid, provided that layer material and layer thickness remain constant. The applicability of this very simple apparatus is practically without any limitation. In my laboratory, we are using the *Elution-Gradient-TLC* for separation of *nucleic acid derivatives* (9).

I would like to mention that the combined application of this technique with a *discontinuous gradient in the stationary phase* has been found very useful in the analysis of *nucleic acid derivatives*. Other applications are, e.g. separation of lipids, amino acids, mucopolysaccharides according to the findings of Niederwieser (7). As I had already pointed out, it is also possible to carry out the separation with *gradients in the stationary phase*. I feel that at the present time, *discontinuous gradients in the stationary phase* are more important than *continuous* ones. *Discontinuous gradients* in the stationary can be produced by :

a) *Treatment of certain regions* of a coated plate (e.g. spraying or immersion in a solution)

b) *Simultaneous coating* of different adsorbent onto the plate.

I would like to give three examples : Honegger (10) for instance, produced an *activity gradient* by dipping an activated Silica Gel G plate several times into 5 per cent *water in aceton* and *decreasing the immersion depth* each time. Another technique has been usd by Berger (11) who used *Stahl's applicator divided in two segments,* which were filled with two different adsorbents. Berger had shown that *monoiodo-tyrosine, diiodo-tyrosine* and *thyroxine* can be separated, without difficulties, on *Dowex- 1×2 (OH⁻)-layers.* However, if the ^{131}*iodine-labeled* compounds are to be separated in the presence of ^{131}*iodine, thyroxine* as well as ^{131}*iodine* remain together at the starting point. If one wishes now to measure the radioactivity of labeled-thyroxine, which is especially important in clinical chemistry, ^{131}iodine must be removed, whilest thyroxine should not be touched. This problem induced the *French group* (11) to search for new methods in TLC. They used as mentioned the Stahl-applicator *divided into two segments.* The first part of the layer contained silver chloride which is capable of holding back the ^{131}*iodine ions,* thus *thyroxine* is now separated and the activity measurement can be done without any difficulties. In my laboratory two adjacent layers of different composition have also been found suitable in the separation of *nucleotides, nucleosides and nucleic acid* bases (12). It is well known that *nucleotides* on the one hand, and *nucleosides* as well as *nucleic acid* bases on the other, can be separated on *PEI-cellulose layers* using *water* as solvent (13). In this case *nucleotides remain at the origin,* while *nucleosides* and *nucleic acid bases travel.* The first layer therefore consists of *PEI-cellulose,* the second one of *plain cellulose.* After the *nucleosides and nucleic acid bases* have reached the *second (cellulose) layer,* both parts of the plate are to be developed with different solvents. It is very practical to carry out this chromatography not *on plates* but on *plastic sheets,* which can be cut, and the parts can be developed separately. The combined application with *gradient-elution* is especially useful. The *nucleosides* and *bases* which are transferred into the *cellulose part* can be separated in two-dimensions in the usual manner, while the *nucleotides* may be chromatographed using *different gradients* in *both directions.* Although, this *gradient*-technique did work satisfactorily, the preparation of multiple layers was somewhat difficult, due to the special technique needed for the pretreatment of PEI-cellulose. In order to overcome these problems. the *Autotransfer-TLC* has been developed (12). The mixture, consisting of nucleic acid bases, nucleosides and nucleotides, was spotted to a piece of plastic sheet coated with PEI-cellulose. The sheet was irrigated with water, to concentrate the nucleic acid bases and nucleosides in the upper part. This part of the layer was then cut out, and transfer was made to a cellulose plate (or sheet). The nucleosides and nucleic acid bases transferred to the cellulose can be separated as a group, while the nucleotides, which did not move, can be transferred to PEI-cellulose, where the separation of these compounds, as a second group, takes place. It should be pointed out that the *Autotransfer* technique may also be applicable to other separation problems, e.g. group separation of amino acids, or separation of basic, acid and neutral compounds, in general.

In the last three years great progress has been achieved in the combined application of TLC with other separation techniques. So, the separation of complex mixtures of *nucleotides, amino-acids* or *peptides* by electrophoresis combined with TLC has been described by different auhors (cf. 1). Without any details, I would like to mention the so-called « *fingerprint* »-*technique* where *enzymatic cleavage products of proteins* and *polypeptides* are separated by means of electrophoresis and chromatography (2, 3).

It has been shown that the separation on thin-layers is superior to those on papers. A micropreparative version has also been described, after which the isolated peptides can be used for sequence analysis. The combined use of TLC and gas-chromatography, either *continuously* or *discontinuously,* represents other very important advances.

One of the major problems in TCL is, as I feel, the *quantitative evaluation* of the chromatograms. A quantitative evaluation can be made *directly* on the layer (these are the so-called « *in situ* » techniques) or *after elution.* The substances must be spotted as carefully as possible for any quantitative determination. Suitable instruments are precision pipetts or micro syringes. Every operation is, of course, subject to certain error. This error can be considerably reduced by increasing the volume applied. The application error can also be reduced with a smaller volume by introducing an internal standard. Although the *elution-techniques* have been most frequently used, we shall limit ourselves to the discussion of the *in situ-techniques.* Among some existing techniques described in the literature, I shall discuss here two methods which have been using in *my laboratory* (cf. 14).

The first technique is the *fluorometric direct evaluation,* the second one is the *spectrophotometric direct evaluation of thin-layer chromatograms.*

First we shall discuss the fluorometric techniques. Fluorometry is a particularly advantageous method among the procedures available for direct quantitative scanning of thin-layer chromatograms.

A) « Fluorescence » measurements are applicable for such compounds which fluoresce on their own or which can be rendered fluorescent by suitable reagents.

B) The « quenching » method, i.e. fluorescence quenching, is applicable for compounds absorbing ultraviolet light without simultaneously emitting secondary light. In order to make use of the UV-adsorbance it is, in general necessary to make the background of the plate fluorescent. Suitable « fluorescence indicators » are well known. In same cases, and with a highly sensitive instrument, the small amount of natural fluorescence of the adsorbent layer is sufficient. The quenching is, in our experience, applicable for compounds showing UV-adsorption up to 400-500 nm.

The measurement of the fluorescence of a compound is probably the most advantageous one of all optical scanning techniques. The advantages are as follows :

1) Measuring the fluorescence is a direct method, the higher the concentration, the higher the intensity of emitted light. Therefore, within the limits of the instrument used, it is possible to adjust the sensitivity of the measurement according to the compound to be scanned. All other optical techniques are based on indirect measurements, whereby the background of the plate determines the sensitivity.

2) Due to the fact that the background appears dark to the instrument, most of the sources of error which are typical for the transmission and reflectance methods are eliminated.

3) The fluorescence method is particularly versatile. Undoubtedly that there is a larger number of compounds showing native fluorescence than there are substances which absorb light of wave lengths above 400 nm as is preferred for photometric measurements. At least this is so with scanning techniques not requiring a spray reagent, which always incorporates an additional source of error.

The versatility of the fluorometric technique is still greatly increased when the fluorescence quenching is included. However, here the same interfering factors apply which are typical for the reflectance technique. If a compound can be scanned either by fluorescence or by fluorescence quenching, the fluorescence method should always be preferred.

Basically the following procedures for fluorometric *in situ* scanning of TLC plates can be applied :

a) *The CF-technique.*

Chromatography (followed by) Fluorescence measurement (the compound shows native fluorescence).

b) *The CRF-technique.*

Chromatography, Reaction, Fluorescence measurement (the separated, non-fluorescent compound must be rendered fluorescent by a spray reagent).

c) *RCF-technique.*

Reaction, Chromatography, Fluorescence measurement (the compound is rendered fluorescent prior to chromatography, whereby the reaction can be carried out before or after sample application).

The same classification can be made for the quenching techniques. (i.e. QF-, CQF- and RCQ-techniques).

For direct fluorometry we are using a TURNER model 111 filter fluorometer modified for TLC scanning (commercially available from CAMAG Muttenz, Switzerland) (cf. Figure 2). As I shall describe later, the Chromatogram-Spectrophotometer PMQ-II from Zeiss can also be used for this purpose. Other instruments have been described in the literature. The primary (excitation) light passes a range selector, then the primary filter. An aperture plate very close to the surface of the layer allows to adjust the excitation slit in width between zero and 3,5 mm. The height of the slit is 15 mm and fixed. The secondary light emitted from either the spot or the layer is fed into the photomultiplier after passing the secondary, the neutral density filter, and the light pipe.

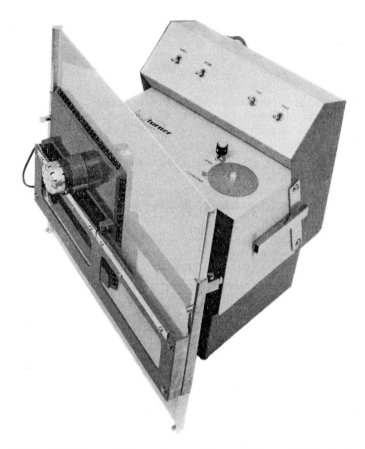

Fig. 2. — The Camag / Turner Scanner (Camag, Muttenz, Switzerland).

The TLC plate is slid with constant speed (10 of 20 mm/min.) past the scanning window. The vertical position of the plate is adjusted by the rack on which the plate holder rides. The direction of scan is, either parallel or perpendicular to chromatography. A low pressure mercury lamp with maximum emission at 366 nm served for all fluorescence measurements. A corning glass filter 7-60 transmittant from 310 to 390 nm was used on the primary side. For the quenching technique a low pressure mercury far-UV-lamp with maximum emission at 254 nm was installed in combination with the primary filter Corning 7-54 with transmittance from 220 to 400 nm. In all cases a sharp cut filter transmittant above 405 nm (Kodak-Wratten 2A) served as a secondary filter. It should be mentioned that, if necessary, the selectivity can be increased by a proper selection of the secondary filter. Other light sources can also be used. The sensitivity of the fluorometer is in the most cases too high in the lowest range. Therefore, neutral density filters (Kodak-Wratten 96A) with transmittances ranging from 1 to 50 % were inserted in addition to the secondary filter. If the chromatographic pattern permits, the scanning should be directed perpendicularly to chromatography. This will give a better base line, particularly with the quenching technique. If the spots on the plate are in sufficient distance from each other, the slit aperture at the instrument should be set to fully open. Resolution of the scan, however, is increased when the slit is adjusted to only 1 mm (14).

For sample application 2 µl self-filling micropipettes were used. The pipetting error plus the error in the actual fluorometric measurement, i.e. the total error attributed to factors other than chromatography and scanning, were estimated and showed a standard deviation of about 1,5 %, which of course is contained in the total variations.

Our investigations have shown that beside fluorescent compounds practically all substances showing absorption up to about 400-500 nm can be scanned (14). The reproductibilities of the fluorescence measurements are better than those of the quenching scans. Scanning examples are given in reference (14).

A linear relationship was found for the fluorescence of DANS-amino acids, whilst the areas of the quenching peaks of the PTH-amino acids are linear to the *logarithm* of the sample quantity. In order to carry out measurements with reasonable reproducibility, the scanning must be made with care (14) ; high error can be made if the scanner is not adjusted correctly. It was further investigated as to what extent fluorometric scanning can be made reproducible if data obtained from different plates have to be compared. This is necessary for instance in *two-dimensional TLC*. A rather poor reproducibility initially observed with amino acid derivates could be attributed to an *influence of time on the fluorescence,* or better, by an influence of the moisture content of the layer. This influence can be eliminated, *as Seiler (15) found,* by spraying the layer with triethanolamine-isopropanol (1 : 4). The time influence on the quenching peaks of *DNP-proline* and *PTH-proline* is pactically not influenced by the spray reagent mentioned. These results indicate that amino acid derivatives can be scanned by fluorometry with fair reproducibility, provided the time influence is eliminated by the spray reagent or that the time delay between drying the plate after the chromatographic run and scanning is standardized. However, standardization of time is the more accurate method (14). This particular example shows also that it is advisable to check in all cases of in situ scanning of TLC plates whether a time influence on the results has to be taken into account.

The quantitative evaluation of thin-layer chromatograms by measuring the light absorption by the *diffuse reflectance* or by the *transmission* method is particularly interesting in quantitation of substances which do not show *native fluorescence* and cannot be *rendered fluorescent* by chemical reactions. There was no lack of attemps, therefore, in developing suitable instruments I shall speak here about the *Chromatogram-Spectrophotometer PMQ-II* of Zeiss, (14, 16) which shows very great versatility. Figure 3 shows this instrument. For different purposes, two different arrangements are used. In the case of *adsorption measurements* (i.e. reflectance) is the following arrangement, suitable : The chromatogram is illuminated with *monochromatic* light and the *diffusely reflected* light is measured.

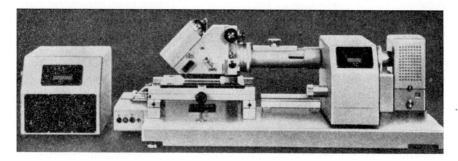

Fig. 3. — The Chromatogram-Spectrophotometer
(Zeiss, Oberkochen, Germany).

We can speak in this case from the *MS-arrangement* (M means mono-chromator ; S means sample). In the case of *fluorescence* or *fluorescence quenching* a second arrangement should be used. This is the so-called *SM-arrangement* (S = sample, M = monochromator). Here a special lamp unit is fitted to the measuring head in place of the detector. For fluorometric determinations a *Hg-lamp* is generally used. The fluorescent light, which is emitted, is *spectrally dispersed* by the *monochromator,* and reaches the detector, which is arranged in this case behind the monochromator. The excitation range is adjusted by suitable filters, the fluorescence maximum is adjusted on the instrument itself. In the case of *fluorescence* it is often possible to use *circular blends*, provided that the spots are round. Using the Chromatogram-Spectrophotometer *colored substances* as well as compounds in the *ultraviolet* region (spectral range : 200-2500 nm) can equally be measured. Moreover, *fluorescence* and *fluorescence quenching* can also be measured. Examples of application are given by Jork (16) and others (cf. 14).

We are using the *Chromatogram-Spectrophotometer PMQ-II* for scanning of different substances (14). Here, I want to give an example. Different quantities of *glycine* (2, 4, 8 and 16 mcg) are chromatographed on Silica Gel and the *optical reflectances* have been measured after coloring with *Ninhydrin.* The scanning was made at *510 nm* perpendicular to the direction of chromatography. There is linear relationship between *square roots of amounts* and the *areas of the peaks.* As already mentioned, the Chromatogram-Spectrophotometer can be used for measurement of *absorption* as well as for *fluorescence quenching.* We, generally, prefer the *absorption method,* owing its better reproducibility. Of course, the peaks which we obtain with *reflectance* spectroscopy depend on the *wavelength,* at which the determination was made, e.g. *PTH-Histine* gives a much higher absorption at *271 nm* than at *254 nm.* The optimal *wavelength* must be therefore determined in each case. Further examples for *in situ reflectance spectroscopy* are nucleic acid derivatives (14). All previous examples mentioned concerned the scanning of substances on *one-dimensional* chromatograms. Of course, two-dimensional chromatograms can also be evaluated.

The *diffuse reflectance* on thin-layer chromatograms depends also on the time-intervall between interruption of the chromatography, drying and start of scanning (14). It is advisable therefore, to standardize the drying time. Considering this fact the *relative standard deviation* of scanning amounts, in the case of different chromatograms, about 6 % (14).

When I began the description of the *Chromatogram-Spectrophotometer PMQ-II,* I drew your attention to the high *versatility* of this instrument. For *qualitative identification* and *characterization* of substances separated on thin-layer chromatograms, it is possible to determine *full spectra,* since the monochromator gives any desired wavelength between *200 and 2500 nm.* These spectra can serve also as guides in choice of the « right » *wavelength* prior scanning as mentioned earlier. There is no doubt that the different spectral characteristics determined by *in situ* methods allow the identification of minute amounts. The *in situ determination* of *absorption spectra* is now being applied in my laboratory in identification of substances in *complex natural mixtures.*

References

1) E. STAHL, Dünnschichtchromatographie, Springer, Berlin, 1967.
2) G. PATAKI, Dünnschichtchromatographie in der Aminosäure und Peptidchemie. De Gruyter, Berlin, 1966.
3) G. PATAKI. Techniques of Thin-Layer Chromatography in Amino Acid and Peptide Chemistry, *Ann. Arbor Science Publ.,* Ann. Arbor, 1968.
4) G. PATAKI. *Chrom. Rev.,* 9 : 23, 1967.
5) K. RANDERATH and E. RANDERATH, in *Methods of Enzymology,* 12, part A 323 (1967).
6) G. PATAKI, in *Advances in Chromatography* (eds. : C. Giddings and R.A. Keller). 7 : 47, 1968.
7) A. NIEDERWIESER. *Chromatographia,* 2 : 23, 1969.
   A. NIEDERWIESER and C.G. HONEGGER, *Advances in Chromatography,* (eds. : C. Giddings and R.A. Keller), 2 : 124, 1966, and private communication.
8) A. NIEDERWIESER and M. BRENNER. *Experientia,* 21 : 50, 105, 1965.
9) G. PATAKI and A. NIEDERWIESER. *J. Chromatogr.,* 29 : 133, 1967.
10) C.G. HONEGGER. *Helv. chim. Acta,* 47 : 2384, 1964.
11) J.A. BERGER, G. MEYNEL, J. PETIT and P. BLANQUET, *Bull. Soc. Chim. France,* 2662, 1963.
12) G. PATAKI, J. BORKO, H. Ch. CURTIUS, F. TANCREDI. *Chromatographia,* 1 : 406, 1968.
13) K. RANDERATH and E. RANDERATH. *J. Chromatograph.,* 16 : 111, 1964.
14) G. PATAKI. *Chromatographia,* 1 : 492, 1968.
15) N. SEILER and M. WIECHMANN. *Z. analyt. Chemie,* 220 : 109, 1966.
16) E. STAHL and H. JORK. *Zeiss Informationen,* 16 : 52, 1968.

Determination of steroids in biological materials by gas chromatography

J. W. A. MEIJER,

Gaubius Institute of the University of Leyden, The Netherlands.

Summary

Gas chromatography is being used more and more extensively for the routine determination of steroids in materials of biological origin. In practice two groups of problems present themselves : firstly the preliminary processing of the sample in order to obtain a specimen suitable for injection and secondly the gas chromatographic separation and detection. In most cases the processing consists of a rather complicated sequence of steps such as hydrolysis, extraction, purification by other chromatographic techniques and conversion of the steroids into derivatives to increase their separability or detection response. The reliability of the analysis depends on the completeness of all of those steps. Eventual losses during this procedure can be compensated by the use of a suitable internal standard. The optimum performance of the gas chromatographic step requires special precautions to prevent a partial adsorption or catalytic conversion of a part of the specimen.

Automatic injection proves to be of great value in routine analyses. It may be expected, that the further development of highly efficient open tubular columns, which begin to find practical application in the steroid field, will allow the purification procedures to be less complicated. Practical applications of gas chromatography in the determination of saturated sterols in lipids and of testosterone in urine are discussed.

I. — INTRODUCTION.

History of gas chromatography of steroids commences in 1959. In the course of that year Eglington and co-workers (1) with the aid of this method succeeded in separating some relatively stable steroid hydrocarbons and ketones. When doing so they still applied the then known techniques which caused retention times to be impractically long. Soon afterwards various other authors demonstrated that, when applying improved circumstances, the greater number of known steroids possess a sufficient degree of thermo-stability to be subjected to this method of analysing. Horning, Sweeley, Van den Heuvel (2,3), Beerthuis, Recourt (4) etc. are among these workers of the first hour.

The total number of publications on gas-chromatography of steroids today has substantially surpassed thousand. From this literature various phases in its development can be recognized. In the initial publications already the most important principles of column technology, namely the application of uncoloured, thoroughly purified supports and of only low concentrations of thermostable stationary phases like SE-30, QF-1 and other silicones had been laid down. As examples of the results which could possibly be achieved mainly chromatograms of synthetic mixtures were produced. The practical applications were at first restricted principally to the examination of the unsaponifiable fraction from animal and vegetable lipids.

The workers in the field were not long in discovering the great amount of regularity prevailing in the relationship between the structure of steroids and their gas chromatographic behaviour (5).

It was found that gas chromatography kept in store considerable promises as a tool for identifying unknown steroids. In the measure the method was being applied on a more extensive scale for practical problems of analysis, refining of the techniques proved to be indispensable, for it was experienced in a lot of cases that the quantitative exactness fell short owing to the circumstance that a certain part of the sample was changed or adsorbed irreversibly at the column (6). As a means of improvement of the peaksymmetry the modification of polar functional groups with the aid of trimethylsilylating agents was introduced (7). The thermostability of steroids proved to be greatly dependent on the kind of materials with which they get in touch in the course of the gas chromatographic procedure (8). At present it is generally assumed that in order to achieve optimum results the construction of the apparatus should be such as to avoid to the greatest possible extent that the steroid vapours get in touch with metals.

The practical usefulness of gas chromatography was only really put to the test when it got a footing in clinical-chemical laboratories. In this field the range of problems proved to be more complicated than in the case of separating arbitrarily composed mixtures or of tolerably pure natural sterol fractions. The very object is the quantitative determination of minimum quantities of frequently very labile compounds such as androgens, estrogens or corticosteroids in rather complex starting materials. The utility of the estimation depends in many cases for a great deal on the time needed to obtain the analytical information.

II. — SOURCES OF ERROR.

While the gas chromatographic step itself can be performed in a rapid way, it often has to be preceded by rather time consuming preparations in order to convert the sample into a suitable form.

Because these procedures are often very extensive, certain partial losses of the components to be determined are likely to occur (Table I). Consequently a method of determination by which the peak area eventually obtained is exclusively related to the injected part of the original sample is doomed to be greatly inexact.

TABLE I

STEPS IN WHICH LOSS OF MATERIAL IS LIKELY TO OCCUR :

1 HYDROLYSIS (INCOMPLETE REACTION , PARTIAL DESTRUCTION)

2 EXTRACTION

3 FRACTIONATION (INCOMPLETE RECOVERY IN TLC ,
                                COLUMN CHROM. , ETC.)

4 PREPARATION OF DERIVATIVES

5 TRANSFER PROCEDURES

6 GAS CHROMATOGRAPHY (PARTIAL ADSORPTION AND/OR
                                CHEMICAL CONVERSION ON THE COLUMN)

When examining urine samples the losses start already at the first step, necessary to convert the steroid-conjugates into the free steroids. When applying acid hydrolysis, conditions should be chosen at least so drastically that the total quantity of the components to be determined is liberated. In the course of this treatment however, a partial disintegration is inevitable. The hydrolysis is effected in a much gentler way with the aid of enzymes such as sulfatase and glucuronidase. This method has as a disadvantage the considerably greater wear of time. Moreover, owing to the presence of inhibiting compounds in urine the enzymes have to be added in rather large portions. Consequently the enzymes should meet high demands of purity in order to prevent the introduction of contaminants which might interfere with gas chromatography. After the hydrolysis has been completed a solvent extraction is performed. Although the losses resulting herefrom need not be large they should not be neglected.

The extract, which in most cases is of a vivid colour, contains a great many steroids and other lipid soluble compounds which will overlap one another in gas chromatography. Direct injection of this mixture makes only sense when one aims at the determination of steroids which are available in such quantities that their detection is scarcely hampered by the other components. In most cases however, first one or more steps of purification have to be made, for instance with the aid of thin layer, paper- or column chromatography.

When applying these fractionations one seldom succeeds in recovering the compounds in their original quantities. Rather they will be partially lost by irreversible adsorption at the chromatographic media, by chemical decomposition or by incomplete transfer.

Prior to introducing the purified extract into the gas chromatograph, in a number of cases it is submitted to a chemical treatment (Table II). With

TABLE II

SUBSTITUTION OF FUNCTIONAL GROUPS

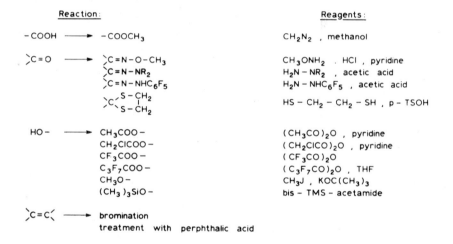

Reaction:	Reagents:
$-COOH \longrightarrow -COOCH_3$	CH_2N_2 , methanol
$\rangle C=O \longrightarrow \rangle C=N-O-CH_3$	CH_3ONH_2 . HCl , pyridine
$\rangle C=N-NR_2$	H_2N-NR_2 , acetic acid
$\rangle C=N-NHC_6F_5$	$H_2N-NHC_6F_5$, acetic acid
$\rangle C \langle {S-CH_2 \atop S-CH_2}$	$HS-CH_2-CH_2-SH$, p-TSOH
$HO- \longrightarrow CH_3COO-$	$(CH_3CO)_2O$, pyridine
$CH_2ClCOO-$	$(CH_2ClCO)_2O$, pyridine
CF_3COO-	$(CF_3CO)_2O$
C_3F_7COO-	$(C_3F_7CO)_2O$, THF
CH_3O-	CH_3J , $KOC(CH_3)_3$
$(CH_3)_3SiO-$	bis-TMS-acetamide
$\rangle C=C\langle \longrightarrow$ bromination	
treatment with perphthalic acid	

a view to improving thermostability and peak symmetry and reducing retention times transfer of hydroxyl groups into trimethylsilylgroups (7) or trifluoroacetylgroups (9) is more and more applied. The gas chromatography of these derivatives seems to make somewhat mitigated demands to the quality of both apparatus and columns. In cases where the quantities available in the starting materials are too small for being detected by means of flame ionization, the application of electron capture detection is the obvious thing to do. For this purpose the steroids should first be transferred into halogen containing derivatives, e.g. chloroacetates (10), trifluoroacetates (9), heptafluorobutyrates (11) or pentafluorophenylhydrazones (12). To illustrate the considerable reduction of retention times, obtainable by substitution of functional groups, figure 1 shows the chromatograms of a mixture of ketosteroids before and after treatment with methoxyamine (13). However simple and smooth the course of the above reactions may be, a reaction running for the full hundred percent in one direction is only wishful thinking.

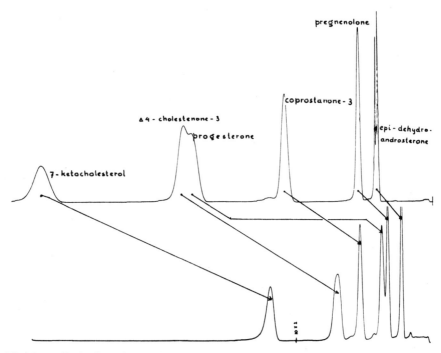

Fig. 1. — Reduction of retention times on QF - 1 by modification of $>C = O$ into $>C = N - O - CH_3$

     Also the last step in the analysis, the proper gas chromatography may, in spite of all precautions, be accompanied by losses, caused by partial decomposition or by irreversible adsorption at the support. The deviations resulting from this phenomenon are increasing in the measure the sample is smaller. In some laboratories the daily gas chromatographic routine is commenced by an injection of reference materials, repeated many times until a so-called pseudostable response plateau has been built up before quantitative determinations are being made.

III. — INTERNAL STANDARDS.

     The sum of the somewhat pessimistic enumeration of losses, stated so far, naturally depends greatly on the skill by which the various steps of the analysis are performed. Regularly, especially in clinical-chemical papers, we find publications in which improvements are recommended. So far, however, the most recommendable way has been to make the outcome of the analysis independent of possible losses. This can be realized by using a select internal standard. This will be of maximum usefulness only if it is added in a stage which lies as early as possible, preferably already at the beginning of the analysis. Putting it ideally, the chemical and physical-chemical properties of this standard should be such that it is participating equally in the adventures of the material in the course of the preparations preceding the final detection. Only in the conclusive step the difference between the chromatographic properties should be sufficient to enable a comparison of quantities.

The use of radio-active internal standards offers the most logical solution to this problem. Various successful applications of this principle have been reported. As a consequence however, the problematics of gas chromatography are enlarged with those related to the accurate counting of radioactivity in the effluent of the column.

A much simpler and less expensive way is to choose as an internal standard a substance of a closely related structure, differing from the sample only to such an extent that a gas chromatographic separation is possible. This difference need not to be greater than one chemically inert group. Of course one should be sure that the standard compound does not occur in the starting material. One could conceive of many steroids which can act as a standard in the gas chromatographic analysis of most natural steroids. Figure 2 gives two examples, which will be discussed in the following.

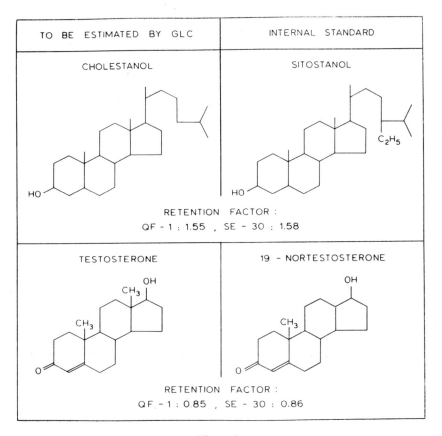

Figure 2.

IV. — DETERMINATION OF SATURATED STEROLS.

The first example is related to the study of saturated sterols in human sterol fractions. An exact determination of the so-called minor components, which constitute only a fraction of a percent of the cholesterol content, for a long time has been a challenge to analytical chemistry. One of these compounds, cholestanol, is a relatively inert, saturated sterol. The absence of a double bond between the carbon atoms 5 and 6 is the only difference on which an analysis of cholesterol-cholestanol mixtures can be based.

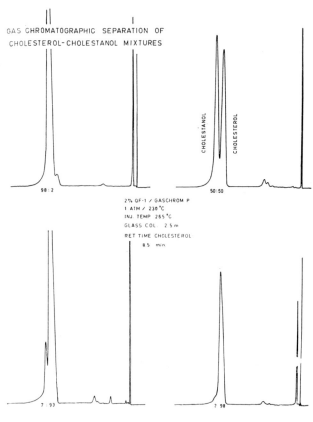

Figure 3.

As shown by figure 3 the gas chromatographic separation of equal amounts of both sterols does not present any problems. However, when the cholestanol content is less than 2 % a quantitative evaluation of the chromatograms is impossible. The resolving-power of even the best conventional packed columns is insufficient to prevent the small cholestanol peak from being submerged by the dominating cholesterol peak. As will be seen afterwards, the development of highly efficient capillary columns may give an answer to this problem. Using conventional columns however, the determination of saturated sterols has to be preceded by the removal of cholesterol. This could be accomplished by means of silvernitrate chromatography. Another way is to convert cholesterol into a compound of greater polarity that can be easily removed from the mixture by means of adsorption chromatography on silicagel. Owing to its double bond cholesterol reacts with bromium, ozone or peroxy acids. The action of perphthalic acid on cholesterol acetate under moderate conditions results in the formation of the epoxides giving a very high yield. Saturated sterolacetates do not participate in this reaction (see Fig. 4). They can be isolated from the reaction mixture by thin layer chromatography and afterwards be estimated by gas chromatography.

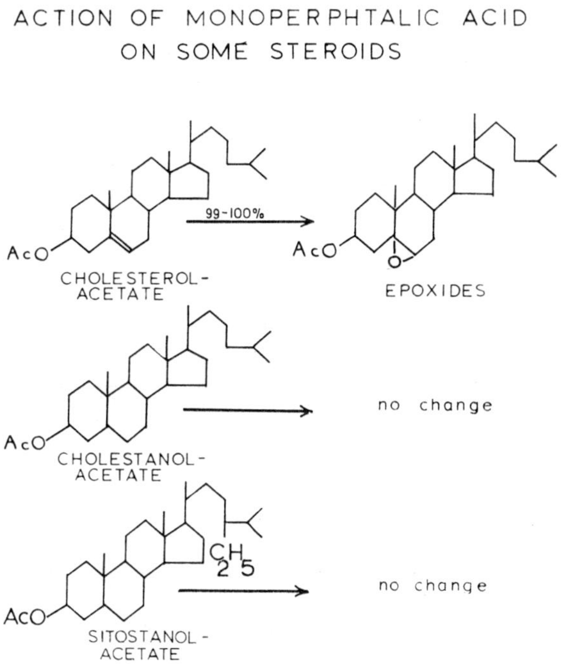

ACTION OF MONOPERPHTALIC ACID
ON SOME STEROIDS

Figure 4.

The above method implies the application of an internal standard by which the areas of the saturated sterol peaks are to be compared. Initially cholestane was used for this purpose. This steroid hydrocarbon has found various applications for similar analytical procedures. However, we conclude from the experiences we have made that cholestane is quite unsatisfactory as a standard in the quantitative analysis of steroids having one or more functional groups. In thin layer chromatography its Rf-value differs considerably from that of the sterol acetates and this makes it necessary to scrape more than one band from the adsorbent layer for the recovery of

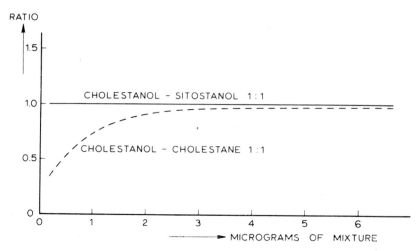

Fig. 5. — Variation of peak area ratio with sample size of mixtures.

the fractionated materials. Furthermore the cholestane peak, having a rather short gas chromatographic retention time, emerges in a region where the base line is still disturbed by low molecular weight impurities from the sample or from the solvents used in the fractionation steps.

TABLE III

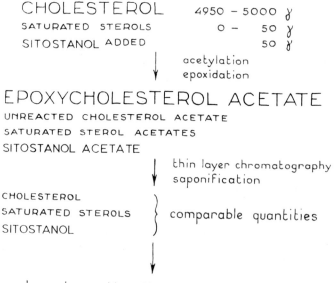

A third objection to the use of cholestane, often overlooked but important as well, is that losses of this hydrocarbon during chromatographic procedures are somewhat smaller than those of more polar steroids. As shown by figure 5 the recorded ratio of cholestane and cholestanol in a mixture of both is dependent on the injected quantity. Considerable errors have to be feared when the sample size is only small. The same figure shows also that this discrepancy is practically zero when another saturated monohydroxysterol is compared with cholestanol. Sitostanol, a substance which under natural conditions is not found in human sterol fractions, is chemically to be regarded as cholestanol which has an extra ethylgroup at C_{24}. Therefore both sterols behave towards a great number of chemical reactions in exactly the same way. Both have the same Rf-value in thin layer chromatography on silicagel, but they are readily separable by gas chromatography. The course of the analysis is schematically shown by Table III. figure 6 shows the chromatogram of a plasma sterol fraction, initially containing less than 0.5 % of cholestanol, after the above fractionation.

A similar procedure can be followed when the study of saturated sterols in plant lipids is aimed at. In this case the standard may consist of cholestanol, a sterol that is absent in vegetable material.

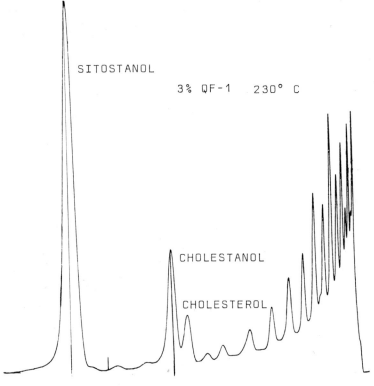

Figure 6.

V. — DETERMINATION OF TESTOSTERONE IN URINE.

Another example relates to the gas chromatographic estimation of testosterone in urine.

In relation to the concentrations in the starting material mentioned this determination causes only difficulties which are inferior to those connected with the determination of testosterone in blood. The particular difficulty is, however, that the testosterone in urine is attended with a great many other steroids, the greater part of which first has to be removed. First step to be made in this case is again the adding of a known amount of an internal standard. Experiments in which $\Delta 1$-testosterone was used for this purpose, failed on account of the instability of this compound. A more suitable standard is nortestosterone, which can be regarded as testosterone, missing the angular methylgroup at C_{19}.

Next step will be the addition of hydrochloric acid, following by boiling of the urine, in order to hydrolyse the conjugates. This treatment is inevitably attended with a certain amount of deterioration. It appears however that the rates of disintegration of testosterone and nortestosterone are practically the same so that their relation is not changed. The extract, obtained by shaking with ether is first treated with diluted alkali, by which procedure a great part of the coloured materials and the acid components (among which the estrogens and bile acids) are removed. Figure 7 shows a chromatogram of the now obtained solution ; the dotted vertical line indicates the site of the testosterone peak. Subsequently the extract is con-

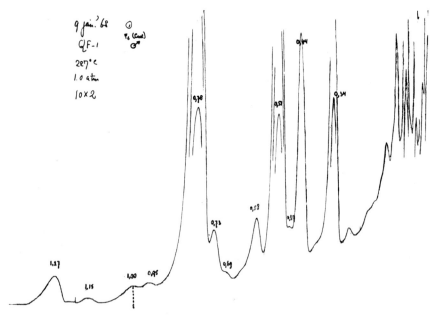

Figure 7.

centrated and applied in the form of a band onto a chromatoplate of Silicagel containing a fluorescent indicator. As a reference it is sided by small quantities of testosterone and nortestosterone. After development with chloroform-ethylacetate (3 : 1) the location of the testosterone-band is verified under a UV-lamp. The Rf-values of testosterone and nortestosterone are the same in this system.

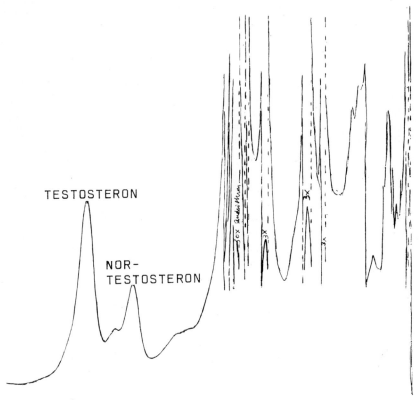

TESTOSTERON

NOR-
TESTOSTERON

Figure 8.

The band in question is scraped from the chromatoplate, extracted and eventually analysed gas chromatographically (Fig. 8). It goes without saying, that in the final calculation of the results the so-called response ratio between standard and material to be determined has to be incorporated.

In literature there exists some discrepancy regarding the normal values of testosterone concentrations in urine. Values found by applying the internal standard method seem to be somewhat elevated owing to the fact that losses during the analysis procedure do scarcely bear on the final results of the analysis.

VI. — INTRODUCTION OF CONTAMINANTS.

As we have dwelt a little long upon the different steps which may cause a partial loss of material, the opposite namely the unwanted introduction of interfering contaminations may as well be mentioned. In almost every laboratory it happens now and then that work on routine samples will loose temporarily its dullness when one suddenly comes across unusual peaks. In too many cases the excitement dies quickly by further investigation when it is revealed that the « new components » originate from impurities introduced during the processing of the sample. Compounds having about the same range of retention times as steroids may be introduced by careless use of lubricants, wax pencils, hand lotions, detergents or even hair preparations. When working on a submicrogram scale, rubber or cork stoppers, plastic bottles and other similar organic materials should be kept far from the samples. The purification of adsorbents, chromatography paper and solvents often offers an almost unsolvable problem. An interesting survey on laboratory contaminants is given by Rouser c.s. (14).

VII. — SOLVENT FREE INJECTION.

In later years nuch attention has been paid to the construction of special sampling systems for steroids and other compounds of low volatility. The principal aim of these new systems is to prevent the introduction of solvents into the analytical column. The smaller volume of the entering vapour packet results in narrower peaks and a better resolution. Moreover, in the analysis of very small quantities the detector sensitivity can be strongly increased without interference from the often very long tail of the solvent peak.

In the communication of Dr. Evrard already an interesting device is shown in which the injected solution is first fractionated at a precolumn. In somewhat less sophisticated methods the samples are directly introduced into the column after evaporation of the solvents. In a technique that has become a routine for years in our laboratory (see Fig. 9) the mixture to be analysed is deposited on a small metal spiral by dipping the latter into a solution and allowing the solvent to evaporate in the air. Subsequently the spiral is suspended via a magnetic device in a cold part of a glass evaporation chamber, which is connected directly to the column via a glass O-ring joint. As soon as the base line is restored, the spiral is dropped into the heated zone. The spirals are made of resistance wire of a special alloy (Kanthal A) which proves to have practically no effect on the thermostability of steroids, provided that its surface is smooth and not etched by acids or flame. The introduction of the spirals can also be performed by an automatic device. The latter method creates the possibility to run series of samples overnight, this increasing the working capacity of the GLC apparatus. More important still is the increase of the reproducibility by the elimination of random variations in manipulation. An apparatus, constructed in our laboratory for routine analyses (20), consists of a cylindrical magazine of plexiglass, provided with 36 small holes to contain the

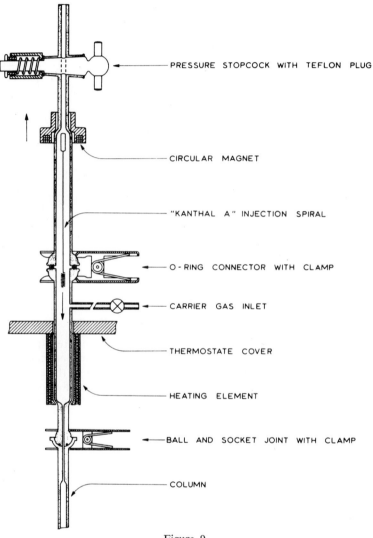

PRESSURE STOPCOCK WITH TEFLON PLUG

CIRCULAR MAGNET

"KANTHAL A" INJECTION SPIRAL

O - RING CONNECTOR WITH CLAMP

CARRIER GAS INLET

THERMOSTATE COVER

HEATING ELEMENT

BALL AND SOCKET JOINT WITH CLAMP

COLUMN

Figure 9.

sample spirals. This magazine, which rotates stepwise by the action of an electromagnet, is enclosed in a gas-tight plexiglass box, connected to the evaporation room of the gas chromatograph. At the activation of the electromagnet, which is controlled by a time clock, a spiral falls through an aperture into the evaporation room. After a series of analyses the spirals can be recuperated from the evaporation room by means of a small magnet. A compact, stainless steel version of the above apparatus, controlled pneumatically, is being manufactured by Carlo Erba (Milan, Italy). Other automatic injection devices are brought onto the market by Pye (England)

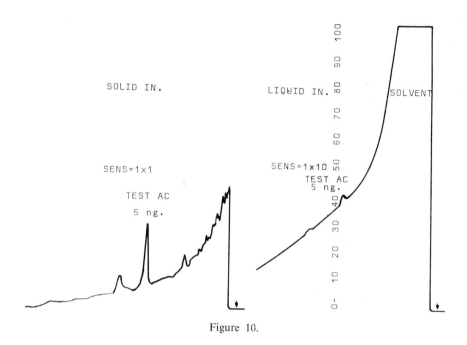

Figure 10.

and by Barber Colman (USA). The advances of solvent free injection are clearly illustrated by figure 10. Using a flame ionization detector, 5 nanograms of testosterone acetate were injected as a solution in the conventional way (b) and without solvent by means of a Carlo Erba automatic sampling system (a). Only in the second case the detector sensitivity could be fully utilized.

VIII. — CAPILLARY COLUMNS.

Automatization can be extended further to the final calculation and interpretation of results by means of electronic integrators and computers. In fact, many features of automatization may be borrowed from the often very ingenious instrumentation, used nowadays in the control of industrial processes.

Up to the present however, the practical profit to be gained by the application of such methods in clinical laboratories is only occasional. As stated already, the time consuming preparation of the samples is still a limiting factor.

A substantial part of these preparations could be omitted if the efficiency of gas chromatographic separation were greater then is practicable with normal packed columns. Therefore great advantages are to be expected in this area when open tubular columns are used. In order to avoid the risk of catalytic conversions on metal surfaces, these columns should preferably be made of glass. Although the idea of using open tubular glass columns in gas chromatography is of older date, the number of investigators who are

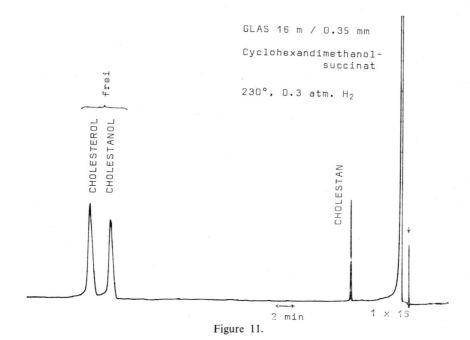

Figure 11.

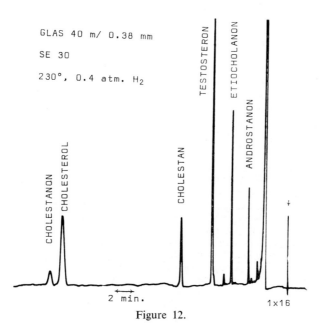

Figure 12.

prepared to cope with the technical difficulties connected with its development has been rather small until now. One of them, Dr. Grob of the Zürich University (15), was so kind to comply with our request by separating a number of steroids on glass capillary columns, not especially prepared for this purpose. The results, shown in Fig. 11 and 12 may act as a stimulus for further adaptation of this method to the analysis of steroids.

IX. — CONCLUSION.

It was endeavoured in the preceding to render a general view of some of the problems, connected with the gas chromatographic determination of steroids in biological materials. As few fields have advanced faster in the past five years, it was not possible to enter into many other practical applications and solutions to technical problems, described in recent literature. Therefore we will not conclude without referring to some sources of further information on the subject. General principles of theory and technique were reviewed by Horning and Vanden Heuvel (16). Many experimental details of clinical chemical applications can be found in the book of Wotiz and Clark (17). The same subject has been treated very recently by Eik-Nes and Horning as a second volume of « Monographs on Endocrinology » (18). Kuppens (19) described some new techniques for the improvement of resolution in steroid gas chromatography. A lot of practical information on the quantitative treatment of microliter samples, use of microcolumns, electron capture detection, reports of investigations on steroids in a variety of biological fluids etc. is contained in a collection of 16 papers presented at the Symposium of the Endocrinological Society (April 4-6, 1966, Glasgow (21)).

Acknowledgements :

The investigations were supported by the Netherlands Organization for Health Research, T.N.O. The author is indebted to Dr Grob for his permission to publish Fig. 11 and 12. The valuable technical assistance of Mr P.W.L. van der Meer and Miss R.P. de Graaf is gratefully acknowledged.

References

(1) EGLINGTON G., HAMILTON R.J., HODGES R., and RAPHAEL R.A.: GLC of natural products and their derivatives. *Chem. & Ind.,* **25** : 995, 1959.
(2) HORNING E.C., SWEELEY C.C., and VANDEN HEUVEL W.J.A.: SE-30 aids separation of complex compounds. *Chem. & Eng.,* July, 40, 1960.
(3) VANDEN HEUVEL W.J.A., SWEELEY C.C., and HORNING E.C.: Microanalytical separations by GLC in the sex hormone and bile acid series. *Biochem. Biophys. Res. Comm.,* **3** : 33, 1960.
(4) BEERTHUIS R.K., and RECOURT J.H.: Sterol analysis by GLC. Symp. on drugs affecting lip. metab., Milan, 2 - 4 June 1960.
(5) CLAYTON R.B.: GLC of sterol methyl ethers. *Nature,* **190** ; 1071, 1961.
(6) HORNING E.C., MADDOCK K.C., ANTHONY K.V., and VANDEN HEUVEL W.J.A.: Quantitative aspects of GLC in biological studies. *Anal. Chem.,* **35** : 526, 1963.

(7) LUUKKAINEN T., VANDEN HEUVEL W.J.A., HAAHTI E.O.A., and HOR-NING E.C.: GLC behaviour of trimethylsilyl ethers of steroids. *Biochim. Biophys. Acta*, **52**: 599, 1961.

(8) MEIJER J.W.A.: Het optreden van omzettingen bij de gaschromatografie van sterolen. *Chem. Weekblad*, **59**: 396, 1963.

(9) VANDEN HEUVEL W.J.A., SJOVALL J., and HORNING E.C.: GLC behaviour of trifluoroacetoxy steroids. *Biochim. Biophys. Acta*, **48**: 594, 1961.

(10) LANDOWNE R.A., and LIPSKY S.R.: The electron capture spectrometry of haloacetates: a means of detecting ultramicro quantities of sterols by GLC. *Anal. Chem.*, **35**: 532, 1963.

(11) CLARK S.J. and WOTIZ H.H.: Separation and detection of nanogram amounts of steroids. *Steroids*, **2**: 535, 1963.

(12) ATTAL J., HENDELES S.M., and EIK-NES K.B.: Determination of free estrone in blood plasma by GLC with electron capture detection. *Anal. Biochem.*, **20**: 394, 1967.

(13) FALES H.M. and LUUKKAINEN T.: O-methyloximes as carbonyl derivatives in gas chromatography, mass spectrometry and nuclear magnetic resonance. *Anal. Chem.*, **37**: 955, 1965.

(14) ROUSER G., KRITCHEVSKI G., WHATLY M., and BAXTER C.F.: Laboratory contaminants in lipid chemistry: detection by TLC and infrared spectrophotometry and some procedures minimizing their occurrence. *Lipids*, **1**: 107, 1966.

(15) GROB K.: Glaskapillaren für die Gas-chromatographie. Verbesserte Erzeugung und Prüfung stabiler Trennflüssigkeitsfilme. *Helv.*, **51**: 718 - 737, 1968.

(16) HORNING E.C., and VANDEN HEUVEL W.J.A.: Quantitative and qualitative aspects of the separation of steroids. *Advances in Chromatography*, **1**: 153 - 198, New York, Marcel Dekker, Inc., 1965.

(17) WOTIZ H.H., and CLARK S.J.: Gas chromatography in the analysis of steroid hormones. New-York, Plenum Press, 1966.

(18) EIK-NES K.B., and HORNING E.C.: Gas phase chromatography of steroids. Heidelberg, Springer Verlag, 1968.

(19) KUPPENS P.S.H.: High resolution GLC in steroid analysis. Eindhoven, Thesis 1968, Technische Hogeschool.

(20) MEIJER J.W.A.: Gas chromatography of steroids. *Planta Medica*, pp. 48-59. Supplement 1967.

(21) GRANT J.K., (Editor). The gas liquid chromatography of steroids. Cambridge University Press, 1967.

COMMUNICATIONS

MEDEDELINGEN

Un nouveau concept en électrophorèse : l'ensemble « Phoroslide » Millipore

par

Jacques JOLLIVET
Millipore S.A. — Paris

INTRODUCTION

Au cours de ces dernières années, l'acétate de cellulose a connu un succès croissant comme support d'Electrophorèse clinique. Ce support permet des séparations rapides et une résolution très grande des différentes fractions, les queues de fraction sont pratiquement inexistantes entre les différents éléments.

Toutefois, il est apparu au cours des manipulations que les propriétés mécaniques de ce support étaient une source de difficultés pour la réalisation des appareils. Il faut noter qu'une membrane d'acétate de cellulose est une mince structure micro-poreuse dont la porosité est environ 80 %. Ce matériel demande beaucoup de soin dans ses manipulations, car il est cassant à l'état sec, souple à l'état humide, mais très fragile dans les deux cas.

En développant le Phoroslide™, Millipore a mis au point un support de séparation qui élimine les limitations d'emploi de l'acétate de cellulose conventionnel. Ce nouveau matériel a permis d'envisager l'électrophorèse sous un nouvel angle. Le système Phoroslide est extrêmement simple et son utilisation précise. Il ne nécessite pas d'entraînement ni de pratique spéciale comme dans le cas des méthodes traditionnelles. Les résultats s'obtiennent facilement, sont sûrs et reproductibles.

MATERIEL

Les principaux éléments du système Phoroslide sont : les bandes-supports, la cuve Phoroslide et les applicateurs, le stabilisateur de tension et le densitomètre Phoroscope™. Tous ces éléments ainsi que les différents accessoires utilisés, sont fournis par Millipore.

Les bandes-support Phoroslide

La Phoroslide possède une structure élastique avec une surface micro-poreuse sur laquelle les migrations électrophorétiques sont identiques à celles obtenues sur l'acétate de cellulose conventionnel. (La Phoroslide est rectangulaire, 25 x 75). On peut faire sur chaque bande deux applications d'échantillon en utilisant l'applicateur Phoroslide. La structure de la bande lui donne une rigidité qui ressemble à celle d'un ressort et qui ne diminue en rien ses performances en électrophorèse, mais, bien au contraire, augmente sa résistance et sa facilité d'emploi. Les Phoroslide peuvent être diaphanisées aisément pour une analyse densitométrique que l'on peut réaliser avec le nouveau Densitomètre Phoroscope de Millipore, soit avec d'autres appareils existants.

La Cuve à Electrophorèse Phoroslide

La cellule Phoroslide peut se diviser en deux parties : une partie inférieure avec les deux bains à tampon et la partie supérieure qui possède, d'une part, les électrodes en platine qui occupent toute la largeur de la cellule et auxquelles toutes les connections électriques sont fixées et un disque rotatif permettant de faire aisément le dépôt de l'échantillon avec les applicateurs Phoroslide. La cuve est faite en un plastique transparent. Un épaulement est moulé dans chaque bain à tampon de telle façon que la Phoroslide une fois en place est automatiquement maintenue en position sans aucun moyen de tension quelconque.

Chaque compartiment de la cuve a une contenance de 5,5 ml et ainsi, le besoin total en tampon s'élève à 11 ml. On peut mettre en série jusqu'à sept cuves à électrophorèse pour un même stabilisateur de tension.

Les Applicateurs Phoroslide

L'extrémité de l'applicateur Phoroslide est formée de deux fils en acier inoxydable maintenus parallèlement dans un corps en plastique. Le capillaire ainsi formé, permet de faire les applications d'échantillons uniformes, rectilignes (environ 4 mm de long) d'un volume constant de 0,3 micro-litre. L'application se fait en mettant les applicateurs chargés d'échantillons dans les trous percés à cet effet au travers du couvercle et du disque rotatif. Les applicateurs sont maintenus en l'air par un épaulement du disque rotatif. Pour réaliser l'application il suffit de faire tourner le disque rotatif qui laisse tomber ainsi les deux applicateurs dont les extrémités viennent en contact avec la Phoroslide et déposent les échantillons.

L'Alimentation

L'alimentation est une source de tension stabilisée, fournissant 100, 200 ou 300 V que l'on peut choisir à l'aide d'un contacteur. Un bouton permet d'inverser la polarité. Ceci permet de faire passer le pôle positif d'un bain à l'autre entre deux électrophorèses successives. La tension à la sortie, ne peut varier de plus de 1 % dans la mesure où l'on n'a pas branché plus de sept cuves en série.

Le Densitomètre Phoroscope ™

Le Densitomètre Phoroscope utilise les méthodes optiques et électroniques les plus modernes pour obtenir une représentation visuelle instantanée de l'électrophorégramme, et permet de faire une mesure des pourcentages représentés par chaque fraction de la séparation. L'électrophorégramme est représenté par une courbe apparaissant sur un tube cathodique. Il est possible d'isoler successivement les différents pics de la courbe par des dispositifs électroniques très simples, ce qui permet d'effectuer les mesures des pourcentages correspondant à chaque fraction.

IMMUNOPHOROSLIDE

Le nouveau système Millipore Immuno-Phoroslide™ pour l'immuno-électrophorèse combine les avantages tels que simplicité, vitesse et économie avec l'obtention d'excellents résultats d'analyse. Ainsi, cette importante méthode d'analyse des protéines devient maintenant un procédé pratique de routine ne nécessitant que peu de temps et un minimum de connaissances spéciales.

La technique ne nécessite que des opérations très simples. Le nouveau et remarquable support Immuno-Phoroslide™ est fourni pour une utilisation immédiate avec des puits de sérum et anti-sérum préformés. Toutes les étapes de la méthode, depuis l'électrophorèse initiale en passant par l'immunodiffusion, le rinçage et la coloration, sont simplifiées et permettent d'obtenir des résultats reproductibles.

Les analyses sont rapides. Tout d'abord grâce à une durée d'électrophorèse réduite et grâce à un rinçage final pour éliminer les protéines qui n'ont pas réagi. La durée totale de cette méthode est d'environ 52 heures alors qu'avec les plaques d'agar cela allait de 72 à 100 heures.

La modicité du prix de ces analyses est due à un appareillage peu cher et à l'élimination de longues heures de travail de techniciens spécialisés. L'équipement est pratiquement le même que celui utilisé dans l'électrophorèse Phoroslide auquel viennent s'ajouter quelques accessoires supplémentaires. Enfin, pour une analyse complète qui durera 52 heures, un technicien ne travaillera vraiment que pendant environ 1 heure.

Des résolutions détaillées de mélanges de protéines complexes sont obtenues de manière routinière. La définition et les détails obtenus sont aussi bons que les résultats donnés par les plaques d'agar.

An improved method for loading high-temperature gas-chromatographic columns

Laboratorium voor Pathologische Biochemie
(Director : Prof. J.V. Joossens),
Akademisch Ziekenhuis St. Rafaël, Leuven, Belgium.

Introduction.

The purpose of the present communication is to relate some improvements of a relatively new method (1) of liquid injection for high-temperature gas-chromatography on packed columns, of incompletely purified biological extracts or similar mixtures. We also propose completing the discussion of the method.

Many gas-chromatographic procedures for biological extracts, are presently reputed requiring considerable clean-up of the original material, prior to chromatography. This is particularly true when the compounds, subjected to determination, at the same time are minor components in the mixture, and are of very low volatility.

High temperature, requisite for partition of low-volatility compounds (e.g. steroids), has an undesirable side-effect on samples that contain large excess of contaminants, the more labile of which, following injection, breaking down thermally. This causes a more or less pronounced trailing of the « solvent » peak on the chromatograms. The future more correct approach eventually might be one where the sample is applied at low temperature, and the temperature then slowly raised up to the point when the low-volatility components migrate.

For the present, injection methods, all based on the concept of isothermal vaporization of the sample, have been perfected. Several efforts were made in the design of high-temperature liquid injectors, to suppress any undesirable effects of solvent back flash. And more recently, attention has focused on practical methods for injection of the dry samples.

A sizable reduction of the « post-injection noise » on high-temperature chromatograms of impure extracts, can be attained, if one prevents the bulk of the vapors composing the front of the chromatogram, from reaching the partition column. This can be done, after taping the column inside the oven at some distance from the top, by momentarily routing the gas stream to atmosphere. Either the isothermal, or the temperature programmed mode, may be used. With the first option, the period of time available for adequate clean-up of the injector space, may be quite short, as determined by the migration characteristics of the more volatile compounds of interest. The programmed mode is more versatile, as rendering possible freezing the compounds of interest, for virtually any period of time, at the top of the vented column segment.

A column by-pass, associated with temperature programming, offers an additional resource. Resistance to the passage of carrier gas of a short length of column being low, it is possible optimizing the injector flow rate under any circumstances, e.g. when the column is long and packed with narrow-mesh particles. It becomes even possible experimenting with extra-large flow rates, obtainable at ordinary inlet pressures. It is then conceivable using unusually large volumes of solvent vehicle, without encountering a back flash effect. Using a large volume of solvent greatly increases the convenience of liquid injection, as a delicate sample concentration procedure, that ruins volumetric precision, is no longer necessary ; and precision syringes in the 0.1 ml range are robust, easy-to-handle devices.

Uptake of heat of vaporization by a large amount of solvent is a cryogenic process, causing a considerable transient lowering of temperature inside the injector space. Far from being undesirable, this effect, with a properly designed system, may play a role of smoothing the thermal shock upon labile sample contaminants.

Modern commercial gas-chromatographs do not permit experimenting thoroughly the resources of a precolumn injection system, because the facility of two juxtaposed, independently programmed ovens, is not standard. Therefore, we had (1) opted for the external addition of a highly simplified precolumn oven, thus escaping from the engineering problem of designing a more conventional oven.

Following our first report (1) the feature of accurate electronic temperature control over the flash heater temperature was added. Another improvement consists in the maintenance of unchanged gas pressure at the top of the partition column during injection, for the primary purpose of preventing any movement of packing particles.

Apparatus and Operation.

A schematic representation of the injection system is seen in figure 1. The reader is referred to the text and figures of our preceding paper (1) for details of the injection steps and wiring of the event programmer.

The precolumn (4) is seen separating flash heater (2) from the column oven (9) of any conventional chromatograph. The venting system, (7), (8), consisting in two swagelock fittings communicating through a piece of capillary, and two lengths of 1/16" stainless steel tubing, is seen located inside the instrument oven.

Flash heater design is seen considerably less sophisticated than is the case with modern conventional liquid injectors. One may note the absence of a complex flow path, and of any provision for preheating the carrier gas. The designation « flash heater» is incompletely descriptive of the two functions of body (2). It supplies the necessary heat of vaporization, and in addition it affords for the precolumn packing an isothermal environment over some distance. Flash heater temperature is a matter of judgement. In our hands it generally lies half-way between the flash heater and column oven (isothermal) temperatures one would use with the classical injection method.

Function of the complex venting system (7), (8), and valves (15) and (16), is to expose the precolumn outlet at atmosphere during injection and injector clean-up, while insuring the absence of pressure surges at the column top. The latter condition is obtained as follows. With the instrument pneumatically in the steady state (i.e. ready for injection, the oven temperature at equilibrium), a reading is made of the pressure at the top of the column (from a precision manometer, seen at the outlet of valve (16) in Fig. 1). This pressure is equal to the pressure at the precolumn top, less the pressure falls across both the precolumn packing and capillary (8). After coarse adjustment of regulator (14) to about the same value, valves (12), (15), and (16) are opened ; regulator (14) is then readjusted to reproduce the first reading (more exacting pressure substitution would require a correction for pressure drop across the piece of tubing past the manometer, or connection of the manometer at the column top by an auxiliary tube).

Injection takes place with the three solenoid valves, (12), (15), (16), open. The role of capillary (8) is mainly to limit to an acceptable value the flow from the column top wasted to atmosphere through valve (15). The latter process has desirable side-effects, as cleaning the capillary bore and exposed surfaces inside the metal fittings, from any low-volatility deposits.

Valve closing, after injection and injector clean-up, may occur simultaneously for (12) and (15), but it is advisable to delay closing of (16) (a pneumatic delay-relay is suitable) for 3-4 sec. until full pressure is restored at the precolumn outlet.

The flow rate through injector and precolumn during injection, with the gas circuitry suggested in figure 1, follows adjustment of pressure regulator (14). Independent control over this parameter is of course possible at the expense of an additional regulator.

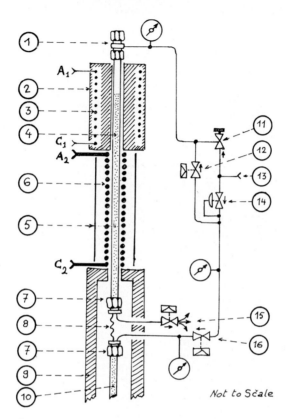

Fig. 1. — Latest version of the precolumn injection system (schematic).

$A_{1,2}$: $C_{1,2}$: Heater terminals, see Fig. 2.

(1) : Septum holder and carrier gas inlet (5/16" × 5/16" Swagelock union).

(2) : Flash heater block, aluminium.

(3) : Heater cartridge (stainless alloy ♯ 304, wire, 7/10 mm, glass-fiber insulated ; resistance at 20°, 15 Ω).

(4) : Packed precolumn, pyrex, 8 mm o.d., 4 mm i.d.

(5) : Screen against air draughts (optional).

(6) : Precolumn miniature oven (bare coil of alloy ♯ 304, wire, 10/10 mm ; resistance at 20°, 5 Ω).

(7) : Swagelock fittings, 5/16".

(8) : Stainless steel capillary, 7 cm- length, 1/16" o.d., 0.01" i.d.

(9) : Column oven.

(10) : Chromatographic column.

(11) : Restrictor valve (constant flow).

(12) : Solenoid valve, normally closed, 1/16"-way (constant pressure).

(13) : Carrier gas inlet, from cylinder regulator, 10 kg/cm².

(14) : Second-stage pressure regulator, outlet range 1-3 kg/cm² (substitution carrier).

(15) : As (12) (precolumn venting).

(16) : As (12) (substitution carrier).

Wiring of the heaters (precolumn heating coil and flash heater cartridge) is seen in figure 2. The heating wire is part of a resistance bridge. As stainless steel is used, instead of traditional nichrome alloys, the heater resistance is temperature-dependent. The system besides is a classical resistance-thermostatic arrangement, where the bridge error signal, after phase discrimination, is used to trigger the power switch K.

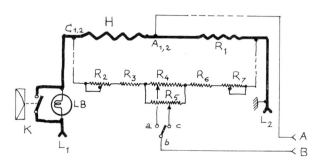

Fig. 2. — Heater wiring (flash heater or precolumn miniature oven) for automatic temperature control.

Components.

H   : Heater wire, stainless steel. For values, see Fig. 1.

R_1 : Constantan wire, 15/10 mm, air-cooled. Approx. 1.5 Ω. (flash heater), or 0,5 Ω (precolumn oven). Connection to H silver-soldered, through short length of copper wire.

R_2 : W.w. adjustable 10 K Ω 2 W (ranging).

R_3 : W.w. 5 K Ω $\pm$ 5 %, 2 W.

$R_{4,5}$: W.w. pot., 200 Ω $\pm$ 5 %, 2 W (temperature settings) ; R_5 not used for flash heater operation ; R_4 then replaced by 100 Ω pot.).

R_6 : W.w. 500 Ω $\pm$ 5 %, 2 W.

R_7 : W.w. adjustable 1 K Ω 2 W (ranging).

K   : S.p. vertical mercury relay, rated 10 A ; or optionally a silicon controlled rectifier. K is triggered by the output of the temperature error amplifier.

LB   : Light bulb, tungsten filament, 60 W ; nominal voltage is that of the power supply (24 or 48 V) (functions as spark suppressor).

Terminals.

L_1, L_2 : To power supply, 24 V a.c., 100 W, or 48 V a.c., 100 W (resp. precolumn oven or flash heater).

a, b, c : Precolumn lower/upper temperature switch-over (to sequence programmer).

A, B : To error amplifier (vacuum tube- or solid state-).

$A_{1,2}$, $C_{1,2}$: Heater terminals, see Fig. 1.

Injection timing.

Figure 1 is nearly self-explanatory from the preceding considerations Precolumn (4) is initially at a temperature low enough for trapping the sample components of interest (e.g. 100-150° C for steroids) and is kept so, until the injection process is completed. The first action consists in opening valves (12), (15) and (16). The sample solution may be injected immediately thereafter, or eventually after cooling the injector space by injection of pure solvent. The following step consists in injector and precolumn clean-up with the valves open. Thereafter the valves are closed, and a pause allows for temperature reequilibration at the different precolumn levels.

Chromatography finally is started under the normal constant flow (from restrictor (11)) by merely raising for some time the precolumn temperature.

Discussion.

The flow rate we use during injection (in the range 1 l/min. S.T.P. [1]) is unusually large. In fact, a claim of the present method is that injection of 0.1 ml of liquid solvent is possible ; the flow rate mentioned is then just optimal for preventing back flash.

It has been occasionally argued that liquid injection might « wash out » stationary phase from the column top. It is quite obvious that this might occur only when solvent in the liquid state penetrates the packing over an appreciable distance. Such condition is ruled out for the case of 0.1 ml of usual solvents injected at a temperature above 200° C.

The excellent trapping capacity of short lengths of packed columns is presently a classical concept (2). In our application however, the occurence of a profound temperature disturbance and the use of a large flow rate, make difficult anticipating the site of grouping of the low-volatility compounds inside the precolumn, or their band width. Due to some difficulty of making direct observations, we used exploiting the different types of information contained in chromatograms of quantitative mixtures (principally steroids) and could conclude the following. 1. Within the range of column loads compatible with quantitative analysis on thin-film columns (0.01 - 20 µg of a single compound), the same peak area for a given mass of a compound is obtained by the precolumn procedure, and by conventional injection. Conventional injections are carried out when the precolumn temperature is maintained equal to that of the column, and the valves are not operated. 2. Quantitative relationships between different peaks are not altered by the precolumn technique. 3. The ratio retention time/peak width, as far as a comparison is possible, is not less favorable with precolumn injection ; considerable band spreading at the precolumn stage is thus ruled out.

References

1. EVRARD, E. *J. Chromatog.,* **27,** 40, 1967.
2. SCOTT, R.P.W. *Proc. 3rd Wilkins Gas Chromatography Symposium,* Amsterdam p. 17, 1965.

Thin-layer chromatography in inorganic analysis

R.C. Hospital, Sittard, The Netherlands.

Summary.

Much work has been published on paper chromatography (PC) of cations. Thin-layer chromatography (TLC) has advantages over paperchromatography, because of its greater sharpness of separation. TLC of cations on silicagel has been worked out by Seiler and some other investigators. In comparison with PC and TLC on cellulose the published results on silicagel are not particularly good. TLC on cellulose is subject to the same conditions which operate in PC. It appears from literature that most investigators, using paperchromatography, have chosen as solvent mixtures : organic solvent (e.g. acetone, methanol, n-butanol)-HCl-H_2O.

By making a study of the correlation between the composition of the solvent mixture and the Rf value, according to Hartkamp and Specker, or the R_M value, as proposed by Lederer and Merkus, it is possible to obtain information about hydration, dehydration, solvation and complexation of the cation and to calculate which composition of the solvent mixture gives the best separation of a given mixture of cations. These theoretical considerations will be put into practice in several examples. The excellent results of the separations of cations on cellulose-thin-layers enable us to conclude that TLC of cations on cellulose is a useful contribution to inorganic micro analysis.

1. Comparison between paper chromatography and thin-layer chromatography on cellulose.

As far as speed is concerned Lederer (1967) is correct in his judgement that there is little or no difference between paper chromatography and thin-layer chromatography, using fast solvent mixtures. In approximately the same time a separation of cations can be achieved, but the sharpness of separation on cellulose-layers is far better.

Figure 1 demonstrates the rather diffuse spots on paper in comparison with the sharp separation using TLC on cellulose. The paper and thin-layer chromatogram were made in the same chamber, with the same solvent mixture and in the same time.

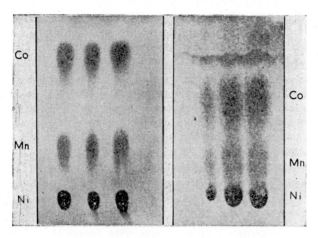

Fig. 1. — Thin-layer chromatogram on cellulose MN 300
(left) and a paper chromatogram (right) on Whatman n° 1
in the same chamber and in the same time.
Spray reagent : pyridylazonaphtol (0.25 % in 96 % ethanol), followed by exposure to NH_3 vapor.
Solvent mixture : acetone-HCl (d = 1.19)-H_2O (90 : 3 : 7).

2. Comparison between TLC on silicagel and TLC on cellulose.

The published results on silicagellayers are rather poor. Extensive
research on separation of cations on silicagel has been carried out by
Seiler (16). However the sharpness of the separation is inferior to the results
on celluloselayers. Many cations bunch together with Rf values in the same
region. Paper chromatography has enabled many problems in inorganic
analysis to be solved. TLC on cellulose layers is subject to the same conditions
which operate in paper chromatography. Therefore TLC on cellulose has to
be preferred.

3. The correlation between the Rf value of a cation and the composition of the solvent mixture.

The theoretical basis of the relation between the Rf value of a cation
in paper chromatography and the composition of the solvent mixture was
first proposed by Hartkamp and Specker (4, 5). They deduced that the
transport of a cation in the mobile phase is dependent on the ionic species
of the cation (aquo-ion, solvo-ion, chloro-complex), moving with the solvent
mixture. Four different transport groups are to be distinguished here. Each
group has a typical sequential order of decrease or increase of the Rf value,
caused by changing the composition of the solvent mixture. In order to investigate the influence of the individual components of the solvent mixture in each
series of experiments the organic solvent concentration is fixed at a constant
level, while the HCl- and H_2O-concentration were varied inversely.

Most investigators using paper chromatography or TLC on cellulose for the separation of cations, have chosen as solvent mixture : organic solvent - $HCl-H_2O$.

Hartkamp and Specker have studied tetrahydrofuran - $HCl-H_2O$. The solvent mixtures, involved in our earlier experiments (12) were acetone-$HCl-H_2O$, n-butanol-$HCl-H_2O$ and methanol-$HCl-H_2O$.

From these investigations can be deduced that the transport of a cation in the mobile phase can be classified into four groups : normal, inhibited-normal, quasi-normal and anomalous transport.

a. *Normal transport :* The transport of the cation is solely dependent on the water concentration in the mobile phase. The cation is moving with the solvent mixture as a hydrated cation (aquo-ion) (e.g. Al, Ni).

b. *Inhibited normal transport :* The hydrated cation is dehydrated by the protons and chloride-ions of the solvent mixture. So, the transport of the cation is strongly inhibited by HCl. This dehydration (decrease of the Rf value) increases with increasing ionic radius (Li < Na < K and Mg < Ca < Sr < Ba) (e.g. Na, K, Ca, Sr, Ba).

c. *Quasi-normal transport :* The cation is moving with the solvent mixture as a solvo-ion. Solvation of the cation by the organic solvent molecules (e.g. methanol) is responsible for the good transport of a cation with a solvent mixture containing too little water to explain the relative high Rf value (e.g. Mg, Li).

d. *Anomalous transport :* The cation is moving with the solvent mixture as a chloro-complex, soluble in the organic component of the solvent mixture.

Increase of the HCl and organic solvent content improves the transport of the cation (e.g. Cu, UO_2, Mn, Co).

The figures 2, 3, 4 and 5 show the Rf curves of these four transport groups.

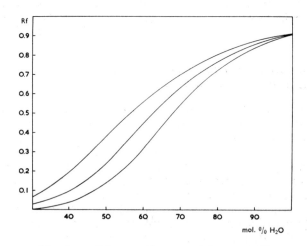

Fig. 2. — Rf-curves for normal transport (5).

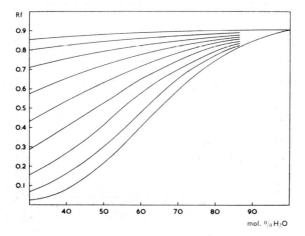

Fig. 3. — Rf-curves for quasi-normal transport (5).

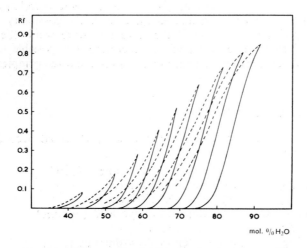

Fig. 4. — Rf-curves for inhibited normal transport (5).

Lederer (9, 10) deduced for a number of cations a linear relationship between the R_M value of the cation and the log $[Cl^-]$ of the solvent mixture. The Lederer equation can be corrected and extended. For anomalous transport we have deduced $R_M = -x \log [Cl^-]$ and for the other transport groups $R_M = x \log [Cl^-]$ (12).

Schematically in figure 6 the R_M values of cations have been plotted against the log $[Cl^-]$ of the solvent mixture. The slope of the straight line (x) indicates the ionic species of the cation, moving with the mobile phase.

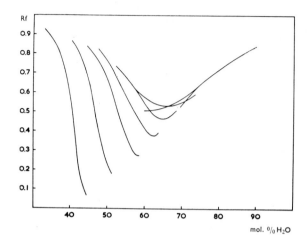

Fig. 5. — Rf-curves for anomalous transport (5).

In short, the most important application for these correlations between composition of solvent mixture and Rf value lies in the possibility of obtaining information which composition of the solvent mixture yields the best separation of a given mixture of cations. Furthermore it is possible to get information about the hydration, dehydration, solvation and the complex-formation of a cation.

4. Detection of the cations.

The best general reagents for the cations, classified according to the H_2S-system, are :

Ag, Pb, Tl	$(NH_4)_2S$; KI ; dithizone ; K_2CrO_4.
Cu, Cd, Bi, Pb, Hg	$(NH_4)_2S$; KI ; dithizone.
As, Sb, Sn	dithizone ; KI ; $(NH_4)_2$ S.
Fe, Al, Cr, Be, UO_2	oxine-quercetin-kojic acid (mixture).
Co, Ni, Mn, Zn	pyridylazonaphtol.
Ba, Ca, Sr, Mg	oxine : tetrahydroxyquinone (2) or glyoxal-bis-(2-hydroxyanil) (14).
Na, K, Li	quercetin, zinc uranyl acetate.

Detailed information on the detection of these cations with a suitable specific reagent may be found in the extensive literature on inorganic paper chromatography (1, 3, 6, 7, 8, 15).

5. Separation of cations with TLC on cellulose.

A sharp separation of Ag, Tl and Pb without tailing is very difficult to achieve. All separations published in the past are insufficient. Also our own method of separation with 10 % NH_4OH is not very good.

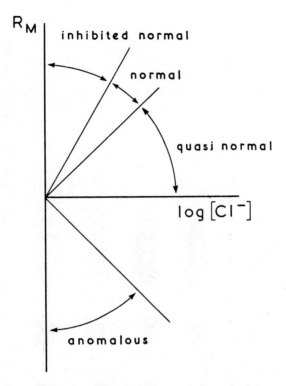

Fig. 6. — Schematic representation of the correlation between the R_M values of cations and the log [Cl⁻] of the solvent mixture for each transport group. The slope of the straight line is an indication of the number of chloride-ions in the chloro-complex of the cation or of the degree of hydration or dehydration of the hydrated cation under the influence of the HCl concentration in the mobile phase.

With NH_4OH as solvent mixture many cations form insoluble hydroxides. This would account for the zero Rf value of Pb. The hydroxide of Tl is more soluble. Therefore Tl is transported over a short distance. The movement of Ag is far better, since the ammine complex of Ag is easily transported by NH_4OH (see figure 7). The variation of the Rf value under the influence of the concentration of the cation, spotted on the cellulose-layer, and the presence of an excessive tailing can be explained if we accept that adsorption of the cation by the cellulose molecules is the major factor in the movement. It must be noted here that if the mobile phase contains free acid there is no adsorption effect, because of the preferential adsorption of the hydroxoniumion by the cellulose. For the separation of the cations Pb, Cu, Cd, Bi and Hg no ideal solvent mixture has ever been found. We have checked all published separations (12). The fast moving mixtures, like acetone-HCl-H_2O or methanol-HCl-H_2O, give rather poor separations,

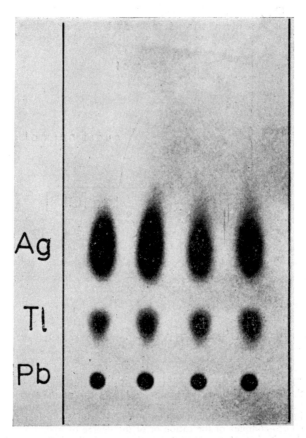

Fig. 7. — TLC on cellulose MN 300 of Ag. Tl, Pb.
Detailed information see text.
Spray reagent : the chromatogram is not dried after deve-
loping and is held above a bottle, containing saturated
ammonium sulphide.
Solvent mixture : 10 % NH_4 OH.

because of the resemblance in transport mechanism of Cd, Bi and Hg. The
best separation (figure 8) can be obtained with n-butanol, saturated with
3 N HCl, which has often been used in paper chromatography. Also As,
Sb and Sn can be separated with this solvent mixture (As < Sb < Sn).

The other cations can successfully be separated with solvent mixtures,
calculated from the data of the experimental and theoretical investigations
of the relation between the Rf value of a cation and the composition of
the solvent mixture, as proposed by Hartkamp and Specker (4, 5), Lederer
(9, 10) and Merkus (12, 13).

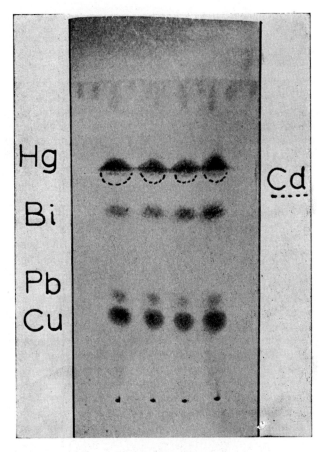

Fig. 8. — TLC on cellulose MN 300 of Cu, Pb, Bi, Cd, Hg.
Spray reagent : ammonium sulphide (see figure 7).
Solvent mixture : n-butanol, saturated with 3 N HCl.

Very good separations can be obtained for Fe, UO$_2$, Be, Al and Cr, for Ni, Mn, Co and Zn and also for Ba, Ca, Sr and Mg with acetone-HCl-H$_2$O. The number of possible compositions of this solvent mixture is enormous. Only a few examples of these separations are presented in the figures 9 and 10. For Ba, Ca, Sr and Mg and for Na, K and Li methanol-HCl-H$_2$O in various compositions can be employed as solvent mixture (see figures 11 and 12).

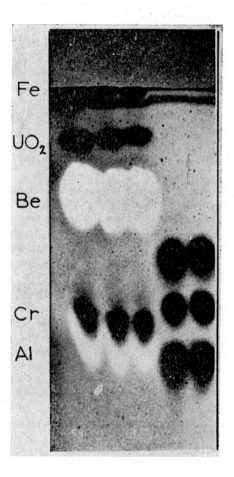

Fig. 9. — TLC on cellulose MN 300 of Al, Cr, Be, UO$_2$, Fe.
Spray reagent : mixture of oxine (0.5 %), quercetin (0.2 %)
and pyridylazonaphtol (0.25 %) in ethanol, followed by
exposure to NH$_3$ vapor. (On the right of figure 9 also Ni,
Mn and Co are visible). Visualization in U.V. light (365 nm).
Solvent mixture : acetone-10 N HCl-H$_2$O (37 : 18 : 12).

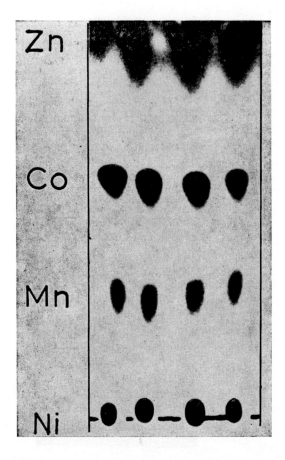

Fig. 10. — TLC on cellulose MN 300 of Ni, Mn, Co, Zn.
Spray reagent : pyridylazonaphtol (0.25 % in ethanol, fol-
lowed by exposure to NH_3 vapor.
Solvent mixture : acetone-10 N HCl-H_2O (90 : 5 : 5).

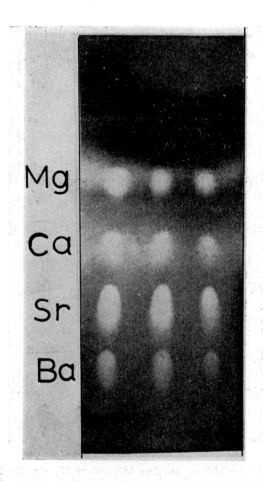

Fig. 11. — TLC on cellulose MN 300 of Ba, Sr, Ca, Mg.
Spray reagent : oxine (0,5 % in 60 % ethanol), followed
by exposure to NH$_3$ vapor. Visualization in U.V. light
(365 nm).
Solvent mixture : methanol-HCl (d = 1,19)-H$_2$O (7 : 1 : 2).

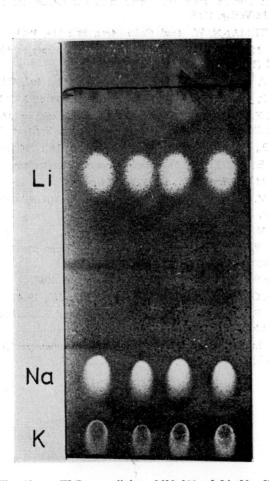

Fig. 12. — TLC on cellulose MN 300 of Li, Na, K.
Spray reagent : zinc uranyl acetate (saturated solution in
N acetic acid). Visualization in U.V. light (365 nm).
Solvent mixture : methanol-10 N HCl-H_2O (80 : 12 : 8).

References

1. BLASIUS, E. Chromatographische Methoden in der analytischen und präparativen anorganischen Chemie unter besonderer Berücksichtigung der Ionenaustauscher, Stuttgart, Enke-Verlag, 1958.

2. BOCK-WERTHMANN, W. *Anal. Chim. Acta,* **28** : 519, 1963.

3. HAIS, I.M., MACEK, K. *Handbuch der Papierchromatographie,* Bd. I (1958), II (1960), III (1963), Jena, G. Fischer Verlag.

4. HARTKAMP, H. SPECKER, H. *Z. anal. Chem.,* **152** : 107, 1956.

5. HARTKAMP, H., SPECKER, H. *Z. anal. Chem.,* **158** : 92, 161, 1957.

6. HECHT, F., ZACHERL, M. *Handbuch der mikrochemischen Methoden,* Bd. III (Anorganisch-Chromatographischen Methoden), Wien, Springer-Verlag, 1961.

7. HEISIG, G.B., POLLARD, F.H. *Anal. Chim. Acta,* **16** : 234, 1957.

8. LEDERER, E., LEDERER, M. *Chromatography,* Amsterdam, Elsevier, 1957.

9. LEDERER, M. *J. Chromatog.,* **1** : 172, 1958.

10. LEDERER, M. In Hecht en Zacherl. : *Handbuch der Mikrochemischen Methoden,* Bd. III, Wien, Springer-Verlag, 1961.

11. LEDERER, M. *Chromatog. Rev.,* **9** : 115, 1967.

12. MERKUS, F.W.H.M. Kwalitatieve analyse van kationen met behulp van dunne-laagchromatografie, Amsterdam, Thesis, 1966.

13. MERKUS, F.W.H.M. *Pharm. Weekblad,* **103** : 1037, 1968.

14. MÖLLER, H.G., ZELLER, N. *J. Chromatog.,* **14** : 560, 1964.

15. POLLARD, F.H., McOMIE, J.F.W. *Chromatographic Methods of Inorganic Analysis,* London, Butterworths, 1953.

16. SEILER, H. In STAHL, E. e.a. : *Dünnschicht-Chromatographie* (ein Laboratoriumshandbuch), Berlin, Springer-Verlag, 2. Ed., 1967.

A study of oxalates as stationary phases
for adsorption chromatography of inorganic ions

by

D.L. MASSART *

Institute for Nuclear Sciences, Proeftuinstraat, Ghent (Belgium).

Summary.

The mechanism of the sorption of Eu on cadmiumoxalate was studied. Eu is sorbed by precipitation on stoechiometrical cadmiumoxalate at sufficiently high Eu concentrations and at low concentrations on cadmium oxalate prepared with an excess of Cd-ions through exchange of Cd^{++}ions bound on the crystal surface againts Eu^{3+}ions. This mechanism contrasts with the metathetical process found for sulfides by other authors. A few practical implications of this mechanism on chromatographic behaviour are discussed.

* * *

Adsorption chromatography was studied as a technique for the separation of metal cations mainly before 1955. The finding of the ion exchange properties and the selectivity of inorganic materials such as zirconiumphosphate or zirconiumoxide has caused however a revival in the interest in this sort of column stationary phases and has led several authors to investigate typical adsorption chromatographic techniques. In this way the study of simple salts as column materials was taken up again more especially in view of developing selective radiochemical separations.

In this context the research effected by Girardi (1) should be mentioned. This author is examining systematically the sorption characteristics of columns of simple salts and has concluded already the study of, among others, cadmium and coppersulfides, copperchloride and leadfluoride. Several interesting separations are possible. Other examples are the sorption of halogenides such as iodide on silverhalogenides such as silverbromide as studied by Eckhardt (2) and the separation of Am^{6+} from Cm^{3+} on CaF_2 as examined by Holcomb (3). The sorption of several metal cations on metalsulfides was studied carefully by Phillips (4) who concluded that the exchange reactions between the metal ions in stationary phase and mobile phase occurs through metathetical reactions such as represented in the following equation :

$$\overline{M_1X} + M_2 \rightleftharpoons \overline{M_2X} + M_1 \quad (1)$$

where M_1 and M_2 represent cations and a bar represents the stationary phase.

---

* Research associate of I.I.K.W.

It was our purpose to investigate in the present work the possibility to use oxalates as stationary phases and to investigate the manner in which reaction (1) goes through.

Experimental.

The oxalates were prepared in 30-50 g batches by precipitation of a 3 % oxalic acid solution with a stoechiometric quantity of the metal ion (« stoechiometric » oxalate) or with a 10 % excess (« non stoechiometric » oxalate). The precipitates were washed six times by decantation with water, filtered off on porous glass filters, washed with $\pm$ 50 ml water and dried over night at 75° C - 80° C. The oxalates were then ready for use except Zr-oxalate which was obtained in very large crystals that had to be crushed.

Batch experiments were performed to determine sorption percentages or distribution coefficients under complete equilibrium conditions. Column experiments were performed in glass columns with internal diameter of 8 mm. In each case the distribution of the cation under study was followed with radioisotopes, prepared by neutron activation in the Thetis reactor of the Institute.

Results and discussion.

Sorption of rare earths isotopes on different oxalates.

The equilibrium adsorption of 1 mg Eu on 1 gram quantities of different oxalates in $O.1N HNO_3$ is shown in Table I.

TABLE I.

Sorption of Eu (1 mg) in $O.1N HNO_3$

% sorption		% sorption	
Cu	50 (pH 4)	Th	0
Co	99.1	Ca	95.1 (pH 4)
Cd	98.5	Sr	91.4 (pH 4)
Zn	96.0	Zr	0
La	100	Pb	56.6

Eu is thus adsorbed to a large extent on all the oxalates investigated, except on ThOx and on ZrOx, both of which are more insoluble than Eu oxalate.

The sorption of different rare earths (R.E.'s) on Cd oxalate was examined. Figure 1, line A represents the losses on precipitation of the R.E.'s using a procedure described by Willard (5). Line B gives the losses on equilibration with Cd oxalate in $O.1N HNO_3$ of 1 mg R.E. and line C the losses on chromatography using 2 gram columns and 5 mg R.E.

It can be seen that the graph of the efficiency of the precipitation procedure is parallelled by the equilibration procedure.

One can thus conclude that the extent to which sorption occurs is determined by the ratio of the solubilities of the oxalates of the two cations taking part in the process.

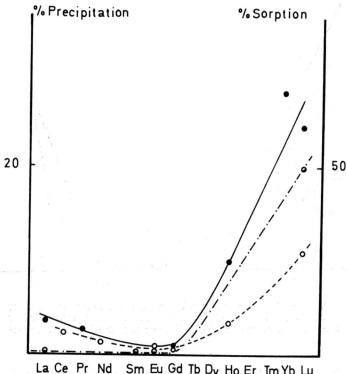

Fig. 1. — % Eu staying in solution.
——————— Equilibration procedure.
— · — · — Chromatographic procedure.
— — — — Precipitation procedure.

Influence of the quantity of ion to be sorbed.

In the case of the adsorption of Eu on 100 mg LaOx a sorption of more than 99 % is obtained for quantities ranging from 2 μg to 10 mg carrier. For higher quantities the sorption decreases gradually. The sorption on CdOx and on PbOx in function of carrier quantity does not show such a simple behaviour. Figure 2 represents the percent Eu staying in solution on equilibration with 100 mg CdOx. On « stoechiometric » CdOx a nearly complete sorption occurs for quantities of Eu between 0.75 mg and 10 mg. While the increasing loss of sorption power at higher quantities is expected, the losses at low loading can not be easily explained on the basis of a

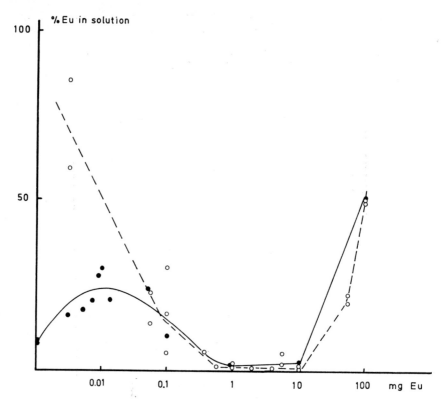

Fig. 2. — % Eu staying in solution on equilibration with 100 mg cadmium in $O.1N HNO_3$ as a function of Eu quantity. Broken line : stoichiometric oxalate. Full line : non stoechiometric oxalate.

metathetical reaction. The behaviour of « non stoechiometric » CdOx is still more complicated. It is identical with « stoechiometric » CdOx at medium and higher loading. At low loading, the loss increases with decreasing loading until the range of 10-20 µg. Eu is reached. From there on the sorption power increases again. A completely analogous behaviour is noted for equilibration on non stoechiometric PbOx. The behaviour of « stoechiometric » CdOx can be explained if one postulates that the sorption of Eu on CdOx and PbOx (and probably on most other oxalates) does not occur through a metathetical reaction, but through precipitation with oxalate anion in solution. This is represented by the following reaction scheme :

$$\overline{M_1X} \iff M_{1 \text{ sol.}} + X \qquad (2)$$

$$M_{2 \text{ sol.}} + X \iff \overline{M_2X} \qquad (3)$$

According to this scheme reaction (3) would only begin to go through as soon as $(M_2)(X) = K_S (M_2X)$.

The behaviour of Eu at very low loading on « non stoechiometric » CdOx can be explained by a secundary process. When a salt is precipitated in the presence of an excess of one of the constituting ions, this ion is adsorbed on the surface. This ion, a Cd^{++}-ion in this case, can probably be exchanged against the Eu^{3+}-cation which forms a less soluble compound with the oxalate. This is certainly the case for Cd-oxalate, precipitated with a 10 % excess of oxalate anion, with which exactly the same sorption results are obtained as with the so called « non stoechiometric » CdOx. As has been shown by Kolthoff (6), the negative charge on the surface is in such a case compensated by the adsorption of positive ions. According to the Paneth-Fajans-Hahn rule the ion which forms the lesser dissociated compound with the anion is preferentially adsorbed, which in this case will lead to the adsorption of Eu. Because it is a surface phenomenon however, the capacity as determined by this reaction alone is very low (in this case it would seem to be 10-20 $\mu g/100$ mg oxalate).

Stoechiometric CdOx is an irreproducible column material for the sorption of μg quantities of Eu. The following sorption percentages were obtained for the sorption of 2 μg of Eu on 1 gram CdOx in O . 1N HNO_3 : 95.2 % - 88.0 % - 77.5 % - 62.2 % - 17.8 %. « Non stoechiometric » CdOx yielded much better results since in a series of 10 experiments no sorption percentage under 99.5 % was obtained.

Influence of the quantity of stationary phase.

When « stoechiometric » CdOx is equilibrated in O . 1N HNO_3 with 1 mg or 1 μg Eu it is found that over the range of 100 mg to 2 gram stationary phase the quantity of stationary phase has no influence on the procentual sorption. The quantity sorbed appears thus to be determined solely by the quantity of free oxalate in solution. The same result is obtained for the sorption of 1 mg Eu on « non stoechiometric » Cd-oxalate but not for the sorption of 1 μg quantities.

Table II summarises results obtained for the sorption of 1 μg Eu in O . 1N HNO_3 on 100 mg or 500 mg CdOx expressed as % Eu in solution.

TABLE II.

Sorption losses on « non stoechiometric » Cd-oxalate.

		100 mg	500 mg
1 μg Eu	A	77.7 %	52.4 %
10 μg Eu		87.0 %	65.2 %
1 μg Eu	B	65.0 %	19.5 %
10 μg Eu		79.5 %	43.4 %

This can be explained by the fact that, if the sorption of small quantities on CdOx is caused by an exchange with adsorbed ions, the extent of the sorption depends on the total quantity of exchangeable cations present.

Group A represents a set of experiments in which the mobile phase was added one week before the equilibration and group B another set in which

the equilibration with Eu was carried out immediately after addition of the
mobile phase. It is clear that the sorption is better for group B. This
is ascribed to a recrystallization process leading to a decrease in surface
From a practical viewpoint, this leads to the conclusion that it is to be
avoided to prepare columns a long time before use. Experimentally, it was
shown that the preparation of « non stoechiometric » columns one week before
use leads to a decrease in efficiency of 10 to 20 % (mean 13.5 %).

Applications.

CdOx presents two favourable characteristics as a column material. The
capacity for example for the rare earths is very high. On a column of
500 mg CdOx 129 mg Sm were sorbed i.e. 50 % of the column was converted
to the Sm form before break-through occurred. Furthermore, the attainment
of equilibrium is fast so that flow rates of 1 ml/cm^2. min were possible.

Complete separation could be obtained of ions that form insoluble
oxalates from ions that do not react, such as Eu from Na. Decontamination
factors of more than 10^4 were obtained by elution with 0.01N HNO$_3$ for
the separation of 1 mg Eu from 200 µg to 500 mg quantities of Na. When
chromatographing two ions that react with oxalate the results depend on
the quantities present as demonstrated in Table III.

TABLE III.

Separations on CdOx columns.

1 mg Eu	100 % sorbed
300 µg Sr	7 % sorbed
1 mg Eu	93 % sorbed
100 mg Sr	1.5 % sorbed
1 mg Eu	100 % sorbed
100 µg Mn	0 % sorbed
1 mg Eu	100 % sorbed
10 mg Mn	30 % sorbed

References

1. F. GIRARDI. Personal communication.
2. W. ECKHARDT, G. HERRMANN, H.D. SCHÜSSLER. Z. anal. Chem. 226 :
   71, 1967.
3. H.P.HOLCOMB. Anal. Chem., 36 : 2329, 1964.
4. H.O. PHILLIPS, K.A. KRAUS. J. Chromat., 17 : 549, 1965.
5. H.H. WILLARD, L. GORDON. Anal. Chem., 20 : 165, 1948.
6. J. KOLTHOFF. J. Phys. Chem., 40 : 1027, 1936.

The trapping of gas chromatographic fractions for infra-red spectroscopy

by

A.S. CURRY, J.F. READ, C. BROWN and R. JENKINS
Home Office Central Research Establishment, Aldermaston, U.K.

Scientific research has a variety of motivations ranging from academic enquiry to commercial exploitation. In Forensic Science the driving force is the practical application of science to the detection of crime and research is designed to obtaining information that will be of use in determining, by analysis and subsequent deduction, the circumstances of a crime and will hence lead to the apprehension of the criminals. The transfer of traces of organic and inorganic compounds from the criminal to the scene of the crime or his removal of these traces from the scene on his person or clothing, with the subsequent demonstration of the materials by analysis is the guiding principle in this particular branch of science. These materials can range from traces of biological debris such as blood spots and single hairs to fragments of paint or glass, small fibres or pieces of plaster, putty or plastic. The analysis of microscopic sized particles is achieved by micro-analysis and gas chromatography either on extracts or of pyrolysis products is the most sensitive general method for the investigation of organic or inorganic-chelate compounds. The identification of a material is however not the main object in an enquiry — the forensic scientist seeks to find other constituents in the material which will give it individual identity and so provide an even better link between the suspect and the crime. In the fields of hair and glass analyses the trace element composition, that is the measurement of up to 40 elements down to the 1 ppm level, in each hair or piece of glass by neutron activation analysis or mass spectrometry, already provides good examples of this type of approach and provides a background, using the inorganic example, of how fundamental research in micro-analysis has to be applied in Forensic Science. The Home Office Central Research Establishment was set up just two years ago to investigate such problems and already has its own spark source mass spectrometer and operates in cooperation with the Atomic Weapons Research Establishment a neutron activation analysis service for the U.K. Such large apparatus has at the present time to be geographically concentrated because of economic considerations but in the field of organic micro-analysis I felt that gas chromatography provided a field of research whose fruits could be applied generally in the existing

Forensic Science Laboratories of which there are 11 in the United Kingdom. Gas chromatography presents us with a very high sensitivity — sometimes less than 10^{-12} gram coupled with high resolving power. Retention times are valuable criteria of identity and if married with retention times of derivatives can lead to absolute identification. Absolute identification is often vital in police investigations. If an old lady who has taken her evening sleeping capsule is cruelly murdered in her bed and in the course of the attack the murderer gets blood splattered from her into his clothing the demonstration that the same barbiturate is present in the blood stains as in the deceased would be vital evidence. This example demonstrates that very high sensitivity is essential — the blood may only contain 1 μg/ml and the total blood on the clothing may only be a few spots ; a vital fact is that no other sample is available for analysis. Sensitivity coupled with absolute identification is therefore vital.

At C.R.E. the toxicology of aircraft accident victims is also investigated in cooperation with the R.A.F. Traces of tranquillisers, stimulants, carbon monoxide and ethanol in the aircrew may be of value in the accident enquiry. Unfortunately in aircraft disasters it is not unusual for bodies to be fragmented and consequently only small pieces of muscle, liver or lung, often putrefying, are available for analysis. The multitude of peaks in toxicology extracts include many decomposition artefacts and make derivative formation on the G.C. column of little value — hence our attention for this and the other reasons detailed above became focussed on the trapping of eluates and also on the combination of GC-mass spectrometry. I now wish to discuss the former in some detail.

Many systems have been described for the trapping of GC eluates and some of these have involved subsequent infra-red spectroscopy but from our search of the literature it became apparent that the vast majority, if not all, this work involved the examination of milligram quantities of material and many stressed the losses that were encountered when smaller quantities were investigated. Our basic aims were therefore to investigate systems which would handle compounds in the microgram range and which required relatively non sophisticated apparatus. Initial excursions into the use of Millipore filters, silver chloride cells, and Multiple Internal Reflectance trapping systems convinced us that the existing commercially available devices were not suitable for our purposes and the use of a stream splitter with a trapping vessel was therefore examined. In these experiments a Pye 104/84 chromatograph was used equipped with a Pye stainless steel stream splitter linked to a Research and Industrial Instruments Co., heated line, this was heated to the temperature of the column. It soon became apparent that when splitting was used not only the quantity but also the flow rate of carrier gas through the detector was affected and the sensitivity of detection was materially altered. Typical results are shown below for equal sensitivity at the detector.

Efforts to increase the flow rate of column gas through the detector thus maintaining the high sensitivity of the flame ionisation detector were not very successful and subsequently a 1 : 10 stream splitter was used which allowed 90 % of the injected material to be collected and the sensitivity of the FID was adjusted by altering the hydrogen flow rate and observing

TABLE 1.

Stream Splitter Approximate split ratio	Min. quantity of each barbiturate in mixture on column	Quantity of each barbiturate passing through the FID detector
No splitter	20 nanograms	20 nanograms
1 : 10	0.4 microgram	40 nanograms
1 : 20	1.8 micrograms	90 nanograms
1 : 100	50 micrograms	500 nanograms

the Hitachi recorder for maximum response. Typical operating conditions for a 10 % SE 30 on silanised Chromosorb W 100/120 for barbiturates and 5 % KOH and 5 W carbowax on Chromosorb W at 220° C for amphetamine type compounds were oxygen free nitrogen at 50 ml/minute and air at 500 ml/minute. No significant column bleed was found using well aged columns of this type.

Several types of trapping vessels were tried and assay of the trapped fractions was undertaken either by re-injection and comparison of relative peak areas or by quantitative ultra-violet spectrophotometry for the barbiturates. As a result of these studies it was found that a small glass tube of length 20 mm and diameter 5 mm fitted with a serum cap passing over the heated needle was suitable provided the needle met the angle of the glass at approximately 30-45°. The tube is cooled in water for relatively high melting point compounds or in acetone/CO_2 or liquid nitrogen for the volatile bases. Provided these conditions are met high efficiency trapping is obtained without aerosol formation at the 1 μg level.

Our next problem was to prepare micro infra-red discs and no doubt many of you have seen beautiful pictures in the manufacturers glossy magazines showing infra-red curves on this quantity of material. We therefore asked manufacturers how they did it only to learn that a milligram or so of drug had been mixed with 300 milligrams of KBr and an aliquot of the mix had been taken to prepare the microdisc! Hence our researches continued! The first thing that was found was that scrupulous attention to cleanliness is essential. The glass trapping tubes had to be cleaned in chromic acid, washed in distilled water and dried at 120° C immediately before use. The potassium bromide used to prepare the disc had to be of optical quality and in the pre-grinding process excessive care to avoid finger and other contamination of agate pestle and mortars and platinum dishes and the discs had to be maintained.

The transference of the trapped eluate to the potassium bromide discs require a steady hand and experience but in principle it is extremely simple. The trapped material is dissolved in 25 μl of dried and distilled chloroform and taken into a Hamilton syringe fitted with a repeating dispenser set at 0.5 μl. On dispensing the first 0.5 μl, a micro drop of chloroform solution forms at the end of the needle tip. Approximately 0.5 mg of powdered K Br can then be carefully taken up and hangs at the end of the tip. Repeated careful dispensation of the chloroform allows transference of the eluate solution onto the suspended KBr with the chloroform being carefully sequentially evaporated by the heat of a table lamp. When all has been transferred the KBr is transferred to a 0.5 mm disc and the disc is prepared by vacuum without the use of pressure. A similar disc prepared from control chloroform from a trapped column fraction when no drug is emerging from the column compensates for impurities and minor column bleed. A R.I.I.C. K Br beam condenser was used in the sample beam with a R.I.I.C. ATO2 attenuator in the reference beam. A Perkin Elmer 225 Infra-Red Spectrophotometer was used in our work but experiments using the same discs in a Unicam SP200 Infra-Red Spectrophotometer showed that excellent curves could be obtained at the 1 μg level. Using the larger machine it was found possible to produce curves which had useful interpretative spectra at less than 500 nanograms and by the use of scale expansion even down to 200 nanograms.

Typical results for barbiturates are shown in the slides. Although this technique is of value for relatively non volatile materials, there are still problems to be overcome in the preparation of micro-discs for less than 10 μg of the amphetamine bases. The volatility of these compounds causes a significant loss during the chloroform evaporation but we have had success in differentiating them from tissue artefacts by trapping to ensure purity and then subsequent rechromatograming on TLC or GLC with derivative formation. For this technique ether can be used as the transferring solution. It unfortunately cannot be used in the preparation of KBr discs because of water interference. The relative simplicity of the GLC-infra-red combination which I have described is of great value in that, for relatively low cost, can be used by adaptation of conventional apparatus already existing in the U.K. Forensic Science Laboratories. The sensitivity obtained is certainly a hundred fold better than previously attained and the technique of micro-disc preparation extends its usefulness to other areas in which prior GLC separation is not necessary.

Chemical effects induced by neutron irradiation of sodium tetrametaphosphate as studied by paper electrophoresis

by

O. Ž. JOVANOVIĆ-KOVAČEVIĆ

Hot Laboratory of the Boris Kidrič Institute of Nuclear Sciences
Beograd, Yugoslavia

In order to study the nature of the changes and defects obtained as a result of the neutron activation of the solid condensed phosphates, the chemical behaviour of ^{32}P in the (n, gamma) irradiated and thermaly treated sodium tetrametaphosphate was investigated. After irradiation in the reactor resp. after subsequent thermal treatment samples were dissolved in water and analysed by use of high-voltage paper electrophoresis. About nine ^{32}P-labelled anions of oxyphosphorus acids were separated and identified. A relatively low percentage of ^{32}P activity in the initial form and in the reduced chemical forms was proved.

Introduction.

Most of the nuclear reactions induce chemical changes and certain defects in solid irradiated systems. Thus, chemical changes are also induced in solid inorganic phosphate systems by their irradiation with thermal neutrons, i.e. by the nuclear reaction $^{31}_{15}$P (n, gamma) $^{32}_{15}$P.

The first studies of the chemical effects of (n, gamma) reactions in alkaline phosphates are related to the name Libby who exposed both solid phosphate systems and their aqueous solutions to thermal neutron irradiation. Using the method of precipitation of phosphorus in the form of magnesium ammonium orthophosphate, Libby found that, actually independently of the target material, about 50 % of the radioactive phosphorus-32 obtained is in the highest oxidation state, i.e. in the form of the initial compound. The presence of the radioactive isotope produced by a specific nuclear reaction in the initial chemical form is referred to in the literature as retention.

The latest works of Lindner, Harbottle, Claridge, Maddock and others, who separated the chemical forms of ^{32}P by using higher resolution methods such as chromatography and high voltage paper electrophoresis, showed that a series of chemical forms of ^{32}P in a lower oxidation state is obtained by the (n, gamma) reaction.

The distribution of ^{32}P formed by (n, gamma) reaction in simple phosphate systems has been extensively studied (1-7). Relatively little attention has been paid to solid alkaline polyphosphates.

In order to have a better insight into the behaviour of recoiled ^{32}P in one cyclic polyphosphate, we exposed sodium tetrametaphosphate — $Na_4P_4O_{12}$ — to thermal neutron irradiation in the reactor and subsequently to a thermal treatment.

Experimental.

The target material — cyclic sodium tetrametaphosphate — was synthesized according to Brauer (8). By recrystallization from water in the presence of ethyl alcohol pure sodium tetrametaphosphate was obtained in the form of a colourless microcrystalline powder.

Besides the target material we also used a series of inorganic phosphorus compounds of the phosphate type (sodium orthophosphate Na_2HPO_4, pyrophosphate $Na_4P_2O_7$, hypophosphate $Na_2H_2P_2O_6$ 6 H_2O, isohypophosphate $Na_3HP_2O_6$. H_2O, sodium tripolyphosphate $Na_5P_3O_{10}$, tetrapolyphosphate $Na_6P_4O_{13}$, and trimetaphosphate $Na_3P_3O_9$), then compounds in which phosphorus is in a lower oxidation state such as : sodium orthophosphite Na_2HPO_3 . 5 H_2O, pyrophosphite $Na_2H_2P_2O_5$, hypophosphite NaH_2PO_2 . H_2O and sodium diphosphite $Na_3HP_2O_5$. Of the above compounds sodium trimetaphosphate and sodium pyrophosphite were synthesized (8), while the other phosphorus compounds and chemicals were of commercial origin (Merck and BDH p.a.) or supplied by different research laboratories. *

Before neutron irradiation in the reactor the thermal and radiation stability of sodium tetrametaphosphate was checked. Before being used for identification all phosphorus compounds were purified by repeated crystallization until their physical constants corresponded to the values cited in the literature.

Samples of sodium tetrametaphosphate in polyethylene or sealed quartz ampoules were irradiated in the VK-2 channel of the RA reactor of the Boris Kidrič Institute at Vinca, at a thermal neutron flux of 3.5×10^{13}n/cm^2 sec (irradiation dose of the order 10^7 rad/h) for 30 minutes or in the BK channel of the same reactor at a thermal neutron flux of 6.5×10^9n/cm^2 sec (irradiation dose of the order 10^5 rad/h) for 10 and 20 days.

To check for the possible appearance of ^{32}P recoil gaseous products (phosphine and others) the quartz ampoules containing the irradiated material were opened in a special apparatus in a nitrogen atmosphere and under a stream of inactive phosphine gas.

To separate the ^{32}P chemical forms, an aliquot of 5-10 λ of the irradiated $Na_4P_4O_{12}$ solution was spotted on the cathode side, 5 cm of the Whatman

* Thanks are due to Prof. Dr. G.W. Wittig, Heidelgerg ; Prof. Dr. L. Horner, Mainz ; Dr. F. Krasovec, Ljubljana ; Dr. J. Speciale, St. Louis ; Dr. M. Halmann, Rehowoth ; Dr. G. Schrader, Wupertal-Elberfeld and Dr. W. Reiss, Braunschweig, for the supply of rare phosphorus compounds.

3 MM paper edge (solvent : 0.1 M lactic acid, 0.1 M pyrophosphoric acid, concentration about 1 mg/100 λ). The separation was performed by high-voltage electrophoresis using a Virus-type chamber with an agar-agar bridge, Pt electrodes, a Cryomate Laude 30D (temp. $0 \pm 0.1°$ C), 60 V/cm gradient and about 5 mA per strip stream. A 0.1 M lactic acid and 0.05 M zinc acetate solution (pH 4.2 - 4.3) was used as the electrolyte (9). Under the above conditions the process lasted about 90 minutes.

After electrophoretic separation of the ^{32}P chemical forms, the paper strips were sprayed with an acid molybdate solution and exposed to a stream of gaseous hydrogen sulphide, thus obtaining characteristic blue spots of the reduced phosphomolybdate complex (10).

The identification was also performed by studying the behaviour of the anions of phosphorus oxyacids in oxidation and hydrolysis (11) and by using neutron activation analysis instead of a colour test. This was done by electrophoresis on paper strips of an active solution of sodium tetra-metaphosphate parallel with a blank containing a chosen inactive phosphorus oxyacid anion which was subsequently irradiated in the reactor.

For neutron activation analysis made after inactive paper electrophoresis, the paper strips (2×1 cm) in Al-foils were irradiated at a thermal neutron flux of 3.5×10^{13}n/cm^2 sec. for 10 minutes. Before activity measurements the irradiated paper strips were cooled for a week so that ^{24}Na would decay.

The activity of the paper strips (2×1 cm) was measured with a well-type scintillation counter. The radiochemical purity was checked by measuring the decrease in activity over nine half-lives. In addition, the beta energy was checked by the absorption method not only in the irradiated materials but also at the activity peaks obtained on paper strips after electrophoresis.

Samples of the neutron irradiated sodium tetrametaphosphate were heated in a thermostat at the temperatures of $60 \pm 0.5°$ C and $180 \pm 0.5°$ C.

Results and discussion.

The series of experiments made to check the possible appearance of gaseous ^{32}P recoil products have shown that in the system investigated no phosphine or any other gaseous products are formed in an amount which exceeds the sensitivity limit of our detection method ($\backsim$ 0.5 % of the total ^{32}P activity).

The results of the chemical and neutron activation analysis, which were in very good agreement, led to the identification of ^{32}P in the following forms (Figs. 1 and 2) :

A	anion of pyrophosphoric acid	$(P_2O_7)^{4-}$
B	anion of hypophosphoric acid	$(P_2O_6)^{4-}$
C	anion of tripolyphosphoric acid	$(P_3O_{10})^{5-}$
D	anion of tetrapolyphosphoric acid	$(P_4O_{13})^{6-}$
E	anion of orthophosphoric acid	$(PO_4)^{3-}$
F	anion of tetrametaphosphoric acid	$(P_4O_{12})^{4-}$
G	anion of trimetaphosphoric acid	$(P_3O_9)^{3-}$
H	anion of orthophosphorous acid	$(HPO_3)^{2-}$
I	anion of pyrophosphorous acid	$(H_2P_2O_5)^{2-}$

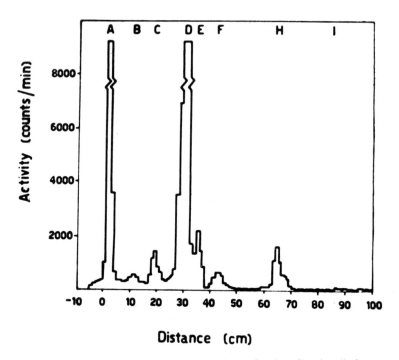

Fig. 1. — Electrophoretic histogram obtained after low flux irradiation
(6.5 × 10⁹ n/cm² sec, 10 d).

Figures 1 and 2 represent electrophoretic histograms which show the activity of particular ^{32}P ionic species as a function of their path on the paper.

As seen in both figures the small activity peak F was identified as the ^{32}P-parent compound whose percentage represents the retention, and the activity peak at the very beginning of the electrophoretic histogram A as the ^{32}P-pyrophosphoric acid.

The results in Table I * represent the relative ^{32}P activity distribution as a function of the conditions of the target material irradiation in the reactor. It is evident that under almost all irradiation conditions the percentage of the retention F is low and one obtains a great yield of the degradation products such as : cyclic trimetaphosphoric acid G which is by one P atom poorer than the parent compound, then the monomer orthophosphoric acid E and the linear polymers - pyrophosphoric acid A, tripolyphosphoric C and tetrapolyphosphoric acid D.

---

* The numerical values given in the tables are average values for five independent determinations. The standard deviation of the yields is 10-20 %.

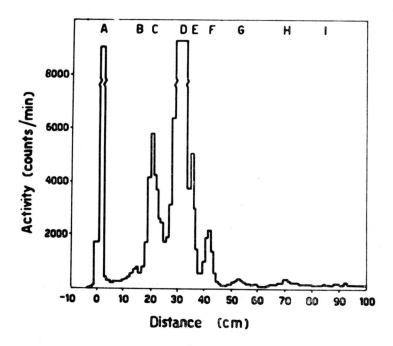

Fig. 2. — Electrophoretic histogram obtained after high flux irradiation
(3.5 × 10^13 n/cm² sec, 30 min).

It should be pointed out that in the above mentioned forms phosphorus is present in the highest oxidation state and even in the mildest conditions of irradiation small yields of reduced ^{32}P in the form of orthophosphorous acid *H* and pyrophosphorous acid *I* are obtained.

TABLE 1.

^{32}P Distribution in $Na_4P_4O_{12}$ under Different Irradiation Conditions

Irradiation conditions		Percentage of total ^{32}P activity								
n/cm² sec	time	A	B	C	D	E	F	G	H	I
6.5×10⁹	10 d	20.7	3.1	6.5	46.7	9.5	4.9	0.3	7.8	0.5
6.5×10⁹	20 d	21.4	2.2	17.3	39.6	10.0	4.8	1.7	2.6	0.4
3.5×10¹³	30 min	23.8	1.3	19.1	40.1	8.9	3.9	1.6	1.3	—
3.5×10¹³ (sample enveloped by Li_2CO_3)	30 min	25.7	2.6	28.1	30.0	9.7	1.4	2.5	—	—

As seen from the results in Table 1, longer irradiation, a higher thermal neutron flux, thus a higher irradiation dose in the reactor, involve reduction of the yields of the cyclic polymers-tetrametaphosphoric acid F and trimetaphosphoric acid G and reduction of the yield of the chain tetrapolyphosphoric acid D in favour of the chain compounds - tripolyphosphoric acid C and pyrophosphoric acid A.

It is evident therefore that, as a result of the recoil reaction in the system investigated, the 4-membered P-O-P ring does break followed by the formation of the 3-membered P-O-P ring, i.e. by chain polymers with shorter P-O-P chains.

Change in the irradiation conditions is also reflected on the yield of the reduced forms of ^{32}P such as orthophosphorous acid H and pyrophosphorous acid I, which is also reduced on transition to more energetic conditions of irradiation.

The irradiation of $Na_4P_4O_{12}$ in the presence of Li_2CO_3, which as a result of the nuclear reaction $_3^6Li$ (n, alpha) $_1^3H$ reduces the thermal neutron flux to about 5 % of its initial value, provides an interesting data about the important role of fast neutrons in the process of the opening and breakage of tetrametaphosphoric acid ring and formation of chain polymers with the characteristic P-O-P chemical bond.

It is an interesting fact that the formation of higher polymerization products could not be detected in the system investigated. In order to check the formation of the higher polyphosphates, which might be masked by the activity peak at the starting point A, we replaced the electrolyte by one containing 96 % 0.1 M acetic acid and 4 % 0.2 M sodium acetate, pH 3.7. Under these conditions the pyrophosphoric acid anion A migrated so that the activity at the starting point disappeared and new activity peaks could not be detected. This fact proved that peak A did not mask any other peak.

The results presented in Table 2 show the ^{32}P distribution after thermal treatment of the neutron irradiated sodium tetrametaphosphate. Comparison of the results presented in Tables 1 and 2 implies a decrease in the yield of ^{32}P in the form of the initial compound F, the tetrapolyphosphoric acid D, orthophosphorous acid H and pyrophosphorous acid I, and an increase in the yield of ^{32}P-tripolyphosphoric acid C. pyrophosphoric acid A and trimetaphosphoric acid G in the case of subsequent thermal treatment. Thermal treatment does not have considerable influence on the yields of ^{32}P-hypophosphoric acid B and orthophosphoric acid E.

These results show that the thermal annealing process is characterized by the opening and breakage of the 4-membered tetrametaphosphoric acid ring, the formation of chain polymers with a characteristic P-O-P chemical bond, degradation of the tetrapolyphosphoric acid into compound with shorter P-O-P chains, and oxidation of orthophosphorous and pyrophosphorous acids into orto- and pyrophosphoric acids.

It may be concluded that the importance of the results obtained so far primarily consists in the identification of the spectra of ^{32}P-labelled compounds formed in the neutron irradiated sodium tetrametaphosphate. We have found

TABLE 2.

^{32}P Distribution in $Na_4P_4O_{12}$ after Thermal Annealing.

Irradiation conditions		Heating conditions		Percentage of total ^{32}P activity								
n/cm² sec	time	temperature	time	A	B	C	D	E	F	G	H	I
6.5×10^9	20 d	180° C	10 min	21.8	2.9	18.2	38.9	9.1	4.3	2.2	2.1	0.5
			30	24.4	2.2	19.7	38.0	7.1	3.5	2.9	2.0	0.2
			60	25.8	2.4	21.1	34.9	7.6	2.9	3.8	1.5	—
			120	27.6	2.1	21.5	34.4	8.4	3.0	2.4	0.6	—
			240	27.9	2.0	22.0	34.7	8.2	2.6	2.3	0.3	—
			2580	27.6	2.2	23.7	33.9	7.9	2.3	2.4	—	—
3.5×10^{13}	30 min	180° C	10 min	24.2	1.5	18.8	39.4	7.8	4.1	2.2	2.0	—
			30	26.0	2.0	20.5	37.9	7.6	3.0	2.5	0.5	—
			60	26.5	1.9	21.7	35.1	8.0	4.3	2.2	0.3	—
			120	28.3	2.3	21.5	34.0	7.2	2.5	4.0	0.2	—
			240	30.1	2.4	20.2	34.3	7.0	2.8	3.2	—	—
			2580	31.0	1.9	23.1	32.5	6.9	0.5	4.1	—	—
3.5×10^{13}	30 min	60° C	10 min	24.7	2.4	19.3	38.6	7.9	3.7	2.3	1.1	—
			30	26.0	2.1	18.9	39.1	7.4	2.7	2.8	1.0	—
			60	27.3	2.2	20.8	36.1	8.1	2.0	2.9	0.6	—
			120	28.1	2.4	21.4	35.2	7.5	1.7	3.4	0.3	—
			240	29.4	2.0	22.3	33.2	7.3	2.0	3.0	0.2	—
			2580	29.7	2.2	23.1	32.9	7.1	1.1	3.9	—	—

that this process leads to the breakage of the initial P-O-P bond and formation of three groups of products : small amounts of cyclic products, largest amounts of chain products with short and long P.O.P. chains and simple inorganic compounds of the phosphate and phosphite type. Besides this, it has been noticed that the irradiation conditions in the reactor and subsequent isothermal annealing have considerable influence on the yield of different ^{32}P species. From the experimental data obtained it may be concluded that in analogy to the situation in the simple inorganic phosphates and in the polyphosphate investigated there exists, during irradiation in the reactor, thermal and probably to a certain extent radiation annealing as well.

ACKNOWLEDGEMENT.

    The author is grateful to the Yugoslav Federal and Serbian Research Funds for financial support and Mrs. B. Pavlovska for her skilful assistance in the experiments.

References

1. W.F. LIBBY. *J. Am. Chem. Soc.,* **62** : 1930 (1940).
2. A.H.W. ATEN Jr., H. VAN DER STRAATEN. *Science* **115** : 267 (1952).
3. P.A. SELLERS, T.R. SATO, H.H. STRAIN. *J. Inorg. Nucl. Chem.* **5** : 31 (1957).
4. T.R. SATO. *Analyst. Chem.,* **31** : 841 (1959).
5. L. LINDNER, G. HARBOTTLE. *Chem. Eff. Nucl. Transf. IAEA,* Vienna, p. 485 (1961).
6. B.F.C. CLARIDGE, A.G. MADDOCK. *Chem. Eff. Nucl. Transf. IAE*A, Vienna, p. 475 (1961).
7. M. HALMANN. *Chem. Rev.* : 689 (1964).
8. P.L. BRAUER, Rukovodstvo po Neorg. Preparativ. Him., *Izdat. Inost. Lit.,* Moskva, str. 273 (1956).
9. M.W. HANNA and L.J. ALTMAN. *J. Chem. Phys.* **36** . 1788 (1962).
10. C.S. HANES, T.A. ISHERWOOD. *Nature* **164** : 1107 (1949).
11. J.R. van WAZER. *Phosphorus and its Compounds.* Interscience, New York (1958).

Paper Electrophoresis in combination
with Polarography

by

J. HOMOLKA
Prague

Summary

The electrophoretic-polarographic method was used for study of blood proteins. In addition to the determination of the relative percentage of individual fractions, i.e. albumins, alfa, beta and gamma globulins, changes in their quality were evaluated. The increase in polarographically active groups expresses the qualitative changes of the proteins which precede the quantitative changes in the composition of the fractions. Thus it is possible to recognise changes in the quality of blood proteins which formerly escaped detection. Examples of the clinical application are given.

**

The blood proteins in clinical practice are usually interpreted, for example in paper electrophoresis, according to the quantity of single fractions. In this way the decrease of albumin and the increase of alfa, beta or gamma globulins is followed.

But from the literature it is known that in the blood proteins even qualitative changes are possible. So in albumin changes in the contents of free base groups (Epstein, 1952), changes in the ability to bind metals and organic dyes (Kusunoki, 1952), serologic aberrant optical activity and viscosity (Jirgensons, 1956), different ultra violet spectra (Rother Sarre, Kluthe, Fischer and Schütte, 1957), different migration in electrical field and many other differences are known.

We supposed that changes of quality in these single fractions could be an early indicator of the state of organism. These changes could be valuable mainly in the case when changes in the quantitative composition of electrophoretic serum proteins fractions are not yet expressed.

Therefore we elaborated a clinically useful method for the estimation of qualitative changes in proteins based on the principle that blood proteins in their denaturation by means of alkali as the result of their quarternary, tertiary or even secondary structure make free till now masked polarographically active groups, and so they raise their polarographic activity. For a better intuition let us say that a polarographic activity of diluted albumin solution by alkali denaturation raises four times in a healthy person, but in a diseased person it may raise only twice.

The principle of this method is the fractionization of serum proteins by means of paper electrophoresis and the isolation of single fractions, as seen in figure 1. These fractions are eluted into physiological saline solution

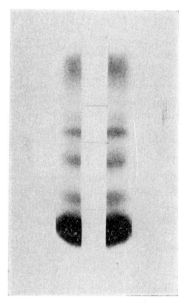

Fig. 1. — The unstained middle strip of the wet paper contains native fractions.

as seen in figure 2. Afterwards they are estimated by means of Brdicka's reaction, i.e. in the solution of a complex trivalent cobalt salt before denaturation and after it. It is very practical and quick to perform polarography by means of dropping mercury cathode with silver anode, as seen in figure 3. The determination before the denaturation gives us the relative percentages of albumin, alfa, beta and gamma globulins. The determination after denaturation shows us how many times these fractions raise their polarographic activity by means of alkali denaturation. We call this multiple relative lability or denaturation capacity as further polarographic active groups are set free.

However it would be pointless not to use for diagnosis generally known electrophoretic quantitative changes in protein fractions and limit oneself only to the mentioned relative lability. Therefore we commonly use a product of relative percentage of protein fraction with the relative lability of this fraction and this we call the maximum lability. The method of determination is accurately described, justified and published by the author (Homolka, 1961, 1964).

Fig. 2. — Eluation of single fractions from the filter paper strips into saline solution.

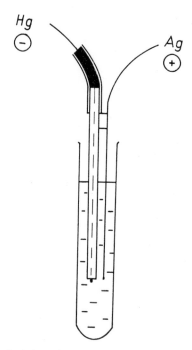

Fig. 3. — Practical dropping mercury cathode with silver anode for quick polarography.

It has been shown in our previous publications that only the albumin fraction is really denaturated by means of alkali. The albumin denaturation has been proved also by other techniques, as by paper electrophoresis, immu-

noelectrophoresis and by ultracentrifugation. The basic is the highest polarographic activity changes of the albumin molecule. Perhaps that is why the
changes in albumins are of main clinical value.

We ascribe the changes in globulins to a higher solubility of globulins
after the alkali addition and this different solubility in health and disease
can be also interpreted as the qualitative changes. But we prefer according
to our clinical experiments to evaluate the albumin denaturation capacity.

Such an example of the electrophoretic polarographic determination can
be seen in figure 4. In the upper part is shown the polarographic registration

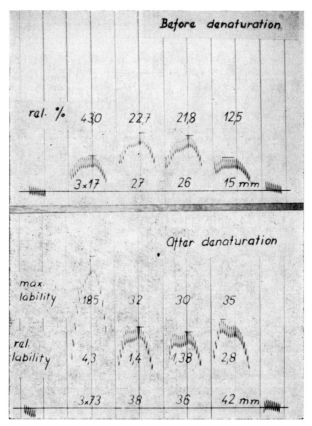

Fig. 4. — Example of the electrophoretic-polarographic determination
of human blood protein fractions.

of the albumin, alfa, beta and gamma globulin serum fractions before the
denaturation with their relative percentage. It is calculated from the wave
heights in mm, the albumin fraction is three times diluted.

In the lower part the same curves are shown, but after the alkaline
denaturation. From the wave heights in mm again the relative lability is

calculated, as can be seen in the middle row. The values of the maximum lability are still higher as they are the products of the relative percentage and the relative lability.

Normal values of the maximum lability in infants and in adults as well for the albumin fractions are 180 to 200. We found out that there were no differences between the newborn, infants and adults.

What is the clinical use of this determination ? We have found it useful in cases where we wished to observe very subtle changes, hardly discernible through other methods or by clinical observations. Such an example is during convalescence after illness. We found (Homolka and Mydlil, 1955) that only after three months of undisturbed convalescence after the diarrhoea of children were the normal levels of the maximum lability attained. Also the convalescence after typical tonsillitis lasted five weeks before the albumin maximum lability was normalised. The long-lasting therapy of rheumatic carditis in children was also followed by means of this method (Homolka and Cenek, 1961). We found, that the application of this method was more convenient than the usual tests such as the sedimentation rate of erythrocytes, the Weltmann and Brdicka filtrate reaction and the estimation of tyrosine and mucoproteins. We have also studied the high sedimentation rate of erythrocytes lasting for many years in these patients and we have gained new facts in this complex problem. The influence of the saline infusions and blood transfusions has been also studied by means of this method in children (Homolka and Mydlil, 1956). Furthermore, the changes in pregnancy have also been studied by this method (Schönfeld, Mach and Herzman, 1960). Now we are interested in the changes in hepatitis as well.

We suppose that the above described determination of qualitative changes in blood proteins and the changes of the polarographic activity of blood albumin especially, by means of denaturation, is a very useful and sensitive method giving evidence even for the subclinical disorders of health. These changes are naturally not specific for certain diseases but we suppose that they are the indicator of the organism reaction in blood proteins in illness. Therefore the application of this method will probably be possible in many different diseases in children and adults.

References

1. EPSTEIN J.A. : Prinjatoj sposob opredelenija kislotnoszczelocznych sootnoszenij w syworotkie krowi. *Biokhimia*, 16, 572-578, 1951.

2. HOMOLKA J. and MYDLIL V. : Blood Proteins in Infants from a Quantitative and Qualitative Point of View. *Ann. paediat.* (Basel), **185**, 129-141, 1955.

3. HOMOLKA J. and MYDLIL V. : The Blood Protein Picture in Diarrhoea of Infants in Comparison with other Diseases. *Ann. paediat.* (Basel) **185**, 142-149, 1955.

4. HOMOLKA J. and MYDLIL V. : The Influence of Saline Infusions on Blood Proteins Composition (in Czech). *Csl. pediat.* **11**, 613-615, 1956.

5. HOMOLKA J. : Die elektrophoretisch-polarographische Quantitäts und Qualitätsbestimmung der Serumeiweiss-Fraktionen.

BÜCHER M. : Moderne chemische Methoden in der Klinik. Leipzig, G. Thieme, 1961.

6. HOMOLKA J. and CENEK A. : The Significance of Changes in Serum Albumin Denaturation Properties in Evaluating the Rheumatic Process during Therapy. (in Czech). *Cas. lék. ces.* **100**, 441-446, 1961.

7. HOMOLKA J. : The Polarography of Proteins and its Clinical Application. (in Czech). SZN Praha, 1964.

8. JIRGENSONS B. : Qualitative Differences Between Preparations of Human Serum Albumin. *Makromol. Chem.*, **21**, 179-192, 1956.

9. KUSUNOKI T. : Binding of Dye by Plasma Proteins in Diseases. *J. Biochem.* (Tokyo), **39**, 349-355, 1952.

10. ROTHER K,, SARRE H., KLUTHE R., FISCHER E. and SCHÜTTE A. : Über immunoserologische und spektralphotometrische Untersuchungen von Serum Proteinen bei experimentalen « Nephrotischen Syndrom ». *Z. ges. exp. Med.*, **129**, 87-110, 1957.

11. SCHÖNFELD V., MACH J. and HERZMAN J. : Polarographic Investigation of Protein Fractions in Pregnant Women Sera. (in Czech). *Cas. lék. ces.*, **99**, 1249-1251, 1960.

12. WUNDERLY Ch. and HÄSSIG A. : Serological Test of Albumin. *Naturwissenschaften*, **39**, 260, 1952.

« KS » une nouvelle chambre chromatographique pour la CCM, à applications multiples

par

F. GEISS et H. SCHLITT
Rapporteur : M. Th. van der VENNE.

I. Définitions.

Nous proposons quelques définitions :

La « *saturation de la chambre* » *chromatographique* est l'état avant et durant le développement où tous les constituants de l'éluant sont en équilibre avec toutes les zones de l'espace gazeux de la cuve chromatographique. Des essais sont en cours pour étudier cet équilibre. Il est à noter que souvent la chromatographie débute loin de l'équilibre et est interrompue avant de l'atteindre.

La « *précharge gazeuse* » est l'expression générale pour la prise par la couche adsorbante « sèche », c'est-à-dire non mouillée par l'éluant, ou située au-dessus du front de molécules gazeuses pouvant influencer le développement chromatographique ultérieur.

La « *saturation sorptive* » est la limite supérieure de précharge gazeuse dans le cas de « saturation de la chambre ».

La « *saturation capillaire* » est le développement proprement dit ; il s'agit du remplissage par une phase liquide, du volume resté libre après la précharge gazeuse.

Quelques mots concernant ces définitions :

Au cas où des constituants étrangers à l'éluant se trouvent dans la cuve chromatographique (par exemple benzène/alcool utilisés pour le conditionnement et benzène comme éluant ; voir *Fig. 2*) l'équilibre pour la « saturation de la chambre » doit s'établir avec tous les constituants présents.

La saturation sorptive est un état peu intéressant : il est difficile à réaliser. De plus, cette saturation provoque une diffusion des spots au point de départ avant le développement.

La saturation capillaire est aussi l'état en fin de développement, la plaque présentant l'aspect mouillé (tout le volume de la couche est rempli).

Ce qui précède n'est qu'un aperçu rapide, tous les détails concernant la « saturation » peuvent être retrouvés dans les publications faites par Geiss et Schlitt (1).

F. GEISS et H. SCHLITT

TABLEAU 1.

Type de chambre	Dénomination abréviée	Saturation de la chambre avant l'élution	Précharge gazeuse	Chambres existant dans le commerce
N chambre large Largeur >> 3 mm				
a) sans revêtement	N non saturée	très faible	restreinte	« Tubes » Desaga « Chromatank » Shandon **) } Tout modèle de cuves carrées ou rondes
b) avec revêtement imbibé par le solvant	N saturée	assez complète	variable *)	
S chambre étroite Largeur < 3 mm				
a) sans contre-couche	S non saturée	nulle	nulle	« S » selon Stahl (Desaga) Type « Sandwich » (entre autres Camag, Kontés, etc.) BN (Desaga) GS-Klimakammer (Desaga) Chromagram (Eastman)
b) avec contre-couche imbibé par le solvant+)	S saturée	complète	variable *)	
c) horizontale avec solutions de conditionnement	KS	au choix	au choix (aussi saturation en gradients)	KS-Variokammer (Camag)

*) Selon la nature de la couche absorbante, de l'éluant, et temps de précharge.
**) Forme de transition vers le type S.
+) ou présence de papier absorbant.

Tableau récapitulatif 1 des chambres chromatographiques le plus couramment utilisées, en fonction de leur forme, le taux de saturation auquel il faut s'attendre.

II. Description de la KS-Variokammer *) *(Fig. 1 a).*

Les précurseurs de la « KS » sont d'une part la chambre BN (2), d'autre part le système décrit par Hesse et Alexander (3). Ces derniers voulaient travailler avec précharge gazeuse tandis que dans la BN on voulait exclure cette précharge.

Selon le cas, les deux conceptions, on le sait actuellement, présentent leurs avantages. La « KS » permet au choix d'utiliser un des deux principes et toutes les formes de transition.

Fig. 1a). — « *KS-Variokammer* ».
Mélange test de colorants : azodérivés, aminoanthraquinoniques, ioniques.

Mixture test dyes : azo-, aminoanthraquinonic and ionic derivatives.

La « KS » se présente comme suit *(Fig. 1b)* : La plaque avec sa couche d'adsorbant tournée vers le bas, peut être mise en contact ou séparée par un volet escamotable de solutions de conditionnement. Il y a deux ouvertures dans le dispositif permettant un balayage gazeux, un bloc chauffant pour la chromatographie continue et un bloc réfrigérant. Nous avons déjà signalé cette chambre antérieurement (4) et lors du 4ᵐᵉ symposium de chromatographie en 1966 (5), mais elle vivait à ce moment sa période artisanale de laboratoire, elle est actuellement commercialisée par Camag**). Les moules de conditionnement à insérer sont de formes carrées.

---

*) Cette dénomination sera utilisée dans ce texte sous sa forme abréviée : « KS ».
**) S.A. Camag, CH 4132 Muttenz (Suisse).

Pour la précharge gazeuse homogène on utilise la forme A, toutes les augettes sont remplies par le même liquide. Pour la précharge gazeuse en gradients discontinus on utilise soit la forme A, soit la forme C selon l'échelonnage du gradient souhaité. Pour réaliser un gradient continu on se sert de la forme B : les augettes latérales sont remplies avec les liquides dont les vapeurs constituent les extrêmes du gradient choisi. Le gradient s'établit par interpénétration de ces vapeurs. Pour la réalisation de précharge gazeuse localement bien délimitée et sélective c'est la forme D, « mosaïque », qui convient.

Fig. 1b). — « *KS-Variokammer* » + *moules de conditionnements.*
Mélange test de colorants : azodérivés, aminoanthraquinoniques, ioniques.

« *KS-Variokammer* » + *conditioning moulds.*
Mixture test of dyes : azo-, aminoanthraquinonic and ionic derivatives.

III. Les possibilités de la « KS ».

– Avec le volet maintenu tout le temps en position fermée, elle peut être utilisée comme une chambre S non saturée.
– Avec l'insertion du moule A rempli d'un seul liquide, le volet étant ouvert on obtient des résultats analogues à ceux fournis par une chambre N ou S saturée.
– Par insertion du moule A ou C on réalise des gradients dans la direction de migration du solvant ou dans la direction perpendiculaire.
– On peut créer une enceinte à humidité contrôlée (ou un gradient d'humidité).
– On peut réaliser une chromatographie continue.
– La couche peut être conditionnée avec des gradients de pH.
– On peut séparer sur une même plaque des constituants de polarité très différente (allant du composé non polaire au composé ionique).

IV. Quelques exemples pratiques.

1. *(Fig. 2).* a) Avec la précharge gazeuse homogène d'un seul constituant identique à l'éluant on obtient une diminution proportionnelle de tous les Rfs, la qualité de la séparation ou l'ordre d'élution ne sont donc pas affectés. La limite de la saturation sorptive se marque par une valeur minimale des Rfs pour les conditions déterminées.

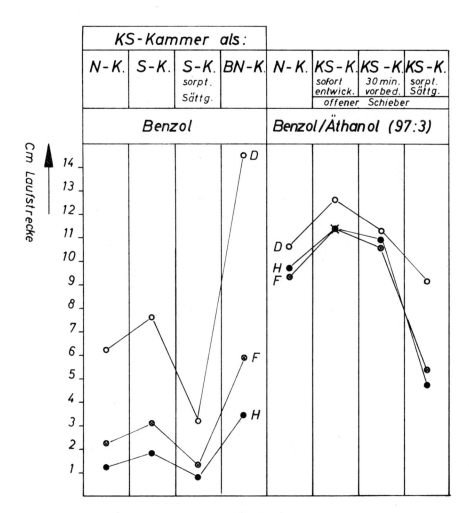

Fig. 2. — *Comparaisons de chambres N - S - KS.*
Pour tous les chromatogrammes : éluant benzène. Humidité relative 60 % Kieselgel G.
Mélange test de colorants : azodérivés, aminoanthraquinoniques, ioniques.

Comparisons of chambers N - S - KS.
For all the chromatograms ; eluant : benzene. Relative humidity 60 % Kieselgel G.
Mixture test of dyes : azo-, aminoanthraquinonic and ionic derivatives.

b) La précharge avec un constituant différent de l'éluant (cas particulier celui de l'humidité ambiante) fut traitée au symposium 1966 (5),
l'étude détaillée figure dans diverses publications faites par Geiss et ses
collaborateurs (4, 6), et Dallas (7).

c) Quant à la précharge par plusieurs constituants, elle donne lieu
à des phénomènes variés : les Rfs décroissent lorsque la précharge croît :
l'ordre de migration du Rouge Soudan (F) et l'indophénol (H) est inversé.
Avec du benzène seul, comme précharge, (H) migre loin derrière (F), avec
une précharge benzène/alcool 97 : 3, et en développant immédiatement, (H)
et (F) sont groupés. Après 30 min. de précharge (H) est devant (F), à la
saturation sorptive l'ordre d'élution est à nouveau inversé.

Pour éclaircir ce phénomène, on a préchargé une autre plaque, perpendiculairement au sens d'élution avec des quantités croissantes d'alcool. Comme
résultat, on a noté que l'inversion F/H ne se produit qu'une seule fois pour
une concentration d'alcool bien définie. A la saturation sorptive il semble
y avoir moins d'alcool sur la couche qu'après 30 min. de précharge.

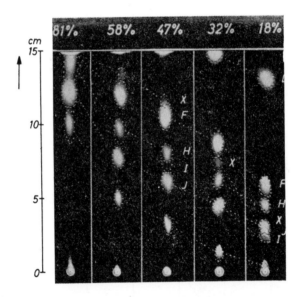

Fig. 3. — *Gradient transversal ; Gradient d'humidité* ; « KS »
kieselgel G ; éluant : benzène. Mélange test de colorants azo-
dérivés aminoanthraquinoniques, ioniques.

Cross gradient ; Humidity gradient ; « KS » kieselgel G ; eluant :
benzene. Mixture test of dyes : azo-, aminoanthraquinonic and
ionic derivatives.

2. *(Fig. 3)*. Cette figure et les suivantes sont des exemples de gradients*)
de précharge. Il s'agit d'un gradient transversal d'humidité, dont l'influence
sur la séparation d'un mélange de colorants est étudiée.

3. La *figure 4* reprend un autre cas de l'influence de l'humidité dans le
cas du comportement du front β dans la chromatographie bizonale (« duo-
zonale DC »). Ce front β se forme à partir d'un mélange de solvants de
polarités différentes (benzène/alcool). Le constituant le plus polaire est
adsorbé préférentiellement sur l'adsorbant, ce qui provoque un retard de
migration de l'alcool par rapport au benzène. Le front β marque un retrait

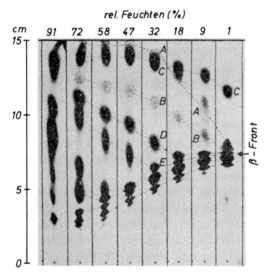

Fig. 4. — *Influence de l'humidité relative sur la position du
front* β. (Gradient transversal).
Chambre « KS » ; mélange de colorants ; kieselgel G ;
éluant benzène/méthanol 90 : 10, volet fermé pendant l'élution.
Mélange test de colorants azo-dérivés aminoanthraquinoniques,
ioniques.

Influence of relative humidity on the position of the β *front.*
(Cross gradient)
« KS » chamber ; mixture of dyes ; kieselgel G ; eluant :
benzene/methanol 90 ⧧ 10, shutter closed during elution.
Mixture test of dyes : azo-, aminoanthraquinonic and ionic
derivatives.

*) L'ensemble des techniques de gradient a été résumé par Niederwieser et
Honegger (8). Stahl (9) a décrit un gradient de sorbant. Le gradient d'éluants est
réalisé soit par démixion spontanée de mélanges de solvants (chromatographie poly-
zonale) soit à l'aide d'un gradient de solvant en profil continu et amené par un
capillaire au contact de la couche adsorbante (10). Tout gradient perpendiculaire à
la direction de l'élution permet par son principe — pour le même mélange — d'étudier
plusieurs conditions chromatographiques sur une même plaque, en vue d'une optimi-
sation. Par contre, un gradient en direction d'élution permet de soumettre *divers*
mélanges au *même* effet chromatographique.

de plus en plus grand par rapport au premier front plus l'humidité augmente. Cela s'explique par une modification de solubilité de l'alcool dans le benzène en présence d'eau. L'alcool séjourne de plus en plus longtemps dans la phase stationnaire et accroît son retard dans la migration. Les tâches qui migrent accidentellement dans la zone du front β, c'est-à-dire dans un gradient spontané, ont une forme lenticulaire et régressent avec ce front, tandis que les constituants au-dessus du front β voient leurs Rfs augmenter en fonction de l'humidité.

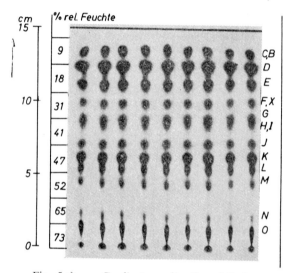

Fig. 5 a). — *Gradient en direction d'élution.*
Mélange de colorants ; chambre « KS », volet fermé, alumine DS-5 Camag ; éluant benzène ; a) activité croissante dans le sens d'élution sous forme de gradient (discontinu) d'humidité de 9 à 72,5 % ; humidité relative ; moule utilisé : C. Mélange test de colorants azo-dérivés aminoanthraquinoniques, ioniques.

Gradient in direction of elution.
Mixture of dyes ; « KS » chamber ; shutter closed ; alumina DS-5 Camag ; eluant : benzene ; a) increasing activity in the direction of elution in the form of a discontinuous humidity gradient from 9 to 72,5 % ; relative humidity ; mould employed : C. Mixture test of dyes : azo-, aminoanthraquinonic and ionic derivatives.

91% 72,5% 58% 47% 31% 24% 18% 9%

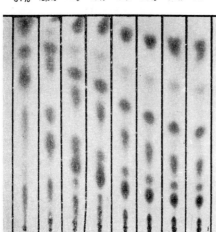

Fig. 5 b). — Comme Fig. 5 a), mais gradient transversal.
Mélange test de colorants : azodérivés, aminoanthraquinoniques, ioniques.

As. Fig. 5 a) but cross gradient. Mixture test of dyes :
azo-, aminoanthraquinonic and ionic derivatives.

4. Maintenant suit un exemple de *gradient dans le sens de migration
du solvant (fig. 5a)*, réalisé avec activité croissante dans le sens d'élution.
On y note une dispersion optimale des taches sur la distance parcourue, mettant en évidence une juxtaposition des zones optimales pour chaque humidité
particulière : *Figure 5b* montre que pour chaque humidité et variant avec
celle-ci il n'y a qu'une fraction du mélange qui soit bien séparée.

5. *(Fig. 6.)* Un gradient dit « de substitution » se forme toujours quand
la précharge est moins polaire que l'éluant. Le front de l'éluant déplace les
molécules adsorbées en formant un espèce de front β, c'est-à-dire un gradient. Ce dernier s'atténue de plus en plus au fur et à mesure que la précharge devient plus polaire (*b*). La partie *c*) de la figure montre comment
on a pu échelonner davantage par chromatographie continue la séparation
des constituants du chromatogramme de l'extrême droite de la partie *b*).

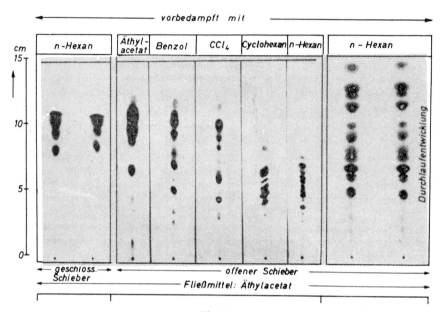

Fig. 6. —

a) « *Précharge gazeuse* » *avec des substances moins polaires que l'éluant.*
   Chambre « KS » ; élution avec le volet fermé pendant l'élution ;
   kieselgel G ; éluant acétate d'éthyle ; humidité relative 20 % ;
   « Précharge gazeuse » 1 h ; mélange de colorants.
b) Influence de la polarité des molécules de précharge sur le gradient
   de substitution.
c) « Précharge gazeuse » 1 h ; élution continue pendant 75 min.
   Mélange test de colorants azo-dérivés, minoanthraquinoniques, ioniques.

a) « *Gas preloading* » *with less polar substances than the eluant.*
   « KS » chamber : elution with shutter closed during elution ; kieselgel G ;
   eluant : ethyl acetate ; relative humidity 20 % ; « Gas preloading » 1 h.;
   mixture of dyes.
b) Influence of the polarity of the preload molecules on the substitution
   gradient.
c) « Gas preloading » 1 h. ; continuous elution for 75 min.
   Mixture test of dyes : azo-, aminoanthraquinonic and ionic derivatives.

6. *Fig. 7* donne un exemple de séparation de constituants allant du composé non polaire jusqu'au composé ionique sur une même plaque chromatographique. Les conditions opératoires furent les suivantes :

Vapeurs de précharge : benzène, (A), éluant : alcool (B), volet : ouvert. Le gradient qui serait trop brutal est atténué par un échange continu au cours de la chromatographie entre A et B. La dilution de B en dessous du front β donne un étalement du profil de concentration dans la phase mobile.

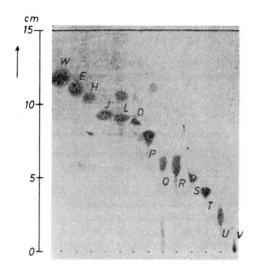

Fig. 7. — *Comment couvrir une large gamme de polarité entre les constituants d'un échantillon.*

Mélange de colorants ; kieselgel G, 55 % d'humidité relative ; chambre « KS » ; 30 min. de « précharge gazeuse » avec du benzène suivie de l'élution avec éthanol, le volet étant ouvert. Mélange test de colorants azodérivés aminoanthraquinoniques, ioniques.

How to cover a wide range of polarity between the constituents of a sample.

Mixture of dyes : kieselgel G, 55 % relative humidity ; « KS » chamber ; 30 min. gas preloading with benzene followed by elution with ethanol, the shutter being open. Mixture test of dyes : azo-,aminoanthraquinonic and ionic derivatives.

V. **Remarques.**

On ne doit pas appeler « phase mobile » le liquide qu'on met dans le réservoir ; ce liquide est dénommé « éluant » (« Fliessmittel »).

Cet éluant, par effet du processus chromatographique entre en contact avec le sorbant, se mélange durant le développement avec des molécules auparavant présentes sur la plaque, et s'y démixe éventuellement. Ce n'est qu'à ce moment que se forme la phase mobile. Son existence est intimement liée à celle de la phase stationnaire.

La phase mobile peut être très variée le long de la couche de la plaque et a une composition qui peut être complètement différente du liquide versé dans le réservoir de l'éluant (sauf dans le cas d'un éluant pur et en l'absence de toute précharge gazeuse). Un exemple simple et parlant est le cas d'une précharge avec des vapeurs d'alcool suivie d'une élution par du benzène. La phase mobile a diverses concentrations benzène/alcool au cours du temps et selon l'endroit considéré.

Par le même raisonnement, le sorbant venant du flacon n'est pas identique à la phase stationnaire du processus chromatographique : la phase stationnaire est constituée par le sorbant, par des molécules de l'éluant et des molécules d'une précharge éventuelle.

Bibliographie.

1) F. GEISS et H. SCHLITT. *Chromatographia*, **1**. 1968.
2) M. BRENNER et A. NIEDERWIESER. *Desaga-Druckschrift zur BN-Kammer*, Heidelberg.
3) G. HESSE et M. ALEXANDER, Journées Internationales d'Etudes des Méthodes de Séparation Immédiate par Chromatographie, Paris, 1961, *Publication GAMS*, Paris, 1962, p. 229.
4) F. GEISS, H. SCHLITT et A. KLOSE. *Z. Anal. Chem.*, **213** : 331, 1965.
5) G. GEISS et M. Th. van der VENNE, dans IVme Symposium Chromatographie-Electrophorèse ; Presses Académiques Européennes, Bruxelles, 1968, p. 153.
6) F. GEISS, H. SCHLITT et A. KLOSE. *Z. Anal. Chem.*, **213** : 321, 1965.
7) M.S.J. DALLAS. *J. Chromat.*, **14** : 57, 1964.
8) A. NIEDERWIESER et C.G. HONEGGER, *Advances in Chromatography* 2, New-York, Ed. J.C. Giddings et R.A. Keller, Marcel Dekker, p. 123, 1966.
9) E. STAHL. *Chem. Ing. Tech.*, **36** : 941, 1964.
10) A. NIEDERWIESER et C.G. HONEGGER. *Helv. Chim. Acta*, **48** : 893, 1965.

Microélectrophorèse en couche mince avec élution continue

par

Maurice SZYLIT

Laboratoire de Biologie physico-chimique
de la Faculté des Sciences (91) Orsay, France.

Introduction.

Le gel d'acrylamide utilisé comme support d'électrophorèse (1) permet d'obtenir d'excellents résultats analytiques : cependant, les méthodes de dosage et de récupération sont parfois aléatoires, que les protéines séparées soient dosées dans le support ou après élution de celui-ci. Dans le premier cas, la concentration de l'échantillon et l'importance relative de ses constituants sont déterminées par une mesure photométrique. Dans le deuxième cas, le gel est découpé en petits fragments, broyé après congélation (2) et les protéines récupérées par centrifugation sont analysées par les méthodes conventionnelles.

Ces divers procédés sont souvent longs et laborieux et parfois imprécis. Il existe en effet quelques restrictions à l'application des lois quantitatives de la spectrophotométrie : la loi de Beer n'est applicable que dans un domaine restreint de concentration et toutes les protéines ne réagissent pas de la même manière avec le colorant (3). L'analyse quantitative est encore plus délicate lorsque la coloration est le résultat d'une réaction enzymatique : les valeurs obtenues dépendent du temps d'incubation, de l'épaisseur du gel et des proportions relatives des diverses fractions (4). Une durée d'incubation trop courte ne fait apparaître que les fractions les plus concentrées, alors qu'une trop longue incubation favorise les bandes mineures. L'élution après le découpage des gels est une opération de longue durée, de faible rendement et provoque une dilution importante des produits (5).

Aussi, avons-nous entrepris la réalisation d'un dispositif de microélectrophorèse préparative qui permet l'élution des protéines au fur et à mesure de leur séparation. Nous sommes partis d'une technique d'électrophorèse en couche mince (4) que nous avons modifiée en nous inspirant d'un travail de Porath (6).

Appareil.

1) *Principe*. Les protéines se déplacent dans une mince plaque de gel disposée verticalement, dans laquelle on crée une cavité obtenue par moulage (chambre d'élution) à une certaine distance de l'origine. Les fractions arrivent successivement au niveau de la chambre d'élution et sont drainées vers un collecteur de fractions par un courant transversal d'électrolyte dont le débit est réglé par une pompe à action péristaltique placée en amont de la cellule. Le circuit d'élution est représenté dans la figure 1 dans le cas d'une élution simultanée à 2 niveaux. L'échantillon est déposé sur une surface parfaitement plane obtenue par moulage (chambre de départ).

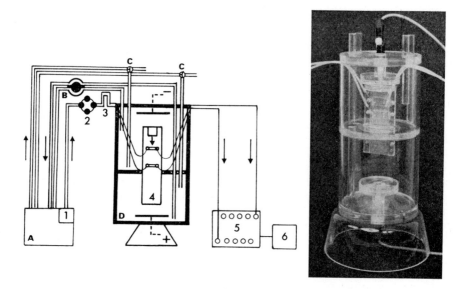

Fig. 1. — Appareil d'électrophorèse avec élution continue : circuit d'élution (1 à 6), circuit de refroidissement (A à D). 1, réserve de tampon. 2, pompe péristaltique. 3, piège à bulles. 4, cellule d'électrophorèse. 5, collecteur de fractions. 6, spectrophotomètre. A, échangeur de calories. B, pompe. C, vidange. D, cuve à électrophorèse.

Apparatus for electrophoresis with continuous elution : elution circuit (1 to 6), refrigeration circuit (A to D). 1, buffer stock. 2, peristaltic pump. 3, bubble trap. 4, electrophoretic cell ; 5, fraction collector. 6, spectrophotometer. A, caloric exchanger. B, pump. C, emptying. D, electrophoresis tank.

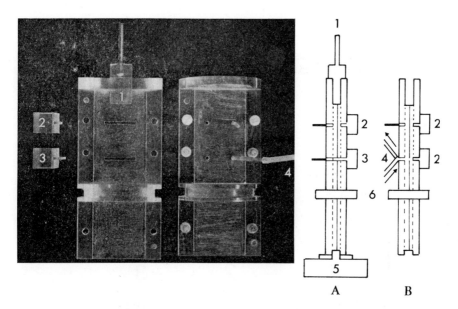

Fig. 2.— Cellule d'électrophorèse utilisant une seule chambre d'élution pendant le coulage du gel (A) et au cours de la séparation (B). 1, cale pour la chambre de départ. 2, cale d'obturation. 3, cale d'élution. 4, capillaire. 5, socle d'étanchéité. 6, joint torique (les proportions ne sont pas respectées).

Electrophoretic cell using a single elution chamber during the pouring of the gel (A) and during the separation (B). 1, wedge for departure chamber. 2, sealing wedge. 3, elution wedge. 4, capillary. 5, tight base. 6, toric joint (the proportions have not been respected).

2) *Réalisation*. Une cellule avec 2 chambres d'élution est représentée dans la figure 2. Elle est constituée par 2 plaques en plexiglas dépoli dont les dimensions sont voisines de celles d'une lame de microscope. Dans l'une des plaques on découpe 2 fentes (1 × 15 mm) à 21 et 38 mm du bord supérieur. Dans l'autre plaque on perce 2 ouvertures (1 mm) à la hauteur de chacune d'elles. La figure 2 (A) représente la cellule dans le cas où une seule chambre d'élution est utilisée. Les 2 plaques sont distantes de 2 mm. On obture l'extrémité inférieure pendant le coulage du gel en plaçant la cellule sur un socle muni d'un joint en caoutchouc (figures 2 (A) et 3). Ce dispositif a en outre l'avantage de maintenir la cellule parfaitement verticale. Dans la fente inférieure on dispose une cale d'élution qui se distingue de la cale d'obturation supérieure par le fait qu'elle vient s'appuyer contre la face interne de l'autre plaque. Les 4 ouvertures circulaires sont obturées. L'étanchéité est assurée par un film de graisse siliconée. Sur le haut de la cuve on place une cale (2 × 10 × 12 mm) pour former la chambre de départ.

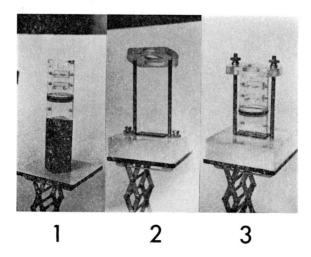

1   2   3

Fig. 3. — Mise en place de la cellule pendant le coulage du gel.
1, cellule. 2, socle. 3, mise en place.

Assembly of the cell during the pouring of the gel. 1, cell. 2, stand.
3, assembly.

Le gel polymérise au bout de 20 minutes à la température ordinaire ; on dégage la chambre de départ et la chambre d'élution, on emmanche 2 capillaires de 0,5 mm de diamètre intérieur. La migration peut s'effectuer librement puisque les 2 cales d'obturation viennent seulement en affleurement. La cellule est prête à l'emploi et peut être placée dans la cuve pour électrophorèse représentée dans la figure 1.

3) *Cuve pour électrophorèse.* Elle comporte 2 compartiments cylindriques de 90 mm de diamètre ; les électrodes sont constituées par un fil de platine disposé en forme d'anneau. La base du réservoir cathodique (supérieur) est percée d'une ouverture de 40 mm de diamètre comportant un joint torique et une butée qui maintient la cellule parfaitement verticale. La cathode est traversée en son centre par un fin tube de verre qui assure la verticalité de l'aiguille de la microseringue servant à délivrer l'échantillon. Le renouvellement des tampons est assuré par 2 circuits indépendants.

4) *Préparation du gel et du tampon.* Nous avons montré (4) que l'on peut renoncer au système discontinu d'Ornstein (2) en réalisant une ligne de départ extrêmement fine. Ceci est obtenu par la formation d'une chambre de départ parfaitement plane et par l'utilisation d'une microseringue de précision (Hamilton 7105). Nous utilisons un tampon continu Tris (10,05 g/l)-

EDTA (0,65 g/l)-Acide borique (0,5 g/l) : à 20° C la conductiblité est égale à 580.10^{-6} ohm^{-1} cm^{-1} et le pH est de 9,1 (4). Nous employons un gel continu à 5,5 % ; on mélange dans l'ordre un volume des solutions suivantes, préparées dans le même tampon qui sert à la séparation : cyanogum 41 BDH (22 %), dimethyl-amino-propio-nitrile (1,6 %), ferricyanure de potassium (0,005 %), persulfate d'ammonium (0,55 %).

5) *Conduite de l'électrophorèse.* Il faut appliquer un champ électrique élevé pour récupérer les bandes les plus lentes dans un laps de temps relativement court. Ceci n'est possible qu'en favorisant au maximum la dissipation de la chaleur dégagée par le passage du courant. Nous réalisons ces conditions en utilisant des gels de faible épaisseur et en maintenant le tampon à une température basse constante (5 à 7° C). Par suite, tous les points du gel sont à la même température. On élimine les bulles d'air en purgeant le tampon et en intercalant un piège à bulles entre la pompe et la cellule. Le débit de l'éluat est réglé à 0,17 ml/min (pompe Desaga). Le capillaire d'élution qui alimente le collecteur de fractions (LKB Utrorac, type 7000) doit être le plus court possible afin d'éviter le mélange des éluats recueillis par fractions de 3 gouttes.

6) *Echantillons et dosages.* La lacticodeshydrogénase du muscle de lapin est une préparation cristalline (Boehringer) reprise dans le tampon Tris après centrifugation. Les zones à activité LDH sont colorées par la réduction d'un sel de tétrazolium (pNBT) à 37° C, à pH 9,1 et à l'obscurité (4). L'activité des éluats est mesurée à 340 mμ à pH 7,6 et à 25° C en présence de NADH 1,6 10^{-4} M et de pyruvate de sodium 1,9 10^{-3} M (8). Les solutions d'hémoglobines A et C sont préparées à partir d'échantillons lyophylisés (Hyland) ; leur concentration dans les éluats est mesurée par photométrie (bande de Soret) et on les détecte dans les gels de contrôle grâce à leur activité peroxydasique (4).

Résultats.

Expérience n° 1. Pour les premiers essais nous avons choisi un colorant de préférence à un mélange protéique : le dépôt du colorant à des intervalles de temps variables permet d'obtenir à volonté des bandes plus ou moins rapprochées et d'évaluer rapidement les caractéristiques principales de l'appareil. Dans la figure 4 (A) nous avons représenté l'élution de 4 bandes successives de bleu de bromophénol séparées par 2 à 3 mm. La distance de migration était de 40 mm. Les bandes recueillies sont bien séparées et relativement peu diluées après un tel parcours (85 % de la bande I sont recueillis en 12 gouttes). L'analyse des aires d'élution montre que le rendement est sensiblement identique pour chaque fraction malgré les différences initiales de concentration et de volume.

Expérience n° 2. La figure 4 (B) montre qu'il n'est pas possible d'éluer en 90 minutes les isoenzymes les plus lents (n° 4 et 5) de la lacticodeshydrogénase lorsque la chambre d'élution est située à 40 mm de l'origine. Le rendement pour les fractions 1, 2 et 3, sensiblement identique pour chacune est voisin de 80 %.

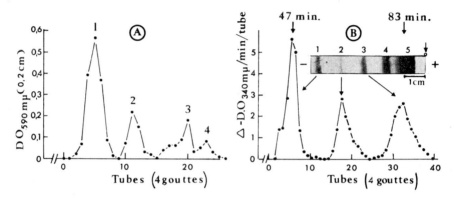

Fig. 4. — Elution de 4 dépôts successifs de Bleu de bromophénol (4_A) et de 55_μ de lacticodeshydrogénase de muscle de lapin (4_B). Acrylamide 5,5 %, 240 volts, 7,5° C, 0,17 ml/min, 40 mm de migration. 4_A : on indique le volume et la concentration relative des échantillons ainsi que les aires d'élution en unités arbitraires (1, 2 μl, $S^2 = 104$) - (2, 0,5 μl, $S^2 = 27,5$) - (3, 2 μl/4, $S^2 = 26$) - (4, 1 μl/4, $S^2 = 12$).

4_B : on indique pour chacune des fractions 1, 2 et 3 les pourcentages calculés d'après l'électrophorégramme (47, 21, 32) et les pourcentages mesurés d'après la courbe d'élution (44, 23, 33). Le rendement pour les 3 fractions est égal à 77 %.

Elution of 4 succeeding layers of bromophenol blue (4A) and of 55 μ of rabbit muscle lacticodeshydrogenase (4B). Acrylamid 5.5 %, 240 volts, 7,5° C, 0.17 ml/min., 40 mm of migration. 4A. The volume and the relative concentration of the samples are indicated as well as the elution surfaces in arbitrary units (1, 2 μl, $S^2 = 104$) - (2, 0.5 μl, $S^2 = 27.5$) - 3, 2 μl/4, $S^2 = 26$) - (4, 1 μl/4, $S^2 = 12$).

4B. For each of the fractions 1, 2 and 3 are indicated the percentages calculated according to the electrophoregram (47, 21, 32) and the percentages measured according to the elution curve (44, 23, 33). The yield for the three fractions is 77 %.

Expérience n° 3. Une distance de migration trop courte ne permet pas une séparation satisfaisante ; trop grande, elle provoque une dilution excessive des fractions lentes. Nous avons résolu ce problème en éluant les fractions lentes et rapides à des niveaux différents favorables à chaque cas d'espèce. L'expérience qui correspond à la figure 5 est conduite de la façon suivante : la 1re chambre d'élution, située à 12,3 mm de l'origine n'est mise en service qu'après le passage de l'hémoglobine A qui sera éluée au niveau de la 2me chambre d'élution située à 29,4 mm. La séparation est rapide, la dilution des 2 fractions est du même ordre et le rendement voisin de 100 %. Une partie de HB_C se retrouve après HB_A après avoir parcouru 29,4 mm ; cela tient à l'existence d'une zone diffuse entre les fractions A et C : l'électrophorèse analytique correspond aux éluats n° 3, 5 $(HB)_A$ 5 $(HB)_C$ 14.

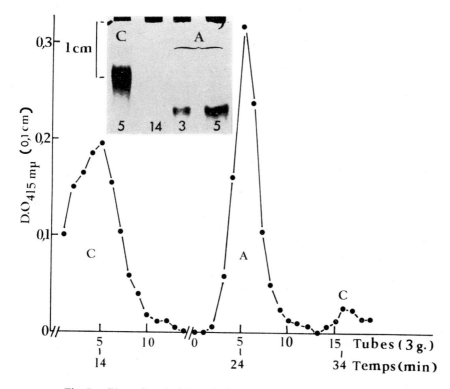

Fig. 5. — Séparation des hémoglobines A et C éluées à 2 niveaux différents (voir texte). Acrylamide 5,5 %, 450 volts, 6,5° C, 0,185 ml/min.
HB$_A$: 0,35 mg-migration sur 29,4 mm-Rend. = 92 %. HB$_C$:
0,35 mg-migration sur 12,3 mm-Rend. = 94,5 et 103 %. L'électrophorèse de contrôle est réalisée par la même cellule pendant
15 minutes à 450 V.

Separation of the hemoglobins A and C eluted at two different
levels (see text). Acrylamid 5.5 %, 450 volts, 6,5° C, 0.185 ml/min.
HB$_A$: 0.35 mg-migration on 29.4 mm - yield = 92 %. HB$_C$:
0.35 mg-migration on 12.3 mm - yield = 94.5 and 103 %. The
control electrophoresis is performed with the same cell during
15 minutes at 450 V.

Discussion.

La technique que nous décrivons (élutions simultanées à plusieurs niveaux) est actuellement appliquée à des protéines non colorées (LDH des
erythrocytes humains). Les résultats sont très reproductibles si les conditions
de la séparation sont soigneusement étalonnées : tension, température, débit
d'élution, etc... Ceci permet de programmer à l'avance les différentes phases
de la séparation. Le rapport des vitesses de migration des hémoglobines A
et C obtenu d'après les courbes d'élution est identique aux valeurs mesurées
sur le gel après coloration par le benzidine. Ces valeurs confirment les résul-

tats présentés par d'autres auteurs (8, 9) et obtenus par des techniques différentes : $V_C / V_A = 0,66$.

Il est donc tout d'abord nécessaire de procéder à une électrophorèse-témoin sans élution pour mesurer les vitesses de migration relatives des composants du mélange. Si l'écart entre les 2 chambres d'élution est constant par construction, on peut par contre modifier la distance de migration en prenant des cales plus ou moins longues pour former la chambre de départ. Le champ électrique est faiblement modifié pour une variation de près de 50 % de la distance entre l'origine et la première chambre d'élution.

Notre dispositif se distingue par de nombreux points des réalisations déjà présentées. C'est un appareil de *microélectrophorèse préparative,* qui assure rapidement le fractionnement de très petits volumes de produits. Les gels plats et minces permettent d'utiliser des champs électriques élevés et par conséquent de récupérer rapidement les fractions lentes avec une dilution minimale. Ces gels très minces dissipent en effet très rapidement la chaleur et nous avons délaissé les dispositifs de refroidissement très complexes qui font usage d'enveloppes concentriques avec une circulation de fluide refroidi. C'est le tampon lui-même que nous faisons circuler et refroidir à l'extérieur de la cuve. Comme la surface de contact du gel avec les plaques est très grande, il n'est pas nécessaire de maintenir le support (9-10-11). Notre chambre d'élution est très simple et extrêmement petite (30 µl) comparée à celle d'autres dispositifs (10-11-12-13) ce qui évite tout mélange des fractions et autorise l'emploi de faibles débits d'élution. Nous avons constaté que les protéines sont éluées avant d'avoir franchi la moitié de la chambre d'élution.

Mais l'avantage essentiel de notre technique est de réaliser l'élution à plusieurs niveaux et sans que les protéines aient à traverser tout le gel. D'une part, nous pouvons éluer les protéines peu mobiles, rapidement et sans trop de dilution. D'autre part, ce dispositif élimine la nécessité d'interposer une membrane hémiperméable entre le gel et les compartiments électrochimiques. Cette membrane complique les montages (9-12-13-14), perturbe les caractéristiques de l'électrolyte (effet Bethe-Toropoff) (5) et provoque une diminution notable du rendement par suite de l'adsorption des protéines (11). Des essais complémentaires ont montré qu'il n'y a pas de déformation du champ électrique au milieu de la cuve. Enfin, le volume des éluats est très petit (3 gouttes).

Toutes les protéines doivent naturellement migrer dans le même sens : cependant, dans le cas contraire, l'échantillon est déposé dans une chambre d'élution et non dans la chambre de départ : ce dispositif d'insertion est particulièrement avantageux dans les électrophorèses sur gradient de pH stabilisé par le gel d'acrylamide (6). La concentration minimale d'échantillon que l'on peut analyser dépend de la nature du mélange et de la sensibilité du dosage. Dans le cas d'un enzyme comme la LDH, on peut mesurer l'activité avec une concentration finale de quelques microgrammes par ml. Par contre, dans le cas d'une protéine sans propriétés catalytiques, une concentration de 100 µg par ml est difficile à mesurer lorsque le volume est de quelques gouttes ; seul un analyseur possédant une chambre de mesure très petite, placé entre la cellule et le collecteur peut résoudre ce problème. Comme la mesure de la concentration protéique tube par tube est une opération fastidieuse, nous vérifions en cours d'électrophorèse la qualité de la séparation. Nous utilisons actuellement des gels comportant 2 pistes, l'une pour l'élution, l'autre de 3 mm de large située sur un côté et qui sert de bande de contrôle. On y dépose l'échantillon et un mélange des hémoglobines A et C. Ces dernières permettent de suivre la séparation et en fin d'électrophorèse on vérifie sur le gel que toutes les bandes de l'échantillon ont bien été éluées. Le tube de concentration maximum de chaque fraction est déterminé par un essai à la touche extrêmement rapide.

Si l'écart entre 2 bandes avant de pénétrer dans la chambre d'élution est inférieur à 1,5 mm, ces 2 bandes ne peuvent être totalement séparées. L'utilisation de champs électriques plus élevés doit permettre d'augmenter la distance de migration sans accroître la dilution. Cette dernière, pour une tension donnée, dépend du débit d'élution, qui doit être maintenu le plus petit possible. Si ce débit, dans le cas de bandes progressant de 2,3 mm par minute est inférieur à 0,075 ml/minute, une partie des constituants échappe à l'élution. Par suite d'une instabilité mécanique des lèvres de la chambre d'élution, nous ne pouvons utiliser des gels de concentration inférieure à 5 %. Des essais préliminaires effectués avec des gels plus minces (1,5 mm) et des chambres d'élution plus petits (15 μl) ont donné satisfaction.

Nous envisageons la possibilité d'utiliser des plaques en quartz montées sur une armature en plexiglas ainsi que des unités de refroidissement par effet Peltier.

M. R. Andouart, par la qualité de son travail et par ses suggestions, a permis la réalisation de cet appareil ; nos remerciements vont aussi à MM. R. Touminet, R. Xerri et J. Cl. Escande.

Références

1. S. RAYMOND et L.S. WEINTRAUB. *Science*, **140** : 711 (1959).
2. L. ORNSTEIN et B.J. DAVIS. Disc electrophoresis ; Eastman Kodak Co (1962).
3. W. GRASSMAN et K. HANNIG. *Hoppe Seyler's Z. Physiol. Chem.*, **290** : 1 (1952).
4. M. SZYLIT. *Path.-Biol.*, **16** : 247 (1968).
5. U.J. LEWIS et M.O. CLARK. *Anal. Biochem.*, **6** : 303 (1963).
6. J. PORATH, E.B. LINDNER et S. JERSTEDT. *Nature*, **182** : 744 (1958).
7. M. SZYLIT et O. IVOREC-SZYLIT. *C.R. Acad. Sc. Paris*, **265** : 1841 (1967).
8. S. RAYMOND et M. NAKAMICHI. *Anal. Biochem.*, **7** : 225 (1964).
9. T. JOVIN, A. CHRAMBACH et M.A. NAUGHTON. *Anal. Biochem.*, **9** : 351 (1964).
10. D. RACUSEN et N. CALVANICO. *Anal. Biochem.*, **7** : 62 (1964).
11. A.M. ALTSCHUL, W.J. EVANS, W.B. CARNEY, E. Mc COUTNEY, H. BROWN. *Life Sciences*, **3** : 611 (1964).
12. J.V. MAIZEL. *Ann. N. Y. Acad. Sc.*, **121** : 382 (1964).
13. H. HOCHSTRASSER, L.T. SKEGGS et J.R. KAHN. *Anal. Biochem.*, **6** : 13 (1963).
14. A.H. GORDON et L.N. LOUIS. *Anal. Biochem.*, **21** : 190 (1967).
15. H. SVENSSON. *Adv. Prot. Chem.*, **4** : 251 (1958).
16. M. SZYLIT. Résultats non publiés.

Gaschromatographic determination of volatile acids in effluent waters

by

B. DECLERCQ, V. BLATON and H. PEETERS

Simon Stevin Instituut voor Wetenschappelijk Onderzoek, Brugge.

Introduction.

The volatile fatty acids in biological material are usually isolated by steam-distillation and recovered as an aqueous solution of their sodium salts.

James and Martin (1) used a micro-partition column of celite and Na_2SO_4 to convert the sodium salts to an anhydrous ethereal solution of free acids and separated them by gaschromatography on a column of Silicon DC 550 and stearic acid on celite.

Further improvements have been reported concerning the isolation by Erwin (2), the removal of water and the final concentration by Gehrke and Lamkin (3).

Efficient columns, reducing reversible adsorption and ghosting peak formation were proposed by Ackman, Metcalfe and Erwin (4, 5, 2).

The flame ionization detector, insensitive to water, has been proposed as the most suitable for this purpose (6).

To detect and quantitate the volatile fatty acids, found in effluent waters, we introduced a few methodological improvements to overcome specific problems related to the extreme dilution of the fatty acid solution, the extraction procedure, the column preparation and the evaluation method.

Experimental.

SAMPLE PREPARATION.

Standard mixtures.

Standard aqueous solutions containing known amounts of acetic, propionic, n-butyric and n-valeric acid are prepared from stock solutions of the individual acids obtained by weighing pure acids. The purity of these acids was controlled by gaschromatography. The final concentration was determined in each case by titration with NaOH.

Effluent sample.

100 ml effluent water is mixed with 6 g $Ca(OH)_2$ and 6 ml $FeCl_3$ (10 %). After 10 minutes the solution is filtered on filter paper. Samples of 20 ml of the clear filtrate are used for further analyses.

Steam distillation.

Twenty ml of a standard solution or effuent water is distilled in the presence of 8 g $MgSO_4 . 7 H_2O$ and 4 ml 10 N H_2SO_4 in a Friedemann apparatus (7) during 2 h at a distillation rate of 150 ml/h. The volatile acids are collected in a 50 % excess of 0.01 N NAOH with phenolphtalein as indicator.

The distillate is concentrated to 25 ml on a free flame and carefully evaporated to dryness in an oven at 80° C to prevent loss by spattering.

Extraction procedures.

In these experiments three extraction procedures are compared.

a) *Extraction with petroleumbenzine, NaHSO₄ and anhydrous Na₂SO₄* (8).

The cooled dry sodium salts are covered with 4 ml petroleumbenzine and the acids are liberated by slowly adding 4 N $NaHSO_4$ solution until complete destaining of the indicator. The top-layer of petroleumbenzine prevents loss of acids by evaporation. The solid is carefully scraped off with a glass rod. To bind the water, anhydrous Na_2SO_4 is added until the top-layer of petroleumbenzine is clear.

One µl of the petroleumbenzine-extract is injected, without further concentration into the gaschromatograph.

b) *Extraction with 1 N trichloracetic acid in acetone* (9, 10).
4 ml 1 N trichloroacetic acid in acetone is added to the dry sodium salts ; the sodiumtrichloroacetate precipitates.

After centrifugation a clear solution is obtained, 1 µl is injected.

c) *Extraction with 1 N Formic acid in acetone* (11).
The acids are extracted with 4 ml 1 N formic acid in acetone.

Gaschromatographic separation.

1 µl of the extract is analysed on a Microtek 2000 MF gaschromatograph. The separation is performed on a 6 ft column of 20 % Tween 80 and 2 % orthophosphoric acid on Anakrom ABS 110-120 mesh.

The acids are detected with a hydrogen flame ionization detector. The temperature of the column, detector and inlet are respectively 165, 220 and 245° C.

Under these conditions acetic, propionic, butyric and valeric acid were completely separated within 15 minutes, as symmetrical peaks and no ghosting were observed. Also isomers are separated. A gaschromatographic pattern of fatty acids in retting liquor is shown in figure 1.

Evaluation.

The identification of the acids is performed by comparing the retention times with those of standard acids.

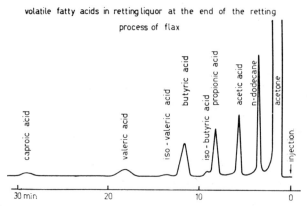

Fig. 1. — Gaschromatographic analysis of volatile fatty acid in retting liquor.

The percentual distribution and the absolute concentrations are calculated by comparing the peak area of each acid with the area of a known amount of n-dodecane as internal standard.

Analysing standard mixtures, the relative response (RR) of equal amounts of each acid and n-dodecane is first established from the equation :

$$RR_A = \frac{Area\ A}{Conc.\ A} \cdot \frac{Conc.\ dodecane}{Area\ dodecane}\ (12)$$

The peak areas are determined by triangulation. The absolute concentration of each acid in a-n unknown mixture can be calculated from the equation :

$$Conc_A = \frac{Area\ A}{RR_A} \cdot \frac{conc\ dodecane}{area\ dodecane}$$

Results and discussion.

Each analytical step has been studied separately in function of optimal recovery.

Steam-distillation.

The recovery of the fatty acids from aqueous solutions is controlled by titration of the distillate with 0.01 N NaOH after 15, 30, 45. 60, 90 and 120 minutes of distillation in CO_2-free medium.

The initial concentration of each acid was $\pm$ 0.1 meq./10 ml. The average distillation rates of the individual and mixed fatty acids are given in table 1 and figure 2. The distillation rate increases from acetic to valeric acid. After 2 hours, the recovery amounts to 99 % for valeric acid and only 78 % for acetic acid. A higher recovery is obtained with a mixture of the acids. After 2 hours of distillation the acids are almost completely isolated.

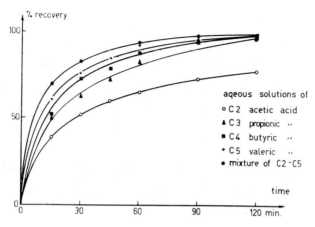

Fig. 2. — Recovery of volatile fatty acids from aqueous solutions by steam-distillation in function of the time.

Table 1.

Percentual recovery of volatile fatty acids and their mixture after steam-distillation in function of time.

Distillation time min.	% Recovery				
	C_2	C_3	C_4	C_5	Mixture
15	39	49	51	61	70
30	51	62	73	77	83
45	60	73	79	87	92
60	65	82	89	93	92
90	73	95	96	97	97
120	78	96	98	99	99
Initial concentration meq/10 ml					
	0.09	0.11	0.10	0.11	0.47

Extraction.

a) *Extraction with petroleumbenzine, $NaHSO_4$ and Na_2SO_4.*

The completeness of this extraction was followed on four samples of 20 ml of a standard mixture evaporated in alkaline medium and extracted. The average values of three gaschromatographic analyses of each sample are given in table 2. With this extraction procedure losses of 50 to 60 % for acetic acid and of 20 % for propionic acid occur.

This agrees with the results of Annison (13), who ascribes the loss of acetic acid to the association with the waterbinding Na_2SO_4. Replicate washing of the Na_2SO_4 with petroleumbenzine should give a higher recovery but the sample becomes too dilute.

Table 2.

Recovery of volatile fatty acids from aqueous solutions by extraction with PE, NaHSO$_4$ and Na$_2$SO$_4$.

Acid	Found		Theoretical	
	meq/l	%	meq/l	%
C$_2$	13.1 ± 0.8	32.6	32.4	50.6
C$_3$	9.6 ± 0.3	29.6	12.4	24.0
C$_4$	9.5 ± 0.3	34.6	10.2	23.5
C$_5$	0.8 ± 0.08	3.2	0.7	1.9
Total	33.0 ± 0.4		55.7	

b) *Extraction with 1 N Trichloroacetic acid in acetone.*

The average values of three analyses of four samples of a standard mixture, after extraction with trichloroacetic acid in acetone are given in table 3.

Table 3.

Recovery of volatile fatty acids from aqueous solutions by extraction with trichloro-acetic acid.

Acid	Found		Theoretical	
	meq/l	%	meq/l	%
C$_2$	28.4 ± 1.4	43.8	32.4	50.6
C$_3$	10.6 ± 0.4	20.1	12.4	24.0
C$_4$	15.1 ± 1.0	34.2	10.2	23.5
C$_5$	0.7 ± 0.1	1.9	0.7	1.9
Total	54.8 ± 0.9		55.7	

The recovery of the acids with this extraction procedure is significantly higher. However, a certain loss of acetic and propionic acid occurs. As demonstrated in table 3 the amount of butyric acid is too high (15.1 meq/l found for 10.24 meq/l theoretical). An abnormal broad peak for butyric acid is found, caused by a contaminant in the trichloroacetic acid-acetone reagent, with almost the same retention time as butyric acid. By decreasing the column temperature to 145° C, the contaminant can be separated partially from the butyric acid peak but a precise calculation of the peak area is difficult. Freshly prepared trichloroacetic acid solutions contain less contaminant. Therefore a third method has been tried out.

c) *Extraction with 1 N formic acid in acetone.*

The average values of three gaschromatographic determinations of four samples of a standard mixture after extraction with formic acid in acetone are given in table 4.

Table 4.

Recovery of volatile fatty acids from aqueous solutions by extraction with formic acid in acetone.

Acid	Found		Theoretical	
	meq/l	%	meq/l	%
C_2	12.6 $\pm$ 0.2	51.0	13.0	50.6
C_3	4.7 $\pm$ 0.1	23.3	4.8	24.0
C_4	4.0 $\pm$ 0.1	23.6	4.1	23.5
C_5	0.3 $\pm$ 0.1	2.1	0.3	1.9
Total	21.6 $\pm$ 0.1		22.2	

This table shows an almost complete recovery of fatty acids and a high reproducibility. This extraction method is selected for further investigation.

Evaluation.

The evaluation was performed against n-dodecane as internal standard. The average relative response factors are calculated from twenty gaschromatographic analyses of a synthetic mixture of known concentrations of acids and n-dodecane (table 5).

Table 5.

The average relative response of acetic to valeric acid against n-dodecane

$$RR_{C_2} = 0.205 \pm 0.010$$
$$RR_{C_3} = 0.357 \pm 0.018$$
$$RR_{C_4} = 0.458 \pm 0.025$$
$$RR_{C_5} = 0.477 \pm 0.032$$

Thirty samples of a standard mixture were then analysed with the modified methods of steam-distillation, extraction and evaluation. The average results obtained for propionic, butyric and valeric acid are in good agreement with the initial concentration (table 6). Only for acetic acid a loss of $\pm$ 25 % occurs, due to the incompleteness of the steam distillation for this acid. However the method has a high reproducibility.

Table 6.

The average results, standard deviations and correction factors of thirty gaschromatographic analyses of an aqueous solution of volatile fatty acids.

Acid	Found		Theoretical		Correction factor
	meq/l	%	meq/l	%	
C_2	10.0 $\pm$ 0.2	45.6	13.0	50.6	1.300 $\pm$ 0.045
C_3	4.6 $\pm$ 0.2	25.7	4.8	24.0	1.051 $\pm$ 0.029
C_4	4.0 $\pm$ 0.2	26.3	4.1	23.5	1.025 $\pm$ 0.049
C_5	0.3 $\pm$ 0.1	2.4	0.3	1.9	0.935 $\pm$ 0.021
Total	18.9 $\pm$ 0.2		22.2		

A correction factor for the distillation error is calculated from the theoretical and experimental values for each acid (table 6). This method is then applied to the determination of volatile fatty acids in retting liquor during the retting process (table 7).

Table 7.

The fatty acid distribution of retting liquor at the end of the retting process.

Fatty acid	meq/l	%
C_2	171.4	54.1
C_3	59.7	23.3
isoC$_4$	trace	—
C_4	45.7	21.2
isoC$_5$	trace	—
C_5	2.6	1.4
C_6	trace	—
Total	279.4	100.0

Conclusion.

After 2 hours of steam-distillation the recovery of butyric and valeric acid from an aqueous solution is almost 100 %. The recovery of acetic and propionic acid is only 75 % and 95 %.

The extraction of the acids from the dry distillate with formic acid in acetone is complete.

Efficient gaschromatographic separations are performed on a 6 ft 20 % Tween 80 and 2 % orthophosphoric acid column, which prevents tailing and ghosting.

Analysis of a sample after steam-distillation and extraction with formic acid in acetone yields reproducible results and becomes accurate by introducing a correction factor for each acid.

This method is efficient for the analysis of effluent water, such as retting liquor.

ACKNOWLEDGEMENT :

This study was supported by a grant of the IWONL. We wish to thank, Bockstaele L., director of the « Onderzoeks- en Voorlichtingscentrum voor Nijverheids-teelten », Provincie Oost-Vlaanderen (Rumbeke - Beitem), for collaboration, and J. Delacourt for her technical assistance.

References

1. JAMES A.T., MARTIN, A.J.P., *Biochem. J.*, **50**, 679-90 (1952).

2. ERWIN, E.S., MARCO, G.J. and EMERY, E.M., *J. Dairy Sci.*, **44**, 1768 (1961)

3. GEHRKE, C.W., LAMKIN, W.M., *J. Agr. Food Chem.*, **9**, 85-88 (1961).

4. ACKMAN, R.G., BURGHER, R.D., *Anal. Chem.*, **35**, 647 (1963).

5. METCALFE, L.D., *Nature*, **188**, 142 (1961).

6. EMERY, E.M., KOERNER, W.E., *Anal. Chem.*, **33**, 146-47 (1961).

7. FRIEDEMANN, T.E., *J. Biol. Chem.*, **123**, 161-84 (1938).

8. KAMER, (van de), J.H., GERRITSMA, K.W., WANSKINK, E.J., *Biochem. J.*, **61**, 174-76 (1955).

9. HUNTER, I.R., HAWKINS, N.G., PENCE, J.W., *Anal. Chem.*, **32**, 1757-59 (1960).

10. SHELLEY, R.N., SALWIN, HORWITZ, W., *J. Assoc. Offic. Agr. Chemists*, **46**, 486-93 (1963).

11. VANDENHEUVEL, F.A., *Anal. Chem.*, **36**, 1930-36 (1964).

12. GREY, T.C., STEVENS, B.J., *Anal. Chem.*, **38**, 724 (1966).

13. ANNISON, E.F., *Biochem. J.*, **58**, 670-80 (1954).

Relation between the structure of organic compounds and their retention time in gas chromatography

by

J.J. WALRAVEN, A.W, LADON and A.I.M. KEULEMANS

Laboratory for Instrumental Analysis,
Technological University Eindhoven, The Netherlands.

Summary.

Logarithmic plots of retention data on two stationary phases have been used by many authors after Pierrotti as an aid to identification. We have plotted retention indices for different classes of compounds on diagrams in which the index on an apolar phase constitutes one axis and the index on a polar phase constitutes the other axis. A remarkable regularity was found. Branched isomers are spread along parallel lines in a repeated pattern. This effect is so general that it is proposed to call it « roofing tile effect ».

On close examination of the apparently random scatter of the points around the « isomer-lines », this was found to be not random but systematic. In the case of the alkanes, cycloalkanes, alkenes and saturatad alcohols, a closer relationship between structure of the compound and position in the diagram was found.

Alkanes can be divided into groups that have equal numbers of primary, secondary, tertiary and quaternary carbon atoms.

The striking thing is that on the basis of this subdivision of isomer groups, further imbrication is found and that the lines are in numerical order.

Introduction.

Gas chromatography today yields so many and so precise results, that for identification purposes it is highly desirable to have available correlations between retention data and structure, that enable us to draw conclusions from the position of a peak, even if the proper reference is lacking.

Martin (1949) found experimentally, that if the logarithm of retention data is plotted against the carbon number in homologous series, a straight line is obtained, in most cases. Logarithmic plots of retention data on two stationary phases will then lead to linear relationships for homologous series (Pierotti, 1956). An example based on recent retention data (Mc Reynolds, 1966) is given in figure 1.

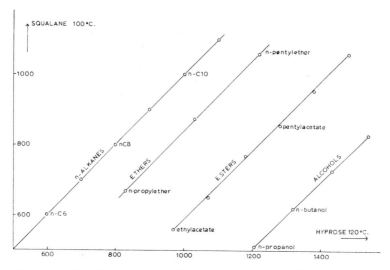

Fig. 1. — Homology lines for alkanes, ethers, esters and alcohols in a diagram of retention indices on squalane and Hyprose.

We have found that two-phase diagrams are especially suited to reveal a new correlation between structure and retention data. This will be described in the following paragraphs.

Correlation between isomers.

1. *Roofing tile effect in a plot of hydrocarbons.*

By plotting the retention index (or the logarithm of the relative retention) of alkanes and alkenes on two stationary phases viz, octadecene at 25° and dimethylsulfolane at 25° (2) we obtain figure 2. An intriguing relationship becomes apparent. Isomers are spread along parallel lines forming imbricated series in the order of the number of carbon atoms. The roofing tile effect is in this example more striking for the class of the alkenes.

2. *Roofing tile effect in a plot of isomeric esters.*

To find out whether the regularity is general we have plotted data (3) of methyl esters of carboxylic acids. The stationary phases are squalane and polypropylene sebacate, (Fig. 3). In this case the points representing isomers can be joined by straight parallel lines with very little scatter.

3. *Roofing tile effect for cycloalkanes and phenyl alkanes of high molecular weight at 200° C.*

Schomburg (4) published retention indices of cyclohexyl alkanes and of phenyl alkanes with 13-19 carbon atoms at 200° C on Apiezon and on polyphenylether. When we plot these data in the manner described before,

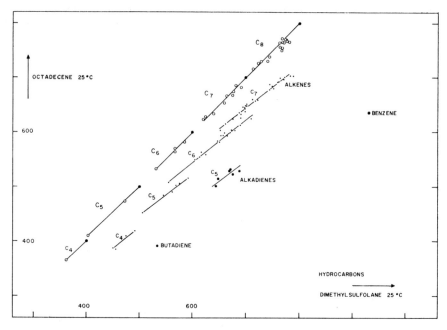

Fig. 2. — Plot of retention index data, showing roofing tile series and scattering not due to measurement errors. (Data based on Cramers, 1967.) (2), (6).

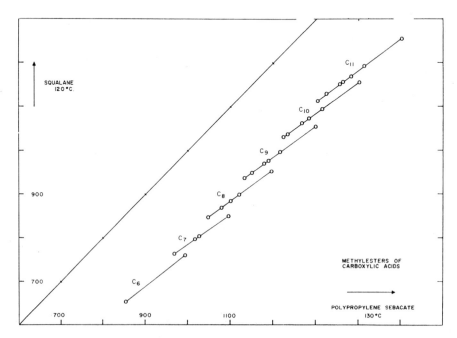

Fig. 3. — Plot of retention index data, showing roofing tile series. (Data taken from Schomburg, 1964). (3), (6).

it is again apparent that isomers can be joined by straight lines and that the lines are parallel, forming an imbricated series. For both classes of compounds, the slope of all these line segments is the same and there is a constant shift due to structural change of cyclohexyl into phenyl. See figure 4.

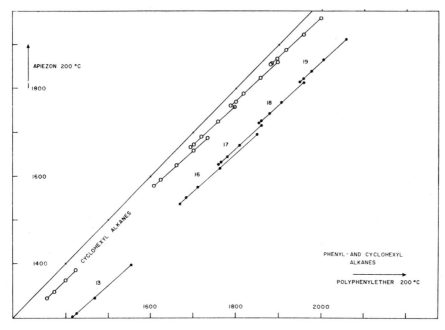

Fig. 4. — Plot of retention index data, showing roofing tile series. (Data taken from Schomburg, 1966). (4), (6).

4. Further imbrication in plots of alkanes.

The scatter of points as illustrated in figure 2 can be accounted for by a numerical structural relationship (6). The basis is that the number of primary, secondary, tertiary and quaternary carbon atoms for each alkane is counted. This is represented in a code. For instance 2,3-dimethyl butane is given the code number 4020. A group of isomers can now be divided into subgroups of compounds having the same code number. We have found that points representing compounds belonging to the same subgroup can be joined by straight lines, that these lines are parallel and that the lines are stacked in numerical order of the code numbers. See figure 5.

5. Further imbrication in plots of cycloalkanes.

As a source of retention data Cramers (2) table of relative retention values of hydrocarbons at 25° C was used to calculate the log $r_{i,s}$ values of cycloalkanes in the C_5-C_8 region. To obtain a relation between retention and structure, the ring must now be taken into account. This is done by first counting quaternary, tertiary and secondary carbon atoms in the ring and then counting the rest of the molecule according to the method described in the previous paragraph.

E.g. ethylcyclohexane   015   110
      propylcyclopentane 014   1200.

It was found that the five-rings and the six-rings are on separate line systems. The lines are parallel and within each isomer group the lines are in numerical order of the core. See figure 6.

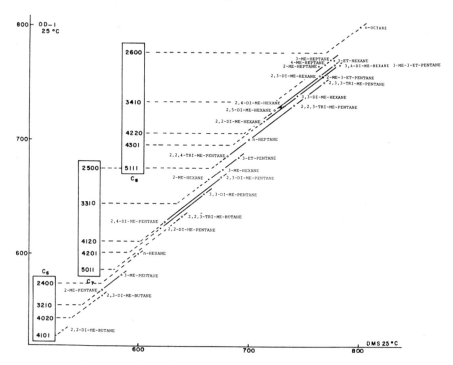

Fig. 5. — Retention index plot of hexanes, heptanes and octanes on octadecene-1 and dimethylsulfolane at 25° C. Alkanes belonging to the same sub-group of isomers are scattered around straight parallel lines. Numbers in rectangles represent code. (Data taken (2), (6).

6. Further imbrication in alkene plots.

The scatter of the alkenes around the isomer lines in fig. 2 can be largely resolved by applying the following code. First the alkenes are coded according to the total number of primary, secondary, tertiary and quaternary carbon atoms without regard to the double bond. E.g. for the hexenes the following subgroups are found : 2400, 3210, 4020, and 4101 (the same as for the alkanes). If the points belonging to one group are considered, e.g. 2400, these can be joined by two parallel lines, one connecting the trans-compounds, the other connecting the cis-compounds. This is applicable to the other subgroups as well. The alkene-1 compounds are found on the cis-line and the 2-methyl-alkene-2 compounds are found on the trans-line. In other words, isomer subgroups according to the four digit code, split into two lines, a trans-line above a cis-line. The subgroups do not overlap and are in numerical order for the butenes, pentenes, hexenes and heptenes. The plot of the hexenes is depicted in figure 7.

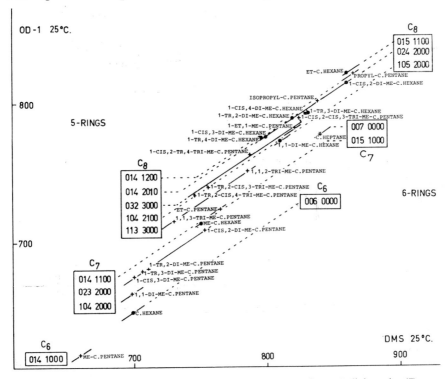

Fig. 6. — Further imbrication in plot of cycloalkanes using a 7-digit code. (7).

7. Further imbrication in plots of alcohols.

The retention indices of isomeric pentanols and isomeric hexanols were taken from Mc Reynolds'collection of retention data (5). A plot on Apiezon- and on sucrose octaacetate at 120° reveals the existence of isomer lines for the alcohols but with considerable scatter. This scatter is largely accounted for by the following code. Firstly the alcohols are divided into primary, secondary and tertiary alcohols, indicated by I, II and III. The groups are defined as follows :

I   C—CH$_2$OH

II $\begin{matrix} C \\ \\ C \end{matrix} \Big\rangle CHOH$

III $\begin{matrix} C \\ | \\ C-C-OH \\ | \\ C \end{matrix}$

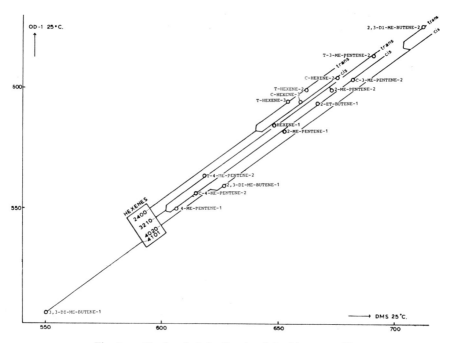

Fig. 7. — Further imbrication in plot of hexenes. (7).

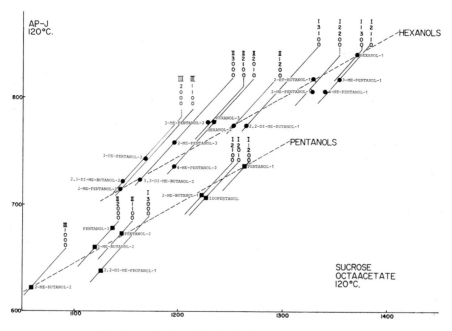

Fig. 8. — Further imbrication in plot of alcohols.
(Data from McReynolds, 1966). (5), (7).

The rest of the molecule attached to these groups, is counted according to the number of primary, secondary, tertiary and quaternary carbon atoms. E.g. 2,3-dimethylbutanol-2 is represented by III-2000. Alcohols bearing the same code, can now be joined by lines. The lines are parallel and in numerical order. See figure 8.

Conclusion.

The numerical codes described in the previous paragraphs are of great value for the correlation of structure and retention parameters. If for instance a compound is found on a certain subisomer line, very definite conclusions about its possible structure may be drawn. Another important advantage of this graphical representation is that a natural classification of volatile organic substances arises (7).

ACKNOWLEDGMENT.

Figures 1-2-3 and 4 are reproduced by courtesy of the publisher of « Chromatographia ». (*Chromatographia*, **1** : 195, 1968.)

References.

(1) PIEROTTI, G.J., DEAL, C.H., DERR, E.L., PORTER, P.E. *J. Am. Chem. Soc.*, **78** : 2989, 1956.

(2) CRAMERS, C.A. *Thesis,* Eindhoven University, The Netherlands, 1967.

(3) SCHOMBURG, G. *J. Chromatog.* **14** : 157-77, 1964.

(4) Ibid. **23** : 18, 1966.

(5) McREYNOLDS, W.O. Gaschromatographic Retention Data, Preston, Evanston, Iliinois, 1966.

(6) WALRAVEN, J.J., LADON, A.W. and KEULEMANS, A.I.M. *Chromatographia.* **1** : 195-198, 1968.

(7) WALRAVEN, J.J. *Thesis,* Eindhoven, Ch. V, 1968.

Le développement multiple en chromatographie de partage [*]

E. SALA

Département de Biologie
Centre d'Etudes Nucléaires de Saclay, France.

En chromatographie de partage sur papier ou sur couches minces, on ne parvient pas toujours à séparer, d'une manière satisfaisante, deux substances déterminées. Ou bien, en effet, les produits à séparer sont de structure semblable. Dans ce cas, ils possèdent des Rf voisins dans la plupart des solvants usuels. Des exemples classiques sont fournis par la leucine et l'isoleucine, le glucose-6-phosphate et le fructose-6-phosphate. Ou bien, d'une façon générale, quelles que soient les substances à séparer par chromatographie, on ne dispose pas toujours de solvants appropriés capables de fournir une bonne séparation, et cela, pour plusieurs raisons qui nous limitent dans le choix des solvants. Parmi ces raisons, citons la bonne volatilité des solvants, s'il s'agit de récupérer les substances, non contaminées par les principes immédiats des solvants. Signalons encore les conditions de pH, s'il s'agit de substances labiles.

Ainsi, on est amené à considérer le cas où deux substances se séparent mal, étant donné leurs Rf trop voisins dans le solvant utilisé. On peut alors avoir recours à l'une des deux méthodes suivantes : la première, bien connue, consiste à effectuer une chromatographie en écoulement continu du solvant, de sorte que les substances soumises à la chromatographie s'éloignent de plus en plus l'une de l'autre : d'où leur séparation possible. Cette méthode n'est pas toujours utilisable, notamment lorsqu'il s'agit de séparer des substances de Rf quelque peu élevés : dans ce cas, les produits entraînés par le solvant risquent de s'échapper du support.

La seconde méthode, moins bien connue, appelée la chromatographie en développements multiples ou successifs, consiste à faire subir au chromatogramme une série de développements séparés les uns des autres par le

---

[*] Les figures ou graphiques correspondant aux diapositives présentées au Ve Symposium International sur la Chromatographie et l'Electrophorèse, à Bruxelles, n'ont pu être, tous, reproduits dans ce texte.

séchage du chromatogramme. Dans le cas le plus simple, c'est un même solvant qui est utilisé. Au cours de cette alternance de développements et de séchages, l'intervalle entre les substances à séparer augmente, dans certaines conditions, avec le nombre de développements, pour atteindre un maximum, puis l'intervalle décroît et a tendance à devenir nul au fur et à mesure que croît le nombre d'opérations. Ce qui est intéressant donc de connaître dans cette méthode, c'est le nombre de développements qui correspond à une séparation optimum.

Ainsi, nous allons exposer maintenant l'étude théorique de cette méthode et ses conditions de validité. Nous verrons ensuite dans quelle mesure les résultats théoriques concordent avec les résultats expérimentaux.

Jeanes et ses collaborateurs, et plus tard, Thoma, ont établi chacun d'eux une théorie approchée du développement multiple. Nous donnons ici la théorie exacte dans ses détails mathématiques. Pour simplifier l'écriture, appelons a le Rf de la substance A migrant le moins vite, b le Rf de la substance B migrant le plus rapidement. Nous supposons dans cette étude théorique, où l'on n'utilise qu'un même solvant :

1° que les Rf : a et b sont constants au cours des développements successifs ;

2° que la distance parcourue par le solvant sur le support est toujours la même.

Dans ces conditions, les chemins successifs décrits par la substance A seront : a, $a(1-a)$, $a(1-a)^2$, ... $a(1-a)^{n-1}$. Nous obtenons ainsi la suite d'une progression géométrique croissante dont le premier terme est a et la raison $(1-a)$.

La distance totale parcourue par la substance A, après n développements sera donc, d'après la formule classique :

$$S = a\, \frac{q^n - 1}{q - 1},$$

égale à $SA = 1-(1-a)^n$.

Pour la substance B, on aura de même :

$$SB = 1-(1-b)^n.$$

Pour un nombre n de développements, l'intervalle entre les deux substances sera donc égal à :

$$D_n = (1-a)^n - (1-b)^n.$$

Les figures suivantes représentent un certain nombre de courbes exprimant la fonction D_n. En abscisses : le nombre de développements. En ordonnées : les intervalles sont exprimés en fractions d'unités de Rf. Ces courbes présentent un maximum pour un nombre de développements qu'on appellera N. Pour ce nombre N, la pente des courbes est nulle. La différentielle de D_n, qui s'annule alors, va nous permettre de calculer N.

La fonction $D_n = (1-a)^n - (1-b)^n$ peut encore s'écrire :

$$D_n = e^{n\,L(1-a)} - e^{n\,L(1-b)}, \text{ (L exprimant les logarithmes}$$

népériens).

La différentielle s'écrira :

$$d(D_n) = e^{n \, L(1-a)} \, L(1-a) - e^{n \, L(1-b)} \, . \, L(1-b)$$

ou encore $(1-a)^n$, $L(1-a) - (1-b)^n$. $L(1-b)$.

Pour le nombre N de développements, la différentielle s'annulant, nous écrirons :

$$(1-a)^n. \, L(1a) = (1-b)^n. \, L(1-b).$$

D'où : $\left(\dfrac{1-a}{1-b}\right)^n = \dfrac{L(1-b)}{L(1-a)}$

Ou encore : $N.\log. \left(\dfrac{1-a}{1-b}\right) = \log. \dfrac{L(1-b)}{L(1-a)}$

Finalement, le nombre N sera égal à, dans le système des log. vulgaires :

$$N = \dfrac{\log. \dfrac{\log. (1-b)}{\log. (1-a)}}{\log. (1-a) - \log. (1-b)}$$

Ainsi, cette formule permet de déterminer le nombre théorique N de développements qui correspond à une séparation maximum pour deux substances de Rf égaux à a et b.

Examinons les conditions de validité de cette méthode chromatographique. Le nombre théorique N peut prendre des valeurs entières ou décimales, selon les valeurs de a et de b. En particulier, il peut être compris entre 1 et 2. Dans ce cas, il est possible de déterminer les conditions pour lesquelles l'intervalle D_1, obtenu après le premier développement, sera égal à l'intervalle D_2, donné après les deux premiers développements. Etablissons ainsi l'égalité entre D_1 et D_2. Après arrangement et simplification, nous aurons :

$$a^2 - b^2 - a + b = O$$
$$a^2 - b^2 - (a\text{-}b) = O$$
$$(a\text{-}b) \, (a + b\text{-}1) = O$$

Cette dernière équation nous montre qu'un deuxième développement n'apporte aucune amélioration dans la séparation de deux substances si la somme de leurs Rf est égale à 1. On démontrerait facilement qu'un deuxième développement n'est avantageux que si la somme des Rf est inférieure à 1. D'où la conclusion suivante, toujours sur le plan théorique : la méthode de chromatographie en développements multiples n'est efficace que si la somme des Rf des deux substances à séparer est inférieure à l'unité.

Expérimentation.

Nous avons supposé dans l'étude théorique, que les Rf sont constants au cours des développements successifs. En réalité, il n'en est pas ainsi ; et dans l'étude expérimentale que nous allons aborder, nous verrons dans quelle mesure la valeur expérimentale trouvée pour N concorde avec la valeur calculée.

Tous les essais qui suivent ont été effectués sur des substances radio-actives marquées au ^{14}C, qui nous ont permis, après développements successifs et autoradiographies, de situer exactement l'emplacement des différentes taches ou des différentes bandes de substances. Nous avons ainsi étudié le comportement de divers amino-acides et de divers sucres, en chromatographie sur papier ou sur couche mince.

Nous présentons, dans les graphiques suivants quelques-uns de nos résultats :

A. Tout d'abord, en chromatographie sur papier, les courbes représentant les intervalles trouvés entre des substances de Rf peu élevés, sont les suivantes :

— séparation de sérine-thréonine sur Whatman 3 en chromatographie ascendante. N théorique égal à 3,86 ; N expérimental : 4 ;

— séparation des mêmes substances en chromatographie descendante : N théorique égal à 4,3 ; N expérimental : 5. La courbe inférieure représente les intervalles sur un chromatogramme effectué, à l'échelle préparative, à partir d'un mélange de sérine et de thréonine à parties égales, le mélange portant sur 0,5 mg par centimètre de longueur de Whatman 3 ;

— toujours, à partir de substances présentant des Rf relativement faibles, voici la séparation de thréonine et d'alanine. N théorique égal à 3,0 ; N expérimental : 3.

B. Les graphiques suivants représentent les intervalles obtenus sur la séparation de substances de Rf plus élevés : ainsi la séparation de valine et de phénylalanine : N théorique : 1,77 ; nombres expérimentaux : 2 et 3. Remarquons, dans ce cas, que la somme des Rf approche de l'unité. Le développement multiple, encore efficace ici, va cesser de l'être pour des substances de Rf plus élevés.

C. Voici un graphique représentant la séparation de valine et de phényl-alanine sur Whatman 3 en chromatographie ascendante. La correspondance entre la théorie et l'expérimentation est encore vérifiée pour la valeur du nombre N, et nous voyons qu'un deuxième développement sera défavorable, conformément à ce que prévoit la théorie : la somme des Rf est en effet supérieure à l'unité.

Si les résultats obtenus en chromatographie sur Whatman 3 donnent une bonne coïncidence entre les valeurs théoriques et expérimentales, relativement au nombre N, les résultats nous ont paru moins satisfaisants en ce qui concerne la chromatographie sur couches minces.

Voici, sur les graphiques suivants, différentes séparations sur cellulose MN300 :

Séparations de substances de Rf plus élevés :

arginine et sérine	: N théorique 4,6 ; N expérimentaux 6-8
lysine et arginine	: N théorique 5,85 ; N expérimentaux 6-8
glucose et fructose	: N théorique 3,4 ; N expérimental 4
thréonine et alanine	: N théorique 3,1 ; N expérimentaux 6-8

Séparations de substances de Rf plus élevés :
valine et phénylalanine   : N théorique 1,4 ; N expérimental   2

En conclusion : d'après les résultats de nos essais, nous pouvons écrire que, dans la technique de développements multiples :

1° en chromatographie de partage sur papier, le nombre expérimental de développements, correspondant à la séparation maximum de deux taches ou deux bandes coïncide très sensiblement avec le nombre théorique N, lorsque le papier est équilibré avec les vapeurs du solvant. Pour les bandes, les quantités déposées ne doivent pas excéder 0,5 mg de mélange par centimètre de longueur de Whatman 3.

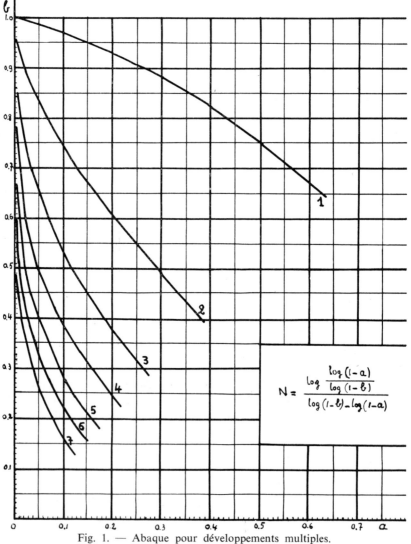

$$N = \frac{\log \dfrac{\log (1-a)}{\log (1-b)}}{\log (1-b) - \log (1-a)}$$

Fig. 1. — Abaque pour développements multiples.
Abacus for multiple developments.

2° en chromatographie de partage sur couches minces de cellulose, le nombre expérimental correspondant à la séparation maximum est sensiblement supérieur au nombre théorique N.

Nous avons cru utile de vous présenter, dans le graphique précédent, une abaque permettant de déterminer rapidement le nombre expérimental de développements qu'il faut faire subir à deux substances de Rf égaux respectivement à *a* et *b* pour obtenir une séparation maximum : il suffit de prendre le point d'intersection des perpendiculaires élevées sur les deux axes, pour l'abcisse *a* et l'ordonnée *b*, et de lire le nombre entier N qui correspond à la courbe la plus proche. Pour la chromatographie sur couches minces, ce nombre N sera augmenté de 1 à 2 unités.

Avant de terminer cet exposé au cours duquel nous avons démontré l'efficacité et la validité de la méthode chromatographique à développements

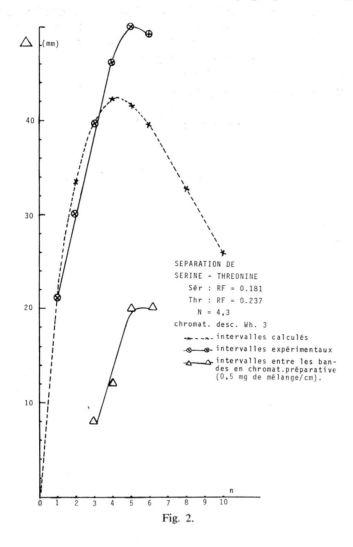

Fig. 2.

multiples, nous devons souligner l'importance de cette technique du point de vue du degré élevé de purification obtenue pour des substances soumises à cette méthode.

Voici un exemple de séparation qui illustre notre pensée : un mélange à parties égales de sérine et de thréonine radioactives, partiellement dégra-dées par radiolyse au cours du stockage, a été soumis à la chromatographie à développements multiples. Un mélange identique a subi une chromato-graphie en écoulement continu. La quantité totale mise en œuvre était de 0,5 mg par centimètre de longueur de Whatman 3. Les autoradiographies montrent sur la figure de gauche les emplacements des bandes obtenues après le premier développement ; au centre, les positions des bandes après le 6me développement ; à droite, les emplacements après trois jours de chro-matographie en développement continu.

Après découpage des bandes obtenues, élution, concentration et analyse effectuée par chromatographie des spots correspondants, nous arrivons aux résultats suivants, exprimés en pourcentage de pureté :

6me développement : sérine   : 99,7 %
                                    thréonine : 99,3 %

Développement continu : (3 jours)
                                    sérine   : 97,7 %
                                    thréonine : 98,0 %.

A gas chromatographic study involving charge-transfer interaction with the stationary phase

C.W.P. CROWNE, J.M. GROSS and M. HARPER
(Department of Chemistry, Sir John Cass College, London E.C.3)
and
P.G. FARRELL
(Department of Chemistry, McGill University, Montreal, Canada)

The use of complexing additives in gas-liquid chromatography (GLC) was first recognised by Bradford et al (1), who found that the addition of silver nitrate to their inert stationary phase (ethylene glycol) selectively retarded the passage of olefins through the column, relative to saturated hydrocarbons. The analytical potential of this technique was extended by Phillips and co-workers (2) who increased the range of cationic complexing agents.

In addition to the purely analytical uses of such systems, Gil-Av, Cvetanovic and others (3) determined stability constants for the Ag^+-olefin complexes in ethylene glycol and the results obtained agreed well with those determined by a conventional distribution method. Activity coefficients for binary systems such as n-hexane in 1,2,4-trichlorobenzene have been measured and the specific retention volume of the solute, Vg move, related to the equilibrium solute concentration present in the column (4). These results also gave values to within 5 % of those determined by static methods. Measurements of excess free energies and other thermodynamic parameters have shown that these are independent of any GLC parameters. This technique is therefore suitable for measuring the extent of any specific interaction such as those mentioned and it has been suggested also that the retention of certain compounds on stationary phases which could act as electron acceptors is in part due to charge-transfer interaction (3, 5).

In an attempt to examine this type of interaction we investigated the substituted aniline-2,4,7-trinitrofluorenone system. In this it is possible to make spectroscopic studies of charge-transfer association in addition to measurements of excess thermodynamic functions by the chromatographic method. Langer et al (5) have shown that the apparent activity coefficient for the solute can be expressed in the form :

$$\gamma = (1\text{-c}) \, \gamma° \tag{1}$$

where c is the fraction of dissolved solute complexed to the stationary phase and γ° is the activity coefficient for uncomplexed solute at infinite dilution. If the charge transfer association constant is represented by K_1 and that for any other association by K_2, it can be shown that :

$$\frac{1}{\gamma} = \frac{1}{\gamma^\circ}(1 + K_1 + K_2) \tag{2}$$

whence $1/\gamma$ may be expressed as a linear function of K_1. Taking the mole fraction association constants for charge transfer complexing measured in dichloromethane at 25° as measures of the relative complexing power of the donors ($\equiv K_1$), these did show linear relationship with $1/\gamma$ (6). These relationships separate into series of primary, secondary and tertiary amines, additional interaction between the stationary phase and the primary and secondary amines being evident. To further investigate the extent, if any, of charge transfer interactions in GLC systems we have measured association constants and other thermodynamic parameters using 2,4,7-trinitrofluorenone dissolved in either tritolylphosphate or polypropylene sebacate as the stationary phase, and anilines as solutes. The low solubility of 2,4,7-trinitrofluorenone and competitive interaction with the solvent influenced the choice of involatile liquid phase. The results obtained for K_1 values are shown in Table I, together with spectroscopic values obtained in dichloromethane or tritolylphosphate solution.

TABLE I.

Values for the association constants K_1 for the substituted aniline-2-4-7-trinitrofluorenone system in the units l. mole.$^{-1}$.

Donor	G.L.C.		Spectroscopic	
	K_1 (PPS, 150°)	K_1 (TTP, 150°)	$K_1{}^a$ (TTP, 23°)	$K_1{}^b$ (CH$_2$Cl$_2$, 25°)
(1) N,N-dimethylaniline	0.72	0.34	0.17	0.84
(2) N,N-diethylaniline	0.50	0.17	—	0.39
(3) N-ethyl N-methylaniline	0.57	0.24	—	—
(4) N,N-dimethyl-p-toluidine	0.92	0.43	—	0.88
(5) N,N-dimethyl-o-toluidine	0.32	0.14	0.03	0.21
(6) N,N-dimethyl-2,4-xylidine	0.29	0.09	—	0.32
(7) N,N-dimethyl-2,6-xylidine	—	0.04	—	0.20
(8) Aniline	0.68	0.30	0.10	0.32
(9) o-toluidine	0.74	0.33	—	0.45
(10) m-toluidine	0.76	0.38	0.12	0.49
(11) p-toluidine	0.78	0.40	—	0.61
(12) p-ethylaniline	0.70	0.35	—	0.50
(13) 2,4-xylidine	0.94	0.41	—	0.74
(14) N-methylaniline	0.71	0.27	0.13	0.61
(15) N-ethylaniline	—	0.22	—	—
(16) N-n-propylaniline	—	0.19	—	—
(17) N-n-butylaniline	—	0.17	—	—

(a) Values determined at λCT, 495 mμ.   PPS = Polypropylene sebacate.

(b) Values taken from COOPER, CROWNE and FARRELL. *Trans. Faraday Soc.,* **62** : 2725 (1966).   TTP = Tritolylphosphate.

The order of the derived association constants is the same in both solvents and the complexing power of the amines can be interpreted in terms of the electronic effects of the alkyl substituents and steric effects of the ortho groups. The GLC values are considerably higher than expected from spectroscopic studies and this is further emphasised by the extrapolation of the spectroscopic data for tritolylphosphate to higher temperatures. It is possible that these high K_1 values arise from surface excess effects; i.e. the polar 2,4,7-trinitrofluorenone may be concentrated at the surface of the stationary liquid and thus give rise to adsorption chromatography (7). To eliminate this possibility we studied different column loadings which should result in a variation of surface : volume ratio for the stationary liquid. No change in specific retention volumes for anilines, both nitrogen substituted and nitrogen unsubstituted, was observed.

It is well known that partition methods measure *all* types of interaction between the components in the system and the K_1 values obtained by the GLC method may therefore be too high. Similar effects have been observed in liquid phase partition experiments with picrates, where distribution results are considerably higher than those obtained by spectroscopic or electro-chemical methods. This is further shown by comparison with the spectros-copically measured complexing orders — that for the GLC measurements with either 2,4,7-trinitrofluorenone or di-n-nonyltetrachlorophthalate does not parallel that for chloranil. Measurements with s-trinitrobenzene as acceptor show far better agreement between GLC and spectroscopic results. Thus in the case of strong charge-transfer interactions the GLC results seem to reflect

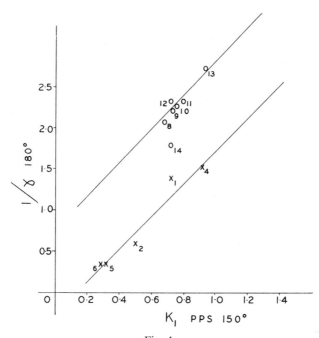

Fig. 1.

those measured spectroscopically. For weak charge-transfer complexing other interactions may be comparable in magnitude with, or larger than, this effect and as one is always measuring positive effects only, the GLC measured K_1 values will be greater than the spectroscopic values.

From earlier experiments (6) using the acceptor itself as the stationary liquid it was found that there was a constant increment in $-\Delta G_e^{\circ}$ for primary amines relative to tertiary amines. This additional interaction ($\equiv K_2$) was suggested as being consistent with hydrogen bonding from aniline to the acceptor, or dipole-dipole interaction which is sterically hindered in the tertiary amines. In this work derived K_1 values at 150° plotted against $1/\gamma$ values for the donors at 180° show similar results, demonstrating the existence of additional associations with the primary amines (fig. 1). The accuracy of K_1 measurement by the GLC is at worst $\pm$ 10 % and in most cases $\pm$ 3 %.

There has been considerable discussion in the literature as to the accuracy of charge-transfer association constants as measured by the Benesi-Hildebrand method, or its modification (8). No account is taken of solvation

TABLE II.

Values for the enthalpy of complex formation $-\Delta H^C$ for the substituted aniline-2,4,7-trinitrofluorenone system, measured in k.cal.

Donor	G.L.C.		Spectroscopic	
	a $-\Delta H^C$ (PPS)	a $-\Delta H^C$ (TTP)	b $-\Delta H^C$ (TTP)	c $-\Delta H^C$ (CH$_2$Cl$_2$, 25°)
(1) N,N-dimethylaniline	4.87	4.37	1.77	2.2
(2) N,N-diethylaniline	5.91	5.84	—	1.4
(3) N-ethyl N-methylaniline	5.64	5.39	—	—
(4) N,N-dimethyl-p-toluidine	6.73	3.86	—	2.6
(5) N,N-dimethyl-o-toluidine	6.51	5.18	7.24	0.4
(6) N,N-dimethyl-2,4-xylidine	8.53	6.23	—	0.9
(7) N,N-dimethyl-2,6-xylidine	—	1.74	—	0.4
(8) Aniline	4.49	4.45	2.35	1.0
(9) o-toluidine	5.38	3.26	—	1.1
(10) m-toluidine	6.41	2.98	1.90	1.1
(11) p-toluidine	5.76	3.44	—	1.1
(12) p-ethylaniline	6.56	4.25	—	1.2
(13) 2,4-xylidine	4.82	3.02	—	1.1
(14) N-methylaniline	5.67	4.40	2.42	1.2
(15) N-ethylaniline	—	2.83	—	—
(16) N-n-propylaniline	—	3.95	—	—
(17) N-n-butylaniline	—	3.58	—	—

The $-\Delta H^C$ values are the mean values for the following temperature ranges :
    (a) 130-150°   (b) 23-52° and (c) 4-34°.
TTP = Tritolylphosphate.   PPS = Polypropylene sebacate.

in these calculations and methods of overcoming this major deficiency have been proposed. The measured K_1 values are also sensitive to changes in the donor : acceptor ratio (9) and are functions of the units used to analyse the data (10). We therefore measured the temperature variation of the GLC K, values to obtain $-\Delta H^C$ and $T\Delta S^C$ data, and these are shown in Tables II and III. The $-\Delta H^C$ values of Table II for the GLC results reflect the data of Table I, illustrating the influence of solvent upon the derived association constants. In accord with other work the spectroscopically derived values are low. The $-T\Delta S^C$ data again show the excepted trends.

TABLE III.

Values for the entropy of complexation $-T\Delta S^C$, for the substituted aniline-2,4,7-trinitrofluorenone system, measured in k.cal.

Donor	G.L.C.		Spectroscopic	
	$-T\Delta S^C$ (PPS, 150°)	$-T\Delta S^C$ (TTP, 150°)	$-T\Delta S^C$ (TTP, 23°)	$-T\Delta S^C$ (CH$_2$Cl$_2$, 25°)
(1) N,N-dimethylaniline	4.73	3.58	0.63	2.10
(2) N,N-diethylaniline	5.52	4.56	—	0.84
(3) N-ethyl N-methylaniline	5.43	4.32	—	—
(4) N,N-dimethyl-p-toluidine	6.85	3.26	—	2.52
(5) N,N-dimethyl-o-toluidine	5.74	3.68	4.90	—0.52
(6) N,N-dimethyl-2,4-xylidine	7.74	4.42	—	2.32
(7) N,N-dimethyl-2,6-xylidine	—	8.81	—	—0.55
(8) Aniline	4.26	3.56	0.84	0.33
(9) o-toluidine	5.33	2.42	—	0.63
(10) m-toluidine	6.00	2.25	0.58	0.68
(11) p-toluidine	6.39	2.77	—	0.81
(12) p-ethylaniline	6.42	3.48	—	0.79
(13) 2,4-xylidine	4.85	2.35	—	0.92
(14) N-methylaniline	5.50	3.44	1.10	0.91
(15) N-ethylaniline	—	1.66	—	—
(16) N-n-propylaniline	—	2.67	—	—
(17) N-n-butylaniline	—	2.22	—	—

PPS = Polypropylene sebacate.   TTP = Tritolylphosphate.

Whereas the association constants may be widely different as determined by GLC and spectroscopy it is generally found that ΔH^C values for charge-transfer complexing are similar whatever the method of measurement employed. This was illustrated by the agreement between $-\Delta H^C$ values determined spectroscopically and the excess partial molar enthalpies determined using 2,4,7-trinitrofluorenone as the stationary liquid. Thus the differences in $-\Delta H^C$ values observed here are due to the solvation differences in the two stationary phases. In the spectroscopic studies the donor component is in considerable excess whereas in the GLC work the acceptor is in almost infinite excess. It therefore seems that although the chromatographic method reflects charge-transfer complex formation (fig. 1), unless one component of the complex is used as the stationary phase itself quantitative data yield only gross solvation effects and other interactions. The alternative suggestion is that the data adds further support to the theories that charge-transfer interaction is no more than random collisional interaction and that the low K_1 values in tritolylphosphate as solvent also imply this. Although possible we believe that organic charge-transfer interactions occur over and above normal collisional interactions and, as part of the solute-solvent interactions, should be evident in suitable GLC systems.

References

1. B.W. BRADFORD, D. HARVEY and D.E. CHALKLEY. *J. Inst. Petroleum*, **41** : 80 (1955).
2. D.W. BARBER, C.S.G. PHILLIPS, G.F. TUSA and A. VERDIN. *J. Chem. Soc.*, 18 (1959).
3. E. GIL-AV and J. HERLING. *J. Phys. Chem.* **66** : 1208 (1962) ; M.A. MUHS and F.T. WEISS. *J. Amer. Chem. Soc.*, **84** : 4697 (1962) ; R.J. CVETANOVIC. F.J. DUNCAN, W.E. FALCONER and R.S. IRWIN. *J. Amer. Chem. Soc.*, **87** : 1827 (1965).
4. C.F. CHUEH and W.T. ZIEGLER. *Amer. Inst. Chem. Eng. J.*, **11** : 508 (1965).
5. S.H. LANGER, C. ZAHN and G. PANTAZOPLOS. *J. Chromatogr.*, **3** : 154 (1960) ; S.H. LANGER and J.H. PURNELL. *J. Phys. Chem.*, **67** : 263 (1963).
6. A.R. COOPER, C.W.P. CROWNE and P.G. FARRELL. *Trans. Faraday Soc.*, **62** : 2725 (1966).
7. D.E. MARTIRE. *Anal. Chem.*, **38** : 244 (1966).
8. See e.g. S. CARTER, J.N. MURRELL and E.J. ROSCH. *J. Chem. Soc.*: 2048 (1965).
9. W.B. PERSON. *J. Amer. Chem. Soc.*, **87** : 167 (1965).
10. I.D. KUNTZ, F.P. GASPARRO, M.D. JOHNSON and R.P. TAYLOR. *J. Amer. Chem. Soc.*, **90** : 4778 (1968).

Continuous-dynode electron multiplier detectors used with a one- (or two-) dimensional vacuum scanner for detection of tritium and radiocarbon in thin-layer radiochromatography

by

S. PRYDZ, B. MELÖ and J.F. KOREN

(Institute of Physics, University of Oslo)

T. Kristiansen (Uppsala)

Summary.

A step scanner for measurement of one- or two-dimensional distributions of radionuclides in thin-layer chromatography (TLC) has been constructed and tested. Different versions of miniaturized channel electron multipliers (to be operated in vacuum) have been applied as detectors. A description is given of the principle of operation, the vacuum step scanner and the way of data presentation.

Counter tests have been performed with calibrated standards of ^{3}H and ^{14}C in the form of polymethylmethacrylate pieces with known electron fluxes (numbers of emitted electrons per second per cm^2). Several TL media have been used with applicated spots of ^{3}H or ^{14}C glucose solutions. Upon drying, spots of activities down to 5 nCi could easily be detected (spot area ca. 0.5 cm^2). A resolution of 1 mm can be achieved in the one-dimensional scans.

Some more results, however preliminary, will be reported and the method discussed to some length. As a method of TL radiochromatography intended not to damage or contaminate the sample (e.g. by admixture of dry or wet scintillator) the scanning by electron multipliers might seem promising, even if the necessary vacuum of 10^{-5} torr is an additional requirement as compared to conventional techniques of measurement.

Introduction.

The purpose of this paper is to draw attention to a new electron detector having, possibly, some advantages in thin-layer radiochromatography detection (TLRC) of tritium (^{3}H) as compared to gas-flow GM detectors. The included discussion will be in general terms and rather short as the system is still under investigation. While a further development of the described system seems possible some disadvantages are obvious.

In principle, a number of methods exist for radionuclide detection. In practice, however, the liquid scintillation methods are still the most sensitive ones. In some cases there might nevertheless be a need for a system of detection rendering the chromatogram undamaged and free from scintillator contaminations.

For the detection of radionuclides other than ^{3}H the sensitivity of ordinary gas-flow GM detectors is fully sufficient. For ^{3}H detection, on the other hand, the smaller energy of the electrons render their path lengths only a few μm in the sample substances or a few mm in air (or in the counter gases). From these facts stems the low counting efficiency for tritium (number of counts per number of disintegrations during the same time only about 0,5 to 5 %). In addition, a preliminary (unpublished) experiment showed a slightly increased ^{3}H-electron exposure of X-ray films when exposed in a moderate vacuum. These considerations lead to the use of an electron multiplier detector operated in a vacuum system.

Is was supposed, also, that the reproducibility in such a system would be much less dependent upon the keeping of a constant gap between the detector and the chromatograms. This was found to be correct. The vacuum requirement, however, excludes the investigation of too volatile substances.

Description of the system.

In figure 1 is shown a diagram of the β-detection equipment. The chromatogram is supported by a 20 × 20 cm table moveable in two perpendicular directions by means of pulse-controlled step motors. The scanner

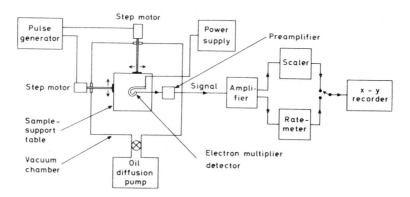

Fig. 1. — Block diagram of the β-detection equipment. The step motor control unit is not shown, neither is the electronic coupling of the recorder position to the detector position.

is contained in a stainless steel housing which may be evacuated to 10^{-5} torr in less than 15 minutes by means of an ordinary oil diffusion pump. The signal from the detector is amplified and recorded by a strip chart recorder (or, in two dimensions, a modified x-y recorder might be used). Synchronized pen movements are obtained by use of two ten-turns potentiometers. The position reproducibility is of the order of $\frac{1}{2}$ mm. The signal, in the form of amplified pulses, may be recorded either as a D.C. output signal from a pulse count-rate meter or in the form of a histogram by use of a special scaler. In this case the height of each column is proportional to the number of counts obtained in the corresponding position during the constant preset step time. This type of recording is free from peak distortion which may easily be introduced by the large time constants necessary when using a ratemeter at low intensities. Further, this type is suitable for simultaneous print out. Figure 2 shows the vacuum scanner and the detector.

Fig. 2. — The electron detector, closely coupled to its preamplifier, may be adjusted in height above the chromatogram which is seen fixed to the three-point support table of the vacuum x-y scanner.

The detector.

Many types of electron multiplier are now available. They are all based on the principle of electron cascading, either within a structured system of dynodes or along a single high-resistance dynode film supported by an insulating material. The factor of multiplication is given very roughly as $A = n^m$, where n is the number of secondary electrons emitted from the dynode surface per incident (primary) electron, and m is the number of multiplying jumps along the surface of the dynode strip (or simply number of dynodes in the first type). 'A' may be found as high as 10^6 (or 10^8 in the first types). We have applied miniaturized detectors of the continuous-dynode type. The dynode is made as a high-resistance coating on the internal wall of small-bore curved quartz tubes. Detector apertures : either the end opening of the tube, if wanted less than 1 mm diameter, or a side slit 1×15 mm cut in the quartz wall of a specially made version ; high voltage (2 - 4 KV) ; and a very low power consumption. (All units are from Mullard).

The small dimension of such detectors might be of importance when high spatial resolution is required. Small dimensions would also be advantageous if a rack of several detectors should turn out to be useful. Likewise, the small dimension of the counter is thought to be of importance for the total noise level. This has contributions due to radioactivity in wall materials, dust contamination from chromatograms (possibly), background radiation, and thermally released electrons from the dynode coating. (Contamination by radioactive dust may to some extent be removed by careful washing in alcohol). Mean noise countrates less than 1 count per 10 minutes have been found in some detectors with no screening other than the walls of the vacuum container made from 2 mm stainless steel.

Preliminary tests and results.

The measurements have been done partly on ^3H and ^{14}C standards in the form of thinplates of polymethylmethacrylate with electron fluxes known from calculation (electrons per second per cm^2). In addition several TL media have been used with active glucose solution applied directly as circular spots (with a tendency to show drying rings).

Counting efficiency :

The overall counting efficiency for ^3H on TL media probably is within the range from one to some few %, obviously somewhat higher on silica than on alumina. The reason for this difference is not known but is supposed to be differences in surface structure of the layers. The response varies linearly with the activity from 5 nCi to more than 5 µCi. This applies to the maximum readings over the centres of the spots as well as to values obtained by weighing the areas of paper under the recorded curves representing the various spots.

Limit of detection :

The smallest amount of tritium that may be detected depends upon the detector noise and the speed of scanning as well as upon the properties of the TL preparations. In figure 3 some combinations of scanning speed (steps per minute, 2 steps equal 1 mm) and time constants are compared for a 50 nCi spot. The drying ring can be recognized. With the best detector we find roughly a signal to noise ratio of 2 for a 5 nCi spot scanned at a speed of 1 mm per minute. This was, however, with an increased noise level due to detector contamination.

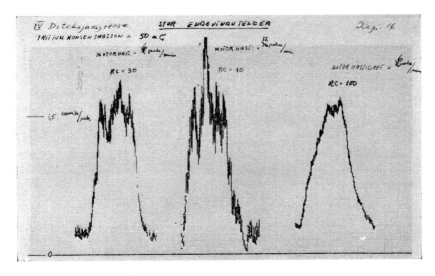

Fig. 3.— Different types of scan over a 50 nCi spot of tritiated glucose on Eastman Chromatogram SiO_2 sheet. At two of the settings the drying ring may be recognized.

Angular field of view :

To obtain a proper estimate of counting efficiency (the overall efficiency as well as that for the detector itself) the angular field of view must be known. The angular variation in sensitivity was investigated by use of the standards in combination with various thin metal frames.

Spatial resolution :

As is to be supposed the resolution of closely spaced spots decrease when the distance to the detector is increased. However, this decrease in resolution (and signal intensity) is much slower than would be expected for spots with no angular variation in electron flux.

In figure 4 are shown the spatial resolutions obtained with three different distances to the counter for the same pair of spots. The high resolution is supposed to be due to a marked angular variation in electron flux, which should in fact be found for low-energy electrons emitted by an active substance covering a highly porous surface of a material with high stopping power.

One additional reason for the observed decrease in resolution when the detector distance is increased is that one component of the noise signal is found to be proportional to this distance. This component might be due to more or less isotropic background radiation inside the vacuum chamber.

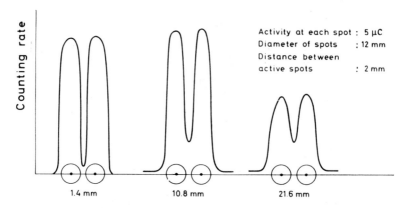

Distances between detector and preparate

Fig. 4. — The pen trace corresponding to a pair of spots is shown
for three different detector distances (1,4, 10.8, and 21.6 mm). At
the same scale are shown the spots with their centres and drying
rings. The distance between the drying rings is 2 mm.

Conclusion.

A further developed equipment of the described type might be found
valuable in TLRC work when destruction or contamination of the chromatogram can not be tolerated. More detailed results will be published elsewhere
within the near future. We are grateful to many members of the staff at
the University of Oslo, Physics Department for help in the construction
and building of the equipment and to B. Maehlum for encouragement and
advice in the initial phase of the work.

Gaschromatografen voor steroïdenbepaling

door

C. VAN HEDDEGEM

Ad. Auriema-Europe, Brussel.

Samenvatting.

Onlangs is de gaschromatografische steroïdenbepaling tot een betrouw-bare analysemethode ontwikkeld geworden en een belangrijk hulpmiddel geworden voor de geneesheer die een diagnose moet stellen (1). Hierdoor heeft men interesse gekregen voor eenvoudige toestellen die geschikt zijn voor deze bepaalde analyse.

De Barber-Colman model 5300 gaschromatograaf is volledig ontworpen in funktie van de steroïdenanalyse. Voor routine- en research-werk werd een automatische gaschromatograaf ontworpen. Voor de biochemist kan een proportionele teller gekoppeld aan een gaschromatograaf zeer nuttige resultaten leveren.

ROUTINEGASCHROMATOGRAAF

Inleiding.

Tot voor enkele jaren waren de meeste toepassingen van de gaschromatografie nog te vinden in de petrochemie. Door de recente ontwikkeling van deze techniek evolueerde deze laatste tot een betrouwbare analysemetode voor de medische laboratoria.

De Barber-Colman model 5300 routinegaschromatograaf (fig. 1) is ontworpen voor de klinische analyst die de gaschromatografie wenst toe te passen voor de routinebepaling van 17-ketosteroïden en estrogenen in urine.

De gaschromatografische steroïdenbepaling is niet zo eenvoudig als bijvoorbeeld de bepaling van koolwaterstoffen. De lage dampspanning, de temperatuurgevoeligheid en de polariteit van deze complexe organische molekulen maken dat de apparatuur met de meeste zorg moet ontworpen worden.

Injektieblok.

Men heeft zich steeds afgevraagd in welke mate metalen de katalytische ontbinding van de steroïden in de hand werken.

Niettegenstaande glazen injektiesystemen zonder twijfel veiliger zijn in het gebruik, is het mogelijk metalen injektiesystemen voor de steroïden-

Fig. 1.

analyse te gebruiken, op voorwaarde dat zij met zorg ontworpen zijn om « hot points » te vermijden.

Kolom.

Eveneens voor de kolom gebruikt men meestal glas. Dit biedt bovendien het voordeel dat men vlug kan vaststellen of de kolom behoorlijk gevuld is.

De temperatuurregeling moet stabiel zijn om « hot points » te vermijden en ook om reproduceerbare analyses mogelijk te maken.

Detektor.

Aanvankelijk gebruikte men voor de detektie van steroïden de argonionisatiedetektor. Stilaan is men overgegaan op de vlamionisatiedetektor die een kleiner uitgangssignaal geeft doch waarvan de verhouding signaal/ruis zeer gunstig gelegen is. Bijgevolg worden hier hogere eisen gesteld aan de elektrometer. De ongevoeligheid van de vlamionisatiedetektor voor water maakt hem bovendien zeer geschikt voor het analyseren van biochemische produkten.

De vlamionisatiedetektor is tevens in ruime mate ongevoelig voor temperatuurvariaties.

Eenvoudige bediening.

Een eis die niet gesteld wordt aan een research-gaschromatograaf, maar wel van belang is voor een routineapparaat is de eenvoudige bediening. Bij de Barber-Colman steroïdenanalysator werden de bedieningsorganen zo eenvoudig mogelijk gehouden, zij werden herleid tot het afregelen van de elektrometer en temperatuurinstelling voor de kolomoven.

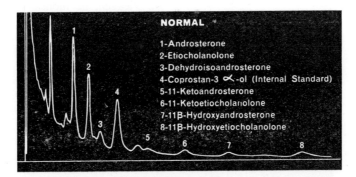

Fig. 2

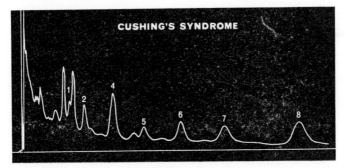

Fig. 3.

Figuren 2 en 3 tonen enkele chromatogrammen die met dit toestel verwezenlijkt werden.

AUTOMATISCHE GASCHROMATOGRAAF

Inleiding.

Het beschreven toestel, dat met 71 monsters kan geladen worden, kan gedurende meerdere dagen zonder toezicht analyses uitvoeren. Dit laat toe een doeltreffender gebruik te maken van de gaschromatograaf zowel voor routineanalyses als voor research waar een grotere hoeveelheid gegevens in een kortere tijd kunnen verkregen worden.

Beschrijving van het toestel.

Het toestel bestaat uit een standaard gaschromatograaf voorzien van een automatisch injektiesysteem. Figure 4 toont een blokschema van dit appa-

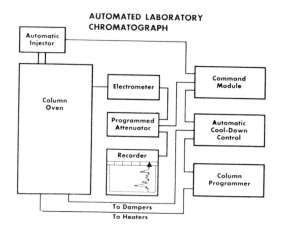

Fig. 4. — Electrical Diagram for Automated
Analytical G.C.

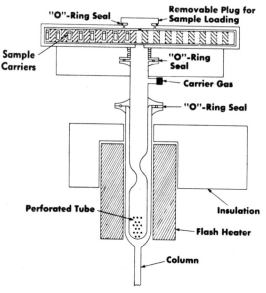

Fig. 5. — Automatic Sample Injector
(cutaway side view)

raat. De operator moet enkel de analysevoorwaarden vastleggen, de monsters laden en het toestel starten. Zonder tussenkomst van buitenuit worden alle monsters één na één geanalyseerd. Als centraal element heeft men de automatische analysekontroller die alle essentiële regelingen kontroleert.

Wanneer de analyse uitgevoerd wordt bij konstante temperatuur, worden de monsters opgebracht op een voorafbepaald tijdsinterval.

Ingeval van temperatuurprogrammatie kontroleert de automatische analysekontroller het volledig programma, het koelen van de kolomoven, het instellen van het termisch evenwicht en het opbrengen van het monster.

Een geprogrammeerde attenuator laat toe, indien gewenst, bepaalde pieken te verzwakken. Automatische korrektie van de basislijn na elke analyse is eveneens voorzien.

Op een impuls van de automatische analysekontroller brengt de automatische injektor (fig. 5) een monster boven de kolom waar het in de « flash heater » valt. Het monster verdampt dan met een efficiëntie die te vergelijken is met die van een konventionele injektie.

Voor niet-vluchtige stoffen, zoals steroïden worden de monsters bewaard op het oppervlak van een inerte drager, onder stikstofatmosfeer.

Toepassingen.

17 Ketosteroïden.

Fig. 6 toont twee analyses van een mengsel 17-ketosteroïden standaarden. 28 stalen werden achtereenvolgens geanalyseerd en de resultaten zijn opgenomen in tabel 7.

De standaarddeviatie is minder dan $\pm$ 0,1 en de standaardvariatie minder dan $\pm$ 2 %.

TMSi 17-KETOSTEROIDS

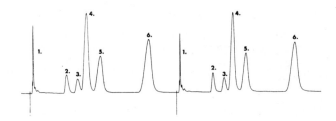

Fig. 6. — Two consecutive Runs of TMSi, 17-Ketosteroids Peak Identification : 1. Solvent, 2. Androsterone, 3. Etiocholanolone, 4. Dehydroisoandrosterone, 5. Epicoprostanol, 6. 11-Ketoetiocholanolone.
Column 2 % NGS at 215° C using a FID at 3×10^{-10} amperes.

TMSi 17-KETOSTEROIDS

Peak (PK) 1. Androsterone
2. Etiocholanolone
3. Dehydroisdandrosterone
4. Epicoprastanol
5. 11-Ketoetiocholanolone

PEAK AREA RATIOS

Run No.	PK3/PK1	PK3/PK2	PK3/PK4	PK3/PK5
1.	5.32	5.68	1.60	0.942
4.	5.28	5.68	1.66	0.938
8.	5.43	5.83	1.62	0.948
12.	5.43	5.77	1.62	0.935
16.	5.27	5.66	1.59	0.936
20.	5.42	5.79	1.63	0.943
24.	5.48	5.88	1.62	0.936
28.	5.27	5.72	1.66	0.942
Standard Deviation	$\pm$0.0974	$\pm$0.0793	$\pm$0.0264	$\pm$0.00447
Coefficient of Variation	$\pm$1.82%	$\pm$1.38%	$\pm$1.63%	$\pm$0.47%

Fig. 7. — Peak Area Ratios for 28 Consecutive Separations of the Same Sample.

Pregnanediol in urine.

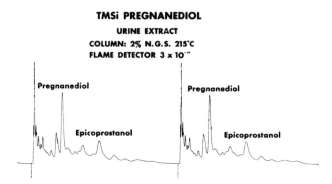

Fig. 8. — Consecutive Runs of Urinary Extract (operating conditions same as Figure 6).

TMSi PREGNANEDIOL

URINARY EXTRACT
PEAK AREA RATIOS

Run No.	Pregnanediol/Epicoprostanol
1.	1.83
2.	1.88
3.	1.82
4.	1.82
5.	1.82
6.	1.82
7.	1.84
8.	1.86
9.	1.83
10.	1.81
15.	1.83

Standard Deviation ± 0.0226

Coefficient of Variation ± 1.23%

Een urinestaal van een zwangere vrouw in de achtste maand werd gehydrolyseerd, de steroïden geëxtraheerd en omgezet in de trimethylsilyeters. Figure 8 toont twee van de 15 opeenvolgende scheidingen. De gegevens van deze scheidingen werden opgenomen in fig. 9. De standaarddeviatie is ± 0,02 en de variatiecoëfficiënt is ± 1,23 %. De estrogenen : estrone, estradiol en estriol werden ook geanalyseerd met gelijkaardige resultaten.

Fig. 9. — Peak Area Ratios for Pregnanediol vs. Epicoprostanol.

DETEKTIESYSTEEM VOOR GEMERKTE MOLEKULEN

Inleiding.

Voor research-werk bestaat een telsysteem dat rechtstreeks na een gas-chromatograaf kan geschakeld worden (fig. 10). Meerdere detektiesystemen werden beschreven door Vanden Heuvel en Horning. De meest gebruikte systemen zijn : de scintillatieteller, de ionisatiekamer en de proportionele teller. Het hieronder beschreven toestel is een proportionele teller.

Fig. 10.

Werking van het toestel.

Het eluent van de kolom wordt in twee delen gesplitst. Een gedeelte gaat naar een conventionele massadetektor, het andere gedeelte naar de proportionele teller.

Figure 11 toont het blokschema van dit detektiesysteem dat bestaat uit drie essentiële delen : de verbrandings-reduktieoven, de proportionele teller en de niveaumeter.

De verbrandingsoven bevat een kwartsbuis gevuld met koperoxide die verwarmd wordt tot $\pm$ 700° C. Hierin wordt het monster verbrand tot CO_2 en water. Voor de ^{14}C-detektie wordt de reduktieoven niet gebruikt. Het water wordt geabsorbeerd in een daartoe bestemde buis en het droge gas stroomt naar de proportionele teller.

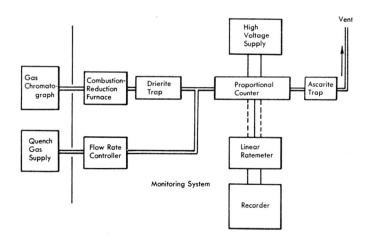

Fig. 11. — System Diagram of the Model 5190.

Monsters die tritium bevatten doorstromen eveneens de verbrandings-oven, doch worden vervolgens door de reduktieoven geleid die een kwarts-buis bevat gevuld met staalwol. Deze oven bevindt zich eveneens op een temperatuur van $\pm$ 700° C. Water wordt hierin gereduceerd en het effluent vloeit na droging door de proportionele teller.

Wanneer de gepaste gasbieten ingesteld zijn wordt een radioaktieve bron tegen de teller geplaatst en een spanning wordt aangesloten.

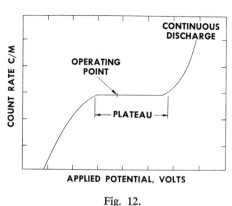

Fig. 12.

Fig. 12 toont de telsnelheid in funktie van de spanning wanneer deze laatste varieert van 1200 tot 2400 volt. Het plateau vertegen-woordigt het werkgebied dat zich gewoonlijk uitstrekt van 1700 tot 1900 volt.

In principe bestaat de teller uit een cylindrische buis met een cen-trale elektrode die geïsoleerd opge-steld is. De bijzonderste voordelen van de proportionele teller zijn : grote gevoeligheid zonder overbe-lasting bij grote monsterhoeveel-heden, goede werking bij de om-gevingstemperatuur en de mogelijk-heid om gebruikt te worden voor een continu-detektie op het effluent van een gaschromatograaf. De niveaumeter integreert het tellersignaal in funktie van de tijd en levert een signaal dat rechtstreeks kan geregistreerd worden.

Toepassingen.

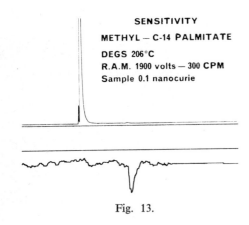

SENSITIVITY

METHYL — C-14 PALMITATE

DEGS 206°C
R.A.M. 1900 volts — 300 CPM
Sample 0.1 nanocurie

Fig. 13.

Fig. 13 illustreert de detektielimiet voor een monster van 0,1 nanocurie methylpalmitaat. Uit het ruisniveau kan men afleiden dat de detiektielimiet voor ^{14}C ongeveer 0,05 nanocurie bedraagt of ongeveer 100 desintegraties per minuut.

Fig. 14 toont de scheiding van een mengsel 17-ketosteroïden met 20 nanocuriedehydroisoandrosterone.

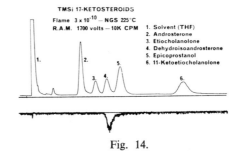

TMSi 17-KETOSTEROIDS

Flame 3 x 10^{-10} — NGS 225°C
R.A.M. 1700 volts — 10K CPM

1. Solvent (THF)
2. Androsterone
3. Etiocholanolone
4. Dehydroisoandrosterone
5. Epicoprostanol
6. 11-Ketoetiocholanolone

Fig. 14.

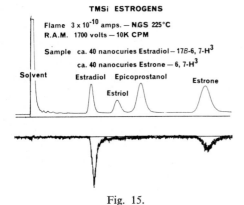

TMSi ESTROGENS

Flame 3 x 10^{-10} amps. — NGS 225°C
R.A.M. 1700 volts — 10K CPM

Sample ca. 40 nanocuries Estradiol — 17β-6, 7-H^3
        ca. 40 nanocuries Estrone — 6, 7-H^3

Solvent   Estradiol Epicoprostanol   Estrone

        Estriol

Fig. 15.

Fig. 15 illustreert de scheiding van estrogenen met 40 nanocurie estradiol en 40 nanocurie estrone.

Referenties

1. G. LEHNERT, H. VALENTIN, W. MÜCKE en H. DEUTZMANN. « Quantitative Bestimmung von Androsteron, Ätiocholanolon und Dehydroepiandrosteron im Harn mit Hilfe der Gas-Chromatographie ». *Ärztl. Lab.* **11** : 143-148 (1965).

2. LANTZ, C.D. and MORGART, J.R. « An Automated Laboratory Gas Chromatograph » presented at the Pittsburgh Conference on Analytical Chemistry and Applied Spectroscopy, March 1967.

3. LANTZ, C.D. and WOODS, R.A. Radioactivity Monitoring System for Gas Chromatography, available as Barber-Colman bulletin F-13225.

Modulation of sorption phenomena on gel filtration matrices

by

R. STROM

(Istituto di Chimica Biologica Università Roma, Italia)

Gel filtration of low-molecular weight substances is largely influenced by sorption phenomena, particularly when dealing with aromatic or heterocyclic compounds. These substances are retarded by an interaction with the gel matrix. This interaction is largely influenced by the composition of the eluent : a high ionic strength enhances it, while the presence in the eluent of substances having double bands and/or carbonyl, amide or aromatic groups depresses it. It is practically independent from the surface tension and from the viscosity of the eluent, as well as from the concentration of the substance being eluted, as shown by the symmetrical shape of the elution curves.

The « sorption » effect is probably due to an interaction between the gel matrix and the — electrons of the substance being eluted, and is therefore affected, apart from variations in the composition in the eluent, by the distribution of the electrons in the substance and by the chemical nature of the gel matrix. The elution volume is determined both by the gel filtration (i.e. by the Stokes'radius of the substance and by the degree of cross-linking of the gel) and by the « sorption » effect. The possibility of modifying at various degrees all these phenomena, by variations in the gel used and in the ionic strength, the pH and the composition of the eluent, makes this method a useful tool for the separation of low-molecular weight substances.

Radiochromatographie gaz-liquide à l'aide de deux compteurs proportionnels installés en série

par

H. LELOTTE, J. FURNELLE, J. WINAND et J. CHRISTOPHE

Laboratoire de Chimie biologique et de la Nutrition,
Faculté de Médecine et de Pharmacie, Université libre de Bruxelles, Belgique.

Introduction.

Nous pratiquons depuis quelques années la radiochromatographie gaz-liquide de mélanges complexes d'esters méthyliques d'acides gras peu radioactifs. Ces acides gras sont marqués au ^{14}C au cours d'incubations de fragments de tissu adipeux et de tranches de foie (1-3). Souhaitant mesurer en continu les radiations β des effluents chromatographiques, nous avons rejeté après quelques tâtonnements la scintillation sur cristaux d'anthracène (4-7), pour adopter la désintégration en comptage proportionnel (8-12), après pyrolyse, de manière à travailler à la température ambiante (10-12).

L'unité de mesure utilisée comporte deux compteurs proportionnels installés en série. Elle permet, en mesure continue, une bonne résolution de la radioactivité d'esters méthyliques d'acides gras de temps de rétention voisin (16:0 et 16:1 par exemple) et la détection de radioactivité aussi faible que 0.15 millimicrocuries dans le 16:0.

Abréviations utilisées :
GLC : chromatographie gaz-liquide ;
THT : très haute tension ; V : volt ; eV : électron volt ;
16:0, 16:1, etc... : esters méthyliques des acides gras désignés par le nombre des atomes de carbone et des doubles liaisons.

Matériel et méthodes.

A. *Chromatographie gaz-liquide et combustion d'une fraction des effluents.*

Les esters méthyliques d'acides gras radioactifs sont séparés au moyen d'un chromatographe Packard de la série 7800 (La Grange, Etats-Unis) dont l'étuve (modèle 802) chauffe à 183° C une colonne hélicoïdale de verre (185 cm × 0.4 cm de diamètre intérieur) garnie d'Apiezon L à 10 % sur Chromosorb W 80-100 mesh. Le débit d'Argon est de 60 ml/min (fig. 1).

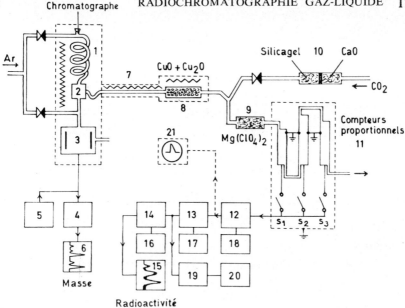

Fig. 1. — Schéma d'ensemble du dispositif de radiochromatographie gaz-liquide.

— Le *chromatographe* Packard modèle 7800 comporte l'étuve modèle 802 (1) ; un diviseur de flux gazeux (1/8 + 7/8) (2) ; le détecteur de masse à l'ionisation de l'argon, modèle 810 (3) ; un électromètre modèle 840 (4) ; une T.H.T. de 500 V (5) ; un enregistreur potentiométrique Honeywell type Electronic 17 (6).

— Gaîne chauffante (7) ; four à pyrolyse Packard modèle 325 (8) ; pièges à eau (9) et (10) ; boîtier en tôle d'acier contenant 3 compteurs proportionnels de 15, 38 et 25 ml et 3 interrupteurs S1, S2 et S3 (11).

— *Electronique de mesure* (MBLE - Philips) comprenant une T.H.T. stabilisée avec changeur d'impédance PW 4220 (500-3000 V) (12) ; un amplificateur de gain maximum 500, à seuil inférieur minimum réglable PW 4270 (13); un intégrateur linéaire PW 4242 (14) ; un enregistreur potentiométrique Honeywell type Electronic 15 (15) ; 3 blocs d'alimentation basse tension PW 4210 (16, 17 et 18) ; une échelle de comptage à 7 décades, avec affichage lumineux type PNC 012 (19) ; une minuterie à moteur synchrone (20) ; un oscilloscope (21).

General diagram of gas/liquid radiochromatography layout.

— The Packard chromatograph type 7800 comprises : the model 802 oven (1) ; a gaseous flux divider (1/8 + 7/8) (2) ; an argon ionisation mass detector type 810 (3) ; an electrometer type 840 (4) ; source of tension of 500 V (5) ; a Honeywell potentiometric recorder type Electronic 17 (6).

— Heating sheath (7) ; Packard pyrolysis oven type 325 (8) ; water traps (9) and (10) ; steel-plate box containing 3 proportional meters of 15, 38 and 25 ml, and three switches S1, S2 and S3 (11).

— Electronic measuring apparatus (MBLE Philips) comprising a stabilized source of tension with PW 4220 variable impedance (500-3000 V) (12) ; a 500 maximum gain amplifier with a PW 4270 adjustable minimum lower threshold (13) ; a PW 4242 linear integrator (14) ; a Honeywell potentiometric recorder type Electronic 15 (15) ; 3 PW 4210 low voltage supply units (16, 17 and 18) ; a counting scale with 7 scales and luminous setting type PNC 012 (19) ; a timer with synchronous motor (20) ; an oscilloscope (21).

A la sortie de la colonne chromatographique, un diviseur de flux gazeux en acier inoxydable fourni par Packard et placé dans le four même, envoie le 1/8 de l'effluent vers le détecteur de masse et les 7/8 restants vers le four à pyrolyse. La première fraction reçoit un complément d'Argon qui ramène le débit vers le détecteur à 60 ml/min, de façon à ne pas perturber les conditions de mesure par ionisation de l'Argon (détecteur de masse Packard, modèle 810).

Les 7/8 de l'effluent, destinés au radiocomptage, sont conduits sous une gaine chauffante à 250° C au four à pyrolyse (Tri-Carb Packard, modèle 325), où ils sont oxydés à 800° C sur CuO dans un tube de silice transparente type 453 de 6 mm de diamètre intérieur (Quartz et Silice, Paris, France). Quatre ml/min de CO_2 séché sur CaO et silicagel sont ajoutés à l'effluent et le tout est déshydraté sur $Mg(ClO_4)_2$.

B. *Les compteurs proportionnels.*

Trois compteurs proportionnels de 15, 25 et 38 ml ont été construits (fig. 2). Leurs constantes de temps respectives, en secondes, pour une vidange à 63 % sont de 14.0, 23.4 et 35.6 et pour une vidange à 99.9 % de 67.5, 118.5 et 171.0. Les temps de comptage sont respectivement de 15, 25 et 38 secondes. Pour les 3 compteurs, la T.H.T. optimum sur l'anode est de 2060 V, le rendement (E) pour le ^{14}C est de 90 %, le background (B) vaut 2.8 cpm/ml et le facteur de mérite ($\dfrac{E^2}{B}$) est de $\dfrac{90^2}{2,8} = 2892$/ml.

Chaque cathode est un tube d'acier inoxydable 18/8 (Sandvik, Suède) de 16 mm de diamètre intérieur et 18 mm de diamètre extérieur. La face interne du tube est très finement polie. L'anode en fil de tungstène de 40 μ de diamètre, est montée entre deux obturateurs isolants en plexiglas (fig. 2).

La concentration du champ électrique entourant l'anode est inversément proportionnelle à la surface de celle-ci, de telle sorte qu'en réduisant le diamètre de l'anode, on diminue la différence de potentiel à appliquer. De ce fait, une anode de 30 μ de diamètre place le plateau de comptage à 1650 V, contre 2000 V pour une anode de 40 μ et 2500 V pour une anode de 100 μ (résultats non publiés).

Par ailleurs, nos cylindres allongés (16 mm de diamètre intérieur) donnent de meilleurs résultats qu'un cylindre large et plat. Ce n'est que dans le premier cas que le gaz s'écoule sans turbulence, que les volumes morts sont presque inexistants (pas de recombinaison spontanée des ions ou d'absorption par les parois) et que le champ électrique est réparti de la manière la plus symétrique possible.

Les trois compteurs peuvent être disposés en série dans un boîtier en tôle d'acier d'un mm d'épaisseur (55 × 150 × 375 mm) relié à la masse. Trois interrupteurs (S1, S2 et S3) isolés à 3000 V permettent d'appliquer la T.H.T. sur l'un des compteurs (fig. 1). De cette manière on peut choisir, à chaque instant, le compteur qui convient le mieux pour une zone chromatographique (voir plus loin), sans devoir interrompre la mesure. L'ensemble assure, à température ambiante, un travail sans fuite et est démontable

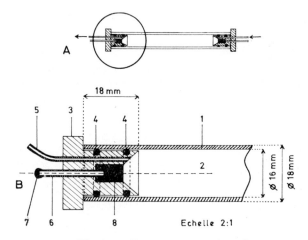

Fig. 2. — *Compteur proportionnel.*

A. Schéma général.   B. Détail des extrémités.

1. Cathode en tube d'acier inoxydable 18/8.
2. Anode en fil de tungstène de 40 μ.
3. Obturateur en plexiglass.
4. « O ring ».
5. Tube capillaire en Teflon d'un $\oslash$ ext. de 1.5 mm.
6. Tube en cuivre d'un $\oslash$ ext. de 2 mm.
7. Soudure Sn - Pb 40/60.
8. Bouchon souple, type 5396730 (Packard Instrument).

Proportional counter.

A. General view.   B. Detail of extremities.

1. 18/8 stainless steel tube cathode.
2. 40 μ tungsten wire anode.
3. Plexiglass shutter.
4. Circular joint.
5. Teflon capillary tube outside $\oslash$ 1.5 mm.
6. Copper tube outside $\oslash$ 2 mm.
7. Sn - Pb 40/60 welding.
8. Flexible plug, type 5396730 (Packard Instrument).

rapidement. L'électronique de mesure, entièrement transistorisée, est décrite dans la figure 1.

C. *Méthode d'utilisation du comptage proportionnel.*

La tension utilisée est suffisante ($>$900 V) pour faire travailler les compteurs en « comptage proportionnel » et non plus en chambre d'ionisation. Elle évite d'autre part d'atteindre les conditions du comptage selon Geiger et Müller ($>$2600 V) (13, 14).

Dans la zone de comptage dit proportionnel, la position du plateau de nos compteurs est déterminée par des injections répétées d'une quantité connue de $^{14}CO_2$ dérivant de la combustion d'un standard d'ester méthylique de [U-^{14}C] palmitate. Les plateaux de comptage se situent pour les trois compteurs entre 1935 V et 2050 V quand le mélange gazeux contient 94 % d'Ar et 6 % de CO_2 et entre 1800 V et 1840 V pour l'Ar pur (fig. 3). Ce décalage du plateau provient du fait que l'énergie moyenne nécessaire pour former une paire d'ions est de 26.4 eV seulement dans l'Ar, contre 32.9 eV dans le CO_2 (15).

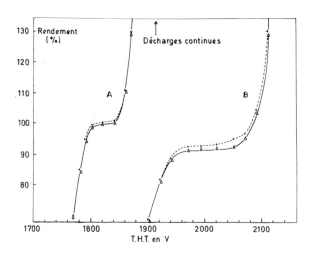

Fig. 3. — Position du plateau de comptage proportionnel et rendement des compteurs de 38 et 15 ml.

A. Argon 100 %.   B. Argon 94 % et CO_2 6 %.

+ — — — — + : compteur de 38 ml

Δ ——— Δ : compteur de 15 ml

Position of the proportional counting plate and yield of the 38 and 15 ml meters.

A. Argon 100 %.   B. Argon 94 % and CO_2 6 %.

+ — — — — + ; 38 ml meter

Δ ——— Δ ; 15 ml meter

La nécessité du CO_2 comme gaz stabilisateur réside dans le fait qu'il allonge le plateau de comptage et que de faibles variations de la T.H.T. n'influencent plus le rendement. En outre, ce gaz est partiellement auto-coupeur, empêchant les avalanches parasites formées par échanges anode-cathode, ce qui réduit le bruit de fond et améliore la reproduction des mesures. Enfin, l'apport de CO_2 provenant de la combustion des molécules organiques devient négligeable par rapport au CO_2 introduit artificiellement.

Résultats et discussion.

A. *Linéarité de la réponse et limite de sensibilité des compteurs proportionnels pour l'ester méthylique du [U-^{14}C] palmitate.*

Une solution standard d'ester méthylique de l'acide [U-^{14}C] palmitique (à 0.15 c/mole d'Applied Science, State College, Etats-Unis) est injectée en concentration croissante.

La figure 4 montre la parfaite linéarité de réponse des 3 compteurs, placés sous 2050 V, pour des injections variant de 490 dpm (0.22 m μC) à 156 800 dpm (70.5 m μC). Les mesures sont reproductibles et stables et les rendements atteignent 91 à 97 %.

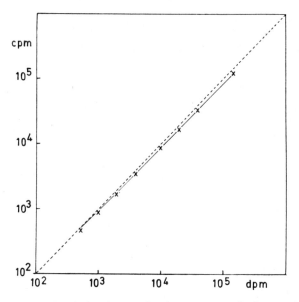

Fig. 4. — Linéarité de la réponse des 3 compteurs, placés sous 2050 V, pour des radioactivités du $^{14}CO_2$ allant de 490 dpm (0.22 m μc) à 156.800 dpm (70.5 μc). La moyenne des réponses (+ —— +) est comparée au rendement théorique de 100 % (— — — —).

Linearity of the response of the 3 meters at 2050 V, for $^{14}CO_2$ radioactivies from 490 dpm (0.22 m μc) to 156,800 dpm (70.5 μc). The average of the responses (+ —— +) is compared with the theoretical yield of 100 % (— — —).

B. *Limites de sensibilité de la radiodétection pour tous les esters méthyliques d'acides gras, en fonction de leurs temps de rétention.*

La limite inférieure de radiodétection de l'ester méthylique du [U-^{14}C] palmitate définie ci-dessus permet de prédire les limites de sensibilité pour d'autres esters méthyliques. Des acides gras de radioactivité spécifique identique, mais de longueur de chaîne croissante ont une limite de radiodétection inversément proportionnelle au temps de rétention, c'est-à-dire à la déviation standard de la gaussienne représentant la masse (16). Autrement dit, si un pic radioactif de [U-¹⁴C] palmitate, de déviation standard 1 à la masse, correspond à 200 dpm, un pic de déviation standard 1/2 présente la même hauteur pour 100 dpm seulement. Dès lors, en posant sur un diagramme logarithmique dans les 2 coordonnées, le point correspondant au temps de rétention et à la radioactivité facilement décelable du standard palmitique, la droite passant par l'origine permet de prévoir la limite de radiodétection de tous les esters méthyliques.

La figure 5 montre deux droites valables pour les compteurs de 38 ml et 15 ml, mis sous tension à 2050 V. Ces abaques permettent de prévoir le moment propice pour passer d'un compteur à l'autre.

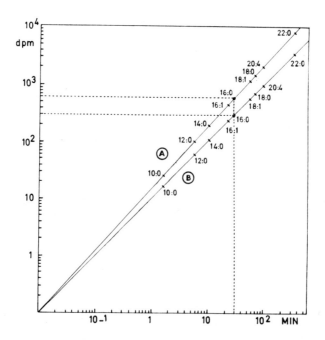

Fig. 5. — Abaques logarithmiques indiquant les limites de radiodétection de 9 esters méthyliques (de 10:0 et 22:0) avec les compteurs de 15 ml (droite A) et de 38 ml (droite B) opérant à 2050 V. La quantité minimum de dpm décelable dépend linéairement du temps de rétention (exprimé en min.).

Logarithmic abacus indicating the radiodetection limits of 9 methylic esters (from 10:0 to 22:0) with the 15 ml meter (straight line A) and the 38 ml (straight line B) operating at 2050 V. The minimum quantity of dpm detectable depends linearly on the retention time (expressed in minutes).

C. *Disposition en série des compteurs de 15 ml et 38 ml.*

La mise en série des compteurs de 15 ml et 38 ml n'exerce aucun effet défavorable sur la performance du grand compteur, en dépit de l'augmentation du volume mort. Dans ces conditions, le petit compteur, dont le temps de vidange à 99.9 % est de 67.5 sec, s'avère utile pour discriminer les acides gras de temps de rétention courts et voisins, par exemple 14:1 et 14:0. Une seconde utilisation du compteur de 15 ml est le passage de pics d'activité spécifique élevée, nécessitant une injection suffisante pour obtenir un enregistrement de masse convenable (cas d'un milieu d'incubation contenant du [U-^{14}C] palmitate p. ex.). Le début et la fin d'un pareil pic tendent à masquer la radioactivité des pics voisins, si ces derniers sont peu marqués et on a intérêt à utiliser le compteur de capacité minimum. Le second compteur (38 ml), qui a un temps de vidange à 99,9 % de 171 sec, convient particulièrement pour la radioactivité d'acides gras longs ou peu radioactifs. Le compteur de 25 ml a des qualités tellement intermédiaires que son emploi ne s'est jamais imposé dans nos travaux.

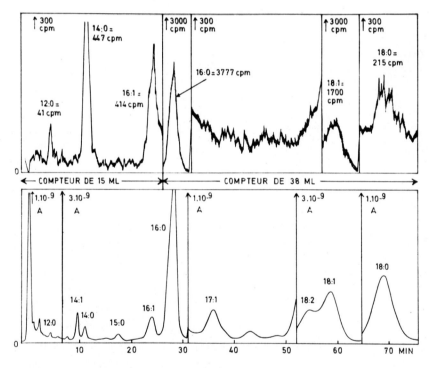

Fig. 6. — Radiochromatographie des esters méthyliques d'acides gras libres présents dans le tissu après incubation de tranches de foie de souris en présence de [1-^{14}C] acétate 0.1 mM (29 c/mole). 3.3 m μc sont injectées en tout.

Radiochromatograph of the methylic esters of free fatty acids present in the tissue after incubation of slices of mouse liver in the presence of [1-^{14}C] acetate 0.1 mM (29 c/mole). 3.3 m μc injected in all.

La fig. 6 illustre un exemple concret : il s'agit des esters méthyliques d'acides gras libres du foie de souris incubé en présence de [1-^{14}C] acétate. La T.H.T. est appliquée d'abord sur le compteur de 15 ml pour mesurer les radioactivités des pics jusqu'à 16:1 et ensuite sur le compteur de 38 ml pour les détections des pics tardifs peu radioactifs. L'intégration des pics montre que l'on récupère plus de 90 % des dpm injectées.

D. Considérations générales.

Le dispositif décrit ci-dessus fonctionne sans arrêt depuis 3 ans. Il a été dédoublé récemment en tirant parti du fait que le chromatographe Packard peut héberger 2 colonnes et 2 diviseurs de flux et que le four peut contenir 2 tubes à pyrolyse.Nous réalisons ainsi 4 analyses complexes par jour. L'alimentation électrique et les courants d'Ar et de CO_2 ne sont jamais interrompus. Tous les matins un standard de [U-^{14}C] palmitate (44 000 dpm = 0.133 µmole) est introduit dans chaque circuit. Les graphiques obtenus avec ces standards permettent à la fois d'exprimer les activités spécifiques des pics expérimentaux en dpm/mg et de vérifier les rendements journaliers des compteurs proportionnels et des détecteurs de masse. Cette précaution est indispensable pour les compteurs proportionnels, car de faibles variations des flux gazeux et de la T.H.T. influencent le rendement. Lorsque la déviation s'écarte trop des valeurs moyennes, les conditions optima sont rétablies avant de lancer un échantillon.

Références

1. J. WINAND, J. FURNELLE et J. CHRISTOPHE. *Bull. Soc. Chim. Biol.*, **49** : 1331, 1967.

2. J. WINAND, J. FURNELLE et J. CHRISTOPHE. *Bull. Soc. Chim. Biol.*, **49** : 1845, 1967.

3. J. WINAND, J. FURNELLE et J. CHRISTOPHE. *Biochim. Biophys. Acta*, **152** : 280, 1968.

4. A. KARMEN et H.R. TRITCH. *Nature*, **186** : 150, 1960.

5. A. KARMEN, L. GIUFFRIDA et R.L. BOWMAN. *J. Lipid Res.*, **3** : 44, 1962.

6. A. KARMEN, I. Mc CAFFREY et R.L. BOWMAN. *J. Lipid Res.*, **3** : 372, 1962.

7. A. KARMEN. *J. Am. Oil Chemist's Soc.*, **44** : 18, 1967.

8. J.T. KUMMER. *Nucleonics*, **3** : 27, 1948.

9. R. WOLFGANG et F.S. ROWLAND. *Anal. Chem.*, **30** : 903, 1958.

10. F. CACACE et INAM-UL-HAQ. *Science*, **131** : 150, 1960.

11. A.T. JAMES et F.A. PIPER. *J. Chromatog.*, **5** : 265, 1961.

12. A.T. JAMES et C. HITCHCOCK. *Kerntechnik*, **7** : 5, 1965.

13. G. NICOLO. L'Electronique dans les appareils de contrôle nucléaire. Paris, Ed. Eyrolles, 1963.

14. J. SHARPE et D. TAYLOR. Mesures et détection des rayonnements nucléaires. Paris, Ed. Dunod, 1958.

15. B.M. TOLBERT. Ionization chamber assay of radioactive gases, Office of Technical Services, Washington, U.S. Department of Commerce, 1956.

16. J.C. BARTLET et D.M. SMITH. *Can. J. Chem.*, **38** : 2057, 1960.

Computer analysis of electrophoretic patterns

by

B.J. HAYWOOD

(U.S.A.)

We have determined by statistical analysis that the area under the majority of protein peaks separated by agar gel or cellulose acetate electrophoresis fit the equation.

$$A = a.\psi \sqrt{\frac{\pi}{\log_e a/y}}$$

when scanned with a spectrophotometric device. Skewed or nongaussian curves usually resulted from poor technique. Using this procedure, it is possible to calculate areas under highly overlapped curves. This type of integration is adaptable to computer calculation. We have designed a program that will calculate peak area, percentage of the total area the peak represents, standard deviation from the normal mean, and deviation from normal mobility. The program is also designed to determine if the intensity of staining is related linearly to the area and percent total protein. A second system involving the direct typing on cards, or recording on magnetic tape, of the signal from the spectrophotometer has been developed. In this case, the signal passes first through a digital converter. This program gives all the information that the other does besides displaying the electrophoretic pattern. These systems have proven valuable in determining if slight alteration of isoenzyme mobility represents genetic deviation, and in studying the assembly mechanisms of isoenzyme systems. It is also valuable in hospital situations where large numbers of electrophoretic analyses are performed.

Synergism in the reversed-phase partition chromatography of uranium and thorium

by

N. CVJETIĆANIN

Hot Laboratory Department, Boris Kidrič Institute of Nuclear Sciences, Beograd (Yugoslavia)

Summary.

Our earlier investigations of the behaviour of [241]Am(III), Ce(III) and La(III) on paper treated with a mixture of thenoyltrifluoroacetone (HTTA) and tri-n-octylphosphate (TOP) have shown that the retention of these ions is higher than if each extractant is used separately. The synergistic effect of the mixture of HTTA and TOP in this case was investigated for U(VI) and Th(IV) by reversed-phase partition chromatography.

The composition of the metallic species was determined by correlating $\log (1/Rf—1)$ with the amount of HTTA and TOP fixed on the paper and with the activity of the hydrogen ion in the mobile phase. $UO_2(TTA)_2TOP$ and $ThCl(TTA)_3TOP$ are found to be formed under the investigated conditions. Conditions for the separation of uranium from thorium and their mutual separation from some tervalent lanthanides are also given.

* * *

Reversed-phase partition chromatography with various organic extractants, especially with liquid ion-exchangers as the stationary phase, has proved to be a very useful method of investigating the behaviour and separation of metallic ions (1—7). Taking into account the analogy between reversed-phase partition chromatography and liquid extraction by using the same extractant, it was interesting to investigate the effect of a mixture of two organic solvents on the behaviour of metallic ions by chromatography.

The great interest in the extraction with mixed solvents is mainly due to the enhancement of the extraction called « synergistic effect » or « synergism » which often appears in the extraction of metallic ions. One of the conditions to obtain this effect is that one of the reagents should act as a complexing or chelating agent, which is capable of neutralizing the charge on the metal ion, while the other is an active donor solvent (8,9). It was found that a mixture of thenoyltrufluoroacetone (HTTA) and ternary organophosphates extracts synergistically various metallic ions from aqueous solutions (8–14).

The synergistic effect of the HTTA and tri-n-octyl phosphate (TOP) mixture was first investigated on ^{241}Am(III), Ce(III) and La(III) (15). On the

chromatographic paper treated with the given mixture all three ions show considerable increase in retention, i.e. decrease in the R_f values compared with individual extractants. The value log $(1/R_f-1)$ for all three ions reaches maximum at a fixed composition of components in the mixture. Experiments have shown that in all three ions a complex of composition $M(TTA)_3(TOP)_2$ is formed, which was in accord with the data obtained by solvent extraction (12,13).

The behaviour of U(VI) and Th(IV) was investigated then in a similar way, using the same mixture of organic solvents. The composition of metallic species in organic phase is determined and conditions for their mutual separation and separation from the lanthanides and actinides, are given.

Experimental

Impregnation of the chromatographic paper with a mixture of HTTA (BDH, London) and TOP (Fluka, Buchs, Switzerland) solution in carbon tetrachloride was done by dipping paper strips in the organic phase as described earlier (6,15). An aqueous hydrochloric acid solution of varios concentrations was used as the mobile phase.

The investigations were carried out with chloride solutions of U(VI) and Th(IV). Aliquots of ~ 0.002 ml which do not contain more than 5.10^{-8} equivalents of each ion were spotted on the paper. Uranyl ion was identified with a 3 % solution of potassium ferrocyanide, and thorium with a 0.05 % aqueous solution of thoron.

The experiments were carried out on Whatman No. 1 paper strips, 2×35 cm, by ascending chromatography. The chromatograms were developed 4–5 hours at the temperature of 23–25°C. The reproducibility of the R_f values was $\pm$ 3 %.

Results and discussion

The preliminary investigations proved that U(VI) and Th(IV) show much higher retention than ^{241}Am(III), Ce(III) and La(III) at the generally same concentration of the extractants in the impregnation mixture. It was for reason that in this case the chromatographic paper was impregnated with a mixture of HTTA and TOP of a total concentration of 0.05 M.

Table 1 gives the Rf values of uranium and thorium, while figure 1 shows the distribution of both ions between 0.1 M HCl and mixture of HTTA and TOP of constant total molarity 0.05.

TABLE 1

Rf values for U(VI) and Th(IV). Paper impregnated with a mixture of HTTA and TOP. The mobile phase is 0.1 M HCl

Mixture of organic extractants		R_f	
		U(VI)	Th(IV)
TOP	0.05 M	0.92	0.57
HTTA TOP	0.0025 M 0.0475 M	0.85	0.48
HTTA TOP	0.005 M 0.045 M	0.70	0.15
HTTA TOP	0.010 M 0.040 M	0.24	0.13
HTTA TOP	0.020 M 0.030 M	0.25	0.12
HTTA TOP	0.030 M 0.020 M	0.34	0.11
HTTA TOP	0.040 M 0.010 M	0.52	0.19
HTTA TOP	0.045 M 0.005 M	0.63	0.38
HTTA TOP	0.0475 M 0.0025 M	0.75	0.40
HTTA	0.05 M	0.97	0.80

The presented curves are rather similar to those obtained by solvent extraction of the above mentioned ions (8).

To determine the composition of the complexes we used a relation which correlates the extraction coefficient with the Rf value in reversed-phase partition chromatography. If the aqueous and organic phases in both processes are the same, the following equation is valid for the given ion

$$\log (1/R_f - 1) = \log E_a^o + \log k \qquad (1)$$

where k is a constant depending on the chromatographic conditions.

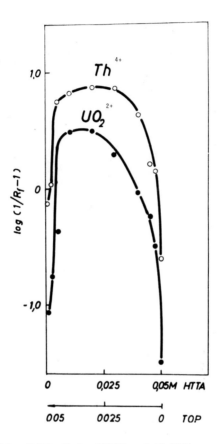

Fig. 1. — Plot of log (1/Rf—1) for U(VI) and Th(IV) as a function of molar
concentrations of HTTA and TOP in the mixture.

The Rf values of U(VI) and then of Th(IV) were investigated as a function
of the activity of H^+ ion and the amount of HTTA and TOP. In these experiments two variables out of three were always kept constant. The activities
of the hydrogen ion in the eluent were calculated from hydrochloric acid
molarities through the tables reported in the literature (16,17). Figure 2 shows
the dependence of log (1/Rf—1) on log a_H^+ for U(VI) for three different
mixtures of HTTA and TOP whose composition in one series of experiments
was constant. The slope of the straight lines is about – 2.

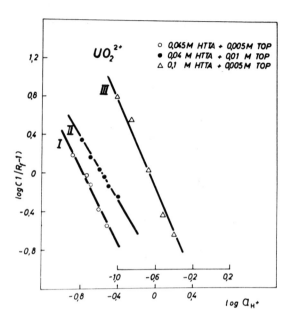

Fig. 2. — Plot of log (1 R — 1) vs, log a_{H}^{+} for U(VI). The slope of line I is
—2, line II —1.7, line III —2.2. The abscissa shifted to the right is for the
straight line III.

The extraction data obtained by investigating the synergistic effect in sexavalent and tetravalent actinides, have shown that the tendency for the two solutes to associate increases with increasing concentration of the ternary organophosphate, which has also been confirmed by measurement in the infra-red spectral region. Such association decreases effective individual concentration of the extractants (8,11). This is why in the course of our further investigations the concentration was HTTA > TOP.

The dependence of log (1/Rf — 1) on the amount of HTTA was then investigated at constant concentration of TOP in organic phase and constant concentration of H^{+} ion in mobile phase. The amount of HTTA, mg/cm², was obtained by weighing the paper before and after impregnation. The weights obtained by treating the paper with a mixture of HTTA and TOP were shown to be additive in comparison with the values obtained by impregnation with HTTA alone or with TOP alone. On the graphs, the HTTA and TOP concentrations are expressed in millimoles.

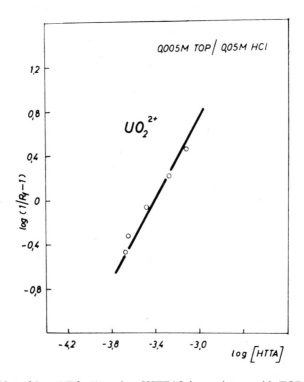

Fig. 3. — Plot of log (1/Rf—1) vs. log [HTTA] in a mixture with TOP for U(VI). The slope of the straight line is 2.

Figure 3 shows the dependence of log (1/Rf — 1) on log [HTTA] for U(VI). The concentration of TOP was always 0.005 M, and the concentration of HTTA in the mixture of impregnation ranged from 0.02—0.1 M. The concentration of the eluent was 0.05 M HCl. The slope of the straight line is 2.

The dependence of log (1/Rf — 1) on log [TOP] was investigated in the same way (Fig. 4). The HTTA concentration was 0.1 M, while the concentration of the eluent was 0.05 M HCl.

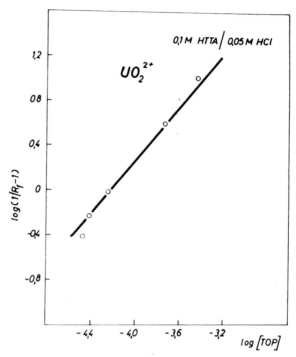

Fig. 4. — Plot of log (1/Rf — 1) vs. log [TOP] in a mixture HTTA for U(VI).
The slope of the straight line is 1.

From the data obtained by extracting uranium with a mixture of HTTA
and neutral organophosphates from chloride and nitrate medium (8,13),
the reaction can be expressed by the following equation :

$$UO_2^{2+} + 2HTTA + S \rightleftarrows UO_2(TTA)_2S + 2H^+ \qquad (2)$$

where S is the ternary organophosphate. The equilibrium constant of the
reaction is

$$K_e = \frac{[UO_2(TTA)_2S]_{org}\, a_H^{2+}}{[UO_2^{2+}]\,[HTTA]^2_{org}\,[S]_{org}} \quad . \quad f(\gamma) \qquad (3)$$

where a_{H^+} is the hydrogen ion activity in aqueous phase, while f (γ) is the
ratio of the activity coefficients of the components. For given experimental
conditions f (γ) may be assumed to be constant. The extraction coefficient
of the metallic complex is :

$$E_a^o = \frac{[UO_2(TTA)_2S]_{org}}{[UO_2^{2+}]} \qquad (4)$$

and taking into account equation (1) we obtain

$$\log (1/Rf - 1) = 2 \log [HTTA] + \log [TOP] - 2 \log a_{H^+} + const \qquad (5)$$

The term const. includes K_e, $f(\gamma)$ and k. The slopes of the straight lines obtained
in our investigations show that the complex $UO_2(TTA)_2TOP$, which is in
accord with the extraction data, is formed.

The behaviour of thorium was investigated in a similar way. Figure 5 shows the dependence of log $(1/Rf-1)$ on log $^aH^+$. The chromatographic paper was impregnated with a mixture of 0.05 M HTTA and 0.005 M TOP. As thorium shows a considerably higher retention than the ions investigated so far, the experiments were carried out in the concentration range from 0.1—1.5 M HCl.

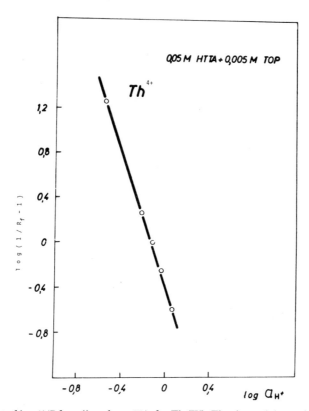

Fig. 5. — Plot of log $(1/Rf - 1)$ vs. log $^aH^+$ for Th (IV). The slope of the straight line is —3.

The dependence on the amount of HTTA is given in figure 6. The concentration of TOP in the impregnation mixture was 0.005 M, while the concentration of the eluent was 1 M HCl. The slope of the straight line was 3. Figure 7 shows the dependence of log $(1/Rf - 1)$ on the amount of TOP, at a constant HTTA concentration of 0.05 M and concentration of the eluent of 0.5 and 1 M HCl. The slope of the lines is 1.

The data obtained by extracting thorium with a mixture of HTTA and ternary organophosphates from chloride medium imply that a complex of the composition $Th(TTA)_4S$, which does not contain anion of mineral acid is formed. These experiments were carried out from 0.1 M HCl (9,13). However, the data obtained from a nitrate medium show that extracted complexes

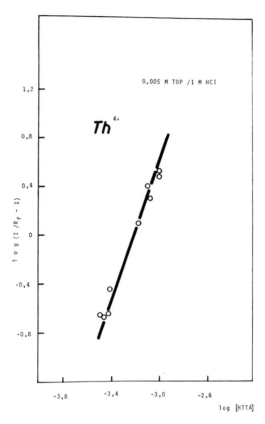

Fig. 6. — Plot of log (1/Rf — 1) vs. log [HTTA] in a mixture with TOP for Th(IV). The slope of the straight line is 3.

include ions of mineral acid, $Th(NO_3)(TTA)_3S$ and $Th(NO_3)_2(TTA)_2 \cdot 2S$ (11). The extraction of the metallic species in these conditions can be expressed by the following general-type equation (18) :

$$M^{z+} + pX^- + nHA + mS \rightleftarrows MX_pA_{z-p}(HA)_{n-z+p}S_m + (z-p)H^+$$

where HA is the chelating organic acid, X^- — the anion of mineral acid and S — the organic electron donor molecule.

The investigations of the behaviour of thorium by chromatography were performed in our case with concentrations from 0.1—1.5 M HCl, which are considerably larger than those used in liquid extraction (9,13). From the slopes of the straight lines shown in Figures 5, 6 and 7 the following reaction is supposed to proceed :

$$Th^{4+} + Cl^- + 3HTTA + TOP \rightleftarrows ThCl(TTA)_3 TOP + 3H^+$$

This implies that a complex thus formed would involve the chloride ion to preserve electroneutrality.

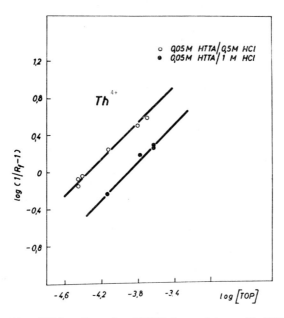

Fig. 7. — Plot of log (1/Rf — 1) vs. log [TOP] in a mixture with HTTA for Th(IV).
The slope of the straight lines is 1.

The paper impregnated with a mixture of given solvents is suitable for mutual separation of metallic ions. By choosing corresponding concentrations of the extractants in the mixture and the concentration of the eluent, we performed separations which are shown in Table II.

Table II. — Chromatographic separations on paper treated with a mixture of HTTA + TOP.
Eluent : HCl.

Elements	Mixture of organic extractants	HCl M	Rf
U(VI)—[241]Am(III)—[144]Ce(III)	HTTA 0.15 M TOP 0.05 M	0.075	U(VI) = 0.0; [241]Am(III) = 0.32; [144]Ce(III) = 0.76
Th(IV)—Ce(III)—La(III)	HTTA 0.15 M TOP 0.05 M	0.05	Th(IV) = 0.0; Ce(III) = 0.40; La(III) = 0.82
Th(IV)—U(VI)—Ce(III)	HTTA 0.08 M TOP 0.005 M	0.05	Th(IV) = 0.0; U(VI) = 0.21; Ce(III) = 0.98
U(VI)—Ce(III)—La(III)	HTTA 0.08 M TOP 0.05 M	0.025	U(VI) = 0.0; Ce(III) = 0.20; La(III) = 0.56

Acknowledgements

The author is grateful to the Yugoslav Federal and Serbian Research Funds for financial support.

References

1. E. CERRAI and C. TESTA, *J. Chromatog.*, **8** : 232, 1962.
2. E. CERRAI in M. LEDERER (Editor), *Chromatographic Reviews*, Vol. 6, Amsterdam Elsevier, p. 129, 1964.
3. E. CERRAI and G. GHERSINI, *J. Chromatog.*, **24** : 383, 1966.
4. J. W. LAUGHLIN, G.J. KAMIN and C.V. BANKS, *J. Chromatog.*, **21** : 460, 1966.

5. N. CVJETIĆANIN, *Bull. Boris Kidrič Inst. Nucl. Sci.*, (Beograd) **15** : 201, 1964.
6. N. CVJETIĆANIN, *J. Chromatog.*, **32** : 384, 1968.
7. S. KAWAMURA, T. FUJIMOTO and M. IZAWA, *J. Chromatog.*, **34** : 72, 1968.
8. H. IRVING and D.N. EDGINGTON, *J. Inorg. Nucl. Chem.*, **15** : 158, 1960.
9. T.V. HEALY, *Nucl. Sci. Eng.*, **16** : 413, 1963.
10. H. IRVING and D.N. EDGINGTON, *J. Inorg. Nucl. Chem.*, **20** : 314, 1961.
11. Ibid, **20** : 321,1961.
12. Ibid, **21** : 169, 1961.
13. T.V. HEALY, *J. Inorg. Nucl. Chem.*, **19** : 314, 1961.
14. Ibid, **19** : 328, 1961.
15. N. CVJETIĆANIN, *J. Chromatog.* **34** : 520, 1968.
16. R.A. ROBINSON and R.H. STOKES, Electrolyte Solutions, London, Butterworths, pp. 491 and 504, 1959.
17. H.S. HARNED and B.B. OWEN, The Physical Chemistry of Electrolytic Solutions. New York, Reinhold Publishing Corporation, pp. 343 and 547, 1950.
18. I.J. GAL and R.M. NIKOLIC, *J. Inorg. Nucl. Chem.*, **28** : 563, 1966.

A review on the chromatography and electrophoresis on the papers impregnated with the inorganic ion exchangers

by

Mohsin QURESHI

Chemical Laboratories, Aligarh Muslim University, Aligarh, India.

Synthetic inorganic ion exchange papers were first reported for the separation of alkali metals using chromatography by G. Alberti and co-workers. They used 0.1N HNO_3 and 0.2N NH_4NO_3 as developers and the separations obtained were of 15 γ Na^+ and K^+ and of 100 γ Li^+, Rb^+ and Cs^+ on ammonium molybdophosphate papers (1). These authors also used zirconium phosphate papers (2) and the separations which were not possible with HCl alone could be achieved with HCl containing NH_4Cl, NaCl or KCl. These papers were also used by them for the rapid separation of the short lived radioactive elements and for the study of absorption equilibria (3). Then the chromatographic behaviour of UO_2^{++} and Pb^{++} at different temperatures was studied and a linear relationship between Rf values and the degree of impregnation was established (4). Slightly higher values were noted at the higher temperatures. The use of these papers were then explored for the spot tests of Tl^+, Hg_2^{++}, Ag^+ and Cs^+ after rapid and selective separation of these ions on ammonium molybdophosphate papers (5). Electrophoretic studies on such papers were also made by G. Alberti and coworkers (6). Thus they separated La, Pr, and Sm by electrophoresis on paper loaded with zirconium phosphate using a 0.5N $HClO_4$ as the electrolyte under an electric field of 4.7 V/Cm.

The papers loaded with zirconium selenite were used for chromatographic studies by Jeronimo and Da Costa (7) and a good separation of Cu^{+2} (Rf = 0.25) and Au^{+3} (Rf = 0.66) was obtained. These authors also used zirconium phosphate papers (8) to see the behaviour of the metals of 1st group. Thus in 0.1N HCl Ag^+ stays at the point of application and Cu^{+2} has an Rf value of 0.70 to 0.80. Separations of the heavy radio-element (9) and of the different valence states of the same elements (10) may also be obtained on zirconium phosphate papers. The alkaline earths may be separated together from caesium on zirconium molybdate papers with 0.5N cl, after which they can be separated (11) from each other with

0.9N HCl. H. Schroeder used ammonium molybdophosphate papers and obtained the following separations (12).

1. Cs^{137}(Rf = 0) from Sr^{90}, Y^{90}, Cs^{144}, Pr^{144} and Ru^{106}(Rf = 1) in 1M HNO_3.

2.   Sr (Rf = 1) from Cs, Y, Ce, and Pr (Rf = 0) in 0.1M HNO_3.

The papers impregnated with the salts of heteropolyacids have also been used by other workers in the field. As a result Li, Na, K have been separated (13) on ammonium tungstophosphate after developing for 3 hours in the solvent system 0.4M NH_4NO_3 — 0.1M HNO_3 — Methanol with the ratios of Methanol to H_2O from 1 : 1 to 5 : 1. Prasilova and Sebesta (14) have defined the optimum conditions for the separations Cs — Rb, Sr — Y, CS — Y, and Rb — Y on these papers using $NH_4 NO_3$ — HNO_3 system. Separations of Li, Na, K, Rb, and Cs were then investigated (15) on the papers impregnated with ammonium molybdo-phosphate using aq. ethanolic 0.6M NH_4NO_3 — 0.1M HNO_3 as solvent system. Thus a satisfactory separation was obtained with solvents having ratios of ethanol to H_2O from 1 : 1 to 5 : 1.

The use of the ion exchange materials having tin for the preparation of the chromatographic papers could not be explored until the behaviour of 28 inorganic cations was first shown on stannic phosphate papers by the Japanese workers Hai Yin and Hsiang Chu (16) in 1966. They used the developers as 0.1 — 4N HCl, HNO_3, H_2SO_4, $HClO_4$ and acetic acid. We also started in our labs. the paper chromatographic studies on these papers and the bevaviour of 21 cations and 2 anions were discussed (17) on stannic phosphate papers with HCl of various concentrations and $HClO_4$ as developers.

Detailed studies were then made in these laboratories on stannic phosphate, stannic tungstate (18), stannic molybdate, stannic selenite, tita-nium tungstate, and titanium molybdate papers. Table 1 gives some selected separations of one metal ion from numerous metal ions on these papers and table 2 shows some important binary, ternary and quaternary separations obtained on them.

A quantitative separation of selenium from metal ions on stan-nic tungstate papers has been achieved (19). Thus selenium can be separated from the substances which may contain Cu^{+2}, Fe^{+3}, Zn^{+2}, Mn^{+2}, Hg^{+2}, Pb^{+2} and Bi^{+3} as impurities. Electrophoretic studies have also been performed for the separation of metal ions (20) on stannic phosphate papers. This technique can therefore be utilized to give a number of inter-esting and difficult separations. To mention a few Tl^+, Hg^{+2}, As^{+3}, Ni^{+2}, Co^{+2}, Mn^{+2}, Cr^{+3}, Zn^{+2}, Mo^{+6}, W^{+6}, Pd^{+2}, Pt^{+4}, Au^{+3}, Ba^{+2}, Sr^{+2}, K^+, Rb^+, Cs^+ can be separated from numerous metal ions. Some binary and ternary separations are also possible. Thus Fe^{+3} — Cr^{+3}, Pt^{+4} — Ru^{+3} — Cu^{+2}, Au^{+3} — Pd^{+2} — Cu^{+2} and Sr^{+2} — Ba^{+2} separations are easily achiev-ed. In the following pages is given a comparative study of the behaviour of 50 metal ions on the papers impregnated with stannic tungstate, stannic molybdate, titanium tungstate and titanium molybdate in three solvent systems.

TABLE 1

Some separations of one metal ion from numerous metal ions.

Cation separated	Solvent system	Paper used	Ions likely to interfere	Time
Au^{+3} (1.00-0.80) from 46 metal ions.	MeOH + 1OM HCl + AcOH (6 : 1 : 4)	Stannic selenite	Nil	40 minutes
Mg^{+2} (1.00-0.93) from 41 metal ions	0.1M NH$_4$Cl	»	Hg^{+2}, Pd^{+2}, Zn^{+2}	20 minutes
Nb^{+5} (1.00-0.90)	0.1M HClO$_4$	»	K$^+$, Pd^{+2}, Ga^{+3}, Hg^{+2} and Be^{+2} or	20 minutes
Or Be^{+2} (1.00-0.90) from 41 metal ions.	0.1M HClO$_4$	»	Nb^{+5}	20 minutes
Zn^{+2} (0.94-0.87) from 41 metal ions.	Ethyl methyl ketone + Acetone + 1M HCl (7 : 3 : 3)	Stannic tungstate	Au^{+3}, Hg$_2$$^{+2}$, Hg^{+2}, Bi^{+3}, and Cd^{+2}	1 hour
Tl^{+3} (1.00-0.91) from 51 metal ions.	Acetone + Acetic acid + n—Butanol + 1M NaCl (1 : 1 : 1 : 1)	Titanium tungstate	Sn^{+2}, Sn^{+4}, Hg^{+2}, Nb^{+5}	1.75 hour
K$^+$ or Rb$^+$ or Cs$^+$(0.77-0.56) from 44 cations.	2M Ammonium formate + 2M NH$_4$OH (1 : 1)	»	Ag$^+$, Ni^{+2}, Mo^{+6}, Se^{+4}, Nb^{+5}	20 minutes
Sb^{+3} (0.46-0.19) from 38 cations	Acetone + Acetic acid + n-butanol + 1M NaCl (1 : 1 : 1 : 1)	Titanium tungstate	Hg^{+2}, Bi^{+3}, Pd^{+2}, Al^{+3}, Be^{+2}, Ga^{+3}, Zn^{+2}, Pt^{+4}, Ca^{+2}, Sn^{+2}, Sn^{+4}, Nb^{+5}, UO$_2$$^{+2}$.	1.75 hour.
Sb^{+5} (1.00) from 37 cations	n-Butanol + HNO$_3$ (8 : 2)	Stannic phosphate	Hg$_2$$^{+2}$, Au^{+3}, Mo^{+6}, Sn^{+2}.	2.5 hour.
Pt^{+4} (0.96)	0.1M Ammonium carbonate	Stannic tungstate	Mo^{+6}, Se^{+4}, UO$_2$$^{+2}$	20 minutes
Or Mo^{+6} (0.94) from 42 cations			Pt^{+4}, Se^{+4}, UO$_2$$^{+2}$	
Se^{+4} (0.92) from 33 cations	0.1M Ammonium tartrate in 4M NH$_4$OH	»	Ag$^+$, Al^{+3}, Cr^{+3}, Ce^{+3}, Ce^{+4}, Mo^{+6}, Ga^{+3}, Zr^{+4}, K$^+$, Rb$^+$, Cs$^+$ and Pt^{+4}	20 minutes

TABLE II

Some important binary, ternary and quaternary separations.

Solvent system	Separations achieved	Paper used	Time
Acetone + Acetic acid + n-Butanol + 4M HCl (1 : 1 : 1 : 1)	$Al^{+3}(0.16-0.34)$ — In^{+3} $(0.42-0.59)$ — $Ga^{+3}(0.66-0.70)$ — $Tl^{+3}(0.93-1.00)$	Titanium tungstate	2 hour
»	$Al^{+3}(0.15-0.32)$ — Be^{+2} $(0.36-0.58)$ — $Ga^{+3}(0.63-0.69)$ — $Tl^{+3}(0.93-1.00)$	»	»
»	$Fe^{+2}(0.18-0.33)$ — $Fe^{+3}(0.64-0.70)$	»	»
Ethyl methyl ketone + Acetone + 50 % HCl (1 : 6 : 1)	$Al^{+3}(0.00-0.08)$ — Be^{+2} $(0.14-0.35)$ — $Fe^{+3}(0.70-0.80)$	»	30 minutes
Acetyl acetone + Acetone + 50 % HCl (6 : 3 : 1)	$Tl^{+}(0.00)$ — $Zn^{+2}(0.25-0.47)$ — $Ga^{+3}(0.74-1.00)$	»	25 minutes
Ethyl methyl ketone + n-Butanol + 50 % HCl (6 : 3 : 1)	As^{+3} — Sb^{+3} or Bi^{+3} $(0.35-0.00)$ — $(0.57-0.50)$ or $(0.60-0.58)$	Stannic tungstate	35 minutes
Dioxan + Satd. soln. of NaF + 1M HCl (3 : 1 : 6)	$Zr^{+4}(0.71-0.52)$ — $Th^{+4}(0.12-0.00)$	»	»
MeOH + 10M HCl + HCOOH (6 : 2 : 2)	Cs^{+} — Rb^{+} or $K^{+}(0.00)$ — $(0.32-0.21)$ or $(0.28-0.20)$	Stannic selenite	40 minutes
0.1M H_2SO_4	Mo^{+6} — Cr^{+3} $(0.10-0.00)$ — $(1.00-0.90)$	Stannic selenite	20 minutes
0.1M $HCLO_4$	Al^{+3} — Cr^{+3} $(0.64-0.37)$ — $(0.17-0.00)$	»	20 minutes
1M $HClO_4$	UO_2^{+2} — V^{+4} — Th^{+4} (0.34) — (0.72) — (0.05)	»	20 minutes
Methanol + 10M HCl + Acetic acid (6 : 1 : 4)	$Au^{+3}(1.00-0.80)$ — Ag^{+} $(0.22-0.00)$ — $Pt^{+4}(0.63-0.45)$	»	40 minutes
n-Butanol + HCl (8 : 2)	$W^{+6}(0.00)$ — $Mo^{+6}(0.56)$	Stannic phosphate	3.25 hours
1N Amm. Formate	$Mg^{+2}(0.83)$ — $Sr^{+2}(0.47)$ — $Ba^{+2}(0.18-0.00)$	Stannic tungstate	20 minutes
n-Butanol - 50 % HNO_3 (6 : 4)	$Ag^{+}(0.00)$ — $Cu^{+2}(0.28)$ — $Au^{+3}(0.98)$	»	2 hours
0.1M Amm. tartrate in 4M NH_4OH	$Se^{+4}(0.86)$ — $Te^{+4}(0.07-0.00)$	»	20 minutes
2N HCl + 2M Ortho phosphoric acid (1 : 1)	$Sb^{+3}(0.07)$ — $Sb^{+5}(0.36)$	Stannic phosphate	25 minutes
Acetyl acetone + Acetone + 50 % HCl (7 : 3 : 1)	$Zn^{+2}(0.78-0.54)$ — $Mn(0.40-0.21)$ — $Ni^{+2}(0.06-0.00)$	Stannic tungstate	20 minutes

EXPERIMENTAL

Apparatus :

Glass jars, 20 by 5 cm. were used to develop paper strips 14 by 3 cm. Whatman N° 1 paper was used throughout the studies.

Reagents :

Chemicals and solvents were either E. Merck (Darmstadt) or British Drug House Analytical grade reagents. Stannic chloride pentahydrate was a Poland product and titanium chloride was used 15 % (w/v) of B.D.H. England.

Preparation of Ion Exchange Papers :

All the papers were prepared using the following standard procedure : The papers were first dipped in 0.25M solution of stannic chloride or titanium chloride for about 5 seconds ; the excess of the solution was drained off and the papers were placed on the paper sheet and dried for 15 minutes. Then they were dipped in 0.25M solutions of sodium tungstate or sodium molybdate, as the case may be for 5 seconds. The excess of the reagents were drained as before and the papers were allowed to be dried at room temperature. Then they were washed in distilled water three times (for 5 minutes each time) to remove any reagent remained in excess. The papers were used as such after drying at room temperature.

Detectors :

Yellow ammonium sulphide was used to detect Ag^+, Pb^{+2}, Hg_2^{+2}, Hg^{+2}, Tl^+, Bi^{+3}, Cd^{+2}, As^{+3}, Sb^{+3} and Pd^{+2}. A fresh solution of cobaltinitrite was used to detect K^+, Rb^+, and Cs^+. La^{+3}, Ce^{+3}, Ce^{+4}, Y^{+3}, In^{+3}, Zr^{+4}, Th^{+4} and Nb^{+5} were detected with 0.1 % alcoholic alizarin red S solution. Te^{+4} and Mo^{+6} were located with the help of $SnCl_2$-HCl reagent. Al^{+3}, Be^{+2}, Ga^{+3} were detected with 1 % alcoholic aluminon. Zn^{+2}, Mn^{+2}, Cr^{+3}, Ir^{+4} were detected by diphenyl carbazide, and potassium ferrocyanide detected Fe^{+3}, UO_2^{+2}, V^{+4}, Cu^{+2} and Ti^{+4}. Dimethyl glyoxime was used for Ni^{+2} and Co^{+2} and Ba^{+2} and Sr^{+2} were located by 5 % aqueous solution of sodium rhodizonate and Au^{+3} by 5 % aqueous hydroquinone. Mg^{+2} was detected with quinalizarin while Ru^{+3} and Se^{+4} were detected with 2N HCl solution of thiourea. Pt^{+4} was detected with aq. KI solution and W^{+6} by 5 % alcoholic pyrogallol solution. The detector for Sn^{+2} was 5 % aq. phosphomolybdic acid, for Sn^{+4} dithiozone and for Fe^{+2} potassium ferricyanide.

Procedure :

Thin glass capillaries were used to spot the test solutions. The chromatograms were conditioned for 10-15 minutes and then the solvent was allowed to ascend 11 cms. on paper in the glass jars. They were then withdrawn and the spots were detected after drying the chromatograms.

Solvent systems used :

(1) Acetone + Acetic acid + n-butanol + 4M HNO_3 (1 : 1 : 1 : 1).
(2) 0.5M HCl + saturated aq. solution of KCl (1 : 1).
(3) Ethyl methyl ketone + acetone + 50 % HCl (6 : 3 : 1).

RESULTS

A comparative study of the chromatographic behaviour of 50 metal ions on the stannic tungstate, stannic molybdate, titanium tungstate and titanium molybdate papers was performed using three solvent systems. The results are summarized in the figures 1 - 12.

DISCUSSION

The plots of Rf values versus metal ions show a comparative study of 50 metal ions on four different ion exchange papers — stannic tungstate, stannic molybdate, titanium tungstate and titanium molybdate in three solvent systems. It was found that in some cases the detection on these papers is not very clear. The difficulty was reported most frequently in the case of titanium molybdate papers. This may be due to the fact that titanium and molybdenum give coloured complexes with most of the detecting reagents. However, the detection is not so difficult on other papers.

The same method of preparation of the papers was followed throughout the studies which helps in comparing the results. The method which we used to follow previously has been slightly modified. Previously we prepared the papers in hot solutions. But when the same papers were prepared in cold solution, the results obtained did not differ much. So it was then thought that the paper may be prepared more easily and in less time by the latter method. It was found practically correct and the results obtained were quite reproducible.

As it is evident from the graphs, the behaviour of the metal ions in one solvent system is the same irrespective of the ion exchange paper we used. The reason may be that all the ion exchangers are cation exchangers so their ion exchange behaviour should be the same. Thus in solvent system 1 (acetone + acetic acid + n-Butanol + 4M HNO_3 (1 : 1 : 1 : 1) Au^{+3} has got the highest Rf value while Ti^{+4}, Fe^{+3}, Zn^{+4}, Nb^{+5}, Mo^{+6}, Te^{+4}, Ba^{+2}, and Hf^{+4} are retained. In solvent system 2 (0.5M HCl + satd. KCl (1 : 1)) the general trend of metal ions is to move. It may be due to the formation of negatively charged chlorocomplexes of these metals. Only a few metal ion such as Ce^{+4}, Zr^{+4}, Nb^{+5}, Sb^{+3}, Te^{+4}, Ce^{+3} and Hf^{+4} have low or zero Rf value probably due to their relatively larger ion radii.

References

(1) ALBERTI, G. and GRASSINI, G. *J. Chromatog.*, **4** : 423, 1960.

(2) ALBERTI, G. and GRASSINI, G. *Ibid.*, **4** : 83, 1960.

(3) ALBERTI, G., DOBICI, F. and GRASSINI, G. *Ibid.*, **8** : 103, 1962.

(4) GRASSINI, G. and PADIGLIONE Claudia, *Ibid.*, **13** : 561, 1964.

(5) ALBERTI, G. *Ibid.*, **31** : 177, 1967.

(6) ALBERTI, G., CONTE, A., GRASSINI, G. and LEDERER, M. *J. Electroanal Chem.*, **4** : 301, 1962.

(7) JERONIMO, M.A.S. and NUNES DA COSTA, M.J. *J. Chromatog.*, **5** : 546, 1961.

(8) NUNES DA COSTA, M.J. and JERONIMO, M.A.S. *Ibid.*, **5** : 456, 1961.

(9) ADLOFF, J.P. *Ibid.*, **5** : 365, 1961.

(10) SASTRI, M.N. and RAO, A.P. *Ibid.*, **9** : 250, 1962.

(11) CABRAL, J.M.P. *Ibid.*, **4** : 86, 1960.

(12) SCHROEDER, H. *Ibid,* **6** : 361, 1961.

(13) ZHU JUN ZHANG, YING BO-HAI, TING ZHEN BANG and SHIH-HIEN SHEN. *Acta Chim. Sinica*, **31** : 218, 1965.

(14) PRASILOVA, J. and SEBESTA, F. *J. Chromatog.*, **14** : 555, 1964.

(15) SHIH NIEN SHEN, ZHU JUN ZHANG, HUEI VEN CHANG. *Acta Chim. Sinica*, **30** : 21, 1964.

(16) PE HAI YIN and YUNG HSIANG CHU. *Hua Hsuch Hsuch Pao*, **32** : 103, 1966.

(17) QURESHI, MOHSIN and QURESHI S.Z. *J. Chromatog.*, **22** : 198, 1966.

(18) QURESHI, MOHSIN, AKHTAR, IQBAL and MATHUR, K.N. *Anal. Chem.*, **39** : 1766, 1967.

(19) QURESHI, MOSHIN and MATHUR, K.N. *Anal. Chim. Acta.*, **41** : 560, 1968.

(20) QURESHI, MOHSIN and ISRAILI, A.H. *Ibid.*, **41** : 523, 1968.

Les particularités de la détermination radiochromatographique des métabolites urinaires de la S^{35} méthionine dans certaines maladies

I. SZANTAY,

(IIIe Clin. Medic. Cluj - Roumanie - Directeur prof. Dr O. Fodor)

Dans nos recherches antérieures (1, 2, 3) nous avons démontré que dans les urines des malades d'hépatite épidémique, d'hépatite chronique et d'ulcère duodénal, après l'administration de la S^{35} méthionine (2 micro Ci/kg corps), une plus grande quantité de S^{35} va être éliminée, par rapport

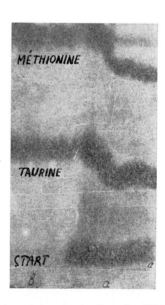

Fig. 1. — Identification des composés de l'extrait S^{35}AA et S^{35}AM obtenu des urines de 24 heures d'un malade d'hépatite chronique active, selon la méthode du « témoin demi-superposé ».

Identification of the constituents of the S^{35}AA and S^{35}AM extract obtained from the 24 hour urines of a patient suffering from active chronic hepatitis, using the « semi-superposed witness » method.

à des valeurs qu'on avait trouvées chez les personnes de contrôle. Les fractions S^{35} amino-acidiques et S^{35} aminique (S^{35} AA et S^{35} AM), extraites des urines de ces malades, présentent également une hausse significative par rapport à la normale (1, 2, 3).

Afin de connaître la composition de l'extrait S^{35} AA et S^{35} AM, celui-ci a été soumis à l'analyse radiochromatographique. A cet effet, de l'extrait de 0,5 ml obtenu sur 20 ml urines on applique une quantité de 0,08 ml sur une bande de papier chromatographique Whatman nr. 1, d'une longueur de 5 cm. L'irrigation du chromatogramme fut effectuée avec une solution de N-butanol, eau, acide formique et alcool étylique 40 : 40 : 10 : 2 (4). Après dessèchement, les chromatogrammes ont été mis en contact avec des films radiographiques Orwo Rapid RF$_2$, de dimensions adéquates. Après un mois, temps d'exposition, les films viennent d'être développés, dans des conditions strictement identiques.

L'identification des fractions sur les radiochromatogrammes a été effectuée selon la méthode des « témoins demi-superposés » (4) (fig. 1).

Ensuite on a procédé au photométrage des composés S^{35} AA et S^{35} AM. Dans la figure 2 nous présentons la courbe d'extinction obtenue des urines de 24 heures chez une personne de contrôle.

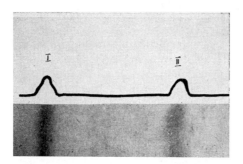

Fig. 2. — Le radiochromatogramme de l'extrait S^{35} AA et S^{35} AM obtenu des urines de 24 heures d'une personne de contrôle.
I. Radiotaurine — II. Radiométhionine.

Radiochromatogram of the S^{35}AA and S^{35}AM extract obtained from the 24 hour urines of a patient under observation.
I. Radiotaurine — II. Radiomethionine

On constate une élimination équilibrée de radiotaurine et de radio-méthionine.

Dans la figure 3 nous présentons un radiochromatogramme obtenu des urines de 24 heures d'un malade d'hépatite épidémique.

Il en ressort que, par rapport à la normale, en ce cas-ci s'élimine une grande quantité de radiotaurine et de radio-méthionine non métabolisée (l'élimination accrue de la radiométhionine est considérée comme un signe de la souffrance hépatique).

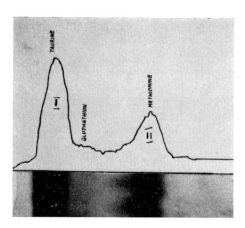

Fig. 3. — Le radiochromatogramme de l'extrait S³⁵ AA et S³⁵ AM obtenu des urines de 24 heures d'un malade d'hépatite épidémique.
I. Radiotaurine — II. Radiométhionine.

Radiochromatogram of the S³⁵AA and S³⁵AM extract obtained from the 24 hour urines of a patient suffering from epidemic hepatitis.
I. Radiotaurine — II. Radiomethionine

Dans la figure 4 on présente un radiochromatogramme obtenu des urines de 24 heures d'un malade l'hépatite chronique active.

On y constate que dans le cas de ce malade l'élimination de la radio-méthionine non-utilisée est également de beaucoup au-dessus du niveau de l'élimination normale, dépassant même celui de l'élimination de la radio-taurine.

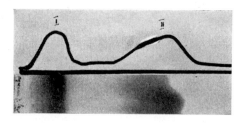

Fig. 4. — Le radiochromatogramme de l'extrait S³⁵ AA et S³⁵ AM obtenu des urines de 24 heures d'un malade d'hépatite chronique active.
I. Radiotaurine — II. Radiométhionine.

Radiochromatogram of the S³⁵AA and S³⁵AM extract obtained from the 24 hour urines of a patient suffering from active chronic hepatitis.
I. Radiotaurine — II. Radiomethionine

Dans la figure 5 on présente deux radiochromatogrammes obtenus des urines de 24 heures des malades d'ulcère duodénal dans la phase active.

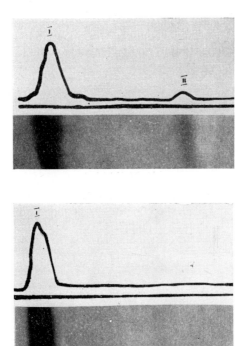

Fig. 5. — Deux radiochromatogrammes obtenus des urines de 24 heures de deux malades d'ulcère duodénal.

I. Radiotaurine — II. Radiométhionine.

Two radiochromatograms obtained from the 24 hour urines of two patients each with a duodenal ulcer.

I. Radiotaurine — II. Radiomethionine

On peut observer que l'élimination urinaire a décerné des fractions S^{35} AA et S^{35} AM en cas d'ulcère duodénal est redevable à l'augmentation de l'élimination de la radiotaurine. Parallèlement, on peut constater aussi la diminution, ou bien l'absence même de la radiométhionine.

Du pourcentage relatif calculé de la courbe d'extinction de la radiométhionine et de la radiotaurine et de la valeur de la fraction S^{35} AA et S^{35} AM, exprimée en pour-cent par rapport à la dose administrée, on arrive à calculer l'élimination de la S^{35} méthionine et de la S^{35}-taurine, exprimée en pour-cent par rapport à la radioactivité administrée.

$$\frac{A.\,B}{100} = C \text{ où :}$$

A = valeur en p. 100 de la radiométhionine, respectivement de la radiotaurine.

B = valeur de l'élimination de la fraction S^{35} AA et S^{35} AM, exprimée en pourcentage par rapport à la dose administrée.

C = valeur de la radiométhionine, respectivement de la radiotaurine, exprimée en pourcentage par rapport à la dose administrée.

Résultats

L'élimination de la radiométhionine, de la radiotaurine, calculée selon la méthode exposée plus haut, chez les malades d'hépatite épidémique, d'hépatite chronique et d'ulcère duodénal est présenté dans le Tableau n° 1.

TABLEAU I

Valeurs de l'élimination urinaire de la radiotaurine et de la radiométhionine chez les malades d'hépatite épidémique, d'hépatite chronique active et d'ulcère duodénal, par rapport aux valeurs normales, exprimées en p. 100 par rapport à la dose administrée.

	Radiotaurine		Radiométhionine	
	% Δ	$\pm$	% Δ	$\pm$
Normal	2,96	1,02	2,31	0,55
Hép. épidémique	8,43	2,12	5,76	1,48
Hép. chronique active	7,68	1,80	8,48	2,11
Ulcère duodénal	7,48	1,90	2,67	0,95

L'analyse radiochromatographique de la fraction S^{35} aminoacidique et S^{35} aminique montre que dans l'hépatite épidémique et l'hépatite chronique active l'augmentation de l'élimination de cette fraction est causée par l'élimination augmentée de la S^{35} méthionine et de la S^{35}-taurine, pendant que dans l'ulcère duodenal, elle n'est redevable qu'à l'élimination élevée de la radiotaurine, associée dans certains cas à une diminution de l'élimination urinaire de la méthionine non-métabolisée.

Bibliographie

(1) SZANTAY I., HOLAN T., FARCASANU M., ABRUDAN RODICA, GIDALI M., ONESCIUC I., prof. GAVRILA I.: Recherches concernant certaines modifications métaboliques de la méthionine dans l'hépatite épidémique. R.I.H. XIV.-155, 1964.

(2) SZANTAY I., HOLAN T., FODOR O., COTUL S.: Quelques aspects biochimiques du métabolisme de la méthionine dans l'ulcère duodénal et gastrique. *Rev. Roum. de Biochim.*, **2**; 67, 1965.

(3) SZANTAY I., COTUL S.: Studiul metabolitilor urinari ai metioninei în hepatita cronică, după administrare de asparagină si acetil-metionină. *Stud. si cercet. de Medic. Internă*, **9**; 35, 1968.

(4) SZANTAY I., K. ALANIA VALERIA, URAY ZOLTAN: S^{35}-tel jelzett metionin anyageseretermékeinek rádiokromatográfiás analizise kapesán nyert tapasztalatok. *Kisérletes orvostudomány*, **15**; 658, 1963.

Etude chromatographique des lipides hépatiques du rat après ingestion de différentes doses d'alcool

par

Prof. E. TURCHETTO, H. WEISS, E. FORMIGGINI et P. BORRI

(Università di Bologna, Italia)

On a étudié les variations des lipides hépatiques du rat après ingestion d'alcool-eau 10 %, 20 %, 30 % (traitement de deux semaines).

En telles conditions les lipides hépatiques changent de 4,5 % (animaux controls) à 7,1 %, 9,8 %, 12,6 %.

Les phospholipides séparés des graisses neutres sur colonne représentent les 59 %, 54 %, 49 %, 44 % des lipides totaux ; les phospholipides calculés comme tissus hépatiques frais montrent une légère augmentation.

Les acides gras des lipides totaux, des phospholipides et des graisses neutres, tendent à augmenter leur saturation.

La lisolecitine et la phosphatidiléthanolamine des phospholipides séparés par T.L.C. augmentent, tandis que la phosphatidilcholine diminue.

Gaschromatographic determination
of ethchlorvynol* in biological material

by

I. SUNSHINE **, R. MAES*** and N. HODNETT **

SUMMARY

A gas chromatographic method for the quantitative determination of ethchlorvynol (Placidyl *) was developed. The method was applied to the biological specimens, obtained from persons who took therapeutic doses of the drug and from two victims of fatal poisonings. The toxicological results obtained by the described technique were compared and discussed with those determined by using Wallace's spectrophotometric method.

INTRODUCTION

Ethchlorvynol (Placidyl ®) is a chlorinated carbinol : ethyl beta-chlorovinyl ethynyl carbinol, with the following structural formula :

According to other tertiary carbinols, ethchlorvynol possesses potent hypnotic and anticonvulsant activities.

Algeri et al. (1) described a colorimetric method for the determination of ethchlorvynol in biological material which is time consuming, not specific and insensitive. Wallace et al. (2) developed a sensitive, specific

---

* Placidyl ,ethyl β-chlorovinyl carbinol) Abbott Laboratories.
** Coroner's Office, Adelbert Road, Cleveland (Ohio) U.S.A.
*** University of Leuven, Pharmac. Institute, Vanevenstraat, Leuven (Belgium).

method which involves the formation of a carbonylderivative of ethchlorvynol, isolation of this carbonyl derivative by steam distillation and its quantitative determination by spectrophotometry.

Kazyak and Knoblock (3) developed a gaschromatographic method for the qualitative detection of ethchlorvynol, but presented no data on the application of this procedure to biological specimens. A gaschromatographic procedure for the qualitative and quantitative determination of Placidyl ® and its application to biological material is the basis of the following research.

EXPERIMENTAL

Biological Specimens

Six human volunteers took 200 mg of ethchlorvynol. Subsequently samples of their blood and urine were obtained at periodic intervals and analyzed on the ethchlorvynol concentration.

Biological specimens were also obtained from two victims of a fatal poisoning due to ethchlorvynol. In one case, the victim, a 51 year old female, was found comatose at home at 8 a.m.

She was admitted to a local hospital and treated with I.V. fluids and penicillin ; suction was applied to clear the bronchial passage. The patient died 20 hours after admission.

Extraction Procedure and gaschromatographic Determination

Homogenize 3 g of tissue with an equal amount of water.

Pipette 5 ml of blood, 5 ml of urine or 5 ml of tissue homogenate into a glass stoppered centrifuge tube or a separatory funnel.

Add 2 ml of ethyl acetate. Stopper and shake gently during 2 minutes. Centrifuge at 2000 r.p.m. for 5 minutes. Remove 0.5 ml of the supernatant fluid into a calibrated centrifuge tube, and slowly evaporate the solvent to 0.2 ml, using a fine jet stream of nitrogen. Rinse the walls of the tube twice with 1 - 2 ml ethylacetate and evaporate to approximately dryness. Add 0.05 ml of a solution of 50 mg of methylpentynol in ethylacetate (= internal standard).

Immediately inject 5 µl aliquots into a gaschromatograph Aerograph 600 C with a flame ionization detector and a 1 mV Minneapolis-Honeywell recorder with a Disc-integrator, and equipped with a 6' x 1/8" o.d. glass column packed with 1 % Hi-Eff 3-BP (Neopentyl Glycol Succinate) on 80/100 Mesh Gas-Chrom Q.

Following chromatographic conditions are found to be optimum :

Temperature : Column   110° C
              Injector port   150° C
              Detector block 130° C

Gas flow rates : Helium (carrier gas)   35 ml/min.
                 Hydrogen   25 ml/min.
                 Air   500 ml/min.

Results and Discussion

The gaschromatographic retention time for methylpentynol and ethchlorvynol was determined to be 3'27" and 3'51" respectively. (Figure 1) The lowest concentration of the drug that could be detected was 0,25 μg/ml.

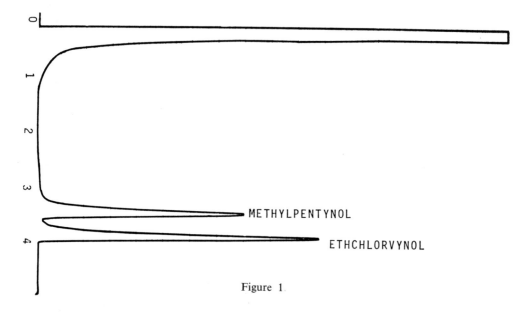

Figure 1

To ascertain the accuracy and the sensitivity of the gaschromatographic procedure, recovery studies were made on aqueous solutions, blood, urine and tissue (Table 1).

The method was furthermore applied to the biological samples obtained from persons who took therapeutic doses of the drug, and from the victim of the fatal poisoning. The concentrations found in these biological specimens are listed in Tables II and III.

When the given procedure was followed, an average of 96 % of the drug was recovered from the aqueous, blood and urine solutions, containing 5 mg and 10 mg of ethchlorvynol per 100 ml respectively. An average recovery of 88 % of the ethchlorvynol added to liver was obtained.

When the ethchlorvynol concentration was 0.25 mg per 100 ml of blood, urine or water, 60 - 70 % of the drug was recovered. This low recovery is probably due to small losses of the compound which are very significant at this low concentration. Corroborating Wallace's report, the highest concentration of the drug in the blood, following therapeutic use of ethchlorvynol, was found to be within 1 - 2 hours after ingestion. In three subjects no ethchlorvynol was detected in the urine during the collection period which was restricted to 8 hours following ingestion. The volume of urine was not recorded in any instance, so no conclusion can be made about the rate of excretion or the percent of the ingested dose that was excreted.

TABLE I

Recovery of ethchlorvynol in biological specimens

Sample	Ethchlorvynol added in MG %	Number of determinations	% Recovery of ethchlorvynol
Aqueous	0.25	4	72 ± 4.2
	1	3	80 ± 4.0
	5	3	94 ± 1.3
	10	3	96 ± 1.3
Urine	0.25	4	61 ± 1.6
	1	3	81 ± 2.3
	5	3	92 ± 2.1
	10	3	100 ± 3.6
Blood	0.25	4	60 ± 2.7
	1	3	79 ± 3.3
	5	3	98 ± 5.3
	10	3	99 ± 5.3
Liver	5	2	90 ± 3.2

TABLE II

Concentration of ethchlorvynol in blood after ingestion of 200 mg

Subject N°	Hours after ingestion			
	1/2	1	2	4
1	0.13 mg %	0.16 mg %	0.09 mg %	0.00
2	0.00	0.15 mg %	0.00	0.00
3	0.00	0.06 mg %	0.15 mg %	0.06 mg %
4	0.16 mg %	0.09 mg %	0.06 mg %	0.00
5	0.00	0.10 mg %	0.00	0.00
6	0.10 mg %	0.18 mg %	0.00	0.00

Concentration of ethchlorvynol in urine after ingestion of 200 mg

Subject Nº	Hours after ingestion		
	0 — 2	2 — 4	4 — 8
1	0.00	0.03 mg %	0.04 mg %
2	0.00	0.00	0.00
3	0.00	0.00	0.00
4	0.03 mg %	0.08 mg %	0.00
5	0.00	0.00	0.00
6	0.00	0.04 mg %	0.00

TABLE III

Concentration of ethchlorvynol in biological specimens obtained
from two victims of a fatal poisoning

Tissue	Concentration of ethchlorvynol	
	Case 1	Case 2
Blood	15.4 mg/100 ml	7.8 mg/100 ml
Urine	1.0 mg/100 ml	0.9 mg/100 ml
Bile	16.0 mg/100 ml	10.3 mg/100 ml
Liver	8.3 mg/100 g	5.8 mg/100 g
Kidney	2.2 mg/100 g	0.2 mg/100 g
Spleen	2.7 mg/100 g	
Brain	4.8 mg/100 g	
Stomach Content	11.3 mg/100 ml	12.7 mg/100 ml

The method described above is applicable to the qualitative and quantitative determination of ethchlorvynol in emergency situations in clinical toxicology, since the complete analysis of a sample can be performed in one hour. This method is slightly more sensitive and faster than Wallace's technique ; but data are comparable.

Acknowledgment : This work was partially supported by a grant 9863-06 from the National Institute of General Medical Sciences (U.S.A.).

References

(1) ALGERI, E.J., KATSAS, G.G., LUONGO, M.A. *Am. J. Clin. Path.,* **38** : 125-130, 1962.
(2) WALLACE, J.E., WILSON, W.J., DAHL, E.V. *J. For. Sci.,* **9** : 342-352, 1964.
(3) KAZYAK, L., KNOBLOCK, E.C. *Anal. Chem.* pp. 1442-1452, 1963.

L'électrophorèse automatique en gradient de saccharose des protéines sériques

Résultats dans une enquête épidémiologique

par

J.R. CLAUDE, J. LELLOUCH, F. CORRE, J.L. RICHARD, E. PATOIS *

L'électrophorèse des protéines sériques fournit en biologie clinique des informations de grande valeur pour établir ou confirmer le diagnostic de différents types d'affections. Sa mise en œuvre systématique au cours d'enquêtes épidémiologiques paraît donc être du plus grand intérêt. Malheureusement, la réalisation en grande série de cet examen pose des problèmes pratiques difficiles, que l'on peut résoudre en accélérant les conditions de migration dans le champ électrique, et en automatisant les opérations. L'électrophorèse automatique en gradient de saccharose, ou Autophorèse ®, mise au point par Skeggs et Hochstrasser (1), utilise à la fois ces deux possibilités ; c'est la raison pour laquelle nous l'avons retenue pour réaliser une étude statistique des fractions protéiques sériques chez des sujets examinés dans le cadre d'une enquête épidémiologique consacrée au dépistage des maladies cardiovasculaires et des maladies par athérosclérose.

Nous rappellerons tout d'abord les caractéristiques techniques et biologiques de la méthode ; nous indiquerons ensuite les résultats qu'elle nous a fournis au cours de l'enquête entreprise.

Matériel et méthodes employés.

1° Nature des sujets ; Examens pratiqués :

On a examiné 358 sujets de plus de 20 ans, de sexe masculin, déterminés par tirage au sort au sein d'une catégorie professionnelle homogène (Gardiens de la paix de la Ville de Paris) ; ils sont tous nés en France et leur âge moyen est de 36 ans ; les prélèvements ont été effectués le matin à jeun.

---

* : Travail du Groupe d'Etudes sur l'Epidémiologie de l'Athérosclérose (GREA), dont la composition est indiquée à la fin de ce travail.

® : Nom déposé par la Compagnie Technicon.

On a mesuré, outre les paramètres protéiques, la cholestérolémie (2), la triglycéridémie (3), le glycémie (4), l'urémie (5), l'uricémie (6), la tension artérielle, le rythme cardiaque, le réflexogramme achilléen, certaines données morphologiques (taille, poids, périmètre thoracique, circonférence iliaque, etc...), divers plis cutanés ; un essai dynamométrique a été réalisé pour chaque bras ; enfin, d'autres éléments, tels que la notion de consommation de tabac sont indiqués par un questionnaire auquel les intéressés ont répondu.

2° *L'électrophorèse en gradient de saccharose des protéines sériques :*

Nous nous limiterons ici à l'énoncé du principe de cette technique et à une description rapide de l'appareillage, les détails pratiques ayant été décrits dans d'autres travaux (Skeggs et Hochstrasser, 1 ; Corre et Claude, 7).

Le principe est le suivant : on établit rapidement dans une cellule de quartz un gradient de densité à l'aide de solutions de saccharose de molarité variable. Le serum sanguin est introduit dans ce milieu et soumis à une différence de potentiel de 100 volts. Le fractionnement est obtenu en 20 minutes en veine liquide, la stabilité de la séparation en diverses zones étant assurée grâce au gradient de densité. Le repérage des fractions protéiques est ensuite effectué par spectrophotométrie dans l'ultra-violet à 280 nm. L'ensemble des opérations est automatisé et programmé.

L'appareillage peut être divisé en deux catégories d'éléments

a) Des éléments du classique Autoanalyzer Technicon :
— Un distributeur d'échantillons.
— Deux pompes proportionnantes assurant l'une le remplissage de la cellule, l'autre sa vidange.
— Un potentiomètre enregistreur muni d'un intégrateur de courbes.

b) Deux éléments propres à l'appareil :
— Un banc optique comprenant la cellule électrophorétique à parois de quartz et le dispositif qui permet le repérage des fractions : pour cela, un rayonnement ultra-violet émis par une lampe à hydrogène est filtré par un monochromateur pour la longueur d'onde choisie (280 nm), puis focalisé en un très mince faisceau qui est transmis par une tête de lecture. Cette dernière balaye verticalement les parois de la cellule de manière à ce que les diverses zones protéiques soient successivement traversées par le faisceau qui, de l'autre côté de la cellule, sera reçu sur un tube photomultiplicateur. Les informations électroniques sont transmises au potentiomètre sur lequel les courbes représentatives de chacune de ces fractions sont directement tracées et leur surface intégrée.
— Un système de programmation commande le déroulement des opérations, qui se succèdent de façon cyclique (remplissage de la cellule par les solutions de saccharose, admission du sérum, passage du courant, lecture, vidange de la cuve, nouveau remplissage, etc...).

Les analyses se déroulent ainsi de façon discontinue, à raison de 3 échantillons par heure, soumis finalement pendant 20 minutes à une tension de 100 volts.

Pour chaque tracé, la surface des pics correspondants aux diverses fractions protéiques sera multipliée par un coefficient de correction approprié fonction de leur richesse en aminoacides aromatiques, responsables de l'absorption à 280 nm. On calcule ensuite les pourcentages relatifs et les valeurs en poids (la protéinémie totale est déterminée par ailleurs).

3° *Valeur de la méthode :*

La séparation obtenue n'est pas tout à fait analogue à celle de l'électrophorèse sur papier, par exemple. On distingue 5 fractions :

— Un bloc albumines $+$ α-1-globulines non séparable.

— Les α-2-globulines.

— Des α-3-globulines de nature très mal connue.

— Les β- et les γ-globulines, très bien séparées, parfois même dédoublées.

On a comparé la valeur de cette technique à l'électrophorèse sur acétate de cellulose gélatineux ou cellogel, également très pratique et très utilisée en biologie clinique. Les points suivants ont été dégagés :

a) La reproductibilité de la mesure est en faveur de l'Autophorèse, surtout pour les globulines.

b) Il n'existe pas de différences statistiquement significatives pour une probabilité inférieure à 5 % entre les moyennes mesurées pour chacune des fractions communes avec les deux méthodes. En d'autres termes, leur correspondance est excellente.

c) La qualité de la séparation obtenue est meilleure pour les fractions les moins rapides avec l'Autophorèse. On a pû aisément ainsi dépister au laboratoire quelques paraprotéinémies, sans aucune équivoque.

Résultats obtenus au cours d'une enquête épidémiologique.

La qualité de l'Autophorèse, le rendement qu'elle permet, nous l'ont fait retenir pour l'exploration biologique de sujets examinés dans le cadre d'une enquête épidémiologique sur les facteurs de risque dans les affections cardiovasculaires et les maladies par athérosclérose. Le propos du présent travail est d'exposer les résultats obtenus sur un échantillon tiré au sort d'une population examinée de façon systématique. Ces résultats serviront ensuite de base pour évaluer dans l'avenir la probabilité de l'assimilation de certains éléments du protéinogramme à de tels facteurs de risque. On a recherché notamment le mode de distribution des fractions protéiques, leur variation avec l'âge, les corrélations existant entre elles ainsi qu'avec d'autres données cliniques, morphologiques et biologiques.

Les calculs statistiques ont été réalisés pour les valeurs exprimées en pourcentage, ainsi que pour les valeurs exprimées en grammes/litre calculées à l'aide du pourcentage et de la protéinémie totale. Les tableaux qui seront présentés mentionneront généralement les résultats obtenus avec ces deux formes d'expression ; cependant, l'expression en grammes/litre sera fréquemment préférée pour la discussion statistique, car elle est moins complexe que l'estimation en pourcentage où intervient la notion des divers rapports des fractions entre elles.

1° *Distribution de la protéinémie et des fractions protéiques :*

Les valeurs, exprimées en g/l, varient pour la protéinémie totale de 58 à 86, avec une moyenne de 69,4 et un écart-type de 3,5 g/l. La distribution est assez régulière, d'allure sensiblement normale (fig. 1). La présence de deux pics distincts pour 69 et 71 g/l est vraisemblablement accidentelle et ne semble pas devoir être interprêtée comme une bimodalité de la courbe.

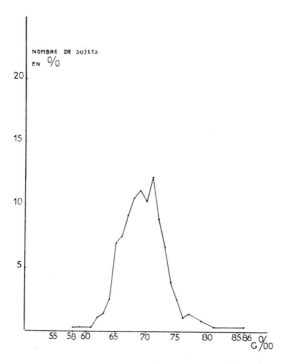

Fig. 1. — Courbe de distribution des protides totaux.

Distribution curve of the total protides.

Pour les fractions séparées par électrophorèse, les valeurs extrêmes, les moyennes et les écarts-types sont présentés au Tableau I, et exprimés en pourcentage et en poids de protéines. L'aspect des courbes de distribution (selon l'évaluation pondérale) (figures 2, 3, 4, 5) est toujours assez régulier, avec néanmoins certaines différences

— On observe une assez bonne symétrie générale pour le bloc albumines-α-1-globulines, les α-3 et les γ-globulines.

— Une tendance à la dissymétrie qui se manifeste pour les α-2 et les β-globulines en faveur des plus fortes valeurs.

TABLEAU I.

Valeurs extrêmes, moyennes et écarts-types des diverses fractions protéiques séparées par Autophorèse dans la population étudiée.

	Alb. + α 1		α 2		α 3		β		γ	
	%	g/l	%	g/l	%	g/l	%	g/l	%	g/l
Valeurs extrêmes	49-73	33-53	3-11	3-8	1-4	1-3	6-14	4-12	9-26	6-21
Moyennes	62,55	43,36	6,31	4,34	2,82	1,90	9,87	6,83	18,44	13,01
Ecarts-types	3,42	2,71	1,33	0,96	0,54	0.47	1,52	1,22	2,79	2,23

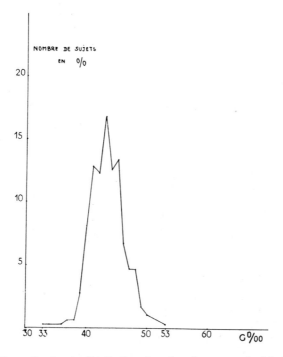

Fig. 2. — Courbe de distribution des albumines + α 1 globulines.

Distribution curve of the albumins and α-1 globulins.

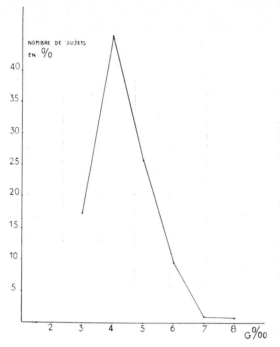

Fig. 3. — Courbe de distribution des α 2 globulines.

Distribution curve of the
α 2 globulins.

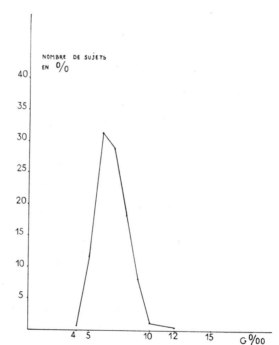

Fig. 4. — Courbe de distribution des β globulines.

Distribution curve of the
β globulins.

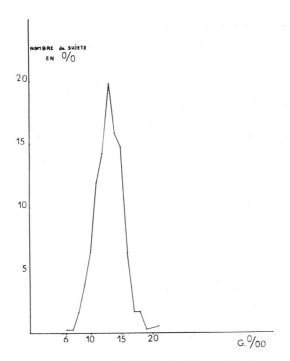

Fig. 5. — Courbe de distribution des γ globulines.

Distribution curve of the γ-globulins.

L'expression en pourcentage des résultats conduit à des courbes comparables. On a calculé également le rapport Alb. $+ \alpha$-1-glob./α-2 $+ \alpha$-3 $+ \beta + \gamma$-glob., qui correspond sensiblement au classique rapport sérine/globuline : il a été trouvé en moyenne à 1,69 avec un écart-type de 0,26.

2° *Variations avec l'âge.*

Alors que la protéinémie totale ne varie pas significativement avec l'âge des sujets (Tableau II), toutes les fractions protéiques y sont liées, sauf les γ-globulines. Ce lien avec l'âge est négatif, assez important et très significatif pour le bloc Alb. $+ \alpha$-1 glob., qui diminuent donc avec le vieillissement ; il est positif, mais d'importance et de signification plus ou moins grandes pour les α-2, α-3 et β-globulines (par ordre croissant d'importance), qui augmentent donc avec le vieillissement. L'expression des résultats en % et en poids donne également des résultats comparables. Le rapport Alb. $+ \alpha$-1 glob./α-2 $+ \alpha$-3 $+ \beta + \gamma$-glob. est aussi corrélé négativement à l'âge, de façon équivalente aux albumines, mais non davantage.

TABLEAU II.

Corrélations des protides totaux et des fractions électrophorétiques avec l'âge.

	Proti-des totaux	Albumine + α 1		α 2		α 3		β		γ		Alb.+α 1 / α2 + α3+ β+γ
	g ‰	g ‰	%	g ‰	%	g ‰	%	g ‰	%	g ‰	%	
Age	—0,03	—0,25	—0,25	0,13	0,13	0,15	0,19	0,21	0,24	0,06	0,07	—0,25
	***	***	*	*	**	***	***	***				***

Ces variations sont particulièrement importantes, car les corrélations avec d'autres données étant susceptibles de passer ainsi par l'âge, il sera par conséquent indispensable de les calculer à âge constant.

3° *Corrélations entre les fractions protéiques :*

Les corrélations entre les fractions exprimées en poids sont portées sur le tableau III. Il est évident qu'une forte corrélation positive existe entre

TABLEAU III.

Corrélations des fractions protéiques entre elles (résultats exprimés en g/l).

	Protides Totaux	Alb. + α 1	α 2	α 3	β	γ
Protides totaux		0,55 ***	0,19 ***	0,14 *	0,48 ***	0,54 ***
Albumines + α 1			— 0,15 **	—0,07	—0,08	—0,22 ***
α 2				0,07	0,13 *	—0,03
α 3					0,08 *	0,02
β						0,25 ***

la quantité de chacune des fractions et la protéinémie ; on relèvera surtout la corrélation négative générale entre le groupe Alb. + α-1 et les autres globulines, très significative notamment pour les γ-globulines ($p<0,001$), ainsi que le lien positif dans presque tous les cas à l'intérieur du groupe des globulines, très marqué entre les β- et γ-glob. ($p<0,001$). L'âge ne modifie pas ces données.

4° *Corrélations entre protéinémie et fractions protéiques et les autres paramètres cliniques, morphologiques et biologiques :*

Les résultats, calculés à âge constant et pour les valeurs exprimées en g/l, sont présentés sur les tableaux IV, V, VI. On retiendra plus spécialement les points suivants :

a) Les corrélations des protéines totales reflètent globalement celles de l'ensemble des fractions qui les composent.

b) Le groupe Alb. + α-1 est corrélé avec l'uricémie et la tension artérielle.

c) Les α-2-globulines sont liées à la tension artérielle mesurée au cours de l'examen clinique, c'est-à-dire au repos.

d) Les α-3-globulines sont reliées au poids, à des données morphologiques qui paraissent reflèter surtout la musculature et l'ossature, et à l'uricémie et la tension artérielle (de façon moins nette).

e) les β-globulines sont aussi corrélées avec le poids et des éléments du morphogramme, qui sont cependant assez différents des précédents, et surtout avec divers plis cutanés. Cet ensemble paraît reflèter davantage un lien avec l'importance du tissu adipeux. Ces globulines sont correlées aussi positivement avec la cholestérolémie et la tension artérielle diasto-lique, négativement avec l'urémie.

f) Très curieusement, les γ-globulines ne sont liées avec aucun paramètre.

La comparaison des tableaux VI et VII démontre l'influence de l'âge. On remarquera par exemple que le lien entre le taux d'albumines et la tension artérielle, inapparent si l'on ne tient pas compte de l'âge, devient significatif lorsque ce dernier est tenu constant. Ceci s'explique par l'existence d'une corrélation de ces deux éléments avec le vieillissement, mais de signe inverse (les albumines diminuent ; la tension artérielle augmente) ; cette évolution masque donc la relation sur l'ensemble de la population.

En outre, certains paramètres ne sont reliés à aucun des éléments du protéinogramme ; c'est le cas de la taille, des essais dynamométriques, du réflexogramme achilléen, de certaines mesures morphologiques, du rythme cardiaque et de la glycémie.

TABLEAU IV.

Corrélations du protéinogramme avec la corpulence, le réflexogramme achilléen, l'essai dynamométrique et la consommation de tabac, évaluées à âge constant.

	Taille	Poids	Poids maximum	Reflex. Achil.	Dynamomètre		Consommation de tabac		
					Droit	Gauche	Cigarette	Pipe + Cigare	Total
Protides totaux	−0,02	0,12*	0,11*	0,08	0,07	0,07	0,003	−0,15	−0,03
Albumines + α 1 glob.	−0,03	0,04	0,04	0,10	0,03	0,03	0,08	−0,14	0,05
α 2	−0,05	−0,01	−0,01	0,03	−0,04	−0,004	0,11	0,27	0,10
α 3	0,05	0,15**	0,14**	0,09	−0,01	0,00	−0,09	0,06	−0,07
β	−0,05	0,16**	0,16**	−0,21	0,01	0,02	−0,004	−0,26	−0,02
γ	0,04	0,02	0,02	0,08	0,08	0,06	−0,11	−0,12	−0,13*
Albumines + α 1 glob.	−0,02	−0,08	−0,08	0,09	−0,04	−0,04	0,09	−0,01	0,10
α 2	−0,03	−0,02	−0,02	0,01	−0,06	−0,03	0,12	0,43*	0,14*
α 3	−0,002	0,10	0,09	0,17	−0,01	0,05	−0,21***	0,04	−0,18**
β	−0,02	0,14**	0,15**	−0,20	−0,01	0,002	−0,02	−0,14	−0,02
γ	0,05	0,01	0,01	−0,06	0,08	0,05	−0,13*	−0,12	−0,14*
Alb. + α 1/ α 2 + α 3 + β + γ glob.	−0,03	−0,09	−0,08	0,05	−0,07	−0,05	0,06	0,02	0,06

Résultats en g/l.

Résultats en %

TABLEAU V.

Corrélations du protéinogramme avec les données du morphogramme, évaluées à âge constant.

	Périm. Thorac.	Hauteur Grand-troc.	Diam. Bitroc.	Diam. Bidelt.	Circonf. Bras	Pli Cut. Bras Ext.	Pli Cut. Bras Post.	Pointe Omoplate	Hauteur Thorax	Circonf. Iliaque	Hauteur Tronc
Protides totaux	0,10	−0,04	0,09	0,01	0,11	0,10	0,10	0,16*	−0,04	0,18**	0,04
Alb. + α1 (Résultats en g/L)	0,10	0,01	0,02	0,03	0,05	0,11	−0,01	0,10	0,07	0,10	−0,03
α2	−0,01	−0,01	−0,10	−0,07	0,04	0,02	0,07	0,09	−0,14*	0,05	−0,08
α3	0,16**	0,03	0,11*	0,16**	0,14*	0,06	0,12	0,12	0,04	0,15*	0,10
β	0,13*	−0,06	0,08	0,06	0,21**	0,19**	0,15*	0,22***	0,01	0,27***	0,09
γ	−0,05	−0,06	0,11*	−0,05	−0,03	−0,13*	0,04	−0,03	−0,08	−0,01	0,07
Alb. + α1 (Résultats en %)	−0,004	0,05	−0,08	0,02	−0,07	0,04	−0,11	−0,06	0,10	−0,07	−0,09
α2	−0,03	0,02	−0,13*	−0,07	0,05	0,04	0,05	0,07	−0,11	0,03	−0,06
α3	0,13*	−0,02	0,07	0,13*	0,16*	0,08	0,12	0,14*	0,09	0,13*	0,10
β	0,10	−0,03	0,06	0,04	0,18**	0,18**	0,15*	0,18**	0,001	0,23***	0,09
γ	−0,06	−0,05	0,11*	−0,03	−0,05	−0,16*	0,03	−0,06	−0,09	−0,05	0,08
Alb. + α1/α2 + α3 + β + γ glob.	−0,02	0,04	−0,09	0,004	−0,06	0,05	−0,10	−0,05	0,08	−0,07	−0,09

TABLEAU VI.

Corrélations du protéinogramme avec divers paramètres biologiques, la tension artérielle et le rythme cardiaque, *évalués à âge constant.*

		Cholest.	Triglyc.	Urémie	Glycémie	Uricémie	Avant prise de Sg.		A l'examen clinique		
							Tension Art. Syst.	Tension Art. Diast.	Tension Art. Syst.	Tension Art. Diast.	Rythme cardiaque
Résultats en g/l.	Protides totaux	0,13*	0,18***	—0,14**	0,10	0,21***	0,18***	0,16**	0,21***	0,19***	0,10
	Alb. + α 1	0,11*	0,12*	—0,06	0,09	0,20***	0,18***	0,13*	0,15**	0,13*	0,10
	α 2	0,06	0,12*	0,05	0,09	0,07	0,08	0,07	0,20***	0,22***	0,11*
	α 3	0,02	0,09	0,05	0,07	0,19***	0,13*	0,16**	0,12*	0,13*	0,01
	β	0,21***	0,07	—0,20***	0,02	0,03	0,09	0,03	0,13*	0,14**	0,03
	γ	—0,07	0,05	—0,09	0,001	0,01	—0,04	0,02	—0,03	—0,06	—0,03
Résultats en %	Alb. + α 1	0,00	—0,03	0,06	0,01	0,01	0,01	—0,02	—0,02	—0,02	0,01
	α 2	0,02	0,07	0,09	0,05	0,01	0,02	0,01	0,15**	0,15**	0,10
	α 3	—0,02	0,02	0,09	0,06	0,20***	0,06	0,13*	0,06	0,12*	—0,03
	β	0,20***	0,02	—0,15**	—0,02	—0,02	0,07	—0,01	0,11*	0,12*	0,01
	γ	—0,12*	—0,02	—0,05	—0,04	0,05	—0,07	—0,006	—0,11*	—0,13*	—0,06
	Alb. + α 1/α 2 + α 3 + β + γ glob.	—0,01	—0,05	0,08	0,004	0,01	0,02	—0,02	—0,03	—0,03	0,03

TABLEAU VII.

Corrélations du protéinogramme avec divers paramètres biologiques, la tension artérielle et le rythme cardiaque. (Il n'est pas tenu compte de l'âge.)

	Cholest.	Triglyc.	Urémie	Glycémie	Uricémie	Avant prise de Sg.		A l'examen clinique		
						Tension Art. Syst.	Tension Art. Diast.	Tension Art. Syst.	Tension Art. Diast.	Rythme cardiaque
Résultats en g/L										
Protides totaux	0,10	0,17**	—0,14***	0,09	0,21***	0,17**	0,15**	0,18***	0,16**	0,10
Alb. + α 1	0,004	0,04	—0,05	0,03	0,17***	0,12*	0,06	0,05	0,03	0,07
α 2	0,11*	0,15**	0,05	0,11*	0,08	0,10	0,10	0,23***	0,25***	0,13*
α 3	0,07	0,12*	0,04	0,09	0,20***	0,16**	0,18**	0,17**	0,17**	0,03
β	0,27***	0,12*	—0,19***	0,06	0,05	0,13*	0,08	0,19***	0,20***	0,05
γ	—0,04	0,07	—0,09	—0,01	0,01	—0,03	0,03	—0,01	—0,03	—0,03
Résultats en %										
Alb. + α 1	—0,10	—0,10	0,06	—0,04	—0,01	—0,04	—0,07	—0,11*	—0,11*	—0,01
α 2	0,07	0,11*	0,09	0,08	0,02	0,05	0,04	0,18***	0,19***	0,11*
α 3	0,06	0,08	0,08	0,10	0,21***	0,09	0,17**	0,13**	0,17***	—0,01
β	0,28***	0,08	—0,15**	0,03	0,01	0,11*	0,05	0,19***	0,19***	0,04
γ	—0,08	0,002	—0,05	—0,02	—0,04	—0,06	0,02	—0,08	0,09	—0,05
Alb. + α 1/α 2 + α 3 + β + γ glob.	—0,10	—0,12*	0,08	—0,04	—0,01	—0,03	—0,07	—0,12*	—0,11*	0,004

Commentaires.

L'étude épidémiologique et statistique du protéinogramme n'a pas été réalisée jusqu'à présent de façon systématique et l'ensemble des résultats précédents pourrait donner lieu à de très nombreux commentaires et pourrait amener à formuler nombre d'hypothèses sur le plan biologique. Nous nous bornerons à évoquer ici les aspects qui nous paraissent les plus notables :

1° Sur le plan technique, l'électrophorèse automatique en gradient de saccharose s'est remarquablement prêtée à notre discipline particulière de travail.

2° La population étudiée montre une distribution globale comparable à celle d'une population sensiblement normale (Girard et coll. 8); on insistera plus particulièrement sur la stabilité de la protéinémie dans les tranches d'âge envisagées, s'opposant aux variations des fractions constituantes : en d'autres termes, le vieillissement provoque ainsi un certain remaniement des fractions protéiques (abaissement des albumines et accroissement de la plupart des globulines), sans retentir sur la somme de celle-ci. Des causes métaboliques doivent être évidemment responsables de ce phénomène. On retiendra cependant le caractère très « neutre » des γ-globulines, qui se comportent comme une véritable variable indépendante des paramètres étudiés. En raison de la potentialité considérable de ces protéines (qui comportent notamment les immunoglobulines), cet aspect peut étonner.

3° L'étude des corrélations existant entre les fractions, tout comme l'évolution avec l'âge, confirme la présence de deux populations différentes de protéines au sein du sérum sanguin sur le plan biologique, comme sur le plan physicochimique d'ailleurs : les albumines, corrélées négativement avec toutes les globulines, et les globulines toutes corrélées positivement entre elles. Le lien important entre β- et γ-globulines n'est pas sans évoquer leur augmentation conjointe dans certaines circonstances pathologiques.

Il n'est pas vraiment sans intérêt de constater que toutes les fractions sont corrélées positivement à la protéinémie : ceci prouve qu'il n'existe aucun élément invariant parmi ces fractions.

Il est bien évident que toutes ces informations sont valables au sein de la population envisagée d'une part, et que d'autre part, elles deviennent erronées pour des sujets atteints par des affections perturbant considérablement le protéinogramme (néphrose lipoïdique par exemple).

4° Les corrélations observées avec les autres paramètres sont parfois surprenantes et certaines sont peu interprétables dans l'état actuel de nos connaissances :

a) La corrélation entre protéines et tension artérielle est fréquente ; elle doit être due à des causes purement physiques.

b) Il est satisfaisant de constater que les β-globulines, qui sont constituées en partie de β-lipoprotéines, sont liées à l'obésité, à la cholestérolémie, aux plis cutanés (pour lesquels intervient la quantité de tissu adipeux). Il est par contre étonnant de ne pas retrouver de liens avec la triglycéridémie, d'autant qu'il en existe entre cette dernière et la protéinémie.

c) Les α-3 globulines, de nature pratiquement inconnue, possèdent des caractères communs avec les β-globulines (liens avec le poids et le morphogramme) ; l'éventualité d'une structure lipoprotéique de cette fraction a donc été envisagée. Malheureusement, l'absence de corrélations comparables à celles des β-globulines, avec les données biologiques (cholestérolémie notamment) ne va pas dans ce sens.

d) La relation entre l'uricémie, la protéinémie, le groupe Alb. + α-1-glob. et les α-3-glob. est assez curieuse. On expliquera difficilement aussi la corrélation négative entre l'urémie, la protéinémie et la quantité de β-globulines.

Conclusion.

Le but du travail qui vient d'être présenté était d'explorer la distribution et les facteurs de variation des éléments du protéinogramme sur une population déterminée. Ces informations étant recueillies, l'intérêt de la détermination systématique de celui-ci, et sa valeur en tant que facteur de risque éventuel, dans une enquête épidémiologique concernant les maladies cardiovasculaires et l'athérosclérose, seront recherchés de deux manières :

— En comparant les résultats présentés ici avec ceux obtenus avec une population dite « à haut risque ».

— En comparant ces mêmes résultats à ceux que l'on obtiendra dans le cadre d'une enquête de morbidité se déroulant sur un nombre d'années suffisant.

C'est la réalisation conjointe d'examens systématiques de dépistage, cliniques et biologiques, liée au traitement des informations recueillies, qui permet sur une échelle suffisamment grande de pratiquer ce type de recherches, où la notion de Groupe interdisciplinaire s'avère indispensable pour les mener à bien.

ANNEXE :

Composition du Groupe d'Etudes sur l'Epidémiologie de l'Athérosclérose (GREA).

Par ordre alphabétique : Dr. G. ANGUERA† ; Dr. A. BARILLON ; Pr. Agr. J.L. BEAUMONT ; Dr. V. BEAUMONT ; Dr. G. BONNAUD ; Dr. BUXDORF ; Pr. Agr. J.R. CLAUDE ; Mme F. CORRE ; Mr. DUCIMETIERRE ; Dr. I. ELGRISHI ; Dr. E. ESCHEWEGE ; Dr. J. GELIN ; Dr. B. JACOTOT ; Dr. H. LAFFONT ; Mr. E. LELLOUCH ; Dr. NAHMANI ; Mlle E. PATOIS ; Dr. J.L. RICHARD ; Pr. D. SCHWARTZ ; Mme TRAN.

Organismes représentés :

Clinique Médicale de l'Hôpital Boucicaut à Paris (Pr. J. LENEGRE).
Direction de l'Hygiène Sociale de la Préfecture de Paris.
Institut National de la Santé et de la Recherche Médicale :
— Groupe de Recherches sur l'Athérosclérose.
— Unité de Recherches Statistiques.
— Section de Cardiologie.
Groupe d'Etudes sur la Fumée de Tabac (GEFT).

Bibliographie

1. SKEGGS, L.T. et HOCHSTRASSER, H. *Ann. N.Y. Acad. Sci.,* **102** : 144, 1962.
2. ETIENNE, G., PAPIN, J.P. et RENAULT, H. *Ann. Biol. Clin.,* **21** : 851, 1963.
3. VAN HANDEL, E. et ZILVERSMIT, D.B. *J. Lab. Clin. Med.,* **50** : 152, 1957.
4. Méthodologie n° 2 : Dosage du Glucose par Autoanalyzer. Cie Technicon, France.
5. Méthodologie n° 1 : Ibid.
6. Méthodologie n° 3 : Ibid.
7. CORRE, F. et CLAUDE, J.R. *Ann. Biol. Clin.,* **25** : 677, 1967.
8. GIRARD, M., DECHOSAL, J. et ROUSSELET, F. Pratique d'Electrophorèse sur papier en biologie clinique, Paris, DOIN éd., 1958.

Purification of beef heart cytochrome C by chromatography on an amberlite XE-64, polyacrylamide bio-gel P-60 and bio-gel P-300 « tandem » column

Eugenia SORU and Karin RUDESCU

Biochemistry and Immunochemistry Department
« Dr. I. Cantacuzino » Institute - Bucharest, Romania

Summary

A new chromatographic procedure for purification of beef heart cytochrome C is described. The method involves as an efficient purification step leading to a high degree of purity chromatography on an Amberlite XE-64, Bio-Gel P-60, Bio-Gel P-300 « tandem » column. Re-chromatography in the same conditions as well as re-chromatography on a Bio-Gel P-300 column gave a Cytochrome C preparation with a still higher degree of purification. The so purified beef heart Cytochrome has ratios A_{550} red/A_{280} = 1.22 — 1.29 ; A_{550} red/A_{550} ox. = 3.69 and shows a single band on polyacrylamide gel electrophoresis and a single peak on chromatography on a Sephadex G-100 column.

$$* \overset{*}{} *$$

A review of the cytochrome C was recently published by Margoliash and Schejter (1) and Palleus (2). Cytochrome C (Cyt. C) has been purified by different chromatographic procedures, (3-9) by molecular sieve chromatography on Sephadex G-75 (10,11) or Sephadex G-100 (12) and recently by ion exchange chromatography on Carboxymethylcellulose (13).

The present paper describes a procedure comprising in the final purification step the chromatography on a column constituted by three different « tandem » layers.

Materials. Cation-exchange resin Amberlite XE-64 (purchased from Serva - Heidelberg) converted into its ammonium form. Bio-Gel P-60 (50-100 mesh) and Bio-Gel P-300 (50-150 mesh) (Bio-Rad Laboratories, purchased from Calbiochem). All chemicals used were of analytical grade.

Methods

Protein was determined by Miller (14) modified Lowry's method with bovine serum albumin (5xcryst.) as standard *nitrogen* by Microkjeldahl. *Iron* according to the sulphosalicylic method (15). *Spectrophotometric* assays were performed in phosphate buffer (pH 6.8) using a Unicam recording spectrophotometer (SP 800 model). Absorbancy was determined at 550 mμ, 535 mμ and 280 mμ (quarz cells of 1 cm light pass). The ratio A_{550} red/A_{280} was followed during the purification procedure. The absorbancy at 550 mμ was measured after solid dithionite was added until complete reduction was obtained. The catalytic capacity of the purified preparation of Cyt. C was assayed in a cytochrome oxydase system according to Stotz (16).

Purification procedure

All procedures were carried out at 4°C.

Step 1. Extraction of Beef Heart Cytochrome C

Fresh beef-hear muscles were trimmed of fat and connective tissue, minced, washed and run through a meat grinder. The mince (3 kg) was stirred with cold water (2400 ml) and adjusted to pH 4.0 with 2.5 NH_2SO_4 (17). After standing for one hour under occasional stirring the pH was adjusted to 6.5 by dropwise addition and under vigorous stirring of 3 N-NH_3. After standing for approximately 18^h the mince was filtered through muslin, re-extracted with cold water (1 kg/300 ml) for one hour and again filtered. The combined filtrate containing the crude Cyt. C was centrifuged and utilized for the further purification of Cyt. C.

Step 2. Adsorption on an Amberlite XE-64 column (NH₄⁺ form) of the crude Cyt. C

Amberlite XE-64 resin (200-400 mesh) was treated and converted for use into its ammonium form as indicated by Lotfield (18). After equilibration with a 50 mM ammonium phosphate buffer (pH 6.4) referred to as « adsorption buffer » the resin was introduced in a chromatographic tube and washed with the same buffer until complete settling (6 cm in high and 3 cm in diameter).

The crude Cyt. C solution was poured on the top of the column and washed in with the « adsorption buffer ». The red Cyt. C solution is retained adsorbed as a solid red band near the top of the resin. The column is washed with the same buffer until the eluates are colourless. The well delimited red resin band containing the adsorbed Cyt. C is carefully removed from the surface of the resin and poured into a small filter glass-fritted funnel. The adsorbed Cyt C is eluted from the resin with the aid of a 401 mM ammonium phosphate buffer pH 7.5. This buffer will be referred to as « desorption buffer ».

Step 3. Dialysis.

The concentrate Cyt C solution is dialysed until ammonium free, against large volumes of bidistilled water. During the dialysis an abundant colourless precipitate is formed and will be removed by centrifugation. A great majority of balast proteins are thus eliminated.

Step 4. Chromatography on an Amberlite XE-64, Bio-Gel P-60,
Bio-Gel P-300 « tandem column »

The column utilized in this purification step is composed by three superposed different layers as follows : a first one at the bottom is an Amberlite XE-64 layer (3x6 cm), the second one superposing the resin layer is a Bio-Gel P-60 layer (3x20 cm), the third one superposing the Bio-Gel P-60 layer is a Bio-Gel P-300 layer (3x3 cm). Every layer before the introduction of the next gel slurry is allowed to settle till no modification of the height could be noted. Care must be taken not to disturb the top surface of the layer during the introduction of the next one. The entire column is equilibrated with the « adsorption buffer ».

TABLE I.

Characterization of beef heart Cytochrome C purified by chromatography on an Amberlite XE-64, Bio-Gel P-60, Bio-Gel P-300 » tandem » column.

Properties	Values
A_{550} red/A_{550} ox.	3.69
A_{550} red/A_{280}	1.24 - 1.29
A_{550} red/A_{535} red	4.00
Fe %	0.46
N %	16.00
Homogeneity	one band in polyacrylamide gel electrophoresis one peak on Sephadex G-100 and Sephadex G-200 respectively Gel filtration

The dialysed Cyt. C preparation (step 3) is carefully layered on the top of the column and washed in with the same buffer. 5 ml effluents were collected (0.5 ml/minute) Cyt. C moved slowly down the Bio-Gel P-300 and respectively Bio-Gel P-60 layer and is retained by adsorption as a homogenous zone on the Amberlite XE-64 resin layer. When the adsorption is completed, the two gel layers are removed from the column, and the remaining resin layer containing the adsorbed Cyt. C is washed with the « adsorption buffer » until the effluents are free of protein. The packed resin thus washed is transferred on a glass filter funnel and the Cyt. C is displaced with the « desorption buffer » (401 mM ammonium phosphate buffer pH 7.5) ; 5 ml fractions were collected (0.5 ml/minute). Each effluent fraction was analysed for protein and spectrophotometrically for A_{550} reduced, A_{280} and A_{535} values. The course of the chromatographic separation is illustrated in figure 1. Re-chromatography in the same conditions of the main peak yield a higher purification of the Cyt. C preparation (Fig. 2).

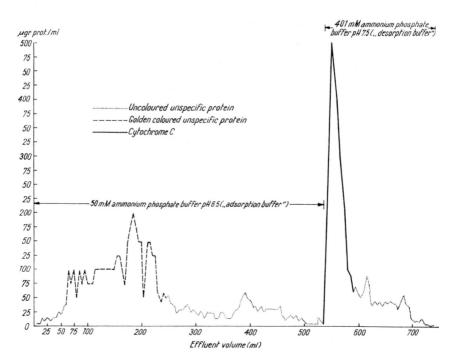

Fig. 1. — The elution pattern when the crude C preparation was subjected to chromatography on an Amberlite XE-64 (3×6 cm), Bio-Gel P-60, (3×20 cm), Bio-Gel P-300 (3×3 cm) « tandem » column equilibrated with the « adsorption buffer » (50 mM ammonium phosphate buffer pH 6.4) developped successively with the same buffer and final elution of Cyt. C from the resin layer with the aid of the « desorption buffer » (401 mM ammonium phosphate buffer pH = 7.5) 5 ml fractions ware collected (0.5 ml/minute).

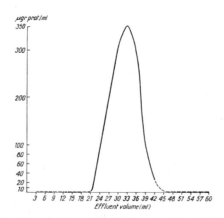

Fig. 2. — Re-chromatography on Bio-Gel P-300 of the Cyt. C purified preparation.
Equilibration and elution with the « adsorption buffer » (50 mM ammonium phosphate pH 6.4).

Step 5. Dialysis and Re-chromatography

The effluents fractions containing the most pure Cyt. C preparation were pooled and submitted to dialysis against a large volume of bidistilled water, any precipitate formed was removed by centrifugation. The clear Cyt. C solution is submitted to re-chromatography on the « tandem » column in identical experimental conditions, or on a Bio-Gel P-300 column equilibrated and eluted with the « adsorption buffer ». The effluents corresponding to the main peak are pooled and dialyzed. For the biological assays on animals the Cytochrome C solution can be sterilized by passing through a bacteriological filter and can be conserved after lyophilization as an intensely red powder.

Results

Figure 1 shows the elution pattern in a typical experiment when the partially purified Cyt. C (step 3) was subjected to chromatography on Amberlite XE-64, Bio-Gel P-60 ; Bio-Gel P-300 « tandem » column as described.

In this special chromatographic condition Cyt. C, because of his low molecular weight, is delayed in his flow through the two gel layers of the column and thus efficiently separated from the great majority of balast proteins having higher molecular weights. These contaminants move ahead of the Cyt. C peak and are eluted when using as development the « adsorption buffer » which in the same time continues to retain adsorbed Cyt. C on the top of the resin layer. The golden coloured impurities, very difficultly removable from the crude Cyt. C preparation (11), as well as the colourless proteins are thus mostly eluted. After elution of these balast proteins the gel layers are removed from the « tandem » column leaving behind only the resin layer containing adsorbed Cyt. C. The resin, thoroughly washed with the « adsorption buffer » till the washings are protein free, retains only the almost pure Cyt. C.

Using as a development the « desorption buffer » Cyt. C is eluted as a nearly symmetrical peak (fig. 1). A small quantity of colourless proteins are eluted behind Cyt. C.

The red coloured effluents are pooled together and dialysed against distilled water. Some uncoloured precipitate formed is removed by centrifugation. The clear intensely red coloured solution of Cyt. C may be conserved as a lyophilized red powder.

For a higher purification re-chromatography on a « tandem » column or even on a Bio-Gel P-300 column can be successfully applied.

Characterization of the purified Cytochrome C

Spectral properties

The ratio A_{550} red$/A_{280} = 1.24 - 1.29$ (Fig. 3), ratio A_{550} red$/A_{550}$ ox. $= 3.69$ ratio comparable to those observed for the monomer form of Cyt. C purified by molecular sieve chromatography on Sephadex G_{75} (11) or Sephadex G-100 (12). *Iron* content : 0.46 % *Nitrogen* content : 16 %.

Electrophoresis on polyacrylamide gel using the slab method (19) and the disc method respectively (20) show a single band for the higher pure preparation (Fig. 4 (c)).

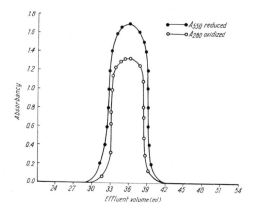

Fig. 3. — Absorbancy at A_{550} red (•———•) and A_{280} (o———o) of the effluents of re-chromatographied Cyt. C on Bio-Gel P-300 column.

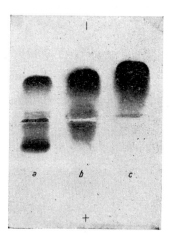

Fig. 4. — Polyacrylamide gel electrophoresis of the crude Cyt. C preparation (a) of the purified on « tandem » column Cyt. C (b) and of the re-chromatographied Cyt. C (c).

As is shown in figure 4 chromatography on a « tandem » column removes from the crude preparation of Cyt. C (Fig. 4 (a)) the great majority of contaminant proteins (Fig. 4 (b)). Re-chromatography removes the last traces of contaminants (fig. 4 (c)).

Chromatography on a Sephadex G-100 column of the pure Cyt. C effected under experimental conditions indicated by Flatmark (11) and Havez et al. (12) as well as chromatography on a Sephadex G-200 column effected under the experimental conditions recommended for molecular weight determination (20) display a single peak.

The catalytic properties controlled in a cytochrome oxydase system (16) are maintained in the Cytochrome C preparation purified with the method described.

References

1. MARGOLIASH, E. and SCHEYTER, A., *Advance Protein Chem.,* **21** : 113, 1966.
2. PALEUS, S., *Bull. Soc. Chem. Biol.,* **49** : 917, 1967.
3. PALEUS, S. and NEILANDS, J.B., *Acta Chem. Scand.,* **4** : 1024, 1950.
4. MARGOLIASH, E., *Nature,* **170** : 1014, 1952.
5. MARGOLIASH, E., *Biochem, J.,* **56** : 529, 1954.
6. HAGIHARA, B., MORIKAWA, J., SEKUZU, I., HORIO, T. and OKUNUKI, R., *Nature,* **178** : 630, 1956.
7. HAGIHARA, B., TAGAWA, K., SEKUZU, I., MORIKAWA and OKUNUKI, K., *J. Biochem* (Tokio) **46** : 11, 1959.
8. YAMANAKA, T., MIZUSHIMA, H., NOZAKI, M., HORIO, T. and OKU-NUKI, K., *J. Biochem.* (Tokio), **46** : 121, 1959.
9. MORRISON, M., HOLLOCHER, T., MURRAY, R., MARINETT, G. and STOTZ, E., *Biochem. Biophys. Acta,* **41** : 334, 1960.
10. PORATH, J., *Biochem. Biophys. Acta,* **39** : 193, 1960.
11. FLATMARK, T., *Acta Chem. Scand.,* **18** : 1517, 1964.
12. HAVEZ, R., HAYEM-LEVY, A., MIZOU, J., BISERTE, G., *Bull. Soc. Chim. Biol,* **48** : 117, 1966.
13. DIXON, H.B.F., THOMPSON, C.M., *Biochem. J.,* **107** : 427, 1968.
14. MILLER, G.L., *Anal. Chem.* **31** : 464, 1959.
15. LORBER, L., *Biochem. Zeitsch.,* **181** : 391, 1927.
16. STOTZ, E., *J. Biol. Chem.,* **131** : 555, 1939.
17. THEORELL, H., *Bioch. Zeitsch.,* **279** : 463, 1935.
18. LOTFIELD, L., Methods in Enzymology, vol. II, p. 752.
19. URIEL, J., *Bull. Soc. Chem. Biol.,* **48** : 969, 1966.
20. TOMBS, M.P., AKROYD, P., Shandon Instrument Applications, n° 18.
21. ANDREWS, P., *Bioch. J.,* **91** : 222, 1964.

A sensitive and non-inhibitory catalytic gel stain for urease

by

W. N. FISHBEIN, M.D., Ph. D.

Chief, Biochemistry Branch
Armed Forces Institute of Pathology, Washington, D.C., USA.

An appreciation of the molecular biology of urease has long suffered from the lack of a sensitive and non-inhibitory catalytic gel stain for this enzyme. A number of investigators in the past have demonstrated multiple molecular forms of this enzyme by ultracentrifugation. However the evidence that these multiple forms are catalytically active has been unconvincing, because the demonstration of multiple forms with Schlieren optics in the ultracentrifuge requires an enzyme concentration some 10,000 fold greater than that used in solution assays of activity. What has been lacking is a specific catalytic stain sensitive enough to demonstrate, in combination with modern gel electrophoretic procedures, the enzymatically active molecular species of urease at concentrations approximating those used in solution assay.

The two critical parameters of a catalytic stain are sensitivity and localization. The gel stains for urease in the past have sacrificed one of these criteria for the other. Sensitivity requires a non-inhibitory stain, and this, in principle, eliminates all of the heavy metal ions which are the common precipitating agents for many histochemical stains ; since urease, due to its large complement of sulfhydryl groups, is extremely sensitive to heavy metals.

The histochemical procedure of Sen (1), using cobalt to precipitate the carbonate product of urea hydrolysis, provided precise localization, but was quite insensitive because urease was markedly inhibited by the cobalt. Similar arguments apply to the use of Nessler reagent to stain urease (2), since the enzyme is markedly inhibited by mercury; and in this case the mercury ammoniate complex does not even form a primary precipitate stain.

The alternate approach has been to take advantage of the alkaline nature of the ureolytic products, as employed by Blattler, Contaxis, and Reithel using a pH indicator to locate areas of alkalinity on the gel (3). This stain is non-inhibitory, but suffers from very poor localization, since the alkalinity diffuses throughout the medium, and the staining reagent itself is diffusible and soluble.

What was needed for urease, therefore, was a non-inhibitory pH indicator which changed color and precipitated above pH 7 to 8. Since such a compound is, to our knowledge, non available, we sought the best of both possible worlds by adapting a tetrazolium salt to fulfill this role. The tetrazolium salts provide a nearly ideal type of biochemical stain because in oxidized form they are yellow, soluble, and non-inhibitory to most enzymes, while upon reduction they become blue and markedly insoluble, thus providing precise localization for dehydrogenase enzymes.

Their value in staining dehydrogenases lies in the fact that they do not reduce non-enzymatically below pH 8. However, we took this latter fact and used it in reverse to develop a tetrazolium stam for urease. Above pH 8 the tetrazolium dyes will accept hydrogens from reducing agents such as thiols. Therefore, in principle, if an enzyme produces an alkaline product, such as urease does, one can use pH to control the reduction of tetrazolium salts by an added thiol, which at the same time serves to protect urease from denaturation. By carrying out the hydrolysis in a mildly acid buffer, alkalinity is localized to the site of urease in the gel, and precise localization is afforded. The principle of the method is shown in figure 1, which describes the sequence of chemical events involved.

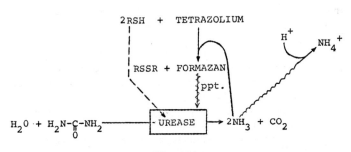

Figure 1.

Verification of the principle was first sought in solution assays, using para-nitro-blue tetrazolium, and dithiothreitol as the mercaptan. The reduction of this tetrazolium salt caused the appearance of a blue formazan which gave a broad absorption plateau centering about 600 nm, which was therefore used for measuring the rate and quantity of reduction. The final absorption level was proportional to the concentration of para-nitro-blue tetrazolium when dithiothreitol was in excess at pH 9. At this pH dithiothreitol reduction of the tetrazolium was quite rapid, and the lowest concentration of thiol producing maximal reduction was easily determined by varying the thiol concentration. As anticipated, the reaction became much slower as the pH was lowered, until at pH 6 reduction was negligible after 24 hours.

Having thus demonstrated by solution assay that pH can be used to control tetrazolium reduction, the next step was to determine the best conditions for the catalytic staining of gel-bound urease. Solutions of commercial urease were prepared in 10 % gelatin and streaked on glass microscope slides which were then immersed in variously compounded staining media.

A corresponding streak of 10 % gelatin without added urease was used as a background control. The staining of the urease-impregnated and control gelatin were graded visually from 0 to 4+, and were verified by densitometry. The results indicated that the best discrimination was provided by an 0.035 citrate or maleate buffer at pH 6.0. Phosphate buffer was inferior because it tended to promote a strong background stain, and Tris-maleate buffer was also unsatisfactory.

The tentative procedure was then applied to urease samples electrophoresed on acrylamide gels, where it was also successful, requiring only a few minor changes in concentration of the mercaptan and the tetrazolium salt to achieve the best sensitivity with the least background.

Table 1.

Urease stain solution.

38.0 ml Distilled water.
3.6 ml 0.5 M Citrate Buffer, pH 6.0.
5.2 ml M Urea.
5.0 ml 0.5 % p-Nitro blue tetrazolium.
14.0 ml Dithiothreitol (add last, with stirring).

Procedure.

(1) Immerse gel in 0.05 M citrate, pH 6.0 and rock gently for 1-2 hours.
(2) Drain buffer off and pour urease stain over gel (*no* water wash).
(3) Incubate gel with stain at 25° (*no* rocking).
(4) After staining, drain off stain solution, cover gel with N HCl (*no* water wash) and rock 1-2 hours.
(5) Gel may be preserved indefinitely in air tight cellophane wrap, and photographed by transillumination.

The components of the catalytic stain in its final form, are shown in Table 1. Since electrophoresis on acrylamide gel is generally carried out under alkaline conditions, the gel is first equilibrated for 1-2 hours with 0.05 M citrate buffer, pH 6. The buffer is then drained off and the staining solution poured over the gel, without any intermediate water wash. The catalytic reaction is stopped by pouring off the staining solution and covering the gel with N HCl, after which the gel may stored for several months in cellophane wrap without any fading of the stain. The gel expands slightly in N HCl but remains transparent and pliable, and may be photographed easily by transmitted light.

At a concentration of 0.1 Sumner unit per cm of slot, clear staining begins within one hour, and over-staining occurs by 3-4 hours. A good quality tetrazolium dye is essential to keep background staining at a minimum. We have found the product manufactured by Cyclo Chemical Co, (Los Angeles, Cal., U.S.A.) to be the best of those we have tried so far. Using this dye, gels can be left in the stain for 4-5 days at room temperature without significant background developing. One can gain a 4-fold increase in the staining rate, although at some sacrifice of background clarity, by doubling the quantity of both the tetrazolium dye and the dithiothreitol in the staining solution.

The specificity of the stain for urease is demonstrated by the absence of staining after 24 hours of incubation when (a) urea is omitted (b) protein denaturants are added (c) the specific urease inhibitor acetohydro amate is added and (d) when the gel slot contains 0.7 mg of serum proteins instead of 0.002 mg urease.

Dithiothreitol may be replaced by 2-mercaptoethanol at 20 times the molar concentration, but this tends to give a heavier background. Para-nitro-blue tetrazolium may be replaced by tetrazolium blue, neo-tetrazolium chloride, or triphenyl-tetrazolium chloride, where economy is essential. These dyes take much longer to produce a perceptible stain, and provide less background clarity over long incubation intervals. The amido-black stain for proteins requires about 50-100 times as much urease as the catalytic stain in order to obtain dark bands.

We have now used the catalytic stain on more than 300 preparations of urease, and it has dramatically altered our appreciation of the molecular complexity of this enzyme. Only a summary of the findings can be presented here : more than a dozen urease isozymes have been demonstrated repeatedly, and these can be grouped into three main categories : (1) a polymeric series of six isozymes named alpha through zeta, with an unresolved ω-component remaining at the slot origin. The fastest-migrating member of this series, α-urease, is the dominant single form in most urease preparations, and has an estimated molecular weight of 450,000 which identifies it as the « traditional » urease. (2) Alpha and β sub-band izozymes have been demonstrated which are characteristic of the particular source of jackbean meal used. Their identity is maintained throughout purification procedures, suggesting that they may be genetically determined. (3) Additional α-sub-bands occur corresponding with successive re-crystallizations, and correlate with marked changes in specific activity. These isozymes have no detectable molecular weight difference, and may be the first functionally significant conformeric isozymes.

The lowest molecular weight form of active urease so far identified is 220,000 and no simple integral series can be fitted to the estimated molecular weights of the various species present.

It is believed that the variability and complexity of urease catalysis noted in the past has been due to the selection and interaction of the particular isozymes with which each investigator was working.

References

1. SEN, P.B., A method of locating urease within tissue by a mecrochemical method. *Indian J. Med. Res.*, **18** : 79, 1930.
2. NIKOLOFF, T., Elektrophoretische Untersuchung der Magen urease. *Z. physiol. Chem.*, **314** : 1, 1959.
3. BLATTLER, D.P. CONTAXIS, C.C. and REITHEL, F.S, Dissociation of urease by glycol and glycerol. *Nature*, **216** : 274, 1967.
4. FISHBEIN, W.N., The structural basis for the catalytic complexity of urease : interacting and interconvertible molecular species. *Ann. N.Y. Acad. Sci.*, (in press).

Separation of hydrolytically active components of cellulase from *Myrothecium verrucaria* by starch gel electrophoresis

by

F.J. RITTER, P.Y.F. PRINS-van der MEULEN and T. van der MAREL

Summary

Using starch gel electrophoresis according to Smithies, desalted crude cellulase from *Myrothecium verrucaria* was separated into at least 12 protein zones. These were tested on their activity towards p-nitrophenyl-β-D-glucoside, sodium carboxymethylcellulose and α-cellulose. They were all hydrolytically active.

When the separate proteins were subjected to re-electrophoresis, they appeared at the same position as that, at which they appeared upon electrophoresis of the crude preparation. In some of the protein zones reducing sugars were found. The results obtained are indicative of a multiple nature of the cellulase from *M. verrucaria,* but are not contradictory to the conception that one, or a limited number of proteins, would be associated with other substances, thus yielding conjugated proteins with different physical and enzymatic properties.

INTRODUCTION

Several investigators, working with a number of different cellulolytically active micro-organisms, including *M. verrucaria,* have in the past obtained cellulase preparations. These mostly seemed to contain a number of different cellulases (1 - 15). Sometimes, however, they were found to contain only one cellulolytic component (16 - 22).

The separation techniques applied, very often included column chromatography and gel filtration (1, 3, 5, 9, 10, 12, 15), but also electrophoresis on starch blocks (6, 7, 8), paper (22, 23), Pevikon powder (2) or starch gel (18) and, recently, also iso-electric focusing (24, 25).

The starch gel electrophoresis of cellulase from *M. verrucaria,* described by Whitaker et al. (18), was simultaneously and independently applied in our laboratory. Our results, however, indicate a multiple nature of this cellulase, whereas Whitaker found no major indications of heterogeneity. Two main differences between our and Whitaker's experimental conditions are the amounts of crude enzyme applied in the electrophoresis (about ten times as much in our experiments) and the character of the buffers used. Moreover, Whitaker applied a number of other purification methods before carrying out the final electrophoresis.

In the following description of our experiments it will be shown that starch gel electrophoresis of crude cellulase from *M. verrucaria* under our experimental conditions gives a great number of cellulolytically active protein zones. Moreover, upon electrophoresis of cellulase fractions, obtained after chromatography of the crude cellulase over Sephadex columns, all protein zones are found again.

DESCRIPTION OF ELECTROPHORETIC TECHNIQUE

The experimental arrangement was essentially the same as the vertical starch gel electrophoresis developed by Smithies (26). The gel, prepared from hydrolysed starch (Connaught) was 255 mm long, 120 mm wide and 6 mm thick, with on each end a thicker part, where the connections with the electrode vessels were made. At a distance of 75 mm from the upper side, 7 slots of each 12 x 1 x 5 mm were present. In general 0.05 ml of a 10 % cellulase solution was applied to these slots. Albumin, when used for comparison, was used as a 2 % solution (0.05 ml). The gel surface was coated with paraffin. The lower side was placed on some layers of Whatman 3 MM paper, in one compartment of a divided electrode vessel, a platinum electrode being placed in the other compartment. Connections were made by cotton wool. On the upper side, the connection between gel and electrode compartment was obtained by a triple layer of Whatman 3 MM paper.

An electric field was applied of 2.5 to 5 Volts/cm, depending on the buffer used. The time of electrophoresis was 16 - 18 hours at 4° C. After termination, the paraffin layer was removed and the gel was cut overlength in two parts, each 3 mm thick, using a strained steel thread. One part was coloured with Amido-black to determine the position of the protein zones. During this procedure some shrinkage occurs. To establish the position of the protein zones in the uncoloured part, the position of the zones on the coloured part was marked onto elastic band, which was then stretched to the size of the untreated part, and the zones were cut out according to the positions indicated on the band.

PRODUCTION OF CELLULASE AND ASSAYS OF
ENZYMATIC ACTIVITY

Details on the methods employed for the production of cellulase and on the enzymatic determinations are published elsewhere (27). The filtrates of the submerged cultures, obtained after 7 days at 26° C on a rotatory shaker, were concentrated about 10-fold. They were desalted over a column of Sephadex G-25 and freeze-dried.

Enzymatic activities were determined towards :

(a) *sodium carboxymethyl cellulose* (NaCMC) ; by definition (10) an enzyme solution contains 1 unit of cellulase activity per ml, when 1 ml of it, incubated with 5 ml NaCMC (0.7 %) produces 0.4 mg glucose equivalent per ml, the reducing sugars being determined by dinitrosalicylic acid ; culture filtrates usually contained about 10 units of cellulase activity per ml.

(b) *α-cellulose ;* the mixture of 1 ml enzyme solution and 5 ml of a 1 % suspension of α-cellulose is, after incubation, centrifuged and the

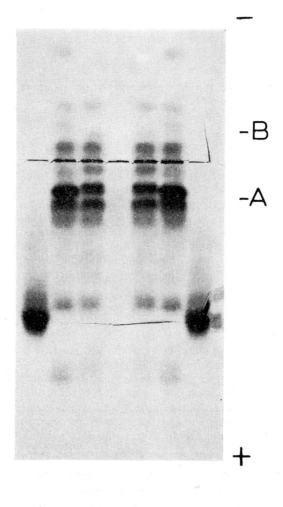

Fig. 1. — Starch gel electrophoresis of various cellulase preparations in Tris-citric acid buffer pH 7.

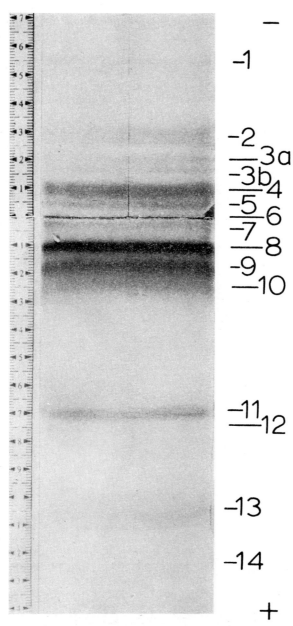

Fig. 2. — Starch gel electrophoresis of 30 mg cellulase
in Tris-citric acid buffer pH 7.0.

reducing sugars are determined according to Somogyi ; unfortunately this very sensitive method cannot be applied to NaCMC, due to formation of a precipitate.

(c) *niphegluc (p-nitrophenyl-β-D-glucoside)* ; the mixture of 1 ml enzyme solution and 0.5 ml niphegluc is incubated and the amount of p-nitrophenol formed is read from a calibration curve.

RESULTS

The best separations were obtained when Tris-citric acid buffers, having a pH between 6.9 and 7.2 and a molarity of 0.02 (in the gel) were used. The buffer around the electrodes was of the same pH, the molarity, however, was 0.05.

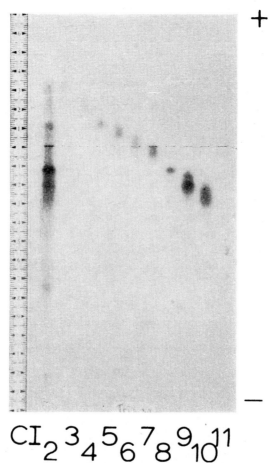

Fig. 3. — Re-electrophoresis in Tris-citric acid buffer pH 7.0 of proteins eluted from zones cut out of a semi-preparative electrophoresis of cellulase in the same Tris-citric acid buffer.

Several cellulase preparations were tested. They showed a similar overall picture but were not completely identical. Figure 1 shows the results for the batches C-I and C-II. Zone A, the first strong band at the (+) side, for example, was much weaker in C-II and the same is true for zone B the second weak zone at the (—) side. The zone of C-I moving fastest to the (—) side and that moving fastest to the (+) side, were absent or much weaker in C-II. A third batch of cellulase, not shown in the figure, gave a pattern which was almost identical to that of C-II.

To investigate the possibility that some of the proteins in the zones were artefacts, originating from proteins in other zones, and to check the reproducibility of the electrophoretic behaviour of these proteins, the proteins were subjected to re-electrophoresis. Therefore, a larger amount of C-I (30 mg in 0.3 ml) was pipetted into a slot of 70 x 1 x 5 mm. The electrophoresis in Tris-citric acid buffer and the establishment of the positions of the protein zones was carried out as described before. The result is shown in figure 2. The zones were cut out, eluted by repeated freezing and defrosting according to Smithies (26) and centrifuged. The gel was washed out three times with buffer and the fractions were concentrated by means of the apparatus of the Membranfilter Gesellschaft in collodion hulls, to a volume of 0.05 to 0.1 ml. Of each fraction 0.02 ml was used for gel electrophoresis. Eleven very small slots (5 x 0.75 x 5 mm) were now used simultaneously in one experiment.

In the Tris-citric acid buffer the proteins were found at their original position (Fig. 3); only some contamination with proteins from adjacent zones was observed.

Upon electrophoresis in borate buffer (pH 8.5), the same order of positions from negative to positive pole was found, but several of the fractions were found to split up into two or three zones. In acetic acid-acetate buffer (pH 4), however, the relative positions were completely different; e.g. fraction 8 ran faster than fraction 4 (Fig. 4).

It was, of course, of interest to find out which of the protein zones contained cellulolytic enzymes. Therefore the preparative electrophoresis of 30 mg cellulase C-I in Tris-citric acid buffer was repeated twice. The zones were again cut out and eluted as described before, but, in order not to dilute the fractions, the gel was not washed. In the combined eluates of each pair of corresponding zones the enzymatic activities towards NaCMC, α-cellulose and niphegluc were determined as well as the amount of protein according to the Folin procedure. The possibility had to be taken into account that some of the extracts contained sugars or other reducing groups, even before the incubation with the enzyme (which would influence the determinations of the hydrolytic activity towards NaCMC and α-cellulose) and also, that they could contain light-absorbing components (this would influence the determination of the activity towards niphegluc). Therefore in each enzymatic determination, when a certain amount of extract was mixed with the substrate, the same amount was also mixed with distilled water and treated identically to the incubated samples. The results are shown in Tables IA-ID. The number of the zones correspond with those of figure 2.

The extinctions of the blank $E_{bl.}$ were substracted from those of the incubated samples $E_{inc.}$ to give $E_{corr.}$.

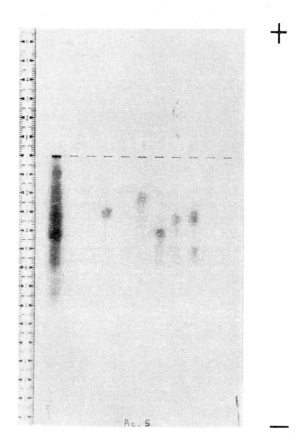

Fig. 4. — Re-electrophoresis in acetic acid-acetate buffer pH 4.0 of proteins eluted from zones cut out of a semi preparative electrophoresis of cellulase in Tris-citric acid buffer pH 7.0.

The blanks of Tables IB and IC are within the experimental error, but in the determinations of the hydrolytic activity towards α-cellulose (Table ID) appreciable $E_{bl.}$ values were found. In this series the reducing sugars were determined by the Somogyi method, which, as mentioned before, is much more sensitive than the DNS method, which was applied for the activity determinations towards NaCMC.

In fractions 6 and 7 the highest amount of reducing sugars were found. In fraction 6 the amount was equivalent to 0.24 mg in 0.05 ml. In the determination of the activity towards NaCMC only 0.01 ml of this fraction

TABLE I A

Protein determination (Folin)

Zone	ml	E
1	0.2	0.148
2	0.2	0.223
3a	0.2	0.244
3b	0.2	0.274
4	0.2	0.519
5	0.1	0.272
6	0.1	0.375
7	0.2	0.594
8	0.2	0.794
9	0.2	0.575
10	0.2	0.332
11	0.2	0.172
12	0.2	0.177
13	0.2	0.157
14 is blank	0.19	0.164

TABLE I B

Hydrolytic activity towards Na-CMC

Zone	ml	$E_{inc.}$	$E_{bl.}$	$E_{corr.}$	mg equ. glucose in reaction mixture (1.1 ml)
1	0.1	0.019	—0.010	0.029	0.09
2	0.1	0.715	—0.010	0.725	0.47^5
3a	0.05	0.349	—0.016	0.365	0.27
3b	0.1	1.550	—0.006	1.556	0.90
4	0.01	0.998	—0.005	1.003	0.63
5	0.01	0.580	—0.006	0.586	0.40
6	0.01	0.685	0.000	0.685	0.45
7	0.01	0.870	0.000	0.870	0.55^5
8	0.01	0.248	—0.012	0.260	0.21
9	0.02	0.118	—0.015	0.133	0.14
10	0.1	0.695	—0.005	0.700	0.46
11	0.1	0.000	—0.010	0.010	<0.08
12	0.1	—0.012	0.015	0.003	<0.08
13	0.1	0.065	—0.005	0.070	0.11
14 is blank	0.1	0.007	0.000	0.007	<0.08

TABLE I C
Hydrolytic activity towards niphegluc

Zone	ml	$E_{inc.}$	$E_{bl.}$	$E_{corr.}$	m. mol. p. nitrophenol formed (5 ml)
1	0.2	0.004	0.001	0.003	—
2	0.2	0.000	0.000	0.000	—
3a	0.2	0.010	—	—	—
3b	0.2	0.028	0.000	0.028	0.005
4	0.2	0.500	0.000	0.500	0.110
5	0.2	0.714	0.000	0.714	0.157⁵
6	0.2	0.650	0.003	0.647	0.142⁵
7	0.2	0.345	—0.001	0.346	0.075
8	0.2	0.055	0.002	0.053	0.012⁵
9	0.2	0.010	0.000	0.010	—
10	0.2	0.012	0.000	0.012	—
11	0.2	0.008	—0.005	0.013	—
12	0.2	0.010	—0.005	0.015	—
13	0.2	0.005	0.000	0.005	—
14 is blank	0.2	0.000	0.000	0.000	—

TABLE I D
Hydrolytic activity towards α-cellulose

Zone	ml	$E_{inc.}$	$E_{bl.}$	$E_{corr.}$	mg equ. glucose in reaction mixture (6 ml)
1	0.5	0.192	0.142	0.050	0.018
2	0.5	0.350	0.138	0.212	0.078
3a	0.3	0.385	0.216	0.169	0.060
3b	0.3	0.661	0.280	0.381	0.141
4	0.05	0.799	0.159	0.640	0.234
5	0.05	0.829	0.313	0.516	0.189
6	0.05	1.18	0.662	0.518	0.189
7	0.05	1.09	0.595	0.495	0.180
8	0.05	0.830	0.047	0.783	0.285
9	0.05	0.476	0.036	0.440	0.162
10	0.3	0.901	0.188	0.713	0.261
11	0.5	0.425	0.127	0.298	0.108
12	0.45	0.278	0.134	0.144	0.054
13	0.5	0.238	0.184	0.054	0.021
14 is blank	0.3	0.164	0.122	0.042	0.015

was available, which would contain about 0.24/5 mg glucose, which is not detectable by the DNS method.

It would be quite conceivable that some reducing sugars can be eluted from starch gel after several times of freezing and thawing. This was tested with a blank gel, subjected to the usual electrophoresis. The amount of reducing sugars present in the protein zones 4, 5, 6 and 7 was, however, far higher than that found in any of the eluates of the blank gel. This suggests, that in these zones, reducing sugars are associated with the proteins. The low $E_{bl.}$ values of zones 8 and 9 in Table ID, can, however, easily be attributed to sugars eluted from the gel. These two most heavy bands, therefore, appear not to contain any reducing sugars.

Fractions 6 and 7 were hydrolysed with dilute sulphuric acid and were analysed chromatographically with respect to the possible presence of specific reducing sugars. The results were compared with those obtained for comparable eluates of blank gel (Fig. 5).

The hydrolysate of the eluate from the blank gel (g) showed a number of spots giving the normal sugar reaction with anisaldehyde. The most prominent of them was probably glucose. These spots were also present in

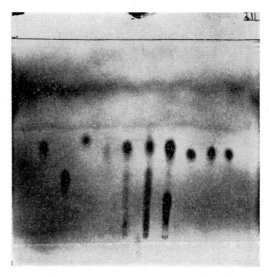

a b c d e f g h i j

Fig. 5. — Thin-layer chromatogram of hydrolysates of eluates from zones 6 and 7 and from a blank gel in comparison with some sugars.
Adsorbent : Silicagel H.
Solvent : Butanol-acetone-water (4+5+4).
Detection : Anisaldehyde reagent.
a = glucose, b = fructose, c = rhamnose, d = xylose, e = hydrolysate zone 6, f = hydrolysate zone 7, g = hydrolysate blank gel, h = galactose, i = mannose, j = maltose.

the hydrolysates (e and f) of the zones 6 and 7, but these contained at least two spots which were not observed in the blank. The identity of these spots could not be established with certainty, but they were evidently not glucose. Therefore, zones 6 and 7 may indeed contain characteristic glycoproteins, that are responsible for the high blank values in the determination of the activity towards α-cellulose.

From the $E_{corr.}$ values of the Tables IA - ID it appears that a hydrolytic activity is associated with most of the zones. Only for zone 1 the activity is doubtful.

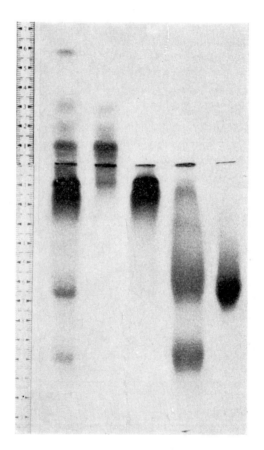

C-I   1   2   3 alb   +

Fig. 6. — Starch gel electrophoresis in Tris-citric acid buffer pH 7.0 of protein fractions obtained on chromatography of cellulase over a column of DEAE-Sephadex A 50.

It appears that the ratio of the activities towards the three substrates is different for different zones. This is most obvious when comparing fractions 4 to 8, where the same amounts of the fractions are taken for the measurements of the activities. According to Table ID for example, fraction 8 has the highest activity towards α-cellulose, whereas, according to tables IB and IC, it has a much smaller activity towards niphegluc and NaCMC than fraction 4. This fraction has the highest activity towards NaCMC, whereas fraction 5 is most active towards niphegluc. This suggests that enzymes with different enzymatic characteristics are present in the various zones.

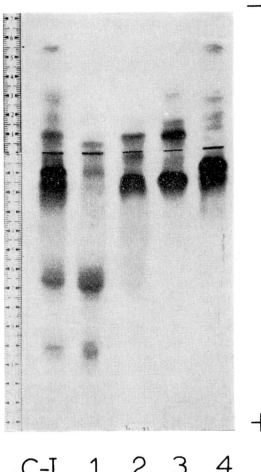

Fig. 7. — Starch gel electrophoresis in Tris-citric acid buffer pH 7.0 of protein fractions obtained on chromatography of cellulase over a column of CM-Sephadex.

The cellulase batch C-I has also been chromatographed in our laboratory over a column of DEAE-Sephadex, using a phosphate buffer pH 8 and 0.01 M. Part of the enzyme was not absorbed (fraction 1). The absorbed enzyme was eluted, using a linear gradient from 0 up to 0.5 M NaCl in the same buffer. The eluate was collected into the fractions 2 and 3. All three fractions contained enzymatic activity towards the three substrates ; only in fraction 3 very little activity towards niphegluc was present. These fractions were subjected to the usual starch gel electrophoresis in Tris-citric acid buffer.

The same cellulase batch was also chromatographed over CM-Sephadex C-50 medium, in phosphate-citrate buffer pH 5.0 and 0.01 M. The absorbed enzyme was eluted with a linear gradient from 0 up to 0.3 M NaCl in the same buffer. Four fractions were obtained, the first one representing unabsorbed enzyme. These fractions too were subjected to the same kind of electrophoresis. The results of the two separations are shown in figure 6 and 7. It appears that all the fractions found in the original cellulase preparations are found again in the fractions obtained from the columns. Some proteins which were not absorbed on CM-Sephadex (Fig. 7, fraction 1), were strongest absorbed on DEAE-Sephadex (Fig. 6, fraction 3) and vice versa. Fraction 3 of figure 6 showed little activity towards niphegluc. This corresponds with the data of table IC, showing that the proteins of the electrophoretic zones 8-13 showed hardly any activity towards niphegluc.

DISCUSSION

When crude cellulase from *M. verrucaria* is subjected to starch gel electrophoresis, either as such or after fractionation on substituted Sephadex columns, a number of separate zones are found, which all contain hydrolytic activity. Some are more active towards α-cellulose, others towards NaCMC or niphegluc. It would be conceivable that some of the enzymes in these zones contain the same protein moiety, but are associated with various other substances (e.g. sugars). Especially the recent communications of Eriksson, Pettersson and Björndal (28, 29) would suggest such a possibility. They found for the fungus Stereum sanguinolentum two chromatographically separated cellulase peaks. After dialysing to zero carbohydrate content, however, the mixed enzymes gave only one homogeneous cellulase peak, upon rechromatography as well as on electrophoresis. Their results suggest, that combination of a cellulase with a carbohydrate can change the properties of the enzyme considerably. The data we obtained in the determination of the reducing groups in some of the eluted electrophoretic zones would be in agreement with such a conception and the results of thin-layer chromatography of hydrolysates of two fractions support the idea that some of the proteins are associated with sugars. Whether the protein moieties in some or all of the hydrolytically active zones are identical or not, is a question that cannot be answered definitely from our data.

There seems to be no doubt, however, that upon electrophoresis of crude cellulase from *M. verrucaria* a large number of hydrolytically active substances can be separated from each other. Some of them may be simple proteins and others are probably conjugated with sugars.

These conclusions may seem to be contradictory to those of Whitaker (18), who also applied gel electrophoresis. However, he carried out a number of purification methods before the electrophoresis and also worked under slightly different conditions.

ACKNOWLEDGEMENT

The cellulase fractions obtained from column chromatography on substituted Sephadex were provided by Ir. J.J. Kannegieter of our laboratory. The thin-layer chromatography work was done by Miss G.M. Meijer.

References

(1) PETTERSSON, G., COWLING, E.B. and PORATH, J. *Biochim. Biophys. Acta*, **67**: 1, 1963.
(2) PETTERSSON, G. and PORATH, J. *Biochim. Biophys. Acta*, **67**: 9, 1963.
(3) LI, L.H. and KING, K.W. *Appl. Microbiol.*, **11**: 320, 1963.
(4) KING, K.W. and SMIBERT, R.M. *Appl. Microbiol.*, **11**: 315, 1963.
(5) LI, L.H., FLORA, R.M. and KING, K.W. *Arch. Biochem. Biophys.*, **111**: 439, 1965.
(6) MILLER, G.L., BLUM, R. and HAMILTON, N.F. *J. Chromatog.*, **3**: 576, 1960.
(7) MILLER, G.L. and BIRZGALIS, R. *J. Chromatog.*, **7**: 33, 1962.
(8) MILLER, G.L. and BIRZGALIS, R. *Arch. Biochem. Biophys.*, **95**: 19, 1961.
(9) SELBY, K. and MAITLAND, C.C. *Biochem. J.*, **94**: 578, 1965.
(10) PRINS-van der MEULEN, P.Y.F. and SCHURINGA, G.J. *Nature*, **187**: 695, 1960.
(11) REINOUTS VAN HAGA, P. *Nature*, **182**: 1232, 1958.
(12) SELBY, K. and MAITLAND, C.C. *Arch. Biochem. Biophys.*, **118**: 254, 1967.
(13) TOYAMA, N. and OGAWA, K. *J. Ferment. Technol.* (Japan), **44**: 741, 1966.
(14) SELBY, K. and MAITLAND, C.C. *Biochem. J.*, **104**: 716, 1967.
(15) PETTERSSON, G. *Biochim. Biophys. Acta*, **77**: 665, 1963.
(16) WHITAKER, D.R., *Arch. Biochem. Biophys.*, **43**: 253, 1953.
(17) WHITAKER, D.R. and THOMAS, R. *Can. J. Biochem. and Physiol.*, **41**: 667, 1963.
(18) WHITAKER, D.R., HANSON, K.R. and DATTA, P.K. *Can. J. Biochem. and Physiol.*, **41**: 671, 1963. 1968.
(19) PETTERSSON, G. and EAKER, D.L. *Arch. Biochem. Biophys.*, **124**: 154,
(20) PETTERSSON, G. *Arch. Biochem. Biophys.*, **123**: 307, 1968.
(21) JERMYN, M.A. *Austr. J. Biol. Sci.*, **15**: 769, 1962.
(22) THOMAS, R. and WHITAKER, D.R. *Nature*, **181**: 715, 1958.
(23) JERMYN, M.A. *Austr. J. Sci. Res.*, **5**: 433, 1952.
(24) BUCHT, B. and ERIKSSON, K.E. *Arch. Biochem. Biophys.*, **124**: 135, 1968.
(25) AHLGREN, E., ERIKSSON, K.E. and VESTERBERG, O. *Acta Chem. Scand.* **21**: 937, 1967.
(26) SMITHIES, O. *Advances in Protein Chem.*, **14**: 65, 1959.
(27) PRINS-van der MEULEN, P.Y.F., van der MAREL, T. and RITTER, F.J. *TNO-Nieuws*, **23**: 275, 1968.
(28) ERIKSSON, K.E. and PETTERSSON, B. *Arch. Biochem. Biophys.*, **124**: 142, 1968.
(29) BJORNDAL, H. and ERIKSSON, K.E. *Arch. Biochem. Biophys.*, **124**: 149, 1968.

A method for quantitative amino acid estimation by gas chromatography

by

J. R. COULTER.

Before beginning I should like to point out three things : the work I am describing has been a dual effort by Mr C. Hann of Adelaide and myself, and has been published *. The method which I shall describe is identical to that described in the paper except as regards the column packing, and those who are interested in a practical way should take note of these differences when I get to them. Thirdly, that this work has been supported by the N.H. and M.R.C. of Australia.

The determination of the structure of proteins is an important study in most biochemical laboratories of the world and for the elucidation of protein structure it is necessary that the precise amount of each amino acid present in a protein be known. As you all know, at the present time, most laboratories are using some modification of the Moore and Stein procedure based on the elution of amino acids from a Dowex 50 type column. This procedure is often automated and this method requires something of the order of 10^{-7} — 10^{-8} moles of material. Furthermore, when all amino acids are present, it is unusual to have this analysis done in less than two hours.

The theoretical advantages of gas chromatography are :

1. It is inherently far more sensitive. It should be possible to analyse 10^{-12} — 10^{-13} M. Our method has achieved quantitation down to 10^{-12}M.

2. The use of gas phase separation rather than liquid phase makes the method much faster and times of less than half an hour should be possible.

3. Cost- and this is the advantage which caused us to look seriously at the G.L.C. method. In Australia a conventional amino acid analyser costs $A 14,000 while the gas chromatograph we are using cost $A 4,000. It should also be noted that by change of column, a gas chromatograph can then be used for other analyses.

Why then has gas chromatographic analysis of amino acids not been used before ? There are three principal problem areas :

1. The amino acids are far too polar to be analysed in the normal gas chromatograph. The use of a temperature which is high enough to

* Journ. Chromat., **36** : 42-49, July 1968.

volatilise an amino acid causes its degradation. Therefore both the COOH and NH_2 groups must be modified in some way to make the compound more volatile. The first problem then, is to modify these groups so that either 100 % of all amino acids are converted to the derivatives or at least a constant percentage conversion over all analyses must take place. It should be noted here that the amino acids are not a single homologous series, and in all laboratories arginine and histidine in particular, have caused a great deal of trouble.

2. Resolution. Many proteins contain up to 20 different amino acids ; a large number of peaks to resolve well enough so that quantitation to within ± 2 % can be achieved ! These are the generally regarded desirable outer limits for quantitative amino acid estimation.

3. Handling very small samples. It is on this point that I think most criticism of published methods can be made. If one is to fully exploit the inherent sensitivity of G.L.C., then the methods of derivative synthesis must handle these incredibly small amounts without loss. It is therefore essential that these syntheses be carried out in a single vessel without any transfer steps. Most published methods have worked up very large quantities of amino acids and then taken a very small sub-sample for G.L.C. analysis. As all of you who have worked with labile proteins know, it is one thing to prepare 2 µg. of pure material but it is quite a different thing to prepare 2 mg. For the smaller amount, most of the standard analytical laboratory methods are adequate.

The conversion of the amino acids to volatile derivatives has generally involved the esterification of the COOH group and the methyl, ethyl, propyl, butyl and amyl alcohols have been used. The NH_2 group has been acetylated, trifluoroacetylated and converted to the 2 : 4 dinitrophenol or P.T.H. derivative. Finally, both the COOH and the NH_2 groups have been modified to trimethyl silyl groups. Our method depends upon conversion to the n-propyl, N-acetyl derivative and esterification is carried out first, followed by acetylation.

The method for derivative manufacture is identical with that which we have published, and those who want particular details are referred to that paper. I should point out here however, that it is essential that all the apparatus be very carefully cleaned and dried and that all the reagents be thoroughly purified and dried. Pyridine, which is used in acetylation, is difficult to purify. Many samples of pyridine contain unidentified substances which, on reaction with acetic anhydride, give three peaks which appear in important parts of the chart record. Nitrogen used in drying must also be very carefully dried.

The instrumental details have been given and I should only like to stress the need for all-glass columns. The support and coating differ from those given in the Journal of Chromatography. This variation has been adopted, not because it gives any better resolution, but because greater reproducibility of resolution and longer column life have been achieved.

Chromosorb W.H.P. 100/120 is now used and before coating the fines are removed. We remove about 1.2 % of fines from this chromosorb. The liquid phase is 0.85 % P.E.G. 6000 mixed with 0.05 % T.C.E.P.E. and the coating solvent is 10 % methanol and 90 % chloroform. If you care to work out the area of this chromosorb and the thickness of the liquid layer, you will realise that the latter is a monomolecular layer only. It is assumed that the function of the very small amount of T.C.E.P.E. is to block remaining polar sites on the chromosorb. All amino acids except isoleucine, leucine and glycine can be resolved without it, but it is essential for the resolution of these three closely running peaks.

The amino acids are esterified by reaction with n-propanol made 8N with respect to Hcl. The reaction is conducted in an apparatus flushed with dry nitrogen at a temperature of 100° C for ten minutes. Two periods of esterification are used and between these the sample is dried with nitrogen. 8N Hcl is required for valine and isoleucine. Lower concentrations will not give 100 % esterification within the time and temperature used. The two periods of esterification are required for lysine and histidine. As tryptophan is being destroyed during esterification it is essential for the quantitation of this amino acid that the concentration of Hcl, and the time and temperature of the reaction be fully reproducible.

The dried esters are acetylated with a mixture of pyridine and acetic anhydride 4 : 1 respectively. The reaction takes two minutes at room temperature for most amino acids but to acetylate the second amino groups of ornithine and lysine, and the hydroxyl groups of threonine and tyrosine, five minutes is required and this latter period is used in the standard procedure.

After drying, the derivatives are dissolved in either chloroform or ethyl acetate and injected into the gas chromatograph.

This standard propylation and acetylation procedure is suitable for all amino acids with the exception of histidine and arginine. Derivatives of these amino acids can not be prepared by this method because the imidazole and guanidine groups respectively are converted to the Hcl salt during esterification. The pyridine of the acetylation mixture is not a strong enough base to remove this Hcl. Reversal of the procedure with acetylation being followed by esterification has proved impossible because of hydrolysis. Arginine is therefore converted to ornithine with arginase enzyme (Worthington) and this reaction proceeds to 100 % when only 1/9000 — 1/27000 moles of enzyme are used. Furthermore, the reaction is not inhibited by a large excess of other amino acids. Histidine is converted quantitatively to aspartic acid with ozone and this reaction is very simply done while the amino acids are dissolved in 0.1N Hcl. If both arginine and histidine are present, two samples are prepared, one treated with arginase and one with ozone.

For further details and for a standard chromatogram, readers are referred to the paper in the Journal of Chromatography. All amino acids except isoleucine and leucine have been completely resolved ; these two still showing a slight degree of overlap. The use of electronic integration has allowed the accurate quantitation of these two peaks and has also allowed an increase in temperature program rate. The average time from the start of esterification to print-out is now about 25 minutes. The method has been used in the analysis of a number of proteins of known structure and has given results comparable with automated resin analysis. It is now in routine use in our laboratory.

Identification of degradation products of amino acids and proteins on exposure to heat, light, alkali, or reducing agents

by

R.S. ASQUITH *

Summary

The methods of identification by chromatographic and electrophoretic techniques of some un-common amino acids found in partially degraded or modified proteins are described. The electrophoretic behaviour of some degradation products of cystine, produced by exposure to ultraviolet light, and their structure is also discussed.

INTRODUCTION

Modern chromatographic and electrophoretic techniques have materially assisted the protein chemist in elucidating the structure of proteins and in identifying many rare, naturally occuring amino acids. Changes in amino acid composition and formation of new amino acids, on treatment of proteins with reagents, can equally be followed by these techniques. However such changes have not received as much attention as the elucidation of structure. Thus, although lanthionine was identified in alkali treated wool as early as 1941 (1), it was not until 1964 that lysinoalanine (2, 3) was identified in alkali treated ribonuclease. Ziegler (4) showed that this amino acid was also present in borax treated wool and explained that the reasons for previous failures to observe it were probably due to the fact that it elutes from amino acid systems using the Hamilton technique (5) as a single peak with lysine. It was later shown that lysinoalanine can be easily detected and estimated by high voltage electrophoresis techniques (6). Later Ziegler (7) found the ornithine analogue of lysino alanine in alkali treated silk fibroin. The purpose of the present paper is to show how a number of unusual amino acids have been identified in our laboratories as degradation products of proteins which have been exposed to alkali, ultraviolet irradiation and reducing agents, using a combination of amino-acid analysis, chromatography and high voltage electrophoresis.

* School of Colour Chemistry and Colour Technology, The University, Bradford 7, England.

UNUSUAL AMINO ACIDS IDENTIFIED

(i) From alkali and heat treatments of wool : —
    β -amino alanine (2,3-diamino-propionic acid)
    2,4 diaminobutyric acid

(ii) From ultraviolet irradiation of cystine : —
    C-methyl-C-carboxy-djenkolic acid
    Cysteine-S-sulphonic acid
    alanine-3-sulphinic acid
    bis-(2-amino-2 carboxyethyl)-tetra-sulphide
    bis-(2-amino-2 carboxyethyl)-tri-sulphide

(iii) From thiol reduced keratin : —
    S-ethyl-2-amino cysteine
    3-(2-amino-ethyl-disulphenyl)-alanine
    3-(2-carboxy-methyl disulphenyl)-alanine

PRODUCTS OF ALKALI OR HEAT TREATMENTS

The importance of lysinoalanine as a crosslinking amino acid in alkali treated keratin has been studied, showing conclusively that this amino acid (8) and lanthionine (9) are involved in crosslinking. During this work it was observed that the acid hydrolysates of alkali treated wool always contain a small amount of a strongly basic amino acid (on electrophoresis at pH 5.2/75 volts per cm. μ arg = 1.37). Further, if wool was treated with alkali and subsequently oxidised with peracetic acid prior to hydrolysis a second basic amino acid (μ arg = 1.28 at pH 5.2/75 volts per cm.) was also formed. The positions are shown on figure 1. On the Technicon amino acid analyser the more basic constituent can clearly be seen (Fig. 2) but

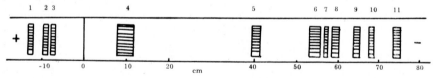

Fig. 1. — Electrophoretic patterns of hydrolysed β-keratose, obtained from alkali-treated wool. (at pH 5.2 and 75 V/cm).

1. Cysteic acid
2. Aspartic acid
3. Glutamic acid
4. Neutral amin acids
5. Lysinoalanine
6. Arginine
7. Histidine
8. Lysine
9. Ornithine
10. 2,4-diaminobutyric acid
11. β-aminoalanine

the less basic constituent tends to elute with ornithine. The U arg value for lysine (four methylene groups) and ornithine (three methylene groups) on electrophoresis under the same conditions are 1.10 and 1.19 respectively, indicating an increase in mobility of 0.09 units for loss of one methylene group. The mobilities found, therefore, for the unknown amino acids are the same as those which would be expected for 2,4-diamino-butyric acid

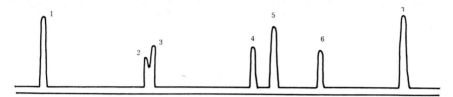

Fig. 2. — Relative positions of basic amino acids obtained from alkali-treated wool on the « Technicon Autoanalyser ». (pH gradient from 2.87 to 5.00 ; temperature 60° C ; 24 hour system).

1. Ammonia
2. 2.4-diaminobutyric acid
3. Ornithine
4. β-aminoalanine

5. Lysine and lysinoalanine
6. Histidine
7. Arginine

(two methylene groups calculated μ arg $=$ 1.28) and β-amino-alanine (one methylene group calculated μ arg $=$ 1.37) respectively. Subsequent comparisons with authentic samples by electrophoresis and ion exchange chromatography confirmed the identity of these compounds.

Most mechanisms for the formation of lanthionine and lysino-alanine postulate the primary breakdown of cystine to give α-amino-acrylic acid residues, subsequent addition to the double bond then giving the end product

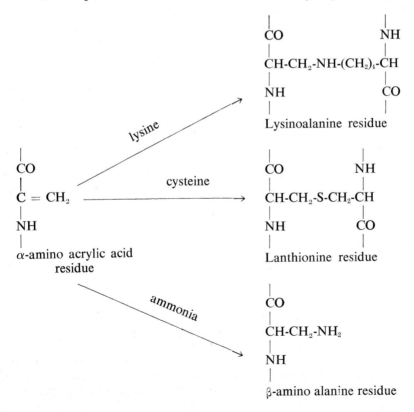

(see below). Undoubtedly therefore, the presence of β-aminoalanine in alkali treated wool can be accounted for as an addition product formed by the interaction of ammonia (from amide side chains).

The presence of 2,4, diamino butyric acid, which only appears if the wool is oxidised prior to hydrolysis, cannot be explained by this mechanism. As the amount is very small, however, it may be that during hydrolysis of the oxidised protein a small proportion of the methionine sulphone decomposes to give this product (10).

Undoubtedly three competitive reactions yielding lanthionine, lysinoalanine and β-amino-alanine respectively, can occur on treating cystine-containing proteins with alkali. It has since been shown (11) that treatment of other proteins with alkali results in the formation of lysinoalanine and β-amino-alanine if cystine is present. The proteins examined were wheat gluten, β-lactoglobulin, ribonuclease, lysozyme, insulin and myoglobin. It has since been shown that the proportions of these products are very dependent on the relative position of lysine in the protein chain. Ribonuclease and lysozyme tend to form large proportions of lysinoalanine preferentially. When ammonia is used as the alkaline treatment medium the proportion of β-amino-alanine formed rises progressively in all cases with increasing ammonia concentration.

Little work from the point of amino acid degradation has been done on the effect of dry heat on proteins. Workers on nutrition proteins (12, 13) have suggested that the lysine content may be reduced by excessive heating. These conclusions have been reached by observations of the decrease of nutritional properties of the heated proteins on feeding to rats.

High voltage electrophoresis at pH 5.2 is an excellent separative technique for basic amino acids. Examination of the amino acid contents of wools which had been heated at various temperatures up to 180° C (when loss of fibre structure occurs) have shown quite clearly that heat treatments above 130° C cause appreciative fall in the lysine contents. At the same time heating at temperatures up to 180° C brings about the formation of some lysinoalanine and some β-amino-alanine. Whilst the amounts formed are small (lysinoalanine maximum 10 mol/gm. of wool and β-amino-alanine maximum 15 mols/gm. of wool), they increase progressively with increasing temperature of treatment up to 180° C. Heat treatments above this temperature bring about total decomposition of most of the amino acids (14).

PRODUCTS OF ULTRAVIOLET IRRADIATION OF CYSTINE

Irradiation of proteins and amino acids in solution with ultraviolet light has shown that cystine is extremely susceptible to decomposition. Much work on this subject has appeared and has been well summarised by McLaren and Shugar (15). In view of the possible influence of this decomposition on the yellowing of wool by light (16) the degradation of cystine has been comprehensively studied in our laboratories (17). The free radical processes which take place produce a large number of products, some of which are of interest here as they have rarely been characterised. In particular the sulphur containing amino acid products are relatively rare. These on electrophoresis strips tend to give a greyer colour with ninhydrin than other amino acids. Amongst the neutral amino acids formed, four bands were observed on high voltage electrophoresis (at pH 1.85/75 volts per centimetre) as probably containing sulphur bridges. Two of these were easily identified as lanthionine and unchanged cystine. The μ gly values of these two amino acids under these conditions were respectively 0.568 and 0.518. This indicates a decrease in mobility of 0.05 units for the addition of one sulphur atom. Therefore the calculated μ gly values for similar substances containing a trisulphur bridge and a tetra sulphur bridge are 0.468 and 0.418 respectively. The values found for the unknown bands were 0.455

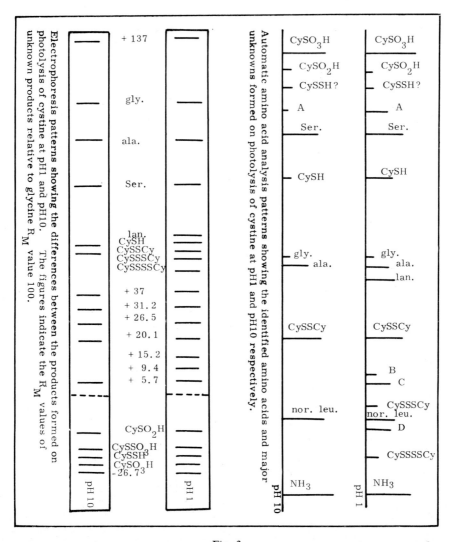

Fig. 3.

and 0.410, sufficiently close to the calculated values to justify comparison with synthesised bis-(2-amino-2-carboxyethyl)-trisulphide and bis-(2-amino-2-carboxyethyl)-tetrasulphide (18). Comparison of the synthesised material, by both paper electrophoresis and ion exchange chromatography, confirmed the identity of these unknowns. As can be seen from figure 3 the order of elution on a single column amino-acid analyser is a reverse of the order of mobility on electrophoresis.

Several acidic sulphur-containing products were formed during the degradation of cystine. These products separated well on high voltage electrophoresis at pH 1.85 (see Fig. 3). On the amino acid analyser the majority of these products eluted as one band with cysteic acid, the exception being alanine-3-sulphinic acid which follows it closely. The latter amino-acid was easily identified by comparison with the previously found position (5) and its identity confirmed by electrophoretic comparison with a synthetic sample. On electrophoresis at pH 1.85/75 volts per cm., the μ gly values of cysteic acid and alanine -3-sulphinic acid are -0.258 and -0.134 respectively. Of the three unknown acids also formed, the one with the mobility (μ gly = -0.267) greater than cysteic acid has not been identified. Attempts to correlate the μ gly values of the other two unknowns (-0.209 and -0.196) gave no indication of their structure. Due to the small amount of materials the structures were only proved on trial and error synthesis based on the theoretical knowledge of the possible products which may be formed by free radical reactions. The substance with μ gly value -0.196 proved to be identical with synthesised cystine -S-sulphonic acid (19). The substance with μ gly value -0.209 proved to be identical to C-methyl-C-carboxy-djenkolic acid synthesised by reaction of pyruvic acid with cystine (20). One other unusual amino acid was detected, and this could not be detected on electrophoresis, possibly because of its ease of oxidation, but appeared as a peak on the amino-acid analyser closely following cysteine-sulphenic acid. It appeared to be identical with S-thiol-cysteine synthesised in an impure form by reaction of cysteine with sodium sulphide (2).

PRODUCTS FROM THIOL REDUCED KERATIN

Treatments of wool with reducing agents such as thioglycollic acid have been used as a basis for permanently setting the fibre (22). It was first assumed that reduction of cystine follows the following sequence (23).

CO   CO

CH-CH$_2$-S-S-CH$_2$-CH + 2HS-CH$_2$-COOH

NH   NH   CO-

⟶ 2 CH-CH$_2$-SH

NH- + HOOC-CH$_2$-S-S-CH$_2$-COOH

It was later postulated (24) that the reaction proceeded via the inter-mediate mixed disulphide. Zahn and Gerthsen (25) on examination of hydrolysates of thioglycollic treated wool, were able to identify a new spot on the paper chromatograms, believed to be a mixed disulphide 3-(2 carbo-xy methyl disulphenyl) alanine. In our laboratories it was necessary to study the quantities of mixed disulphides formed under different conditions of treatment. The synthesis and examination of the mixed disulphide showed that it was impossible to estimate it under standard conditions on the autoanalyser, the peak being completely covered by that for glutamic acid. High voltage electrophoresis at pH 5.2/75 volts per cm. gave a clear separation of the product from other amino acids, enabling it to be estimated colorimetrically by the technique of Atfield and Morris (26). At the same time, irrespective of the pH at which wool had been treated with thioglycollic acid, a band of lysinoalanine (Fig. 7) always appeared on

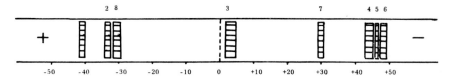

Fig. 4. — Relative position of amino acids obtained by the hydrolysis of thioglycollic acid treated wool on electrophoresis. (pH 5.2 acetic acid/ pyridine buffer : 6 K. V/75 m.A. at 8° C).

1. Aspartic acid
2. Glutamic acid
3. Neutral amino acids
4. Arginine

5. Histidine
6. Lysine
7. Lysinoalanine
8. 3-(2 carboxyl methyl disulpheyl alanine)

the electrophoresis strip. Whilst this is to be expected under alkaline conditions, it is surprising that the presence of a reducing agent will bring about the formation of lysinoalanine under acid conditions. Replacing thio-glycollic acid as a reducing agent by 2-amino-thio-ethanol resulted in no new amino acid being detected by high voltage electrophoresis. However examination of the hydrolysis products on the Technicon autoanalyser

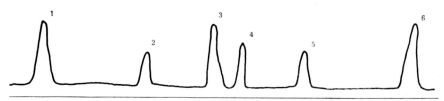

Fig. 5. — Relative position of basic amino acids obtained by the hydrolysis of 2-amino thioethanol treated wool on the « Technicon Autoanalyser ». (pH gradient from 2.875 to 5.00 ; temp. 60° C ; 24 hour system).

1. Ammonia
2. S-ethyl 2-amino cysteine
3. Lysine

4. Histidine
5. 3-(2 amino ethyl disulphenyl) alanine
6. Arginine

(Fig. 5) showed the presence of two new basic materials. One of these was easily identified as the expected mixed disulphide 3-(2-amino-ethyl-disulphenyl)-alanine by comparison with the synthesised materials. Electrophoretic examination under standard conditions showed this substance to have the same mobility as lysine.

The position of the other unknown material on the autoanalyser is identical to that of β-amino alanine, but this latter amino acid was not present on the electropherograms of the mixture. Further the colour developed with ninhydrin by the unknown acid has a relatively high absorption at 440 mu. This is often characteristic of sulphur containing amino acids. It seemed probable therefore that the material could be S-ethyl-2-amino-cysteine. Synthesis of this amino acid (27) and comparison confirmed the structure. Examination on electrophoresis showed that S-ethyl-2-amino-cysteine had the same mobility at pH 5.2 as arginine, explaining the failure to isolate it from the protein hydrolysate by this technique.

These examples of unusual amino acids identified in treated proteins show the versatility of standard chromatographic and electrophoretic techniques in identification. They also show the weaknesses of these techniques (which do not provide sufficient material for classical organic analysis). In some cases therefore it is impossible to make any deductions regarding structures. In these cases the investigator has still to fall back on speculative synthesis. In conclusion I would like to thank my many research colleagues whose work has formed the basis of the paper.

References

(1) HORN, M.J., JONES D.B. and RINGEL, S.J., *J. Biol. Chem.*, **138**: 141, 1941. 1964.

(2) PATCHORNIK, A. and SOKOLOVSKY, M., *J. Amer. Chem. Soc.*, **86**: 1860, 1964.

(3) BOHAK Z., *J. Biol. Chem.*, **239**: 2878, 1964.

(4) ZIEGLER, K., Proc. Internat. Wool Textile Res. Conf. Paris (CIRTEL), **2**: 403, 1965.

(5) HAMILTON, P.B.,*Anal. Chem.*, **35**: 2055, 1963.

(6) MIRO, P. and GARCIA-DOMINGUEZ, J.J., *Melliand Textilber.*, **47**: 676, 1966.

(7) ZIEGLER, K., *Nature*, **214**: 404. 1967.

(8) ASQUITH, R.S. and GARCIA-DOMINGUEZ, J.J., *J. Soc. Dyers and Colourists*, **84**: 211, 1968.

(9) ASQUITH, R.S., MIRO, P., GARCIA-DOMINGUEZ, J.J., *Text. Res. J.*, (in press Oct).

(10) ASQUITH, R.S. and GARCIA-DOMINGUEZ, J.J., *J. Soc. Dyers and Colourists*, **84**: 155, 1968.

(11) ASQUITH, R.S., BOOTH, A.K. and SKINNER, J.D., *Biochem. Biophys., Acta* (in press).

(12) BEUK, J.K. et al., *Symposium Div. of Biol. Chem. of Amer. Chem. Soc. N.Y.*, Sept. 17, 1947.

(13) MECHAM, D.K. and OLCOTT, M.S., *Industr. and Eng. Cham.*, August 1947.

(14) ASQUITH, R.S. and OTTERBURN, M.S., Unpublished Work.

(15) McLAREN, A.D. and SHUGAR, D., *Photochemistry of proteins and Nucleic Acids*, Pergamon Press, 1964.

(16) LENNOX, F.G., *J. Text. Inst.*, **51** : T. 1193, 1960.
(17) ASQUITH, R.S. and HIRST, L., Unpublished Work.
(18) CAVALLINI et al., *Enzymologia*, **22** : 161, 1960.
(19) SORBO, B., *Acta. Chem. Scand.*, **12** : 1990, 1958.
(20) ASQUITH, R.S. and HIRST, L., Unpublished Work.
(21) ZAHN, H. and FAHRENSTICH, R., Private Communication.
(22) DAVIDSON, A.N. and HOWITT, F.O., *J. Text. Inst.*, **53** : 62, 1962.
(23) BERSIN, T. and STUDEL, P., Ber., **71 B** : 105, 1938.
(24) SPRINGEL, H.P., *Text. Res. J.*, **28** : 874, 1958.
(25) ZAHN, H. and GERTHSEN, T., *J. Soc. Dyers and Colourists*, **75** : 604, 1959.
(26) ATFIELD, G.N. and MORRIS, C.J.O.R., *Biochem. J.*, **81** : 606, 1961.
(27) ASQUITH, R.S. and PURI, A.K., *Text. Res. J.*, in press.

The homogeneity and stability of phosphatidylcholine separated by column chromatography

by

D. VANDAMME, V. BLATON and H. PEETERS
Simon Stevin Instituut voor Wetenschappelijk Onderzoek, Brugge.

The isolation of phospholipids from egg — yolk by column chromatography was already studied by several investigators. Lea (1) and Maclagan (2) described separation methods for phospholipids on silicic acid columns, while Hanahan (3, 4) and Singleton (5) used Al_2O_3 as adsorbent.

Only the latter was able to purify egg — yolk phosphatidylcholine by one simple column elution on Al_2O_3. Hanahan isolated the choline from the non choline containing components, but needed a second elution on SiO_2 for the separation of phosphatidylcholine from its lysoderivates. The use of Al_2O_3 as adsorbent is preferable to SiO_2, because Al_2O_3 has a greater separation capacity and is less time consuming.

Modifications of synthetic phosphatidylcholine to lysophosphatidylcholine and other degradation products have been described by Camejo (6) for separations on SiO_2 and by Renkonen (7) on Al_2O_3, but they did not study the isolation of pure phospholipid fractions. Using a slightly modified method of Singleton phosphatidylcholine was eluted, chemically characterised and controlled on homogeneity and stability by GLC, TLC and ion exchange chromatography on DEAE-celluloses.

MATERIALS

All samples investigated in this study are fresh egg — yolks of commercial eggs, immediately used after separation of the whites.

The total lipid concentration of egg — yolk is 30 g % with triglycerides and phosphatidylcholine as the most important lipid fractions, respectively 18 and 7 g % or 58 and 23 percent of the total lipids.

METHODS

1. - **Extraction of lipids.**

Neutral lipids of 15 g egg — yolk are extracted with 50 ml acetone. The precipitate containing proteins and lipids is washed three times with 10 ml aceton and suspended in 100 ml 96 % ethanol to dissolve the polar lipids. After mixing during 1 hour the proteins are filtered on a S x S 589² weissband filter paper. The filtrate is concentrated under N_2 — atmosphere at 70° C. The dry extract is suspended in 5 ml petroleumbenzine and mixed with 60 ml acetone at 4° C. To obtain a maximum precipitate, temperature is reduced to 4° C and the ratio petroleumbenzine to acetone is decreased to 1/12 instead of 1/5 used by SINGLETON at a temperature of 15° C. After one hour the supernatant is decanted and the polar lipids are dissolved in 5 ml chloroform.

2. - **Al_2O_3 - column chromatography.**

Chromatography is performed on basic aluminium — oxyde (Woelm) activity grade I, without any pretreatment. A suspension of 25 g Al_2O_3 in 60 ml chloroform is poured into the column (40 cm height, 2 cm ID, provided with a sintered glass — plate PO). The Al_2O_3 column of 8.5 cm height is washed with 100 ml chloroform, before applying the crude phosphatides. Remained neutral lipids in the extract are removed by elution with 150 ml chloroform. Phosphatidylcholine is eluted with 200 ml chloroform / methanol 19/1 V/V.

The chloroform / methanol 19/1 V/V ratio is preferable to 9/1 used by Singleton because the higher polarity of the solvent may elute traces of sphingomyelin and lysophosphatidylcholine. Chromatography is performed at room temperature and at 2° C in a cooled column, to reduce the modification of phosphatidylcholine.

3. - **Thin layer chromatography on SiO_2.**

The homogeneity of the isolated phosphatidylcholine is controlled on Silicagel G thin layer plates of 0.25 mm thickness, with different solvent systems. The solvent PE/EE/HAc 80/20/1 V/V/V (8) separates cholesterol-esters, triglycerides, free fatty acids, free cholesterol, diglycerides and phospholipids. The whole phospholipid fraction can be separated in phosphatidylethanolamine, phosphatidylserine, phosphatidylcholine, sphingomyelin and lysophosphatidylcholine by developing in the following solvent systems : chloroform / methanol / water 80/35/5 (V/V/V) (9) chloroform / aceton / methanol / HAc / water 100/40/40/20/10 (V/V/V/V/V) (10) and chloroform / methanol / water /NH_3 130/70/8/0.5 (V/V/V/V) (10).

Before development of the chromatograms, the plates are activated during 30 minutes at 110° C, washed in chloroform / methanol / H_2O 80/35/5 V/V/V and reactivated during 30 minutes.

The TLC plates are developed at room temperature in chromatographic chambers lined with solvent saturated Whatman 3 MM paper.

Chromatograms are detected either by a 10 % phosphomolybdic acid in ethanol at 110° C during 10 minutes, or by 0.7 % potassium-dichromate in 55 % H_2SO_4 at 200° C during 30 minutes.

4. P- and N - evaluation.

Phosphor is determined according to a slightly modified method of ROUSER (11). After mineralisation of the organic to inorganic phosphor by perchloric acid, the phosphate is quantitative determined as molybdene blue, on a Zeiss PMQ II spectrofotometer at 825 nm. The molar extinction coefficient of molybdene blue at this wave lenght is 27.10^3.

Nitrogen is determined after Kjeldahl destruction either by titrimetry or by spectrophotometry at 630 nm as phenolindophenol according to the method of MANN (12), the molar extinction coefficient is 15.10^3.

5. - Fatty acid analysis.

Fatty acids are analysed on a Microtek 2000 MF gaschromatograph with double flame ionization detector and temperature programmator. The stationary phase is 10 % DEGS on Anakrom ABS 110 - 120 mesh. The carrier gas is purified nitrogen at a flow of 40 ml/min and a pressure of 1.5 kg/cm².

Column temperature is programmated from 195 to 225° C at 2° C / min after an isothermal period of 15 minutes. The fatty acids are separated within approximately 1 hour.

The phosphatidylcholine samples are transesterified from glycerol to methylesters by a mixture of BF₃/Methanol (125 g/l) during 24 hours at room temperature. The methylesters are extracted with petroleumbenzine. The percentual distribution and the absolute concentrations are calculated by comparing the peak areas, determined by triangulation, to each other and to the area of a known amount of margaric acid (17 : 0) as internal standard.

RESULTS

1. - Isolation of phosphatidylcholine.

A thin layer control of the extraction is given in figure 1. Aceton extracts most of triglycerides and free cholesterol and practically all free fatty acids and cholesterolesters but also 30 % of the egg - yolk phospholipids.

To reduce the modification of phosphatidylcholine, during column chromatography, the isolation is performed with an elution flow of 20 ml/ min equal to the free flow of the column, if a driving head of nearly 20 cm solvent height is maintained over the bed of the alumina. Triglycerides and free cholesterol are eluted with chloroform while phosphatidylcholine is eluted with chloroform / methanol 19/1 V/V. If the solvent polarity is increased, elution of sphingomyelin and lysophosphatidylcholine may occur while phosphatidylethanolamine remained adsorbed on the alumina (5). Quantitation of the isolated phosphatidylcholine by phosphor analysis proves a recovery of 60 % of the original phosphatidylcholine concentration of the egg — yolk sample. Thin layer separations of the column eluates in PE/EE/HAc 80/20/1 V/V/V and in chloroform / methanol / H₂O 80/35/5

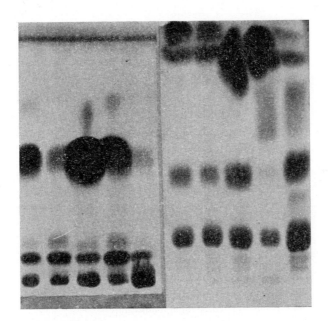

Fig. 1. — TLC control of the lipid extraction in :

	a. PE/EE/HAc	80/20/1
	b. CHCl$_3$/CH$_3$OH/H$_2$O	80/35/5.

Spot 1. Chloroform/methanol extract of egg - yolk lipids.

Spot 2. Chl/Meth. extract of the protein precipitate obtained by the aceton extraction.

Spot 3. First aceton extract.

Spot 4. Second aceton extract.

Spot 5. Crude PL-extract used in Al$_2$O$_3$ chromatography.

V/V/V are given in figure 2. Both solvents demonstrate the homogeneity of the phosphatidylcholine fraction isolated as well at 23 as at 2° C. Similar results are obtained with the other mentioned solvent systems.

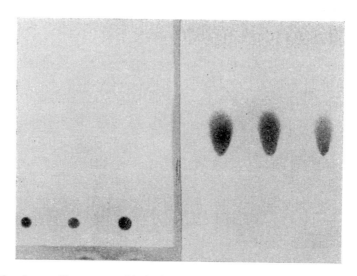

Fig. 2. — Chromatographical homogeneity of phosphatidyl choline controlled by
TLC in a. PE/EE/HAc   80/20/1
          b. CHCl₃/CH₃OH/H₂O   80/35/5.
Spot 1.   Fraction isolated in a cooled column.
Spot 2+3. Fractions isolated at room temperature.

2. - Chemical characterisation.

Phosphatidylcholine isolated from fresh egg — yolk by Al_2O_3 column chromatography contains 30 % palmitic and oleic acid, 15 % stearic and linoleic acid and 55 to 60 % unsaturated fatty acids (table 1)... Singleton on the other hand notes 37 mol % palmitic acid, 33 % oleic, 17 % linoleic and 9 % stearic acid. The small differences between both results may be due to the origin of the lipid samples.

The experimental P/N and P/FA ratios, given in table 2, approximate with a maximum deviation of 8 %, the theoretical values of respectively 2.214 and 0.50. From the theoretical structure of phosphatidylcholine and the P-, N- and fatty acid content, an experimental molecular weight of 777 is calculated, if a mean molecular weight of 350 is taken for all un-identified higher fatty acids. This agrees with the usual accepted value of 775 or 25 x P (table 2).

TABLE I

Percentual fatty acid distribution of phosphatidylcholine
in weight percent.

Fatty acid	Column temperature		
	2º C	23º C	23º C
14 : 0	0.2	0.2	0.1
15 : 0	0.1	0.1	0.1
16 : 0	27.5	29.4	28.5
16 : 1	1.5	1.5	1.5
18 : 0	13.6	13.5	14.0
18 : 1	29.4	30.4	29.8
18 : 2	17.5	16.9	17.0
20 : 0	0.2	0.1	0.1
18 : 3	0.2	0.2	0,2
20 : 3	0.3	0.4	0.4
20 : 4	4.5	3.7	4.2
24 : 0	0.2	0.1	0,2
24 : 1	0.2	0.2	0,3
unidentified	4.6	3.3	3.6

TABLE II

P/N, P/FA and molecular weight of phosphatidylcholine.

Column temperature in º C	N - determination method	P/N ratio	P/FA ratio		Molecular weight	
			absolute	molar	FA	P. Ch.
2	colorimetric	2.227	59.10^{-3}	0.53	279.19	779
23	colorimetric	2.095	53.10^{-3}	0.46	277.54	776
23	titrimetric	2.100	59.10^{-3}	0.53	278.32	777
Theoretical value		2.214	56.10^{-3}	0.50	277	775

3. - Stability control of the isolated phosphatidylcholine.

Isolated phosphatidylcholine kept at different temperatures (23° C, 4° C and — 20° C) in chloroform, in methanol and in a dry state, were controlled on stability by TLC and GLC.

The change in lipid composition after 3 hrs, 1 and 10 days storage is followed by TLC with $CHCl_3/CH_3OH/H_2O$ 80/35/5 V/V/V as developing solvent. The lipid composition is unaltered if the sample is stored in $CHCl_3$ or in methanol as well at — 20° C, and 4° C as at room temperature. Storage, in closed testtubes, of dry phosphatidylcholine during 10 days results in the formation of different degradation products (Fig. 3). Singleton notes breakdown products even after 24 hours exposure to air at room temperature.

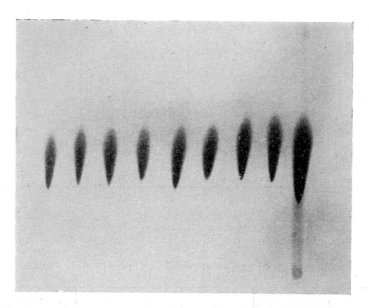

Fig. 3. — TLC control in $CHCl_3/CH_3OH/H_2O$ 80/35/5 of the modification of phosphatidylcholine after 10 days storage in :

1-2-3 chloroform	1-4-7 at - 20° C
4-5-6 methanol	2-5-8 at - 4° C
7-8-9 dry	3-6-9 at 23° C

Gaschromatographic control of the fatty acid composition of samples kept in a dry state shows also a change in fatty acid distribution after 10 days storage at room temperature (table 3). The total fatty acid concentration is decreased with 30 % caused by a decrease of linoleic, arachidonic and all unidentified higher acids.

TABLE III

Stability of the percentual fatty acid distribution
of phosphatidylcholine kept in a dry state.

Fatty acid	fresh	Storage time and temperature					
		1 day			10 days		
		-20° C	$+4^o$ C	$+23^o$ C	-20° C	$+4^o$ C	$+23^o$ C
16 : 0	28	30	29	29	31	28	40
18 : 0	14	12	14	12	14	15	18
18 : 1	30	30	31	28	29	30	34
18 : 2	17	18	15	16	16	17	4
20 : 4	4	4	4	5	4	4	trace
higher FA	5	5	5	5	4	4	1

4. - Influence of the column packing material on the modification of phosphatidylcholine.

Formation of breakdown products of phosphatidylcholine during chromatography on Al_2O_3 and DEAE cellulose was followed by TLC control and phosphor determination.

If a 5 mg sample, applied on 0.5 g adsorbent, is eluted with 10 ml methanol, the recovery of phosphatidylcholine amounts 100 % in the case of ion exchange chromatography and 80 % in adsorption chromatography. After Al_2O_3 separation 2 % of the original sample is recovered as lysophosphatidylcholine, while 18 % remained bound as degraded material to the alumina.

TABLE IV

Percentual phosphor distribution in the methanol eluates
of pure phosphatidyl choline after DEAE and Al_2O_3 chromatography.

Adsorbent	DEAE			Al_2O_3		
Contact solvent	M	C/M	M	M	C/M	M
Contact time in hours	0	24	24	0	24	24
Phos. choline	100	87	20	80	52	36
Lysophosph. choline	—	8	15	2	12	20
Non eluted material	—	1	—	18	36	44
Non identified material	—	4	65	—	—	—

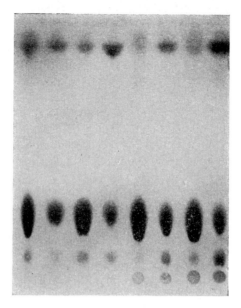

Fig. 4. — TLC control in $CHCl_3/CH_3OH/H_2O$ 80/35/5 of the modification of phosphatidylcholine by contact with Al_2O_3 and DEAE.
1. Contact with Al_2O_3 in C/M 19/1 8 h
2.   »   »   Methanol 8 h
3.   »   »   C/M 19/1 24 h
4.   »   »   Methanol 24 h
5. Contact with DEAE in C/M 19/1 8 h
6.   »   »   Methanol 8 h
7.   »   »   C/M 19/1 24 h
8.   »   »   Methanol 24 h

TABLE V

Percentual distribution of the methylesters formed by contact of phosphatidylcholine with DEAE and Al_2O_3.

FA	DEAE	Al_2O_3
16 : 0	33	34
18 : 0	15	16
18 : 1	32	30
18 : 2	10	10
20 : 4	4	3
unidentified	3	4

If the methanol elution is performed after 24 hours contact of sample and adsorbent, either in chloroform/methanol 19/1 or in methanol, the total phosphor recovery of the eluted sample remains 100 % after DEAE chromatography. However thin layer control of the eluates demonstrates a degradation of 13 % using chloroform/methanol and of 80 % using methanol. The degraded material is recovered as lysophosphatidylcholine, methylesters and as some polar phosphor containing material which remains at the starting point of thin layer chromatograms. Chromatography on Al_2O_3 gives a recovery of 52 % if the contact is performed with chloroform/methanol and of 36 % if pure methanol is used (table 4 and Fig. 4). Percentual distribution of the methylesters formed by contact of sample and adsorbent is given in table 5. Similar results are obtained after Al_2O_3 and DEAE chromatography. The percentual concentration of linoleate found in the methylesters is lower than the linoleic acid concentration in phosphatidylcholine. This demonstrates that linoleic acid is less converted than the other acids.

CONCLUSION

The isolation of egg — yolk phosphatidylcholine by means of Al_2O_3 — column chromatography is possible after a preliminary separation of the neutral and polar lipids by aceton extraction. The chromatographical homogeneity is confirmed by TLC as well in polar as in apolar solvent systems. The recovery of phosphatidylcholine by this method is 60 % of the initial concentration. The phosphatidylcholine contains 30 % palmitic and oleic acid, 15 % stearic and linoleic acid and an unsaturation of 55 %. Chemical characterisation with the P/N and P/fatty acid ratios yields an average molecular weight of 777 which corresponds with the theoretical value.

Stability control of the isolated phosphatidylcholine shows that the chromatographical homogeneity is unchanged if the sample is stored in solution. Storage of dry phosphatidylcholine during several days at room temperature results in the formation of degradation products detectable by TLC and in a loss of linoleic, arachidonic and all higher fatty acids. Contact of phosphatidylcholine with Al_2O_3 must be reduced to a minimum, to prevent the formation of lysophosphatidylcholine, methylesters and other polar breakdown products.

Similar results were obtained after contact of the sample with DEAE cellulose. From these results DEAE and Al_2O_3 can be considered as a catalyst in the methanolysis of phosphatidylcholine. Therefore chromatographic separations must be controlled on interactions of lipid sample and stationary phase as well in adsorption as in ion exchange chromatography.

ACKNOWLEDGEMENT

The authors wish to express their thanks to the NFWGO -grant nr. 1056 which supported this work and to M. Gabriël and N. Van Landschoot for their technical assistance.

References

(1) LEA, C. et al. *Biochem. J.*, **60** : 353, 1955.
(2) MACLAGAN, N.F., BILLIMORIA, J.D., HOWELL, C. *J. Lip. Res.*, **7** : 242, 1966.
(3) HANAHAN, D. et al. *J. Biol. Chem.*, **192** : 623, 1951.
(4) HANAHAN, D. et al. *J. Biol. Chem.*, **234** : 466, 1959.
(5) SINGLETON, W.S. et al. *J. Am. Oil Chemists' Soc.*, **42** : 53, 1965.
(6) CAMEJO, G. *J. Chromatog.*, **21** : 6, 1966.
(7) RENKONEN, O. *J. Lip. Res.*, **3** : 181, 1962.
(8) WOOD, P. *J. Lip. Res.*, **5** : 225, 1964.
(9) SKIPSKI, V.F. *J. Lip. Res.*, **3** : 467, 1962.
(10) PARSONS, J.G. *J. Lip. Res.*, **8** : 535, 1967.
(11) ROUSER, G. *The Lipids*, **1** : 86, 1966.
(12) MANN, L.T. *Anal. Chem.*, **35** : 2179, 1963.

La myoglobine de poule :
isolement et caractérisations préliminaires

par

S. PEIFFER, M. DECONINCK, A.G. SCHNEK et J. LÉONIS

Laboratoire de Chimie Générale I, Faculté des Sciences,
Université Libre de Bruxelles, Belgique.

Résumé.

La myoglobine du muscle cardiaque de poule a été isolée par précipitation au sulfate d'ammonium, puis purifiée par chromatographie sur DEAE-cellulose et/ou sur gel de dextrane Sephadex G-50. Toutefois, il faut signaler que l'électrophorèse décèle encore des composants mineurs dans ces préparations.

Dans l'ensemble, la myoglobine de la poule présente beaucoup de similitude avec les myoglobines de mammifères déjà étudiées. C'est ainsi que des mesures de coefficients de sédimentation et de diffusion, encore qu'elles soient préliminaires, tendent à indiquer une masse molaire inférieure à 21.000 environ. En outre, cette myoglobine se caractérise elle aussi par une grande résistance à la dénaturation alcaline, à la différence des hémoglobines correspondantes.

Les techniques d'étude des résidus terminaux ont indiqué la présence de glycine en position N-terminale, celle de sérine à l'extrémité carboxylique. La composition totale en acides aminés a également été déterminée, sur des échantillons de l'apo-protéine ; les valeurs obtenues sont proches de celles déjà connues pour les autres myoglobines.

La présence de deux résidus de méthionine, par mole de protéine, a rendu possible l'utilisation du bromure de cyanogène comme agent d'hydrolyse spécifique. Les fragments résultant de ce traitement ont été séparés par filtration moléculaire sur gel de dextrane Sephadex. Le peptide N-terminal et le peptide médian peuvent ainsi être comparés, sous l'angle de la composition en acides aminés, à leurs homologues dérivés des myoglobines de cachalot et de cheval.

Méthodes et résultats.

Isolement.

Les cœurs, congelés sitôt après le sacrifice, sont dégelés à 4° C et lavés à l'eau. Ils sont ensuite broyés en présence d'eau distillée (1 l/kg) ; l'extrait obtenu est centrifugé puis filtré. La solution clarifiée est portée à 60 % de la saturation en sulfate d'ammonium puis centrifugée, le précipité étant rejeté. Le liquide surnageant est alors amené à la saturation en sulfate d'ammonium ; le précipité, isolé par centrifugation, contient toute la myoglobine ainsi qu'un peu d'hémoglobine et de cytochrome. Il est fractionné une nouvelle fois au sulfate d'ammonium, entre 60 et 100 % de la saturation. La préparation ainsi obtenue est enfin dialysée et lyophilisée. On notera que toutes les étapes du traitement sont effectuées à 4° C et à pH proche de 8,4.

Chromatographie sur DEAE-Cellulose.

La chromatographie sur DEAE-Cellulose a été réalisée suivant la technique préconisée par Brown (1) ; la figure 1 montre un chromatogramme typique. La fraction 1 se compose de cytochrome et de protéines sans hème ; la fraction 2 contient la myoglobine, dans laquelle on peut reconnaître deux pics correspondant sans doute à la forme acide et à la forme alcaline décrites par Perkoff, Hill, Brown et Tyler (3) ; enfin, la fraction 3 est formée essentiellement d'hémoglobine. Ce type de chromatographie a été mis en œuvre pour les opérations à l'échelle préparative.

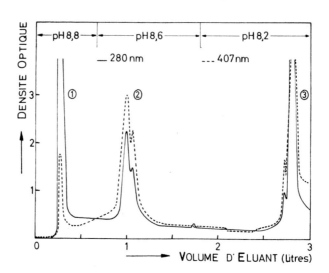

Fig. 1. — *Purification de la Myoglobine de Poule sur DEAE-Cellulose.*

L'échantillon traité est la fraction insoluble recueillie entre 60 et 100 % de la saturation en sulfate d'ammonium. Colonne de 30 × 3 cm, maintenue à 4° C ; élution par tampon Tris 0,05 M, entre pH 8,8 et 8,2 ; la nature des trois fractions majeures est discutée dans le texte.

Purification of chicken myoglobin on DEAE cellulose.

The sample is the insoluble fraction taken between 60 and 100 % of ammonium sulphate saturation. Column of 30 × 3 cm kept at 4° C ; elution with 0.05 M Tris buffer between pH 8.8 and 8.2 ; the nature of the three major fractions is discussed in the text.

Chromatographie sur Sephadex G-50.

La filtration sur gel de dextrane Sephadex G-50 Fine, selon la technique de Gondko, Schmidt et Leyko (2), a été utilisée à des fins analytiques plus que préparatives.

Sédimentation et diffusion.

L'ultracentrifugation des préparations purifiées par chromatographie montre un pic unique et symétrique, dont le coefficient de sédimentation a approximativement la valeur 1,97 S (20° C ; pH $= 8$; solution à 0,6-0,7 %). Evalué dans les mêmes conditions, le coefficient de diffusion a été trouvé égal à $0,87 \times 10^{-6}$ cm²/sec. Ces résultats, malgré leur caractère très préliminaire, sont proches de ceux obtenus par Perkoff *et al.* (3) pour la myoglobine humaine ($s^\circ_{20,w} = 1,815$ S et $D^\circ_{20,w} = 0,96 \times 10^{-6}$ cm²/sec). Ils indiqueraient une masse molaire de quelque 20 % supérieure aux valeurs usuelles pour les myoglobines (17 à 18.000) ; cet écart pourrait cependant résulter d'une certaine association, déjà perceptible aux concentrations utilisées ici.

Electrophorèse.

Après chromatographie, les échantillons de myoglobine ont été soumis à l'électrophorèse sur acétate de cellulose et sur gel d'amidon, à pH $= 7.4$. La première méthode ne permet de déceler la migration que d'une seule bande, alors que, dans les mêmes conditions, la seconde en met trois en évidence. L'une d'entre elles, très intense, représente approximativement 90 % de la préparation.

Ces résultats sont assez semblables à ceux obtenus dans le cas de différents mammifères. Les diverses hypothèses envisagées pour expliquer cette hétérogénéité ne peuvent se fonder sur aucune variation dans la composition de la globine, sinon au niveau des fonctions amides ; plus vraisemblablement, il s'agirait de différences localisées à hauteur du groupe prosthétique.

Dénaturation alcaline.

La résistance de la myoglobine de poule à la dénaturation alcaline a été comparée à celle de l'hémoglobine totale du même oiseau. La réaction est aisément suivie au spectrophotomètre, grâce à la diminution de l'absorption à hauteur de la bande de Soret (412 nm); elle est effectuée à pH $= 12,9$ et à 25° C. La figure 2 montre à quel point les deux hémoprotéines diffèrent, la myoglobine étant pratiquement insensible à l'action du pH en comparaison de l'hémoglobine. Ce test a permis de distinguer sans difficulté la nature des deux pics hémoprotéiques obtenus lors des chromatographies.

Résidu N-terminal.

Pour la détermination du résidu N-terminal par la méthode de dinitrophénylation de Sanger, la technique préconisée par Biserte, Holleman, Holleman-Dehove et Sautière (4) a été utilisée. Après traitement, le dinitrophénylaminoacide a été identifié par chromatographie sur couche mince de silice. La myoglobine de poule purifiée sur DEAE-cellulose ne possède qu'un seul

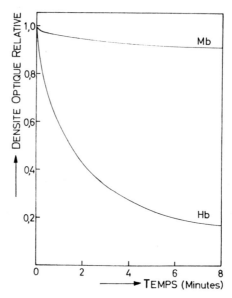

Fig. 2. — *Dénaturation Alcaline de la Myoglobine (Mb)*
et de l'Hémoglobine (Hb) de Poule.

La densité optique relative, mesurée à 412 nm, représente la fraction de
la protéine qui est sous forme native. Il s'agit de la densité optique
instantanée, corrigée pour l'absorption résiduelle de la forme dénaturée,
et ramenée à une variation unitaire entre les états natif et totalement
dénaturé.

Alkaline Denaturation of the Myoglobin (Mb)

The relative optical density measured at 412 nm, represents the protein
fraction which is in native form. It is the instantaneous optical density,
corrected for the residual absorption of the denatured form and reset
to a unitary variation between the native and completely denatured states.

acide aminé N-terminal, la glycine. Dans le cas des mammifères examinés
jusqu'ici, on a également identifié la présence de glycine ou, exceptionnelle-
ment, de valine.

Résidu C-terminal.

L'acide aminé C-terminal a pu être déterminé par hydrolyse à l'aide
des carboxypeptidases A et B, selon la méthode décrite par Ambler (5).
L'étude cinétique de la composition de l'hydrolysat a montré la libération
successive de sérine et de glycine ; spécifions toutefois que l'expérience n'a
pas encore été répétée. Les résultats obtenus ici diffèrent, notamment par
la présence de sérine, de ceux relatifs aux myoglobines de mammifères.

Composition globale en acides aminés.

L'analyse complète a été effectuée à l'aide du dispositif automatique
Beckman Unichrom, pour des échantillons soumis à diverses durées d'hydrolyse. Le tableau I récapitule la distribution des acides aminés dans l'apomyoglobine de poule, et permet la comparaison avec le cas d'autres espèces.

TABLEAU I.

Composition de la Myoglobine de Poule en Acides Aminés.

L'analyse a porté sur des échantillons de globine débarrassée de l'hème. Les résultats sont exprimés en nombre de résidus par mole, de manière telle que le total soit voisin de celui des myoglobines déjà connues ; d'une manière générale, l'incertitude expérimentale est de l'ordre de ± 1 résidu. Dans certains cas, les valeurs obtenues après 24, 48 et 72 heures d'hydrolyse ont été extrapolées au temps zéro (o) ou à une durée infinie (∞) ; habituellement il s'agit de la moyenne arithmétique.

	Poule	*Homme* (a)	*Cheval* (b)	*Cachalot* (c)
Acide aspartique	14	12	10	8
Thréonine	7 - 8	4	7	5
Sérine	7 - 8 (o)	8	5	6
Acide glutamique	18 (∞)	21	19	19
Proline	5	6	4	4
Glycine	13	15	15	11
Alanine	13	11	14	17
Valine	8 - 9 (∞)	8	8	8
Méthionine	2	3	2	2
Isoleucine	8 (∞)	7	9	9
Leucine	14 (∞)	16	17	18
Tyrosine	3	2	2	3
Phénylalanine	8	7	7	6
Lysine	15 - 17	19	18	19
Histidine	9 - 10 (o)	9	10	12
Arginine	4 - 5 (o)	3	2	4
Total	148 - 155	151	149	151

(a) *cf* référence (3).

(b) DAUTREVAUX, M. et BERNARD, S. *Bull. Soc. Chim. Biol.* **44** : 965, 1962.

(c) EDMUNDSON, A.B. *Nature,* **205** : 883, 1965.

En outre, le contenu en tryptophane (2 résidus) et en tyrosine (3 résidus) a été déterminé spectrophotométriquement, grâce à la méthode de
Edelhoch (6) modifiée par Brygier, Schnek et Léonis (7). On notera que les
nombres de résidus de tyrosine obtenus par deux méthodes indépendantes
sont parfaitement concordants.

Enfin, il semble intéressant de relever l'absence de cystine et de cystéine
dans nos échantillons, ainsi que la teneur relativement élevée en acide
aspartique.

Hydrolyse au bromure de cyanogène.

Le traitement de l'apomyoglobine de poule au bromure de cyanogène
a été effectué dans les conditions suivantes : à 25° C, durant 24 heures, dans
l'acide formique 70 %, et en utilisant 45 moles de réactif par mole de
méthionine. Après ce traitement, les peptides résultants ont été séparés par
filtration sur gel de dextrane (fig. 3), puis purifiés en répétant l'opération.

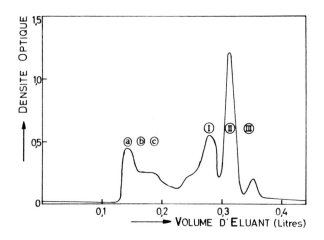

Fig. 3. — *Séparation des Peptides Obtenus par Action
du Bromure de Cyanogène.*

Colonne de 120 × 2 cm, maintenue à 5° C ; gel de dextrane Sephadex
G-50 Fine ; élution par l'acide acétique 0,2 M ; fractions successives
de 4 ml. Les zones a-b-c correspondent à la globine non clivée ou
imparfaitement clivée ; les pics I-II-III renferment les trois peptides
résultant du clivage complet.

Separation of the peptides obtained by action of cyanogen bromide.

120 × 2 cm column, kept at 5° C ; Sephadex dextran gel G-50 Fine ;
elution by acetic acid 0.2 M ; successive fractions of 4 ml. The zones a-b-c
correspond to the non-split globin or the imperfectly split globin.
The vertices I - II - III enclose the three peptides resulting from the
complete split.

Les tableaux II et III comparent les compositions des peptides median (I) et N-terminal (II), obtenus ici, à celles des peptides homologues identifiables dans le cas du cheval et du cachalot.

TABLEAU II.

Composition du Peptide Médian (I) en Acides Aminés.

Les valeurs ont été déterminées pour des durées d'hydrolyse de 24 et 72 heures, sauf pour le tryptophane qui est dosé photométriquement. L'incertitude expérimentale est de l'ordre de ± 1 résidu. Les résultats, exprimés en nombre de résidus par mole, sont calculés en prenant comme base l'acide aminé le moins abondant (tyrosine).

	Poule	*Cachalot*	*Cheval*
Acide aspartique	8	3	4
Thréonine	2-3	3	3
Sérine	2	4	4
Homosérine	Pas déterminé	1	(+)
Acide glutamique	9	7	7
Proline	2	3	3
Glycine	6	6	7
Alanine	5-6	10	8
Valine	3-4	3	3
Méthionine	0	0	0
Isoleucine	3	6	6
Leucine	8-9	8	8
Tyrosine	1	1	1
Phénylalanine	5	2	2
Lysine	7-8	10	11
Histidine	3-4	8	7
Arginine	3	1	0
Tryptophane	0	0	0
Total	67-73	76	74

TABLEAU III.

Composition du Peptide N-terminal (II) en Acides Aminés. *

	Poule	*Cachalot*	*Cheval*
Acide aspartique	7	3	4
Thréonine	3	2	3
Sérine	2	2	2
Homosérine	Pas déterminé	1	(+)
Acide glutamique	8	8	9
Proline	2	1	1
Glycine	6	3	6
Alanine	4	4	4
Valine	4	4	4
Isoleucine	3	2	2
Leucine	6-7	7	6
Tyrosine	1	0	0
Phénylalanine	3-4	3	3
Lysine	5-6	5	6
Histidine	3	4	4
Arginine	2-3	2	1
Tryptophane	2	2	2
Total	61-65	53	57

* Pour les remarques particulières, consulter la légende du tableau II.

Conclusion.

Dans l'état actuel de nos travaux, il apparaît de nombreuses analogies entre la myoglobine de la poule et celles de divers mammifères. On relèvera notamment ce qui concerne la masse molaire, la résistance à la dénaturation, le comportement à la chromatographie et à l'électrophorèse, ainsi que la composition globale en acides aminés, voire le résidu N-terminal et la distribution des peptides constitutifs.

L'établissement de la séquence complète des acides aminés dans la myoglobine de poule est en voie de réalisation. La comparaison avec les structures actuellement élucidées permettra certes une analyse plus fine des rapports phylogéniques entre la classe des oiseaux et celle des mammifères. En outre, nous mettons actuellement à l'étude l'isolement et la caractérisation de la myoglobine du manchot ; ce qui permettra sans doute d'étendre la comparaison au sein de la même classe. Enfin, signalons que l'étude de l'hémoglobine de poule est en voie d'achèvement, l'établissement de cette structure étant susceptible de fournir d'utiles rapprochements entre deux hémoprotéines distinctes issues du même animal.

* * *

Nous tenons à remercier vivement le Dr. J. Dumont, du Centre de Médecine Nucléaire de l'Université Libre de Bruxelles, ainsi que le Dr. J. Close, Directeur du Laboratoire de Biochimie à la Division Pharmaceutique de l'Union Chimique Belge. C'est à leur obligeance que nous devons les multiples analyses d'acides aminés discutées ici.

Nos vifs remerciements vont également à Madame Hannecart, de l'Institut Pasteur du Brabant, pour son aimable coopération dans les mesures de sédimentation.

Le présent programme de recherche n'aurait pas été possible sans l'aide généreuse du Fonds de la Recherche Fondamentale Collective.

Références

1. BROWN, W.D. *J. Biol. Chem.* **236** : 2238, 1961.
2. GONDKO, R., SCHMIDT, M. et LEYKO, W. *Biochim. Biophys. Acta,* **86** : 190, 1964.
3. PERKOFF, G.T., HILL, R.L., BROWN, D.M. et TYLER, F.H. *J. Biol. Chem.* **237** : 2820, 1962.
4. BISERTE, G., HOLLEMAN, J.W., HOLLEMAN-DEHOVE, J. et SAUTIÈRE, P. *Chromatographic Reviews,* **2** : 59, 1960.
5. AMBLER, R.P. *Methods in Enzymology,* **11** : 155, 1967.
6. EDELHOCH, H. *Biochemistry,* **6** : 1948, 1967.
7. BRYGIER, J., SCHNELK, A.G. et LÉONIS, J. Federation of European Biochemical Societies, 5th Meeting. Abstract 567. (Prague, 1968).

Rapid measurement of erythrocyte carbonic anhydrase isozymes by means of cellulose acetate membrane electrophoresis

by

Dr. Guillermo RUIZ-REYES and Maria de Jesus RAMIREZ-ZORRILLA

(Puebla, Mexico)

A rapid method for the quantitation of carbonic anhydrases B and C is described. The procedure is based on cellulose acetate electrophoresis with tris-EDTA-borate buffer, staining by Ponceau S and quantitation by elution. Separations obtained are reproducible. The range of carbonic anhydrases B and C concentration in 63 normal subjects was 0.65 ± 0.46 % and 0.22 ± 0.10 2 S.D respectively. Twenty patients with megaloblastic anemia showed a range of 1.2 to 3.47 % for the B band and 0.15 to 1.06 % for the C one. Band B was found elevated in 20 cases of megaloblastic anemia studied and isozyme C was high in 16 cases (80 %). This finding suggested that carbonic anhydrase B elevations are more specific for megaloblastosis than the C variations, when electrophoretic techniques are used.

L'électrophorèse comme méthode d'exploration de l'effort physique et du degré d'entraînement

par

ROTARU C. *

Le problème des modifications biochimiques et physiologiques dans l'effort musculaire a suscité l'intérêt de nombreux auteurs qui lui ont consacré leurs travaux. Parallèlement, on a cherché à élaborer des méthodes adéquates pour l'investigation de ce phénomène.

L'effort physique entraîne d'importantes modifications biochimiques du sang, conséquence de l'augmentation du métabolisme et des processus d'adaptation aux nouvelles conditions d'activité de l'organisme.

Certains résultats que nous avons obtenus antérieurement (6, 8) démontrent la manière particulière de se comporter des protéines sérosanguines et leur évolution caractéristique dans l'état d'effort physique et de fatigue par rapport aux valeurs au repos.

L'existence de certaines modifications proportionnelles des fractions électrophorétiques pourrait donc constituer un moyen de préciser la valeur du séroprotidogramme électrophorétique dans l'évaluation informative de l'intensité de l'effort musculaire et de l'état d'entraînement physique.

Les investigations ont été effectuées dans des conditions de laboratoire sur des adolescents soumis à un effort dosé, d'intensité moyenne, avec un ergomètre mécanique à manivelle.

Les déterminations pour apprécier la dépense d'énergie ont été faites par la méthode des échanges respiratoires en circuit ouvert (Douglas-Haldane) et ont eu en vue le repos, l'effort pour le travail effectué et la période de restitution après l'effort. L'évaluation des fractions du séroprotidogramme électrophorétique a été faite sur du papier filtre avec des prises de sang récolté avant et après l'effort. Dans certains cas les prises de sang ont été plus grandes car on a pratiqué des récoltes aussi après la période de restitution, quand le pouls artériel et la ventilation pulmonaire sont revenus aux valeurs ayant précédé l'effort.

---

* Lab. de Physiol., Inst. Medico-Pharmac., Universitatii n° 16, Iasi, Roumania.

En examinant les résultats de la dynamique des processus énergétiques dans l'effort ainsi que l'évolution des protidogrammes électrophorétiques on constate que les variations, tant du métabolisme énergétique que de la composition protidique du sérum sanguin ne sont pas les mêmes chez le sujet en état de repos que chez celui fatigué ou presque épuisé. Donc, les répercussions de l'effort musculaire généralisé sur le métabolisme général et sur les caractères physico-chimiques du sang sont incontestables.

Bien que le travail exécuté à l'ergomètre ait été réglé à une valeur constante (244,4 Kg.m./min.) on observe que par rapport à l'état d'entraînement physique de chaque sujet l'effort jusqu'à la fatigue s'exécute dans des durées de temps différentes. Si nous considérons qu'à mesure que l'organisme s'adapte à l'exercice physique, il y a amplification de la synergie respiratoire ce qui assure l'amélioration de la force de travail et une moindre augmentation des échanges gazeux respiratoires à l'effort, nous pouvons établir une relation entre la capacité d'adaptation physiologique du sujet à l'effort et l'augmentation du métabolisme gazeux correspondant à un travail défini. Dans un exercice physique bien précisé — tel que le travail à l'ergomètre à effort dosé — on peut donc prendre comme indice d'appréciation du travail effectué la dépense d'énergie, et comme indice d'aptitude de l'organisme à l'effort, la consommation d'oxygène.

Dans le tableau électrophorétique des séroprotidogrammes nous constatons des modifications qui peuvent être relevées aussi par l'examen comparatif des bandes électrophorétiques du repos et de celles correspondantes obtenues après l'effort.

L'évaluation des fractions électrophorétiques au cours de l'effort physique chez l'homme ont donné lieu à des appréciations controversées.

R. De Lanne et coll. (1) constatent, après un effort intense au véloergomètre une augmentation des alfa-2-globulines ; J. Nitescu et coll. (4), après un effort d'intensité modérée et de grand volume, une augmentation particulièrement des gamma-globulines, tandis que M. Gaglio et R. Mineo (2), P. Lotti et coll. (3) obtiennent après l'effort de durée, l'augmentation des fractions alfa-2 et bêta-globulines. Dans des recherches antérieures nous avons signalé, tant à la suite d'investigations expérimentales (7) que chez l'homme (5), à la cessation de l'effort (dans la période de restitution), en variations, proportions les plus importantes et constantes au niveau des alfa-globulines.

Sur ces aspects, controversés dans la littérature, on pourrait obtenir certaines précisions en prenant en considération concomitamment les modifications du tableau électrophorétique séroprotidique et l'évolution du métabolisme énergétique.

A l'analyse quantitative électrophorétique des séroprotidogrammes on remarque, à l'examen global, l'existence de certaines modifications nettes de la composition pour-cent dans l'état d'effort de l'organisme. On constate premièrement un changement du rapport S/G (déduit par électrophorèse) qui marque une diminution des sérines en faveur de l'augmentation des fractions globuliniques. L'hypoalbuminémic n'est que proportionnelle, car la valeur fondamentale des protides totaux est légèrement augmentée ou non modifiée. Cette augmentation plus ou moins légère du contenu en protéines totales du sang pendant l'effort, peut être considérée entre autres comme une conséquence de la concentration du sang à la suite de l'entraînement des liquides plasmatiques dans le tissu musculaire. Ce phénomène est d'autant plus intense que le muscle est moins entraîné à l'exercice physique. Bien que la quantité de protéines totales augmente ou demeure en apparence normale, on remarque un déséquilibre dans la répartition électrophorétique des sérines-globulines. En analysant les valeurs obtenues on constate que les globulines acquièrent une proportion considérable en particulier au niveau des fractions alfa-2, qu'elles s'accompagnent d'une augmentation médiocre des fractions alfa-1 et d'une augmentation limitée des bêta-globulines, tandis que les gamma-globulines présentent autour des valeurs normales de très petites oscillations.

Nous pouvons donc considérer que si dans le tableau électrophorétique sero-sanguin parmi les modifications déterminées par l'effort musculaire l'hypoalbuminémie est caractéristique, l'augmentation proportionnelle des alfa-2 globulines l'est aussi tout particulièrement, car elle présente des valeurs incontestables et manifestées de manière constante.

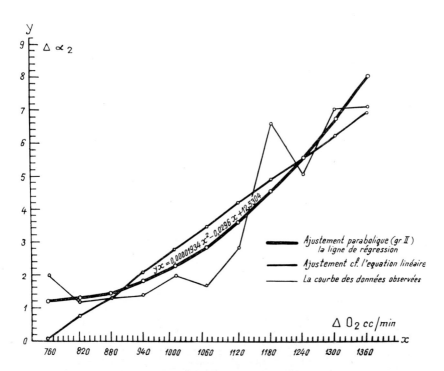

Fig. 1. — L'augmentation des alfa-2 globulines par rapport à l'augmentation de la consommation d'oxygène.

(Δ = la différence du repos à l'effort)

Increase of the α-2 globulins with regard to the increase of oxygen consumption.

(Δ = difference between rest and effort)

292 ROTARU C.

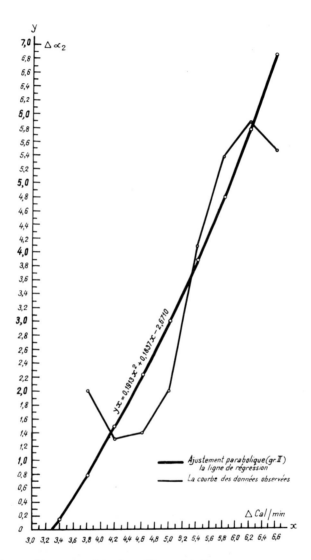

Fig. 2. — L'augmentation des alfa-2 globulines par rapport à l'augmentation de la dépense d'énergie.
(Δ = la différence du repos à l'effort)

Increase of the α-2 globulins with regard to the increase of
energy expended.
(Δ = difference between rest and effort)

Afin de mettre en évidence les relations entre le métabolisme énergétique et le protidogramme électrophorétique, on a recherché une éventuelle corrélation tant avec la consommation d'oxygène qu'avec la dépense d'énergie.

On a considéré comme une variable indépendante, d'une part : la série de valeurs se rapportant soit à la consommation d'oxygène, soit à la dépense d'énergie (en valeurs absolues), ou la différence entre l'effort et le repos ; et d'autre part, comme une variable dépendante, soit les globulines alfa-1, soit les globulines alfa-2 (en valeurs absolues), ou la différence entre l'effort et le repos. Dans ce qui suit, nous notons par delta (Δ) la différence de l'effort au repos.

En analysant les résultats des recherches de la corrélation entre la consommation d'oxygène et les globulines alfa-1 et alfa-2, on remarque les faits suivants : la corrélation entre delta-consommation d'oxygène (cc/min.) et delta-globulines alfa-1 est faible et statistiquement non assurée. La corrélation entre delta-consommation d'oxygène (cc/min.) et delta-globulines alfa-2 est forte ; la corrélation linéaire ($r = 0,85$) est statistiquement assurée. En fait, les données observées sont encore mieux représentées par une courbe parabolique de IIme degré et, dans ce cas, la corrélation est très forte (le rapport de corrélation $\gamma = 0,91$). La figure 1 montre : courbe des données observées (on a groupé les 20 observations en 11 variantes) et concomitamment l'ajustement linéaire de même que celui parabolique.

L'analyse des résultats obtenus en ce qui concerne la recherche de la corrélation entre la dépense d'énergie et les globulines alfa-1 et alfa-2 mettent en évidence des aspects semblables : la corrélation entre delta-dépense d'énergie (cal/min.) et delta-globulines alfa-1 s'avère faible et statistiquement non assurée, tandis que la corrélation entre delta-dépense d'énergie (cal/min.) et delta-globulines alfa-2 est forte ($r = 0,75$) et statistiquement assurée. Dans ce cas aussi, les données observées sont mieux représentées par une courbe parabolique de IIme degré. Le rapport de corrélation empirique de même que celui théorique est pratiquement égale ($\gamma = 0,80$) ; la vérification de la régression parabolique par le test Fischer-Snedecor, conclut à la confirmation de l'hypothèse nulle (respectivement, nos données ne diffèrent pas significativement de la courbe parabolique). La figure 2 montre : courbe des données observées (on a groupé les 20 observations en 8 variantes) et concomitamment l'ajustement parabolique.

En résumé, nos recherches mettent en évidence une corrélation nette de type parabolique entre delta-consommation d'oxygène ou dépense d'énergie et les delta-globulines alfa-2, mais ne nous permettent pas de conclure à une corrélation avec les delta-globulines alfa-1.

Nous mentionnons qu'on n'a pas pu mettre en évidence une corrélation significative entre les valeurs absolues des alfa-1 ou alfa-2 globulines d'une part et la dépense d'énergie pour l'effort total, de durée, et pour le travail extérieur exprimé en Kg/m. d'autre part. En échange, quand on a pris comme variable indépendante la delta-dépense d'énergie exprimée en cal/min. on a constaté une corrélation évidente. De même, on a observé aussi une relation avec la consommation d'oxygène.

Nous pensons qu'on pourrait interpréter la corrélation mentionnée comme reflétant la relation entre les alfa-2 globulines et l'effort effectué ; et si nous nous rapportons à l'installation du phénomène de fatigue métabolique, plutôt la relation avec l'intensité de l'effort qu'avec le degré de fatigue.

Si nous suivons en même temps le rendement du travail et l'évolution du tableau séroprotidique, on constate des relations entre la dépense d'énergie de l'organisme dans l'effort et la restitution, et l'importance des modifications pour-cent des fractions électrophorétiques après un effort. Les variations du tableau électrophorétique du sérum sanguin sont plus accentuées chez les sujets pour lesquels la quantité nécessaire d'oxygène par minute pour un même travail est plus élevée, tandis que chez ceux à consommation d'oxygène plus réduite par minute pour un même travail, on ne met en évidence que de faibles modifications, insignifiantes et même indécelables.

Ces données peuvent donc exprimer une traduction biochimique de la notion d'aptitude physiologique de l'organisme à l'effort, mise en évidence par la consommation d'oxygène par minute des sujets pour fournir un travail musculaire défini.

D'après ces considérations, on peut discuter la différence de comportement des sujets ayant des degrés différents d'entraînement dans le travail physique. De ce point de vue, on a effectué une répartition approximative des personnes, étudiées en deux groupes. Le premier groupe pour lequel la durée de l'effort a été courte, donc le travail effectué a été réduit, tandis que la consommation d'oxygène et la dépense d'énergie atteignaient des valeurs élevées. Le second groupe, avec une durée de l'effort prolongée, donc un travail effectué important, et où la consommation d'oxygène ainsi que la dépense d'énergie étaient plus réduites.

En étudiant la série de valeurs delta-globulines alfa-2 chez les sujets entraînés (IIme groupe) et chez ceux non entraînés (Ier groupe) on a constaté une différence fortement significative (le test t Student : t = 10,1^{+++}). On a pu mettre en évidence aussi une correspondance en fonction de l'état d'entraînement du sujet ; se traduisant par une diminution proportionnelle des modifications du tableau électrophorétique à mesure que l'organisme est plus adapté à l'exercice physique. Ainsi, tandis que chez les personnes examinées du Ier groupe, le rapport sérines/globulines présente des différences importantes, accompagnées d'une augmentation des alfa-2 globulines, avec des différences appréciables, chez les personnes du IIme groupe, avec un degré d'entraînement plus grand, le rapport sérines/globulines présente de faibles différences et on remarque une augmentation médiocre des alfa-2 globulines, avec des différences beaucoup plus réduites.

Cette observation est confirmée par le fait suivant : la répétition du travail à l'ergomètre à des intervalles de 24 heures, prolonge progressivement pour chaque séance la durée de l'effort maintenu, et augmente en même temps le travail effectué ; à mesure que cette adaptation à l'exercice physique a lieu, un redressement des modifications du tableau électrophorétique se produit aussi, particulièrement en ce qui concerne les fractions globuliniques alfa-2.

En conclusion, nous sommes en droit de considérer que l'investigation électrophorétique du sérum sanguin, afin d'apprécier l'évolution des fractions alfa-globuliniques du repos à l'état surchargé de l'organisme par le travail peut constituer un guide de recherche dans l'effort physique, donc dans l'évaluation de son intensité ; et, en même temps, cette investigation constitue un auxiliaire pour caractériser l'état et le degré de l'entraînement.

Bibliographie

1. DE LANNE R., BARNES J., BROUHA L. *J. Appl. Physiol.* n° 13, p. 97, 1958.
2. GAGLIO M., MINEO R. *Méd. Sport.,* n° 14, p. 547, 1960.
3. LOTTI P., PERUZY A., CAPONE M., MONTERVINO C. *Méd. Sport.,* n° 14, p. 660, 1960.
4. NITESCU I., CONSTANTINESCU A., MIHAITA M. *Rev. Fiziol. Norm. Pat., Bucuresti,* n° 6, p. 93, 1961.
5. ROTARU C., STEFAN I. *Rev. Med. Iasi,* n° 1, p. 91, 1959.
6. ROTARU C., PRUTEANU P., ROSCA V. *Rev. Med. Chir. Iasi,* n° 4, p. 1007, 1966.
7. ROTARU C. *Rev. Med. Chir., Iasi,* n° 3, p. 375, 1967.
8. ROTARU C. *Rev. Med. Chir., Iasi,* n° 1, p. 173, 1968.

Ultramicrométhodes d'analyse chromatographique des nucléosides et des oligonucléotides, après marquage externe par le borohydrure tritié

par

S. DESREUMAUX, G. BARBRY et E. SEGARD

Laboratoire de Chimie Biologique, Service de Biochimie Cellulaire,
Faculté des Sciences de Lille.

La mise au point de microméthodes d'analyse des enchaînements nucléotidiques devient nécessaire pour aborder l'étude ponctuelle de la structure des sites d'activité des RNA. La détermination des séquences nucléotidiques présentant une activité précise sur ces molécules, intervient dans les travaux portant sur la structure du site actif des aminoacyl-synthétases pour les t-RNA ou sur celle des sites d'initiation et d'arrêt de la traduction pour les m-RNA polysomiaux ; ce problème se présente également dans les recherches concernant l'importance du site actif de la RNA-polymèrase et son sens de lecture au niveau du DNA. Les quantités très faibles de substrats disponibles dans ce type de travaux imposent de nouvelles recherches purement méthodologiques.

Les techniques actuellement pratiquées dans le cas des t-RNA spécifiques, des RNA 5.S et, plus récemment, des r-RNA (1), font appel à la chromatographie sur colonne (DEAE-cellulose, DEAE-Sephadex A-25 et Dowex 1×2) ou sur papier (électrochromatographie ou chromatographie bidimensionnelle à partir de substrats marqués par le ^{32}P) ; les identifications et dosages, par spectrophotométrie ou autoradiographie et comptage de la radioactivité, nécessitent des quantités initiales de substrat de l'ordre de 5 à 10 mg, pour un seul type de nucléase. Pour atteindre un domaine de sensibilité supérieur, en évitant les causes d'erreur dues aux corrections d'hyperchromicité ou aux marquages statistiques, nous avons retenu le principe des marquages de type « externe » ; ils concernent, par définition, la réaction chimique sélective d'une substance radioactive sur un substrat préalablement isolé et se distinguent donc nettement des marquages *in vivo* et *in vitro* qui se fondent sur une réaction métabolique. En envisageant le cas le plus simple des marquages externes sélectifs des groupements 3'-hydroxyle terminaux des oligo- et polynucléotides, on peut envisager un protocole de microanalyse des enchaînements nucléotidiques dans un domaine de sensibilité allant de la nanomole à la picomole. Nous décrirons les résultats actuels de nos travaux concernant : I - les techniques de marquage externe, de séparation et de dosage des nucléosides et, II - un exemple de leur application à l'étude des produits d'hydrolyse par la RNase-pancréatique du t-RNA de E. coli.

I. — MARQUAGE EXTERNE, SEPARATION ET DOSAGE DES NUCLEOSIDES.

1. Principes.

Parmi les différentes réactions spécifiques de la fonction α-glycol en 2'-3' du ribose, l'oxydation périodique suivie d'une réduction par le borohydrure tritié est la plus intéressante, car elle s'effectue dans des conditions très douces ; elle est, en outre, quantitative et rigoureusement sélective. Le produit de la réaction (fig. 1) est un ether-oxyde substitué

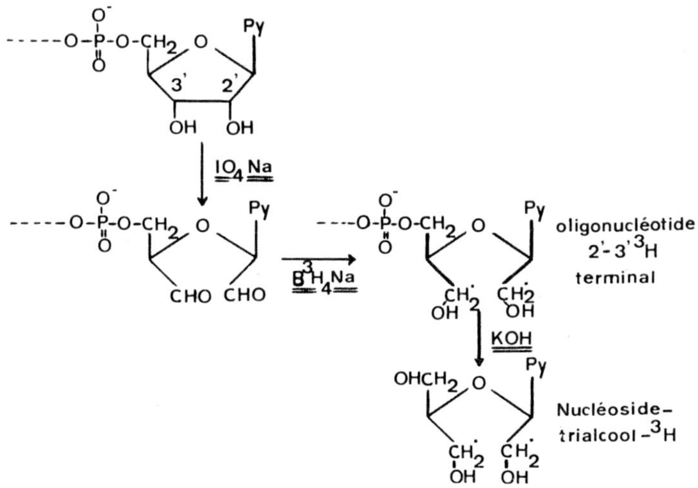

Fig. 1. — Principe du marquage externe des groupements 3'hydroxyle terminaux.

External marking principle of the terminal 3'hydroxyl groups.

portant 3 fonctions alcool primaire, sur les positions initiales 5', 3' et 2' du ribose. Rajbhandary (2) a proposé la dénomination de *nucléoside-trialcool,* avec les symboles C', A', G' et U' correspondant aux 4 nucléosides. Ce principe de réaction dérive de celui décrit, pour les oses et polyols, par Murakami et Winzler (3) ; pour 1 μM de nucléoside, la réaction nécessite 1 μM de métaperiodate de sodium et 0,5 μM de borohydrure de sodium. L'ouverture du cycle furannique du ribose labilise la molécule et la mise au point des techniques d'analyse doit tenir compte des conditions de stabilité de ces dérivés. Nos travaux de mise au point ont été effectués, pour cette raison, à partir de nucléosides témoins.

2. - **Marquage par le borohydrure de sodium tritié.**

Les conditions que nous avons récemment décrites (4) sont dérivées de celles de Rajbhandary (2). Elles s'appliquent aux nucléosides libres (hydrolysats alcalins traités par la phosphatase alcaline) et aux oligonucléotides traités par la phosphatase alcaline. La réaction est conduite dans l'eau distillée ; par suite d'un effet tampon, le pH demeure entre 7,5 et 8,2. Ceci permet l'analyse directe du milieu réactionnel en évitant les surcharges en ions minéraux. Dans le cas des RNA macromoléculaires, la réaction est effectuée dans un tampon de force ionique convenable exempt de fonctions oxydables par l'acide périodique ; la préparation est dialysée avant l'analyse.

A 0,1 ml d'une solution aqueuse, correspondant à une quantité maximale de 0,2 μM de nucléoside 3'hydroxyle terminal, on ajoute 0,1 ml de métaperiodate de sodium 0,04 M et on abandonne 1 heure à 20° C. On ajoute alors 5 μl d'éthylène-glycol, pour détruire l'excès d'acide périodique, et après 15 mn de contact, on abandonne 1 nuit à 20° C en présence de 0,1 ml d'une solution contenant 10 μM de borohydrure de sodium tritié (CEA - 100 mCi/mM), conditionné dans les conditions de Murakami et Winzler (3). En fin de réaction, le milieu renferme, outre le substrat marqué, de l'éthylène-glycol en excès, du méthanol-^{3}H, des iodates et des borates et un excès de borohydrure-^{3}H ; la réaction s'accompagne, en outre, d'une production variable d'eau tritiée provenant de la réduction des fonctions aldéhydes. Nous avons mis au point des conditions d'analyse des nucléosides-trialcools qui éliminent l'interférence de ces produits secondaires et permet une séparation directe, sans aucune purification.

3. - **Analyse chromatographique directe des nucléosides-trialcools.**

a) *Analyse chromatographique sur papier des 4 nucléosides-trialcools normaux.*

L'électrophorèse sur papier est à exclure, car quels que soient le tampon et le pH, le borohydrure-^{3}H et le méthanol-^{3}H interfèrent avec les nucléosides-trialcools. En outre, les solvants acides sont à exclure, car les études de stabilité montrent que ces dérivés sont stables en milieu alcalin, mais qu'ils se dégradent rapidement en milieu acide :

à pH 8, la dégradation est nulle après 48 h. à 20° C et 3 semaines à 4° C ;

dans la potasse 0,3 N, la dégradation est de 5 p. 100 après 48 h. à 37° C ; elle est inférieure à 0,5 p. 100 après 18 h. à 37° C. Le marquage externe est donc parfaitement applicable avant l'hydrolyse alcaline ;

à pH 3,5 et 2,7 la dégradation est de 35 p. 100 après 4 h. à 20° C ; elle est supérieure à 90 p. 100 après 24 h. à 20° C. La dégradation n'est cependant pas supérieure à 5 p. 100 après 24 h. à 0° C dans l'HCl 0,01 N. Les analyses par chromatographie en Isopropanol (—)/HCl (5) et en butanol/acide acétique/eau (6) permettent de caractériser la base et le glycérol libérés.

Les conditions mises au point sont les suivantes (4) : chromatographie bidimensionnelle sur papier Whatman n° 1 avec les solvants indiqués à la figure 2. Le solvant original n° 1 permet d'entraîner le méthanol-^{3}H au front et de séparer le borohydrure en excès. Une électrophorèse à pH 7,5 (ou 3,5 selon le cas) de la partie supérieure du chromatogramme permet de

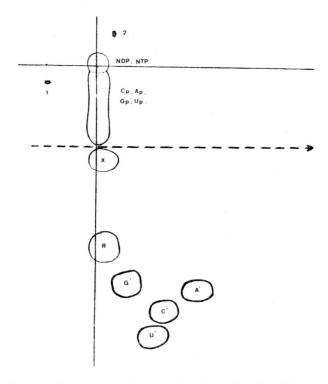

Fig. 2. — Chromatographie sur papier des nucléosides-trialcools.

1 — méthanol/butanol tertiaire/eau (4/4/3) ajusté à pH 10 avec l'ammoniaque - 30 heures -

2 — n-butanol/ammoniaque 6 N. (30/5) - 24 heures -

C', A', G', U' : nucléosides-trialcools — Cp, Ap, Gp, Up : nucléosides 3'phosphates — R : radioactivité du borohydrure en excès — X : composé non nucléique et non radioactif — NDP, NTP : nucléosides di- et triphosphates.

Chromatograph on paper of nucleosides-trialcohols.

1 — Methanol/tertiary butanol/water (4/4/3) adjusted to pH 10 with ammonium - 30 hours.

2 — n-butanol/ammonium 6 N. (30/5) - 24 hours -

C', A', G', U' ; nucleosides-trialcohols - Cp, Ap, Gp, Up ; nucleosides 3'phosphates — R ; radioactivity of excess borohydride — X ; non nucleic and non radioactive compound — NDP, NTP ; nucleosides, di- and triphosphates.

séparer, dans un 2me temps et dans le sens 2, les nucléosides mono-, di-, tri- et tétraphosphates qui ne migrent pas au cours de la chromatographie ; le dosage de ces nucléotides (en mélange (7) ou après séparation), par spectrophotométrie ou par comptage de préparations marquées au ^{32}P, conduit à la détermination rapide de la longueur de chaîne en établissant le rapport nucléotides/nucléoside terminal. Le dosage des nucléosides-trialcools est effectué par comptage des spots de papier dans le mélange PPO, 5 g/POPOP, 0,5 g pour un litre de toluène (Spectromètre Nuclear-Chicago-Mark I). En raison de la radioactivité de fond provenant de l'eau tritiée, il est nécessaire d'introduire de nombreux témoins papier. Dans les conditions opératoires, la réaction est rigoureusement quantitative et le marquage est nul avec les bases, les déoxyribonucléosides et les nucléosides 2'- et 3'-phosphates Enfin, la radioactivité est uniforme pour les nucléosides et les oligo- et polynucléotides quelles que soient les bases constituantes et la longueur de chaîne — ce type de dosage élimine donc les corrections d'hyperchromicité nécessaires en spectrophotométrie et permet une microdétermination simple de la répartition des oligonucléotides libérés à partir de l'enchaînement étudié. Une série de déterminations à partir d'une gamme étalonnée fournira le témoin de dosage. Si l'activité du borohydrure tritié est de 1 mCi/10 μM, on obtiendra des valeurs du témoin comprises entre 400 et 500 DPM par picomole de nucléoside-trialcool (après correction du quenching). En travaillant en routine avec des échantillons de borohydrure dilués au 1/10 (0,1 mCi/10 μM) les dosages sont pratiqués couramment pour des quantités de 10 à 100 picomoles de nucléoside-trialcool.

b) *Analyse chromatographique sur colonne des nucléosides-trialcools normaux et atypiques.*

Une adaptation du procédé décrit par Cohn (8) et modifié par McCully et Cantoni (9) nous permet actuellement de différencier C' et C'^m, A' et A'^m, rT', $I'\psi'$ et U'; une amélioration du calibrage des résines nous permet d'espérer la différenciation de G' et G'^m et de I' et I'^m. Les conditions et les diagrammes d'analyse sont indiqués dans la figure 3. Les profils de radioactivité sont établis après comptage d'aliquotes de 0,5 ml dans le liquide de Bray. La radioactivité résiduelle est éluée avec C'; la présence de ce dernier dérivé doit donc être confirmée par chromatographie sur papier. Le fond est ensuite de 1500 à 1800 DPM et la sensibilité de repérage par la radioactivité est de 50 picomoles de nucléoside-trialcool (soit environ 8000 DPM). La récupération du substrat est de 94/97 p. 100. La technique ne convient pas à l'identification de U^h, car ce dernier se dégrade de façon exponentielle en milieu alcalin ; le noyau réagit, en outre, avec le borohydrure pour former le ribosyluréidopropanol. Sa présence est cependant aisément décelable, soit par différence dans le cas des oligonucléotides libérés par la RNase-pancréatique, soit par la libération spontanée d'oligonucléotides inférieurs à partir des oligonucléotides produit par la RNase-T_1. Enfin, l'absence de marquage de certains oligonucléotides provenant des t-RNA est une indication précieuse de la présence de 2'-0-méthylnucléoside (2'-OmeG ou 2'-OmeU).

4. Analyse chromatographique sur colonne des oligonucléotides marqués.

Les produits d'hydrolyse ribonucléasique des RNA peuvent être marqués par la méthode décrite, après traitement par la phosphatase alcaline. Les oligonucléotides, marqués sur leurs groupements 2' et 3' terminaux, peuvent être séparés par groupes de mono-, di-, tri-, tétranucléotides, etc...

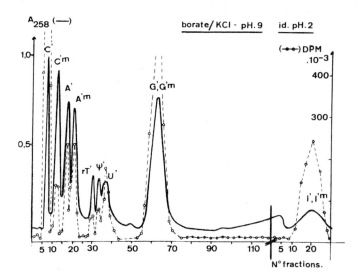

Fig. 3. — Chromatographie sur Dowex 1×8 des nucléosides-trialcools
normaux et atypiques.

Colonne de 0,8×100 cm. — Dowex 1×8 (200/400) — débit : 12 ml/h.
— fractions de 4 ml. — $K_2B_4O_7$-0,02 M./KCl-0,03 M. pH 9 —
nucléosides-trialcools de cytidine (C') — 5-méthylcytidine (C'^{m}) —
adénosine (A') — 1-méthyladénosine (A'^{m}) — ribothymidine (rT') —
pseudouridine (I'ψ') — uridine (U') — guanosine (G') — 7-méthyl-
guanosine (G'^{m}) — inosine (I') — 7-méthylinosine (I'^{m}).

Chromatograph on Dowex 1 × 8 of normal and atypic nucleosides-
trialcohols.

0.8 × 100 cm column — Dowex 1 × 8 (200/400) — flow : 12 ml/h, —
fractions of 4 ml — $K_2B_4O_7$-0.02 M./KCl-0.03 M. pH 9 — nucleosides-
trialcohols of citidine (C') — 5-methylcitidine (C'^{m}) — adenosine (A')
— 1-methyladenosine (A'^{m}) — ribothymidine (rT') — pseudourine (I')
— uridine (U') — guanosine (G') — 7-methylguanosine (G'^{m}) —
inosine (I') — 7-methylinosine (I'^{m}).

(« Isopleths ») en appliquant la méthode de fractionnement sur DEAE-
Séphadex A-25 de Ishikura et al. (10). La figure 4 fournit un exemple de
séparation de quelques oligonucléotides isolés d'un hydrolysat par la RNase-
pancréatique de t-RNA de E. Coli, selon le procédé décrit au II. Un gradient
de NaCl de 0 à 0,5 M dans le mélange Urée-7 M/Tris-HCl-0,02 M. pH 7,6
conduit à la séparation par groupes des mono- aux heptanucléotides. La
radioactivité des réactifs est éluée au volume mort de la colonne et la sensibilité de repérage par comptage comme au 3.b permet d'atteindre des
sensibilités correspondantes à 20 picomoles du nucléoside 3'-hydroxyle terminal. La labilité de ces dérivés en milieu acide pose le problème de la
séparation de chaque « isopleth » sur Dowex 1×2 par élution en Urée/HCl
à pH acide (11) ; les travaux sont en cours sur ce point.

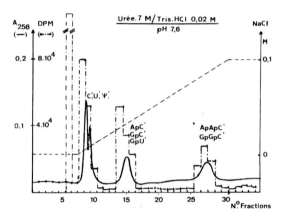

Fig. 4. — Chromatographie sur DEAE Sephadex A-25
d'oligonucléotides marqués.
Colonne de 0,8 × 100 cm — débit 32 ml/h — fractions de 10,5 ml.

Chromatograph of marked oligonucleotides on DEAE Sephadex A-25.
0.8 × 100 cm column — flow 32 ml/h — fractions of 10.5 ml.

II. — APPLICATION A L'ETUDE DES PRODUITS D'HYDROLYSE NUCLEASIQUE DU t-RNA DE E. COLI.

La figure 5 fournit les résultats de la séparation d'un hydrolysat par la RNase-pancréatique de 0,250 mg de t-RNA commercial de E. Coli selon la technique de Singhal (12). Une élution phosphatasique des spots, suivie d'un marquage externe par le borohydrure tritié, permet, en appliquant les méthodes décrites, de déterminer la séquence de la plupart des oligonucléotides à partir de quantités d'oligonucléotides allant de 2 à 50 nanomoles. Les spots sont élués en présence de phosphatase alcaline, puis soumis au marquage externe par la méthode décrite. Le comptage d'une première aliquote permet de préciser la répartition des différents oligonucléotides par rapport au t-RNA total ; l'hydrolysat alcalin d'une 2me aliquote est soumis à l'analyse chromatographique sur papier, selon la méthode 3-a, pour préciser le nucléoside terminal et la longueur de chaîne ; une 3me aliquote est soumise, après hydrolyse alcaline, à une hydrolyse phosphatasique complémentaire suivie d'un 2me marquage, ce qui fournit la composition en bases par analyse sur Dowex 1 × 8 selon la méthode 3-b ; enfin, une 4me aliquote est hydrolysée par la phosphodiestérase de venin et analysée à nouveau par Dowex 1 × 8 après un 2me marquage. On détermine ainsi l'enchaînement des di, tri- et tétranucléotides jusqu'à des quantités limites de 2 à 5 nanomoles. Le détail de ces travaux qui sortent du cadre restreint de cet exposé fait l'objet d'un mémoire en préparation (13).

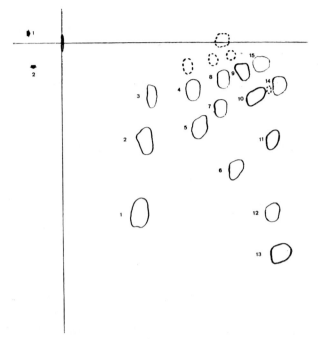

Fig. 5. — Electrochromatographie d'un hydrolysat par la
RNase-pancréatique de t-RNA et de E. Coli.

Papier d'Arches 304. —

1 : électrophorèse en tampon formiate 0,06 M, pH 2,7 - 800 volts ·
8 heures.

2 : chromatographie descendante en butanol tertiaire/formiate d'ammonium 0,06 M, pH 3,8 - (87/63) - 83 heures.

Dépôt correspondant à 250 μg. de t-RNA —

Désignation des spots étudiés : 1 = Cp ; 2 = ApCp ; 3 = ApApCp+
(ApAp)Cp ; 5 = GpCp ; 6 = ApUp + Ap ψ p ; 7 = ApApUp ;
8 = GpGpCp ;   10 = (ApGp)Up ;   11 = GpUp + Gp ψ P ;
12 = rTp + U^h/p ; 13 = Up + ψ p.

Electrochromatograph of a hydrolysate by the RNase-pancreatic of
t-RNA of E. coli.

Arches paper 304 —

1 : Electrophoresis in formate plug 0.06 M, pH 2.7 - 800 volts -
8 hours.

2 - Descending chromatograph in tertiary butanol/formate of ammonium 0.06 M, pH 3.8 - (87/63) - 83 hours.

Sediment corresponding to 250 μg of t-RNA —

Description of spots studied : 1 = Cp ; 2 = ApCp ; 3 = ApApCp +
(ApAp)Cp ; 5 = GpCp ; 6 = ApUp + Apψp ; 7 = ApApUp ;
8 = GpGpCp ; 10 = (ApGp)Up ; 11 = GpUp + Gp ψ p ; 12 =
rTp = U^h/p ; 13 = Up + ψ p.

L'association de ces différentes techniques permet donc, d'ores et déjà, d'envisager de nombreuses applications dans la microanalyse des séquences nucléotidiques. Une exploitation de ces méthodes est actuellement en cours pour préciser le sens de lecture et l'importance du site actif de la RNA-polymérase à partir d'un complexe DNA-RNA, isolé du foie de rat (14).

Bibliographie

1. AMALDI, F. et ATTARDI, G. *J. Mol. Biol.,* **32** : 737-755, 1968.
2. RAJBHANDARY, U.L. *J. Biol. Chem.,* **243** : 556, 1968.
3. MURAKAMI, M. et WINZLER, R.J. *J. Chromatog.,* **28** : 344, 1967.
4. SEGARD, E. et DESREUMAUX, S. *Bull. Soc. Chim. Biol.* (sous presse).
5. WYATT, G.R. *in* « Nucleic Acids », CHARGAFF, E. et DAVIDSON, J.N. Ed. Acad. Press, vol. 1, p. 243, 1955.
6. PARTRIDGE, S.M. *Biochem. J.,* **42** : 238, 1948.
7. NIHEI, T. et CANTONI, G.L. *J. Biol. Chem.* **238** : 3991, 1963.
8. COHN, A.W. *in* « Nucleic Acids », CHARGAFF, E. et DAVIDSON, J.N. Acad. Press, Vol. 1, p. 211, 1955.
9. McCULLY, M.S. et CANTONI, G.L. *J. Biol. Chem.,* **237** : 3760, 1962.
10. ISHIKURA, H., NEELON, F.A. et CANTONI, G.L. *Science,* **153** : 300, 1966.
11. NEELON, F.A., MOLINARO, M., ISHIKURA, H., SHEINER, L.B. et CANTONI, G.L. *J. Biol. Chem.,* **242** : 4515, 1967.
12. SINGHAL, R.P. Lille, Thèse de doctorat, 1967.
13. DESREUMAUX, S., BARBRY, G., SINGHAL. R.P. et SEGARD, E. (en préparation).
14. KRSMANOVIC, V., SERGEANT, A., DESREUMAUX, S., SEGARD, E. et MONTREUIL, J., (en préparation) et KRSMANOVIC, V., SERGEANT, A., DESREUMAUX, S., SEGARD. E. et MONTREUIL, J. *Arch. Intern. Physiol. Biochim..* **76** : 13, 1968.

Dosage de l'oxaloacétate après chromatographie sur papier en présence de glucose

par

J. J. WOHNLICH

attaché de recherche au C.N.R.S., Unité de Diététique, Labor. Chim. Biol.
Hôpital Bichat, Paris - France.

L'oxaloacétate (OAA), métabolite clé de la gluconéogénèse, peut être révélé par la benzidine (80° C, 5 min) en même temps que les glucides (1) après chromatographie sur papier.

Quand il s'agit de séparer un mélange de sucres contenant des oligosides, une chromatographie multiascensionnelle est nécessaire et la tache de glucose interfère avec celle de l'OAA.

Pour rendre possible un dosage simultané sur le même chromatogramme, nous avons développé le procédé suivant : Le glucose peut être dosé directement sur papier à 570 mµ, comme décrit antérieurement (2), sans qu'intervienne la tache jaune citron de l'OAA dont l'absorption à cette longueur d'onde est pratiquement nulle. L'OAA donne une absorption linéaire à 430 mµ dans la zone de 10 à 40 µg pour une gamme d'échantillons de 0,04 ml déposée sur papier (Fig. 4). Mais le glucose absorbe aussi à cette longueur d'onde (Fig. 1).

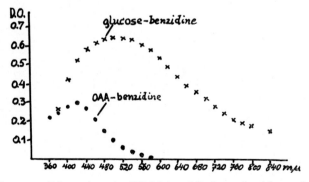

Fig. 1. — Spectres d'absorption de (l'OAA-benzidine) et du (glucose-benzidine) sur papier.

Absorption spectra of (OAA-benzidine) and (glucose-benzidine) on paper.

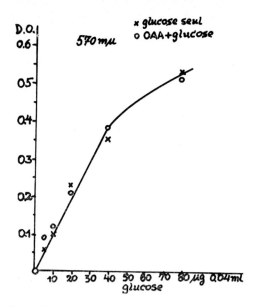

Fig. 2. — Photométrie directe sur papier d'une gamme de glucose avec et sans OAA après chromatographie multiascensionnelle (3 ascensions à 25°C, 14 heures chacune).

Direct photometry on paper of a range of glucose with and without OAA after multi-ascensional chromatography (3 ascensions at 25° C, 14 hours each).

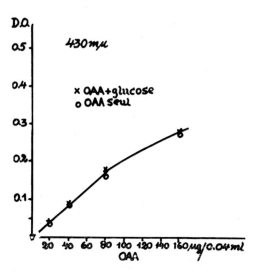

Fig. 3. — Courbes d'étalonnage d'une gamme de l'OAA avec et sans glucose après révélation par la benzidine et élution par l'isopropanol à 30 %.

Calibration curves of a range of OAA with and without glucose after revelation by benzidine and elution by isopropanol at 30 %.

Pour remédier à cet inconvénient, nous découpons la tache qui contient le glucose et l'OAA et nous éluons la couleur jaune citron due à l'OAA par 3 ml d'une solution d'isopropanol à 30 % (30° C, 2 heures) qui ne dissout pas la couleur brune due au glucose. La couleur jaune en solution est lue au spectrophotomètre à 430 mμ contre un blanc, obtenu par élution d'une même surface de papier ne contenant ni glucose ni OAA.

Les résultats sont représentés par les figures 2 et 3. La figure 2, sur laquelle sont données les courbes d'étalonnage de glucose avec et sans OAA, montre une identité pratiquement complète des deux tracés, ce qui prouve que la couleur jaune citron de l'OAA n'intervient pas avec celle du glucose à 570 mμ. La figure 3 donne les courbes d'étalonnage de l'OAA seul et en présence de glucose après élution du papier. Dans ce cas aussi, les courbes sont presque identiques, ce qui montre que la couleur brune due au glucose est insoluble dans la solution d'isopropanol à 30 %.

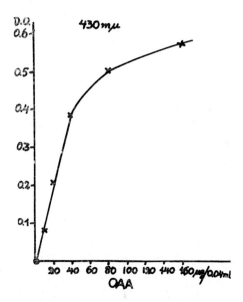

Fig. 4. — Gamme de (l'OAA-benzidine) après chromatographie ascendante (1 seule ascension 14 heures, 20° C) obtenue par photométrie directe sur papier (voir le texte).

Range of (OAA-benzidine) after ascending chromatography (1 single ascension 14 hours, 20° C) obtained by direct photometry on paper (see text).

La séparation de l'OAA et du glucose est pourtant possible dans des conditions bien définies et s'il n'y a pas d'oligosides dans le mélange de glucides à séparer : c'est le cas d'une seule ascension dans le système butanol-pyridine-eau (6 : 4 : 3) effectuée en 14 heures à 20° C. Dans ces conditions, la tache correspondant à l'OAA a une valeur de R_G de 0,85 et elle peut être découpée et lue directement au spectrophotomètre à

430 mμ, comme nous l'avons décrit pour l'acide pyruvique (1) (Fig. 4). Mais dans ces conditions aussi, la lecture directe sur papier est limitée par les concentrations de l'OAA et du glucose (80 μg/0,04 ml environ) à cause d'une interférence de ces deux substances, et le procédé d'élution est nécessaire comme nous venons de le décrire.

Bibliographie

(1) WOHNLICH, J.J. *Bull. Soc. Chim. Biol.,* **49** ; 900, 1967.
(2) WOHNLICH, J.J. *Bull. Soc. Chim. Biol.,* **43** ; 1121, 1961.

The isolation of phytotoxins from liquid cultures of ceratocystis ulmi

by

H. REBEL and C.A. SALEMINK

(Laboratory of Organic Chemistry, State University Utrecht,
the Netherlands)

The Dutch elm disease has made since 1921 many victims under the elm trees in the Netherlands, in the United States and in Canada. The disease is caused by the fungus *Ceratocystis ulmi* (Buisman) C. Moreau, which is transported to these tree species by several kinds of bark beetles.

Fundamental research has been done in Europe as well as in both countries already mentioned and it was *Zentmyer,* who discussed in 1942 that phytotoxic material was present in the filtrate of a nutrient medium in which Ceratocystis ulmi had been grown. This has been confirmed by *Dimond* in 1947, by *Feldman, Caroselli* and *Howard* in 1949, by *Kerling* in 1955 and by *Beckman* in 1956. We started our efforts to isolate the toxic factor or toxic factors in close and indispensable collaboration with *Prof. Dr. L.C.P. Kerling* and *V. Tchernoff,* of the Phytopathological Laboratory Willie Commelin Scholten, State University of Utrecht at Baarn.

In our experiments Ceratocystis ulmi was cultured at 24° in a nutrient medium of the following composition (per liter of tap water)

D-glucose	20	g
L-asparagine-monohydrate	**2**	
KH_2PO_4	1,5	
$MgSO_4\ 7H_2O$	1	
$ZnSO_4\ 7H_2O$	20	mg
$FeCl_3$	10	
pyridoxin	1	
thiamin	1	

The culture liquid — which itself has no phytotoxic activity — was dispensed in 750 or 1000 ml erlenmeyer flasks in layers of about 3 cm in depth and sterilized at 120° C for 20 minutes. After cooling two drops of a concentrated suspension of Ceratocystis ulmi spores were added to the sterilized flasks and these incubated at 23° C during three weeks. The cultures were shaken, during that time, for aeration.

After incubation mycelium and spores has been centrifuged and thereafter the supernatant passed through asbestos sterilizing filters (Carlson, Copenhagen) under pressure or suction.

The phytotoxic activity of this culture filtrate, or of any fraction derived from this, was tested on elm sprouts, originated from the susceptible elm clone *Ulmus hollandica Belgica*, raised according to the method of Tchernoff.

Sprouts of about 15 cm were cut off and placed with their base into water immediately. After a few minutes 2 cm of the sprouts were cut off again under water. These sprouts were placed into flattenedend tubes containing about 12 ml test solutions.

Preparations in water, which contained phytotoxic material, induced distinct and characteristic symptoms of the Dutch elm disease, viz. wilting of the leaves and necrosis of the interveinal leaf zones, as a rule after 24 hours.

Minimum active concentrations to test were about 150-250 mg pro liter.

Moreover the toxicity of pure compounds were tested on rooted, one to two years old, elm cuttings through a special procedure.

The clear light yellow filtrate — with phytotoxic properties — has been treated, in order to isolate the phytotoxic factor or factors, as follows : The solution was concentrated *in vacuo* fifteen times, the temperature kept below 40° C. An equal volume of ethanol was added to the concentrate — with stirring — and rather slimy, fibrous material precipitated. It could easily be removed with a glass rod. This material gave after purification a white hygroscopic product — a polysaccharide with only glucose- units — without phytotoxic activity.

The remaining alcoholic solution is filtered through paper by suction — to remove all polysaccharide particles — and the ethanol evaporated *in vacuo*.

This solution was dialysed to separate material of high molecular weight — if present — from glucose, salt and other compounds of small molecular size. It appeared that the non-dialysable fraction retained the phytotoxicity because the dialysable part was inactive in this respect. This fact indicated that the phytotoxic material must be macromolecular.

Adding acetone or ethanol to the dialysed solution — after concentration *in vacuo* — gave an almost white solid precipitate, that contained all phytotoxic activity and at the same time proved to be the whole solute content of the non-dialysable fraction.

The yield varied between 0,5 — 1,5 g crude toxin/ pro 10 L. culture filtrate.

The crude toxin was not homogeneous, as was indicated by ultracentrifuge sedimentation patterns. Two compounds — to indicate as A and B — are present.

A represents 90 - 95 % of the material and has a molecular weight 25.-30.000.

B, the minor component, has a molecular weight of about 1.000.000 and is probably heterogeneous.

The main problem — now — was to separate the two substances A and B. Before I will indicate the experiments which are carried out, to obtain pure A and pure B, some analytical data of this material are to be given :

The crude toxin has a high molecular weight and is extremely soluble in water, the solutions displaying a remarkable viscosity. The phytotoxic material is heatstable, which — in general — excludes proteins or nucleic acid material.

Infrared spectra of the crude toxin — and later of pure A — clearly show the characteristic pattern of carbohydrate absorptions, which has been confirmed by the presence of monosaccharides after hydrolysis of the material.

Still some amino acids are involved in the structure : elementary analysis indicates nitrogen. UV spectra of the crude toxin or phytotoxic fractions thereof indicate protein material. This is supported by the fact that the culture filtrate of Ceratocystis ulmi and all phytotoxic fractions easily produce stable foams.

Only a trace of sulphur and phosphorus could be detected.

The following experiments were carried out to separate compounds A and B.

Paper chromatography

An attempt to separate compounds A and B (mol. weights ± 25.-30.000 and 1.000.000) by paper chromatography was not successful. The material did not move, even in very polar solvent mixtures. Irregular migration with retardation occurred when water was the developing solvent, but no separation was detected.

Gel permeation chromatography

The molecular weights of A and B differ so widely — as already has been indicated — that it seems very logical to apply gel permeation chromatography for the separation of the two compounds.

The dextran gel Sephadex G 100 (Pharmacia) seemed most suitable for the separation of A and B.

However the elution curves always showed an asymmetrical peak, the elution volume of which did not differ significantly from the column void volume.

The material, represented by the peak, was collected, isolated and centrifuged at 60.000 rpm. Its « schlieren » pattern was identical to that of the original preparation !

We repeated the experiment with a polyacrylamide gel Biogel P 100 (Bio Rad). The eluted material corresponded here too with a « tailed » peak, which yielded a fraction with — in comparison with the crude toxin — unchanged sedimentation characteristics in the ultracentrifuge.

Electrophoresis

Hereafter we tried separation through electrophoresis. It was proved namely that compound B was destructed when the crude toxin was treated with pronase, an aspecific proteolytic enzyme from *Streptomyces griseus*. Substance B probably has predominantly protein character. Compound A — after treatment of the crude toxin with pronase — seemed homogeneous, has the same sedimentation constant of A, collected by preparative centrifuging at 25.000 rpm and was phytotoxic to elm sprouts and trees. The activity, though, seemed to have been weakened, and the yield was rather low.

We subjected the material (A + B) to electrophoresis on paper at pH values between 3 and 9, in order to get information about its electrical properties. At each pH any distinct move of the material could not be detected, only normal band spreading was present.

Ion exchange chromatography

The crude toxin passed Amberlite C9-50 and DEAE-cellulose or DEAE-Sephadex A-50 columns unretarded and unchanged. Dilute buffers were used and the pH was varied stepwise in suitable direction. No separation has been obtained between components A and B.

Preparative ultracentrifuging

Component A — as already has been indicated — could be collected by preparative ultracentrifuging at 25.000 rpm. The procedure suffered from the following disadvantages :

1. Long runs were required : 24 - 27 hours, the time, calculated for complete sedimentation of component B and careful collection of a « safe » volume of a solution with only A.

2. The yield of compound A was low and almost all material, thus collected, ran out in tests, especially tree tests.

3. Furthermore component B could not be obtained in pure form.

Gel permeation chromatography

At this stage the new agarose xerogels Sepharose 2B and 4B (Pharmacia) were introduced. In contrast with other gels, mentioned before, the pore size of the gel Sepharose 4B was large enough to allow at least one of the toxic components (A) to permeate partially into the gel beads. The crude toxin could be separated — at last — into several fractions. The first fraction from the column was substance B. Fraction A came off as two main fractions A_1 and A_2. So far we know now their structures are rather similar. Both compounds A_1 and A_2 possess good phytotoxic properties.

Concerning the chemical properties : Both substances have a high carbohydrate content and show a positive ninhydrin-reaction after hydrolysis. The compounds precipitate in a saturated ammonium sulphate solution (low yield).

This in combination with the results of gel chromatography on Sepharose 4B makes it likely that A_1 and A_2 have protein-polysaccharide character. Further purification of these two fractions is carried out.

Comparison of the in vitro-toxin with the in vivo-toxin

It was not possible — up till now — to prove the identity of the *in vitro* toxins as extracellular factors in the incubated culture filtrate with the *in vivo* toxins in the xylem sap of a diseased elm tree through immunological methods.

An antiserum against the original substance A (A_1 + A_2) could be made, but no reaction with xylem sap could be detected in the experiments so far.

References

BECKMAN, C.H. *Phytopathology,* **46** ; 605, 1956.

DIMOND, A.E. *Phytopathology,* **37** : 7, 1947.

FELDMAN, A.W., CAROSELLI, NE.. and HOWARD, F.L. *Phytopathology,* **39** : 6, 1949.

KERLING, L.C.P. *Acta Bot. Neerl.,* **4** ; 398, 1955.

LAFAYETTE, F. and HOWARD, F.L. *Phytopathology,* **4** : 82, 1951.

SALEMINK, C.A., REBEL, H., KERLING, L.C.P. and TCHERNOFF, V. *Science,* **149** : 203, 1966.

TCHERNOFF, V. *Acta Bot. Neerl.,* **12** : 40, 1963.

ZENTMYER, G.A. *Phytopathology,* **32** : 20, 1942.

ZENTMYER, G.A. and HORSFALL, J.S. *Science,* **95** : 512, 1942.

The separation of nucleotides by electrophoresis and chromatography on ion-exchange paper

by

Prof. G. SERLUPI CRESCENZI

The work that I am presenting has been carried out at the Istituto Superiore di Sanità in Rome with the collaboration of Dr. C. Paolini and Mr. T. Leggio.

We were working for several years on the separation of nucleotides mixtures but the methods used were troublesome and long. Most of these methods are based on ion exchange resins chromatography, paper chromatography and electrophoresis, thin-layer chromatography, etc....

The use of chromatography and electrophoresis on ion-exchange paper gave us a very rapid method to be applied in the determination of these compounds in pharmaceutical preparations as well as in biochemical problems like the determination of adenosine derivatives (mono-, di- and tri-phosphate) for instance in works on oxidative phosphorylation.

Chromatography on ion-exchange paper has the advantage of rapidity because aqueous solvents are generally used ; and because the compounds are linked to the paper by chemical bonds, spots are more compact and the tendency of tailing is reduced. The introduction on an experimental basis of an ion-exchange paper, polyethylene coated, may be another considerable step in avoiding one of the disadvantages of their use — a certain degree of brittleness. Both in chromatography and in electrophoresis the mobility of the compounds to be separated depends strongly on pH in a complicated and unpredictable way : the pH influencing the specific charge, the degree of dissociation of the compounds and of the paper, and the affinity of the compounds for the paper. It has therefore been found useful to study the dependence on pH of the mobility of nucleotides both in electrophoresis and in chromatography. The data obtained can be used to predict the pH at which a particular separation can be achieved. A few examples will be reported, together with the separation by a bidimensional method of twelve derivatives : the 5'-mono, di- and tri-phosphates of adenosine, guanosine, cytidine and uridine.

Ascending chromatography was carried out on 15 × 15 cm. strips (solvent path 13 cm.).

Electrophoresis was carried out with the old technique described by Markham and Smith (1) with water circulating in a glass serpentine immersed in carbon tetrachloride, to cool the latter, A glass support was used to hold the paper and to avoid breakage of the DE 81 paper when wet. No support was used with the polyethylene-coated paper.

For bidimensional separations, electrophoresis was first carried out on a strip of paper. The part of the strip containing the spots was then cut out and the longest edge was overlapped by 1 mm. to the edge of another sheet of paper of dimensions 35 × 45 cm. The two edges were kept together either by a strip of cellulose adhesive tape or by squeezing them between two glass-rods held together by two elastic strings. Since the cellulose tape cannot be used with organic solvents, in this case the two-rod method must be used. In this way spots can be passed from a sheet to another without trouble, thus permitting the consecutive use of different solvents and different types of paper avoiding elution of the spots, concentration, etc. Chromatograms were developed using 1M acetic acid/sodium citrate buffer of the desired pH as solvent. Electrophoresis was carried out in 0.1M acetic acid/sodium citrate buffer of the desired pH.

Table I shows Rf values of the twelve nucleotides at different pH's on normal and polyethylene-coated DE 81 paper.

TABLE I.

Chromatographic *Rf* Values of the Different Nucleotides on Polyethylene-Coated and Uncoated DE 81 Paper
(experimental conditions as described in the text)

Nucleotide	Buffer : 1 *M* acetic acid/sodium-citrate							
	Whatman DE 81				Whatman DE 81, polyethylene coated			
	pH				pH			
	3.5	3.7	3.8	4.0	3.5	3.7	3.8	4.0
CMP	0.60	0.61	0.61	0.73	0.63	0.69	0.72	0.82
UMP	0.39	0.47	0.54	0.71	0.43	0.56	0.61	0.74
AMP	0.41	0.45	0.47	0.58	0.46	0.52	0.55	0.64
GMP	0.29	0.37	0.42	0.53	0.35	0.44	0.49	0.54
CDP	0.29	0.39	0.44	0.58	0.38	0.48	0.51	0.58
UDP	0.15	0.29	0.38	0.63	0.20	0.36	0.44	0.63
ADP	0.17	0.29	0.33	0.47	0.21	0.34	0.39	0.52
GDP	0.08	0.16	0.23	0.35	0.12	0.19	0.28	0.39
CTP	0.08	0.19	0.27	0.43	0.13	0.25	0.31	0.49
UTP	0.03	0.10	0.17	0.56	0.03	0.13	0.20	0.41
ATP	0.03	0.09	0.27	0.34	0.05	0.12	0.19	0.35
GTP	0.01	0.04	0.17	0.26	0.02	0.06	0.09	0.24

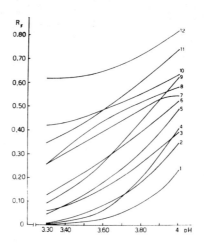

Fig. 1. – Variation of Rf of different nucleotides as function of pH in chromatography on polyethylene-coated DE 81 paper : (1) GTP ; (2) ATP ; (3) GDP ; (4) UTP ; (5) CTP ; (6) ADP ; (7) GMP ; (8) CDP ; (9) UDP ; (10) AMP ; (11) UMP ; (12) CMP.

TABLE II.

Chromatography Rf Values on Polyethylene-Coated DE 81 Paper Prewashed with 1 M Acetic Acid
(experimental conditions as described in the text)

Nucleotide	Buffer : 1 M acetic acid/sodium citrate		
	pH 3.3	pH 3.5	pH 3.7
CMP	0.75	0.78	0.78
UMP	0.51	0.66	0.73
AMP	0.54	0.59	0.67
GMP	0.39	0.50	0.61
CDP	0.42	0.58	0.65
UDP	0.23	0.47	0.59
ADP	0.24	0.41	0.52
GDP	0.14	0.27	0.41
CTP	0.13	0.31	0.50
UTP	0.04	0.16	0.39
ATP	0.05	0.16	0.32
GTP	0.02	0.08	0.21

The general order does not change in the two types of papers, although the Rf values are slightly higher in coated paper an the time required for the solvent to reach the same distance is slightly reduced.

The same results for the coated paper are reported in graphic form in figure 1, which can be used to find a suitable pH for a particular separation to be achieved.

Because of its higher strength the coated paper can be prewashed with the solvent. This treatment reduces to about one-half the time required for a run, but the saturation of ion-exchange groups of the paper by the solvent greatly affects the trend of the curves. The data are reported in table 2 and in graphic form in figure 2. It can be seen that, while with untreated paper a good separation could be obtained at pH 3.7-3.8, a similar separation can be obtained on prewashed paper at pH 3.5.

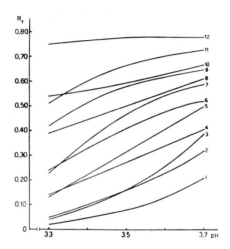

Fig. 2. — Variation of Rf of different nucleotides as function of pH in chromatography on prewashed polyethylene-coated DE 81 paper : (1) GTP ; (2) ATP ; (3) UTP ; (4) GDP ; (5) CTP ; (6) ADP ; (7) UDP ; (8) GMP ; (9) CDP ; (10) AMP ; (11) UMP ; (12) CMP.

It is worth noting that a good separation of adenosine, AMP, ADP, and ATP, and in general of mono-form di- and tri-phosphate of the same base can be achieved in 10 minutes.

Table 3 shows the distances (in cm.) travelled by different nucleotides in 180 min. with a potential gradient of 25 V/cm. The data at pH 4.5 were obtained with a lower potential gradient (19 V/cm.) and referred to the others by calculation. Here again the differences between coated and uncoated paper are irrelevant. The same data are reported in graphic form in figure 3.

It may be interesting to note that, at low pH, on ion-exchange paper, the monophosphates run in front, followed by the diphosphates and the

TABLE III.

Electrophoresis of the Twelve Nucleotides on Polyethylene-Coated and Uncoated DE 81 Paper

(mobilities expressed in cm traveled in 180 min under a potential gradient of 25 V/cm ; experimental conditions as described in the text)

	Buffer : 0.1 M acetic acid/sodium citrate							
	Whatman DE 81				Whatman DE 81, polyethylene-coated			
	pH				pH			
Nucleotide	3.0	3.5	4.0	4.5	3.0	3.5	4.0	4.5
CMP	18.9	21.8	30.0	42.1	17.9	18.5	25.5	36.2
UMP	19.0	22.0	30.2	42.7	18.3	19.4	26.4	37.1
AMP	11.6	21.6	25.0	28.8	12.3	16.7	19.1	26.4
GMP	9.8	20.2	20.0	24.7	11.0	15.5	14.9	22.0
CDP	6.5	18.3	20.8	30.0	7.3	14.1	19.3	31.7
UDP	5.4	18.8	23.4	34.1	7.0	14.0	20.4	31.4
ADP	2.6	11.6	16.0	22.8	3.9	9.6	13.5	22.0
GDP	2.5	9.7	12.1	18.2	3.4	7.9	10.0	17.0
CTP	1.8	8.1	13.7	22.6	2.7	6.2	13.9	27.0
UTP	1.5	7.5	11.0	24.3	2.4	6.2	13.7	24.9
ATP	0.8	4.0	5.8	16.5	1.6	3.4	9.3	17.4
GTP	0.5	3.5	6.6	13.5	1.1	2.5	6.4	12.7

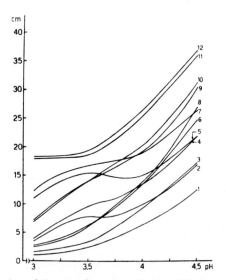

Fig. 3. – Variation of distances in cm travelled by different nucleotides in 180 min with a potential gradient of 25 V/cm as a function of pH in electrophoresis on polyethylene-coated DE 81 paper : (1) GTP ; (2) GDP ; (3) ATP ; (4) ADP ; (5) GMP ; (6) UTP ; (7) AMP ; (8) CTP ; (9) UDP ; (10) CDP ; (11) CMP ; (12) UMP.

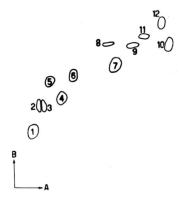

Fig. 4. — Bidimensional electrochromatography of twelve nucleotides on DE 81 paper : (1) GTP ; (2) ATP ; (3) UTP ; (4) GDP ; (5) CTP ; (6) ADP ; (7) UDP ; (8) CDP ; (9) GMP ; (10) UMP ; (11) AMP ; (12) CMP. In direction A, electrophoresis with 0.1 M acetic acid/sodium citrate buffer at pH 3 with a potential gradient of 20 V/cm for 5 hr. In direction B, ascending chromatography with 1 M acetic acid/sodium citrate buffer at pH 3.75 for 24 hr.

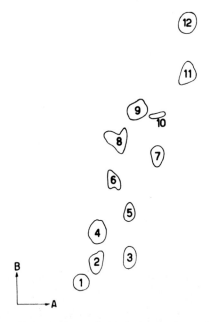

Fig. 5. — Bidimensional electrochromatography by twelve nucleotides on polyethylene-coated DE 81 paper : (1) GTP ; (2) ATP ; (3) UTP ; (4) GDP ; (5) CTP ; (6) ADP ; (7) UDP ; (8) GMP ; (9) AMP ; (10) CDP ; (11) UMP ; (12) CMP. In direction A, electrophoresis with 0.1 M acetic acid/sodium citrate buffer at pH 4.5 with a potential gradient of 19 V/cm for 145 min. In direction B, ascending chromatography with 1 M acetic acid/sodium citrate buffer at pH 3.7 for 20 hr.

triphosphates. At higher pH's the binding effect of ion-exchange groups is less marked until, at pH 5, the three derivatives of the same base run together, and at higher pH's ion-exchange paper behaves like normal paper. On occasion this pH can be used for group separation.

The tables and the data which I have shown have given us a way of choosing suitable pH's to carry out bidimensional electro-chromatography of complex mixtures.

Figure 4 exemplifies the separation of the twelve mono-, di, and tri-phosphates of adenosine, guanosine, cytidine, and uridine by bidimensional electrochromatography on uncoated DE 81 and figure 5 on polyethylene-coated DE 81. paper.

Prewashing the polyethylene-coated paper prior to chromatography, although not changing the general disposition of spots, decreases by half the time required for the separation. A good separation can be achieved in 13 hours.

In conclusion I would say that although theoretical predictions of the dependence of mobility from pH is difficult because it depends on the interaction of many factors, some of which are unknown or are impossible to measure, however, the data presented in this paper have a great practical significance. Chromatography on ion-exchange papers with aqueous buffers is of a rapidity comparable to that of thin-layer chromatography but is much simpler. The separation of simple mixtures, for instance of AMP, ADP, ATP, and adenosine, can be obtained in a matter of minutes. Determinations otherwise impossible can be made by this method, e.g., ATP in the presence of cocarboxylase, or cocarboxylase in the presence of the three adenine nucleotides (9).

References

1. MARKHAM, R. and SMITH, J.D. *Biochem. J.,* **52** : 552, 1962.
2. PAOLINI, C. and SERLUPI-CRESCENZI, G. *Gazz. Chim. Ital.,* 94 : 181, 1964.
3. SERLUPI-CESCENZI, G., PAOLINI, C. and LEGGIO, T. *Ann. Chim.* (Rome), **57**, 438, 1967.
4. SMILLIE, R.M. *Arch. Biochem. Biophys.,* **85** : 557, 1959.
5. JACOBSON, K.B. *Science,* **138** : 515, 1962.
6. LETTERS, R. and MICHELSON, A.M. *J. Chem. Soc.,* 71, 1962.
7. LEVENE, P.A. and SIMMS, H.S. *J. Biol. Chem.,* **65** : 519, 1925.
8. LEVENE, P.A., BASS, L.W. and SIMMS, H.S. *J. Biol. Chem.,* **70** . 229, 1926.
9. POLIZZI-SCIARRONE, M. *Boll. Chim. Farm.,* **104** : 812, 1965.
10. SAUKKONEN, J.J. *Chromatog. Rev.,* **6** : 53, 1964.

Contribution ultérieure à l'étude des Bisalbuminaemies et des Paralbumines

par

Luisa BONAZZI

Instituts hospitaliers de Verona - Département de Biochimie et Hématologie
Médecin en chef, Prof. E. Morelli.

L'objet de cette communication est celui de présenter une nouvelle famille avec la Bisalbuminémie, le trait familier qui se manifeste par la présence de deux albumines dans le sérum, à l'examen électrophorétique à ph 8,6 et un cas de Paralbumine, anomalie identique mais non héréditaire.

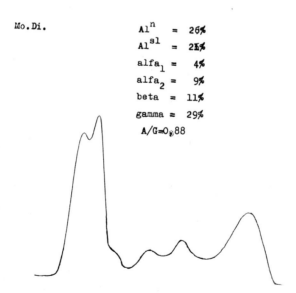

Mo.Di.

Al^n = 26%
Al^{sl} = 25%
$alfa_1$ = 4%
$alfa_2$ = 9%
beta = 11%
gamma = 29%
A/G=0,88

Fig. 1. -- Tracé électrophorétique sur papier du Cas Index de la famille avec Bisalbuminémie.

Electrophoretic pattern on paper of the Index Case of a family with bisalbuminemia.

TABLE I

Cas Index de Bisalbuminémie décrits par :	Année	N. des sujets examinés de chaque souche familiale	N. des sujets avec Bisalbuminémie	Type de Bisalbuminémie	Lieu d'origine de famille
Scheurlen, Whurmann	1955-59	10	4	rapide	Suisse allemande
Knedel	1957	8	5	?	Allemagne
Nennestiel et al.	1957	—	3	rapide	Angleterre
Knedel	1958	3	3	lente	Allemagne
Bennhold et al.	1958	2	2	lente	Suisse allemande
Earle et al.	1958	43	25	lente	Norvège
Weiner	1959	—	9	?	Italie (Salerno)
Wieme	1960	10	7	rapide	Belgique
Franglen et al.	1960	15	6	lente	Angleterre
Miescher	1960	27	8	rapide	Suisse allemande
Adner et al.	1961	17	12	lente	Suède
Sarcione et al.	1962	13	6	lente	Italie (Napoli)
Cooke et al.	1962	—	1	lente	Angleterre
Cooke et al.	1962	—	1	lente	Angleterre
Robbins et al.	1963	9	4	lente	Amérique
Efremov et al.	1964	—	1	lente	Norvège
Tárnoky et al.	1964	11	3	rapide	Angleterre
Galb et al.	1964	—	1	rapide (transitoire)	Allemagne
Drachmann et al.	1965	100	28	lente	Danemark
Terzani et al.	1966	3	3	?	Italie (Roma)
Bell et al.	1967	38	16	lente	Inde

La Bisalbuminémie, (Fig. 1), rencontrée la première fois par Scheurlen en 1955 dans un jeune homme de 25 ans atteint de diabète mellitus, a été décrite avant 1968, en 21 familles, appartenant pour la plupart à des groupes ethniques d'origine norvégienne, anglaise, allemande, danoise ; de ces 21 familles, une seulement est décrite en Italie, deux autres le sont en Amérique, sur émigrants italiens. - Tab. I.

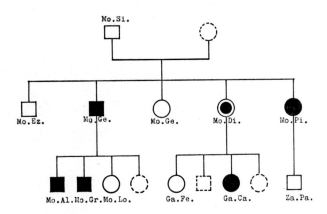

Fig. 2. — Arbre généalogique de la famille avec Bisalbuminémie dont le Cas Index est Mo. Di.

Family tree of the family with bisalbuminemia whose Index Case is Mo. Di.

Quant à nous dans une communication précédente (1968) nous avons présenté cinq familles avec la Bisalbuminaemie, familles qui étaient arrivées à notre observation dans l'exécution routinière d'environ 12.000 tracés électrophorétiques sur papier ; il nous semble pourtant que la présentation d'une sixième famille ne soit pas inutile, soit pour l'illustration de l'anomalie en elle-même et aussi pour la relative fréquence des cas que nous avons constatés en comparaison du très petit nombre de ceux trouvés dans la littérature en général et dans celle d'Italie en particulier. - Tab. II.

TABLE II

Cas de Bisalbuminaemia que nous avons déjà décrits :

Cas Index	N. des sujets examinés de chaque souche familière	N. des sujets avec Bisalbuminaemia
Bo. En.	8	5 sur 3 générations
Ba. An.	5	4 sur 2 générations
Be. Vi.	—	1
Me. El.	+	1
Za. Er.	—	1

Dans la sixième famille, on a pu constater cette anomalie de caractère héréditaire, outre que sur le Cas Index et sur 5 autres membres de la famille pour un complexe de deux générations (Fig. 2). Notre Cas Index est atteint de maladie rhumatismale et les membres de la famille dans laquelle a été rencontrée l'anomalie même, ne présentaient aucun syndrome morbide, confirmant ainsi les données de la littérature.

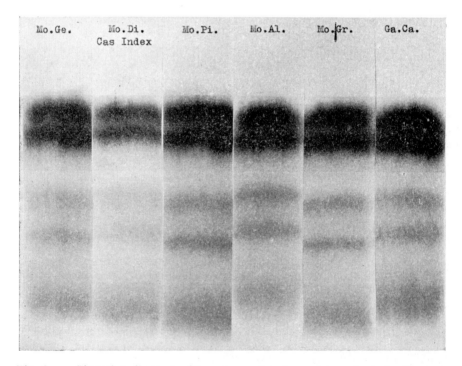

Fig. 3. — Séparation électrophorétique sur papier des membres de la famille avec Bisalbuminémie.

Electrophoretic separation on paper of the members of the family with bisalbuminemia.

Tandis que dans la plupart des cas reportés en littérature la Bisalbuminémie est remarquée lorsque l'on exécute l'examen électrophorétique sur agar-gel ou sur acétate de cellulose, pour nous elle se révèle très visiblement même sur papier (Fig. 3).

Seulement, l'acétate de cellulose nous donne une indication plus sûre de la relative mobilité de deux albumines, comparée à la mobilité d'une albumine normale (Fig. 4).

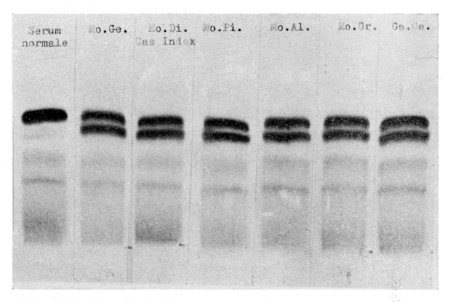

Fig. 4. — Séparation électrophorétique sur acétate de cellulose des membres de la famille avec Bisalbuminémie comparée avec celle d'un sérum ayant albumine normale.

Electrophoretic separation on cellulose acetate of the members of the family with bisalbuminemia compared with the separation of a serum containing normal albumin.

Quant à la mobilité des deux albumines en accord à la classification de Wieme de Bisalbuminémie de type rapide et de Bisalbuminémie de type lent, suivant la migration plus rapide ou plus lente de l'albumine anormale par rapport à l'albumine normale, notre nouveau Cas Index et les membres de sa famille, ainsi que les cinq autres familles que nous avons trouvées dans la région de Vérone, présentent tous l'albumine lente, c'est-à-dire que la fraction anomale est comprise entre l'albumine normale et l'alpha-globuline.

L'éventuelle différence immunologique entre les deux albumines a été recherchée au moyen de l'immunoélectrophorèse.

Nous avons remarqué un arc de précipitation unique plutôt élargi par rapport à l'arc de précipitation provenant d'un sérum avec albumine normale : dans les limites de notre étude, dans tous les cas que nous avons examinés les deux albumines ont démontré avoir la même affinité antigénique. (Fig. 5.)

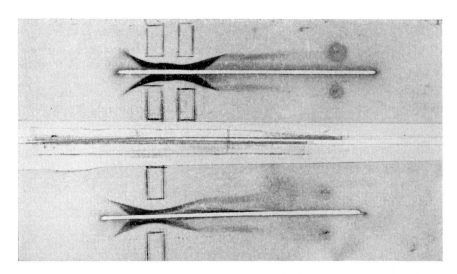

Fig. 5. — Immunoélectrophorèse d'un sérum avec Bisalbuminémie au moyen d'antiserum spécifique antialbumine de lapin de la Behringwerke Ag Marburg/Lahan, (Sérum appartenant au Cas Index Mo.Di.) comparée avec l'immunoélectrophorèse d'un sérum ayant albumine normale.

Immunoelectroporesis of a serum with bisalbuminemia with a specific antialbumin rabbit antiserum from Behringwerke Ag, Marburg/Lahan, (serum of the Mo. Di. Index Case) compared with the immunoelectrophoresis of a serum containing normal albumin.

Les deux albumines sont présentes en quantité à peu près constante et leur rapport est environ 1 : 1 ; de plus la présence des deux albumines n'est liée ni à hyperprotidémie ni à hyperalbuminémie, comme on peut le voir de la Tab III qui montre les valeurs des protéines totales et les pourcentages de chacune des fractions élèctrophorétiques.

Enfin en ce qui concerne la capacité des sérums à fixer des colorants comme marqueurs de l'albumine type, le bleu de bromophénol, nous avons constaté en accord avec les résultats de Earle et coll. que cette substance colore les deux albumines.

A l'encontre de la Bisalbuminémie héréditaire qui vient d'être décrite nous pensons devoir signaler un cas d'observation tout à fait récent de Double Albumine que nous avons rencontré en exécutant l'examen électrophorétique d'un liquide pleurique sans que cette anomalie fût présente dans le sérum du même patient atteint de néoplasie pulmonaire (Fig. 6).

TABLE III

Concentration des protéines sériques des sujets avec Bisalbuminémie.
Le Cas souligné indique le Cas Index de la souche familière.

Cas avec Bisalbuminémie (initiales, sexe, âge)	Protéines totales (biurète)	Fractionnement électrophorétique sur papier						
Mo. Ge, masc., a.48	7,10	AL^n	AL^{sl}	a_1	a_2	beta	gamma	A/G
Mo, Di, fém., a.41	6,80	29	26	4	9	12	20	1,22
Mo. Pi, fém., a.38	7,30	26	21	4	9	11	29	0,88
Mo. Al., masc., a.13	6,90	28	27	3	9	12	21	1,22
Mo. Gr., masc., a.10	6,75	30	28	4	10	12	16	1,38
Ga. Ca., fém., a. 6	7,10	30	29	4	10	11	16	1,44
		28	25	5	8	12	22	1,13
		Al^n (normal) = normale						
		Al^{sl} (slow) = lente						

Ce qui rend la remarque encore plus intéressante c'est que :

— ce pic bifide de l'albumine peut être rencontré seulement à l'exécution de l'électrophorèse sur acétate de cellulose (l'électrophorèse sur papier montre un seul pic anormal en ampleur),

— ce pic bifide de l'albumine est de type rapide, suivant la classification de Wieme, c'est-à-dire que la fraction anomale migre plus rapidement que l'albumine normale,

— ce pic bifide de l'albumine tend à disparaître presque complètement après quelques semaines.

Nous reportons ce cas soit pour la particulière rareté de l'anomalie soit pour la pathogénèse de l'anomalie même.

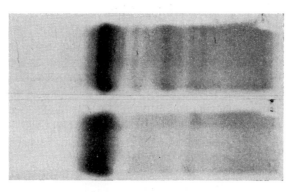

Fig. 6. — Séparation électrophorétique du liquide pleurique avec double Albumine comparée à celle du sérum du même patient ayant albumine normale.

Electrophoretic separation of the pleural liquid with double albumin compared with the electrophoretic separation of the serum of the same patient with normal albumin.

Tandis que nous nous réservons une étude plus approfondie de ce phénomène, qui est peut-être dû au grave trouble métabolique présent en ce patient et tel à modifier perceptiblement quelques propriétés de l'albumine et à varier d'une manière remarquable sa mobilité, nous nous rapportons au cas cité dans la littérature qui décrit la Double Albumine présente dans le liquide amniotique et dans le sérum foetal, mais non dans le sérum maternel (Aharon et coll. 1964) ; Double Albumine transitoire de type rapide non héréditaire (Galb et Coll. 1964) ; Albumines Bifides dues à la capacité de l'albumine même à se lier à des substances étrangères et qui se vérifient quand ces substances (bilirubine, acides gras, urobilinogène, acides biliaires) se trouvent en concentration plus élevée que celle normalement présente dans le sang ; Pics Bifide dus non à l'albumine mais à une préalbumine riche en triptophane.

Ces différents aspects de l'albumine plasmatique convalident l'hypothèse de Sogami et Foster que celle-ci doit être formée d'une population microhétérogène de molécules semblables, mais quelque peu différentes.

Un des plus intéressants développements dans l'étude des protéines dans ces dernières années a été justement la découverte d'une entière série d'aspect biochimique différente dans chacune des protéines plasmatiques, comme conséquence du développement de méthodes sensibles capables de relever et de caractériser des protéines isolées, de mélange hétérogènes excessivement complexes.

L'aspect biochimique de cette microhétérogénéité demande d'abord une discussion sur le mécanisme génétique de la synthèse protéique ; et partant conduit à considérer l'hypothèse que chaque gène exprime ses effets sur une suite d'aminoacides dans une protéine particulière ou chaîne polypeptidique, c'est pourquoi les mutations du gène peuvent être considérées comme la cause d'altération de cette suite, altération qui s'extrinsèque dans la synthèse d'une variante spécifique d'un type particulier de protéine ; et encore demande une étude plus approfondie des problèmes complexes métaboliques.

Bibliographie

ADNER, L. & A. REDNORS. En slakt tva albuminfractioner i serum elektroforesen (Bisalbuminaemia). *Nord. Med.*, **18** : 623, 1961.

BELL H. E., NICHOLS S. F. and THOMPSON Z. R. Bisalbuminemia of fast type with a homosygote. *Clin. Chim. Acta,* **15** : 247, 1967.

BENNHOLD H., OTT H. und SCHEURLEN G. Beitrag zur Frage der genbedingten Bluteiweisstörungen. *Verhandl.. Deut. Ges. Inn. Med.,* **64** : 279, 1958.

BONAZZI L. On a rare genetic variation of plasma albumin : Bisalbuminaemia. *Clin. Chim. Acta,* **20** : 362, 1968.

BONAZZI L. Su di una rara variazione genetica dell'albumina plasmatica : La Bisalbuminemia. Min. Med. 1968 in corso di pubblicazione.

BRZEEINSKI A., SADOVSKY, E. and SHAFRIR E. Protein composition of early amniotic fluid and fetal serum with a case of Bisalbuminemia. *Am. J. Obst. Gyn.,* **89** : 488, 1964.

COOKE K. G., CLEGHORN T. E. and LOCKEY E. Two new families with Bisalbuminaemia : an Examination of possible links with other genetically controlled variant. *Bioc. J.,* **81** : 39, 1962.

DRACHMANN O., HARBOE N. M. G., SVENDSEN P. J. and SOGAARD T. JOHNSEN. Bis or Paralbuminaemia, a genetic alteration in plasma albumin. *Danisch. Med. Boll.,* **12** : 74, 1965.

EARLE D. P., HUTT M. P., SCHMID K. and GLITIN D. A unique human serum albumin transmitted genetically. *Trans. Assoc. Am. Physicians,* **71** : 69, 1958.

EARLE D. P., HUTT M. P., SCHMID K. and GLITIN D. Observations on double albumin : a genetically transmitted serum protein anomaly. *J. Clin. Invest.,* **38** : 1412, 1959.

EFREMOV G. and BRAEND M. Serum Albumin-polymorphism in man. *Science,* **146** : 1679, 1964.

FOSTER J. F. *in* F. W. Putman (Ed.). The plasma proteine, vol. 1°. *Academic Press,* New York, 1960, p. 179.

FOSTER J. F., SOGAMI M., PETERSEN H. A. and LEONARD W. J. The microheterogeneity of plasma Albumins. *The J. of Biol. Chem.,* **240** : 2495, 1965.

FRANGLEN G., MARTIN N. H., HARGRAVES T., SMITH M. J. H. and WILLIAMS D. J. Bisalbuminaemia. A hereditary albumin abnormality. *The Lancet,* **6** : 307, 1960.

FRIEDMAN M., BYERS S. O. and ROSENMAN R. H. The accumulation of serum cholate and its relationship to Hypercholesteremia. *Science,* **115** : 363, 1952.

GALB F. and HBER E. G. Passeggere, nicht hereditäre Doppelalbuminaemie, *Ann. Paediet.,* **202** : 81, 1964.

GLATTHAAR E., SUENDERHAUF H. und WUNDERLY Ch. Eletrophoretische Untersuchungen bei normaler Gravidität und Spathestose. *Schweiz. Med. Wschr.,* **81** : 592, 1951.

GLITIN D., SCHMID, EARLE D. P. and GIVELBERG H. Observations on double albumin. 11. A Peptide difference between two genetically determined human serum Albumins. *J. Clin. Invest.,* **40** : 820, 1961.

GRASSMANN W. K. K. HANNING and KNEDEL M. Über ein Verfahren zur elektrophoretischen Bestimmung der Serum Proteine auf Filterpapier. *Dtsch. Med. Wschr.,* **76** : 333, 1951.

HOCH-LIGETI C., HOCH H. Electrophoretic Studies on Human Serum Albumin *Biochem. J.,* **43** : 556, 1948.

KARJALA S. A. and NAKAYAMA Y. Studies on Fast moving Albumins in human serum. *Clin. Chim. Acta,* **4** : 369, 1959.

KNEDEL M. Die Doppel-Albuminamie, eine neue erbliche Proteinanomalie. *Blut,* **3** : 129, 1957.

KNEDEL, M. Über eine neue vererbte Proteinanomalie. *Clin. Chim. A. Acta,* **3** : 72, 1958.

LEUTSCHER J. A. *J. Clin. Invest.,* **19** : 313, 1940.

MAHAUX J. and DELCOURT R. La réaction de Hanger, la Thimol-Test, la cholestérinémie et le diagramme d'électrophorèse dans les hypothyroïdies chroniques. *Ann. And. Paris,* **10** : 620, 1949.

McLOUGHLIN J.V. Splitting of serum Albumin during Electrophoresis on Paper, *Nature,* **191** : 1305, 1961.

MIESCHER F. Neues Vorkommen der vererbbaren Doppelalbuminämie. *Schweiz Med. Wochschr.,* **90** : 1273, 1960.

NENNSTIEL H. J. and BECHT T. Über das erbliche Auftreten einer Albuminspaltung im Elektrophoresediagramm. *Klin. Wschr.,* **35** : 689, 1957.

ROBBIN J. L., HILLS G. A., MARCUS S. and CARLQUIST J. H. Paralbuminemia : Paper and cellulose acetate electrophoresis and preliminary immunoelectrophoresis analysis. *J. Lab. Clin. Med.,* **62** : 753, 1963.

SARCIONE E. J. and AUNGST C. W. Bisalbuminaemia associated with albumin thyroxine-binding defect. *Clin. Chim. Acta,* **7** : 297, 1962.

SARCIONE E. J. and AUNGST C. W. Studies in Bisalbuminaemia : Binding properties of the two Albumins. *Blood,* **20** : 156, 1962.

SCHEURLEN P. G. Über Serumeiweissveränderungen beim Diabetes Mellitus. *Klin. Wochschr.,* **33** : 198, 1955.

SCHULTZE H. E. Elektrophorese mit isolierten Plasmaproteinen. *Clin. Chim. Acta,* **3** : 24, 1958.

SOGAMI M. and FOSTER J. K. Microheterogeneity as its explanation for resolution of N and F Forms of Plasma Albumin in electrophoresis and other Experiments. *J. Biol. Chem.,* **238** : Pc. 2245, 1963.

TARNOKY A. L. and LESTAS A. N. A new type of Bisalbuminaemia. *Clin. Chim. Acta,* **7** : 551, 1964.

TARNOKI A. L. Varieties of Bisalbuminaemia. *Proc. Of. the Ass. of Clin. Biochem.,* **4** : 12, 1966.

TERZANI G. e NATALIZI G. Su un caso di Bisalbuminemia. *Il Laboratorio nella diagnosi medica,* **9** : 1, 1966.

WEINER L. Citato in Earle P.D. e Coll. *J. Clin. Invest.,* **38** : 1412, 1959.

WIEME, R. J. Studies on Agar gel electrophoresis techniques applications, Brussels, Arscia, 1959.

WIEME R. J. Bisalbuminemia. *The Lancet,* **1** : 830, 1960.

WIEME R. J. On the presence of two albumins in certain normal human sera and its genetic determination. *Clin. Chim. Acta,* **5** : 443, 1960.

WIEME R. J. On the occurence of different types in genetically determined Bisalbuminaemia in H. Peeters (Ed) Protides of the Biological Fluida, Proceedings of the 9th Colloquium Bruges, 1961, Elsevier, Amsterdam, 1962, 221.

WHURMANN F. Albumindoppelzacken als vererbbare Bluteiweissanomalie. *Schweiz. Med. Wochschr.,* **89** : 150, 1959.

Iso-enzymes urinaires de leucine-amino-peptidase et de alanine-amino-peptidase

par

VAN TRIET, A.J. *, VAN DER HEIDEN, D.A. **

Summary.

In some diseases of the Reticulo-Endothelial System (for instance malignancy's like Hodgkin's disease and reticulosis or « benignant » infections like toxoplasmosis) and during pregnancy (from the 10th week till birth) an extra band of L.A.P.- and A.A.P.-activity can be demonstrated after electrophoresis on agarose-gel of concentrated urine.

It will be shown that a good separation of the normal activity (in the α_2-region) from the abnormal activity (in the β-region) only can be obtained on agarose but not on agar-gel slides.

The significance of this extra L.A.P.- and A.A.P.-fraction will be discussed.

* * *

En étudiant les enzymes qui sont en usage pour le diagnostic et le contrôle des maladies diverses, nous fûmes frappés par une qualité spéciale de l'enzyme Leucine-Amino- Peptidase (= L.A.P.).

Dans tous les cas où l'activité sanguine de cette enzyme est augmentée, l'excrétion urinaire est élevée aussi. Mais au contraire, au cours de quelques maladies du système réticulo-endothélial (c'est-à-dire myélome et leucémie) il fut constaté que l'excrétion urinaire est augmentée, malgré une activité sanguine normale.

Les activités sanguines et urinaires de L.A.P. et A.A.P. ont été dosées fluorimétriquement selon la méthode de Roth (1).

---

* Centraal klinisch chemisch laboratorium van het Academisch Ziekenhuis der Vrije Universiteit te Amsterdam, Nederland.

** Klinisch laboratorium van het Ziekenhuis « Sint Jozef », Kerkrade, Nederland.

Pendant nos recherches nous avons confirmé et étendu les résultats déjà obtenus par d'autres auteurs, (2, 3) et nous avons trouvé que ce phénomène concerne aussi d'autres maladies du système réticulo-endothélial. En même temps il fut constaté qu'une autre amino-peptidase, l'alanyl-amino-peptidase (= A.A.P.) montre un paralléllisme parfait avec la L.A.P., soit quant à l'activité totale, soit quant à l'électrophorèse.

L'excrétion urinaire augmentée pourrait être expliquée par une fonction altérée des reins dans toutes ces maladies différentes du système réticulo-endothélial. Cependant, pour nous autres chimistes, cette explication n'était pas acceptable. Il vaudrait probablement mieux penser que dans ces maladies, la synthèse des protéines ne se fait pas normalement, et qu'une protéine anormale caractérisée par une activité enzymatique comparable à celle du L.A.P. et A.A.P. est présente.

Si cette hypothèse se vérifiait, on pourrait compter sur une iso-enzyme spéciale dans l'urine de ces maladies. En en effet, après l'électrophorèse des urines concentrées, on peut mettre en évidence la présence de quelques zones à activité de L.A.P. ou A.A.P., près desquelles l'iso-enzyme normale est trouvée dans la région des α_2 et l'iso-enzyme anormale dans celle des β-globulines.

Les urines concentrées et les sérums sont soumis aux électrophorèses de zone en gel d'agarose et quelquefois en gel d'agar [toutes les deux selon la technique de Wieme (4)]. Les urines ont été concentrées environ 50 fois par ultrafiltration. Pour la révélation des iso-enzymes, les électrophorégrammes sont mis à incuber pendant 30 à 60 minutes à 37° C avec le mélange révélateur. Ce mélange de tampon-substrat contient 20 mg % de leucil- (ou alanyl-)-β-naphtylamide-hydrochlorure et 20 mg % de sel de Fast-Blue-RR dans un tampon trismaléate, pH 7,2. Pendant l'incubation, le β-naphtylamide, libéré par l'enzyme, est précipité comme un pigment rouge dans les zones à activité enzymatique.

Maintenant, je veux vous montrer quelques électrophorégrammes concernant les iso-enzymes de L.A.P. Celles de A.A.P. ont été omis parce qu'elles sont identiques. En employant le gel d'agar, il est très difficile — sinon impossible — d'établir une différenciation entre le normal et le pathologique. Heureusement, cette différenciation est facile en gel d'agarose, dans lequel l'électro-endosmose est différente de celle de gel d'agar (voir la fig. 1). Continuons maintenant par montrer encore quelques maladies appartenant ou non au système réticulo-endothélial (voir resp. les fig. 2 et 3).

Continuons maintenant par discuter encore quelques maladies appartenant ou non au système réticulo-endothélial (voir resp. les fig. 2 et 3).

Jusqu'ici les exemples des maladies du système réticulo-endothélial étaient sans doute malignes. Mais il y a aussi des maladies bénignes de ce système. C'est-à-dire des maladies non cancéreuses, par exemple l'hépatite auto-immune, la maladie de Pfeiffer, l'infection des glandes lymphatiques avec plasmodium vivax (voir la figure 4).

Quelquefois nous avons trouvé une zone anormale, lorsqu'il n'y avait pas de maladie du système réticulo-endothélial. Dans presque tous les cas, le pathologiste a mis en évidence que les tumeurs primaires ont été enkystés par une infiltration plasmacellulaire, par exemple un cas de carci-

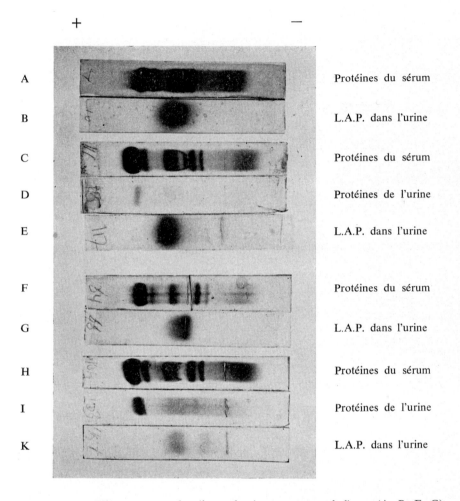

+   —

A   Protéines du sérum

B   L.A.P. dans l'urine

C   Protéines du sérum

D   Protéines de l'urine

E   L.A.P. dans l'urine

F   Protéines du sérum

G   L.A.P. dans l'urine

H   Protéines du sérum

I   Protéines de l'urine

K   L.A.P. dans l'urine

Fig. 1. — Les différences entre les électrophorégrammes en gel d'agar (A, B, F, G)
et ceux en gel d'agarose (C, D, E, H, I, K.).

A — E : infarctus du myocarde, (dans E *une* zone d'activité enzymatique).

F — K : syndrome de Digluglielmo (dans K *deux* zones d'activité de L.A.P.).

Differences between the agar gel electrophoreses (A, B, F, G) and those using agarose
gel (C, D, E, H, I, K).

A — E : myocardial infarction (in E *single* zone with enzymatic activity).

F — K : Di Gluglielmo's syndrome (in K *two* zones with the L.A.P. activity).

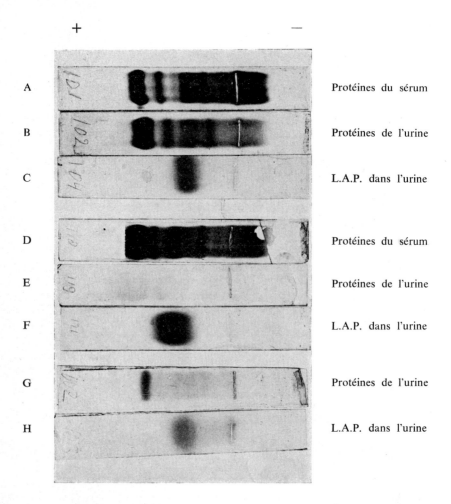

Fig. 2. — Quelques maladies n'appartenant pas au système réticulo-endothélial, par exemple cirrhose (A, B, C), carcinome du colon (D, E, F), et carcinome de l'épipharynx (G, H). Voir la seule zone d'activité enzymatique.

Some diseases not belonging to the reticulo-endothelial system, for example cirrhosis (A, B, C), carcinoma of the colon (D, E, F), and carcinoma of the nasopharynx (G, H). See the one zone of enzymatic activity.

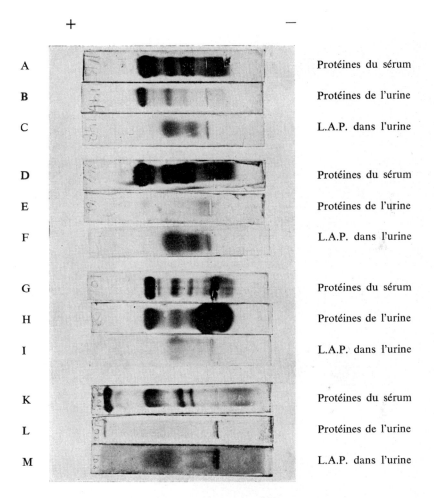

Fig. 3. — Quelques maladies malignes appartenant au système réticulo-endothélial, par exemple réticulo-sarcome et leucémie lymphoïde (A, B, C), maladie de Kahler (D, E, F), maladie de Waldenström (G, H, I) et réticulose maligne (K. L, M).
Voir les deux zones d'activité enzymatique.

Some malignant diseases which belong to the reticulo-endothelial system, for example reticulo-sarcoma and lymphoid leukemia (A, B, C), Kahler's disease (D, E, F), Waldenström's disease (G, H, I), and malignant reticulosis (K. L, M.).
See the two zones of enzymatic activity.

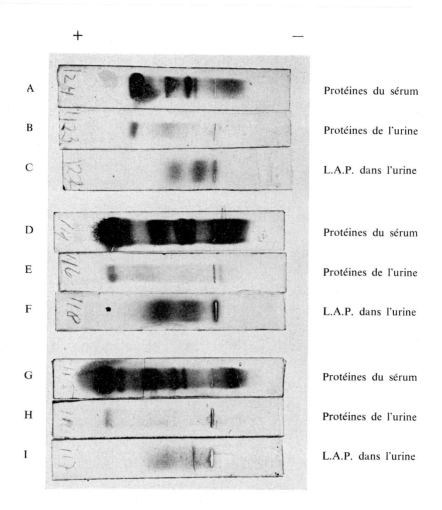

Fig. 4. — Quelques maladies non cancéreuses du système réticulo-endothélial, par exemple hépatite auto-immune (A, B, C), infection des glandes lymphatiques avec Toxoplasmodium vivax (D, E, F) et maladie de Pfeiffer (G, H, I).

Some non cancerous diseases of the reticulo-endothelial system, for example autoimmune hepatitis (A, B, C), lymphatic gland infection with Toxoplasmodium vivax (D, E, F) and Pfeiffer's disease (G, H, I).

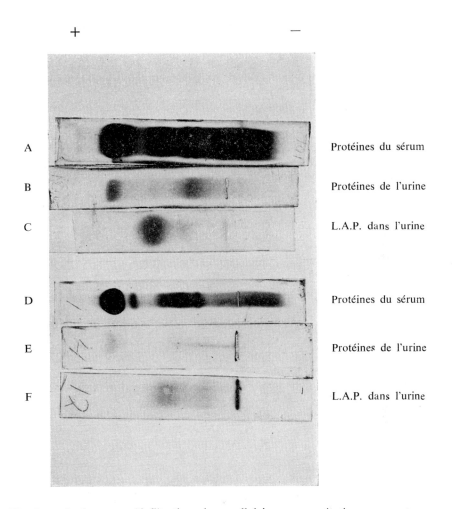

Fig. 5. — Quelques cas d'infiltration plasmacellulaire comme réaction sur une tumeur primaire, par exemple carcinome de la vessie (A, B, C) et Lichen Planus de la bouche (D, E, F).

Some cases of plasmacellular infiltration as reaction on a primary tumor, for example carcinoma of the bladder (A, B, C) and lichen planus of the mouth (D, E, F).

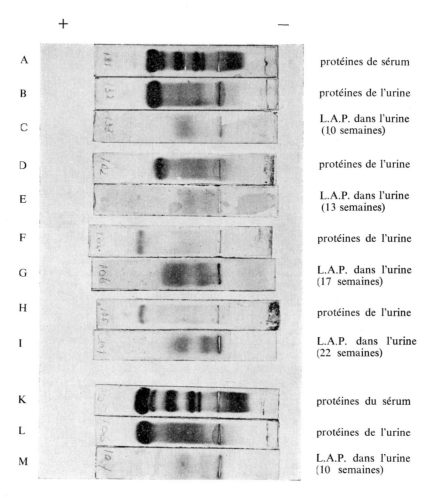

+   —

A — protéines de sérum

B — protéines de l'urine

C — L.A.P. dans l'urine (10 semaines)

D — protéines de l'urine

E — L.A.P. dans l'urine (13 semaines)

F — protéines de l'urine

G — L.A.P. dans l'urine (17 semaines)

H — protéines de l'urine

I — L.A.P. dans l'urine (22 semaines)

K — protéines du sérum

L — protéines de l'urine

M — L.A.P. dans l'urine (10 semaines)

Fig. 6. — Quelques électrophorégrammes pendant la grossesse. A - I grossesse normale ; K, L, M grossesse anormale, c'est-à-dire déjà hypertension pendant la 10me semaine. Voir *une* zone d'activité enzymatique dans c et *deux* zones dans M (les deux pendant la 10me semaine !).

Some electrophoreses during pregnancy. A - I normal pregnancy ; K, L, M abnormal pregnancy, that is hypertension as early as the 10th week. See the *single* zone with enzymatic activity in C and *two* zones in M (both during the 10th week !).

nome de la vessie avec une telle infiltration, et un Lichen Planus de la membrane muqueuse de la bouche avec multiples dégénérations carcino-mateuses (voir la fig. 5).

Il nous reste encore une grande série des électrophorégrammes qui res-semblent à ceux des maladies du système réticulo-endothélial, tel le schéma des iso-enzymes obtenu pendant la grossesse. Nous avons constaté que pendant la treizième semaine environ de la grossesse, des zones d'activité de L.A.P. et A.A.P. aparaissent dans la même région des β-globulines comme chez les réticuloses. Nous soupçon-nons sur base de quelques observations seulement, que les cas de toxicose et d'antagonisme de Rhésus se trahissent par l'intensification ou la production précoce de ces iso-enzymes localisées dans la région des β-globulines (voir fig. 6). Ici, je veux seulement signaler ce phénomène et ne pas le discuter, d'autant plus qu'il y a des irrégularités d'activité totale dans les sérums, c'est-à-dire que le rapport entre les activités de L.A.P. et A.A.P. se change pendant la grossesse, surtout dans les cas pathologiques.

En résumant, nous avons l'impression d'avoir dans les iso-enzymes L.A.P. un critère pour mettre en évidence une maladie du système réticulo-endothélial, soit cancéreuse, soit non cancéreuse. L'inverse, c'est-à-dire la possibilité d'exclure des maladies du système réticulo-endothélial, est une chose de plus grande importance peut-être.

Bibliographie

1. M. ROTH. *Clin. Chim. Acta,* **9** : 448, 1964.
2. M. van RIJMENANT, H.J. TAGNON. *New Eng. J. Med.,* **26** : 1373, 1959.
3. L. LEVITAN, H.D. DIAMOND, L.F. GRAVER. *Am. J. Med.,* **30** : 99, 1961.
4. R.J. WIEME. *Agar Gel Electrophorèse.* 1965.

Procédés chromatographiques d'isolement et d'identification des cétoses

Application à l'urine humaine (*)

Gérard STRECKER et Jean MONTREUIL (**)

Service de Biochimie Cellulaire de l'Institut de Recherches sur le Cancer de Lille (***)
et Laboratoire de Chimie Biologique de la Faculté des Sciences de Lille (****).

Dans des mémoires antérieurs, Montreuil et Boulanger (9) et Montreuil (7) avaient signalé la présence d'un cétose inconnu dans l'urine humaine. Dans une première série de travaux, Montreuil et Goubet (10), se fondant sur le comportement chromatographique et sur les réactions colorées données par ce glucide, étaient parvenus à la conclusion qu'il pourrait bien s'agir d'un céto-hexose : l'*allulose* (ou *psicose)*. Ce sucre étant très peu répandu dans la Nature, nous avons tenu à l'identifier sans ambiguïté en l'isolant en quantité appréciable, de manière à appliquer aux préparations des critères physico-chimiques indiscutables. Au cours de nos travaux nous avons mis en évidence, isolé et identifié huit autres cétoses présents dans l'urine humaine (15).

MATERIEL ET METHODES

I. Préparation des cétoses de référence

Le D-fructose, le L-sorbose et le D-mannoheptulose étaient des préparations commerciales. Le sédoheptulose (D-*altro*-heptulose) a été isolé par chromatographie préparative des extraits aqueux de feuilles de *Sedum* selon le protocole expérimental que nous décrivons plus loin à propos de l'urine. Tous les autres cétoses ont été préparés en utilisant la méthode générale

(*) Ce travail a été effectué grâce à une subvention du Comité Cancer et Leucémie de l'I.N.S.E.R.M. Contrat de Recherche 66-CR-228).
(**) Avec la collaboration technique de Mademoiselle Annick Poitau.
(***) Sac Postal n° 30 ; 59 - LILLE RP (France).
(****) Boîte Postale n° 36 ; 59 - LILLE-Distribution (France).

d'épimérisation des glucides par l'eau de chaux (5 mn à 100° C) suivie d'une destruction des aldoses par le brome (voir ci-dessous) : le D-xylulose, le D-ribulose, le D-allulose, le D-tagatose et D-*allo*-heptulose (*) à partir respectivement du D-xylose, du D-arabinose, du D-fructose, du D-galactose et du D-*altro*-heptulose.

II. Préparation des extraits urinaires

A. *Préparation de l'urine.*

L'urine, recueillie sur un mélange de chloroforme et de toluène, est filtrée après un séjour de 24 h à 4°C, puis concentrée 10 fois par lyophilisation. Le matériel macromoléculaire est précipité par l'addition de 9 volumes d'éthanol absolu et séparé par centrifugation, après un repos de 30 mn. La précipitation est répétée 2 fois et les solutions alcooliques réunies sont évaporées à siccité sous pression réduite.

B. *Purification des solutions glucidiques.*

Le résidu sec est repris par l'eau et la solution obtenue est purifiée par passages successifs sur des échangeurs de cations (*Dowex 50 ; « mesh »* 25-50 ; forme *acide*), puis d'anions (*Duolite A-40 ; « mesh »* 25-50 ; forme *formiate*). La solution effluente renferme une proportion importante d'aldoses et d'osides à groupement réducteur aldéhydique qui sont oxydés sélectivement en acides aldoniques par l'eau de brome à 20°C, en présence de carbonate de calcium. A la fin de la réaction, on élimine le brome en excès par évaporation sous vide et le carbonate de calcium par centrifugation. La solution est ensuite purifiée sur des colonnes d'échangeurs d'ions selon le mode opératoire décrit ci-dessus.

III. Procédés chromatographiques

A. *Chromatographie préparative sur colonne de cellulose.*

La solution précédente est concentrée jusqu'à consistance pâteuse et le résidu est repris par 5 à 10 ml du mélange *n*-butanol/éthanol/eau (1 : 1 : 1). Après centrifugation, la solution obtenue est soumise à une chromatographie sur colonne de cellulose (colonne Chromax L.K.B.) dans le système-solvant *n*-butanol/éthanol/eau (4 : 1 : 1). Des fractions de 15 ml sont recueillies et soumises à l'analyse chromatographique sur papier.

B. *Chromatographie sur papier*

1. *Systèmes-solvants :* n° *1* : n-butanol/acide acétique/eau (4 : 1 : 5) (12) ; n° *2* : acétate d'éthyle/pyridine/eau (8 : 2 : 1) (6) ; n° *3* : acétate d'éthyle/acide acétique/eau (3 : 1 : 3) (4) ; n° *4* : phénol saturé d'eau en atmosphère ammoniacale (1 ml d'ammoniaque pour 100 ml de la phase aqueuse du système-solvant) (12).

(*) Un échantillon d'*allo*-heptulose de synthèse nous a été aimablement fourni par N.K.Richtmyer à qui nous adressons nos vifs remerciements.

2. *Révélateurs spécifiques*. Nous avons utilisé les réactifs à l'oxalate d'aniline de Partridge (13), à l'urée chlorhydrique de Dedonder (1), à l'orcinol de Klevstrand et Nordal (5), à la vanilline de Godin (3), et au *p*-diméthylaminobenzaldéhyde de Svennerholm et Svennerholm (16).

IV. Etude des propriétés physico-chimiques des cétosazones

Il est bien connu que les glucides eux-mêmes se prêtent parfois fort mal à des études physico-chimiques en raison, par exemple, du caractère sirupeux de leurs préparations et les auteurs préfèrent s'adresser à des dérivés qui cristallisent aisément. C'est pourquoi nous avons plus particulièrement étudié les osazones des cétoses que nous avons préparées et purifiées suivant les procédés classiques. La morphologie des cristaux d'osazones a été examinée au microscope et leur point de fusion et leur pouvoir rotatoire (mesuré à l'aide du micropolarimètre Zeiss ont été déterminés.

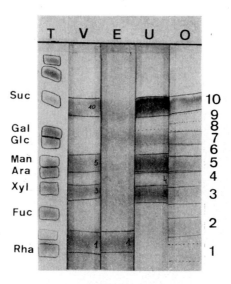

Fig. 1. — Chromatographie monodimensionnelle des cétoses urinaires. Papier Whatman n° 3. Système-solvant : *n*-butanol/acide acétique/eau (4 : 1 : 5). Révélation par les réactifs à la vanilline (V), au *p*-diméthylaminobenzaldéhyde (E), à l'urée chlorhydrique (U) et à l'orcinol trichloracétique (O). T : solution-témoin de sucres purs. 1 : désoxycétose non identifié ; 2 : xylulose + ribulose ; 3 : allulose ; 4 : 3-cétofructose ; 5 : fructose + *allo*-heptulose ; 6 : sédoheptulose ; 7 : *manno*-heptulose ; 8 et 9 : diholosides non réducteurs (voir Fig. 2) ; 10 : saccharose.

Mono-dimensional chromatograph of urinary ketoses. N° 3 Whatman paper. System-solvent : n-butanol/acetic acid/water (4 : 1 : 5). Relevation by the reaction to vanillin (V), *p*-dimethylaminobenzaldehyde (E), with hydrochloric urea (U) and trichloracetic orcinol (O). T : Pure sugar control solution. 1 : unidentified desoxyketose ; 2 : xylulose + ribulose ; 3 : allulose ; 4 : 3-ketofructose ; 5 : fructose + *allo*-heptulose ; 6 : sedoheptulose ; 7 : *manno*-heptulose ; 8 and 9 : non-reducing diholosides (see Fig. 2) ; 10 : saccharose.

TABLEAU I

Comportement chromatographique dans différents systèmes-solvants et réactivité de divers cétoses purs et extraits des urines humaines.

Nature et origine (a) des cétoses	Systèmes-solvants (b)					Réactifs (c)				
	1 R_f	2 R_{MH} (e)	2 R_{FRU} (f)	3 R_f	4 R_f	Aniline	Urée	Orcinol	Vanilline	PDMAB (d)
Erythrulose (P)	—	—	—	—	0,81	brun				
Ribulose (P et U)	0,32	—	2,88	—	0,63	brun		rose		
Xylulose (P et U)	0,32	—	3,16	—	0,58	brun		pourpre		
Allulose (P et U)	0,28	—	1,95	—	0,57	brun	bleu	jaune	bleu	
3-cétofructose (P et U)	0,26	—	—	—	0,59	brun	rose	rouge		
Fructose (P et U)	0,22	—	1	0,19	0,51	brun	bleu	jaune	bleu	
Sorbose (P)	0,23	—	—	—	0,42	brun	bleu	jaune	bleu	
Tagatose (P)	0,23	—	1,30	—	0,48	brun	bleu	jaune	bleu	
Désoxycétose X (U)	0,37	—	2,50	—	0,72	jaune			bleu	violet
Heptuloses (g) :										
Allo- (P et U)	0,23	—	—	—	0,60	brun			orange	
Altro- (P et U)	0,18	1,20	—	0,19	0,46	brun			orange	
Galacto- (P)	—	0,88	—	0,15	0,48	brun			orange	
Gluco- (P)	—	0,93	—	0,16	0,42	brun			orange	
Gulo- (P)	—	—	—	0,21	0,53	brun			orange	
Ido- (P)	—	—	—	0,22	0,50	brun			orange	
Manno- (P et U)	0,15	1	—	0,16	0,40	brun			orange	
Talo- (P)	—	1,38	—	—	—	brun			orange	

(a) P : échantillons purs du commerce ou de synthèse ; U : glucides d'origine urinaire ; (b) (c) : voir texte ; (d) : réactif au p-diméthylaminobenzaldéhyde ; (e) : vitesse de migration par rapport au manno-heptulose ; (f) : vitesse de migration par rapport au fructose ; (g) système-solvant n° 2 : valeurs de Mac Comb et Rending (6) ; systèmes-solvants n°s 3 et 4 : valeurs de Noggle (11).

RESULTATS

I. Chromatographie des cétoses

Nous avons rassemblé dans le tableau I les Rf de différents cétoses dans les systèmes-solvants 1 à 4.

II. Chromatographie des cétoses urinaires

La figure 1 montre que, par *chromatographie monodimensionnelle* sur papier, le nombre de cétoses caractérisés est lié à la nature du réactif utilisé et nous avons précisé dans le tableau I les colorations données par différentes solutions de révélation. Le réactif à l'orcinol est, à cet égard, le plus satisfaisant puisqu'il permet de mettre 10 taches en évidence.

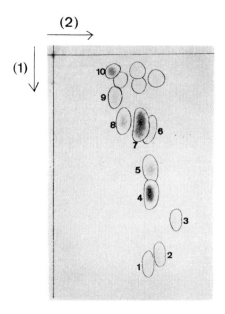

Fig. 2. — Chromatographie bidimensionnelle des cétoses urinaires. Papier Whatman n° 1. Système-solvant (1) : acétate d'éthyle/pyridine/eau (8 : 2 : 1); système-solvant (2) : phénol saturé d'eau (atmosphère ammoniacale). Révélation par le réactif à l'orcinol trichloracétique. 1 : xylulose ; 2 : ribulose ; 3 : désoxycétose non identifié ; 4 : allulose ; 5 : 3-cétofructose ; 6 : *allo*-heptulose ; 7 : fructose ; 8 : sédoheptulose ; 9 : *manno*-heptulose ; 10 : saccharose. Les taches non numérotées correspondent aux bandes 8 et 9 de la Fig. 1 et marquent l'emplacement de diholosides non réducteurs dont 3 ont été identifiés à Glc ←→ Xyl ; Xyl ←→ Xyl ; Ara ←→ Ara.

Bi-dimensional chromatograph of urinary ketoses. N° 1 Whatman paper. System-solvent (1) : acetate of ethyl/pyridine/water (8 : 2 : 1); system-solvent (2) : water saturated phenol (ammoniacal atmosphere). Revelation by the reaction to trichloracetic orcinol. 1 : xylulose ; 2 : ribulose ; 3 : unidentified desoxyketose ; 4 : allulose ; 5 : 3-ketofructose ; 6 : *allo*-heptulose ; 7 : fructose ; 8 : sedoheptulose ; 9 : *manno*-heptulose ; 10 : saccharose. The unnumbered spots correspond to strips 8 and 9 in Fig. 1 and mark the position of non-reducing diholosides, 3 of which have been identified at Glc ←→ Xyl ; Xyl ←→ Xyl ; Ara ←→ Ara.

La figure 2 montre que les 9 cétoses présents dans les urines humaines sont parfaitement séparés par une *chromatographie bidimensionnelle* qui combine l'emploi des systèmes-solvants 2 et 4.

III. Isolement et étude des propriétés physico-chimiques des cétoses urinaires

La chromatographie sur colonne de cellulose (Fig. 3) nous a permis d'obtenir le désoxycétose (1) et l'allulose (3) à l'état pur et d'autres fractions hétérogènes dont la purification a été achevée par chromatographie sur papier.

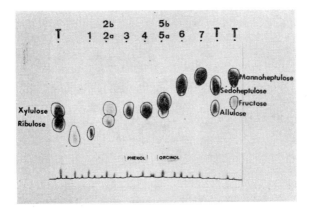

Fig. 3. — Chromatographie sur papier des fractions 1 à 7 obtenues par chromatographie préparative des cétoses urinaires sur colonne de cellulose (colonne L.K.B. ; type Chromax). Papier Whatman n° 3. Révélation à l'orcinol trichloracétique. Système-solvant : phénol saturé d'eau (atmosphère ammoniacale). T : solutions-témoins de sucres purs. Les chiffres désignent les fractions obtenues au cours de la chromatographie sur colonne de cellulose : 1 : désoxycétose inconnu ; 2a + 2b : ribulose + xylulose ; 3 : allulose ; 4 : 3-cétofructose ; 5a + 5b : *allo*-heptulose + fructose ; 6 : *altro*-heptulose ; 7 : *manno*-heptulose.

Chromatograph on paper of fractions 1 to 7 obtained by preparative chromatography of urinary ketoses on cellulose column (L.K.B. column, type Chromax). N° 3 Whatman paper. Revelation by trichloracetic orcinol. System-solvent : water saturated phenol (ammoniacal atmosphere). T : Pure sugar control solutions. The figures indicate the fractions obtained during chromatography on cellulose column : 1 : unknown desoxyketose ; 2a + 2b ; ribulose + xylulose ; 3 : allulose ; 4 : 3-ketofructose ; 5a + 5b : *allo*-heptulose + fructose ; 6 : *altro*-heptulose ; 7 : *manno*-heptulose.

1. En comparant les propriétés physico-chimiques des cétoses ainsi préparés avec celles de cétoses purs de référence, nous sommes parvenus à identifier sans ambiguïté le D-fructose, le D-allulose, le D-*manno*-heptulose et le D-*altro*-heptulose (sédoheptulose) (Tableau II).

2. Pour certains cétoses dont nous ne disposions que de faibles quantités, nous avons dû nous limiter à l'étude du comportement chromatogra-

TABLEAU II.

Propriétés physiques comparées des phénylosazones de cétoses purs
et de cétoses urinaires

Nature des cétoses	Origine	P.F. (a)	$[\alpha]_D^{20}$ (b)		
D-allulose	Pur	159-163	−74 → −68	}	Pyridine
	Urine	156-160	−74 → −66	}	
D-fructose	Pur	207	−72 → −40	}	Pyridine-éthanol
	Urine	200-201	−68 → −35	}	(2 : 3) (v : v)
Heptuloses :					
D-*manno*	Pur	198-199	+80 → − 7	}	id.
	Urine	195-196	+81 → − 5	}	
D-*altro*	Pur	191-192	−64 → −47	}	id.
	Urine	190-192	−64 → −48	}	

(a) en degrés centigrades ; (b) en degrés

phique des oses natifs (*allo*-heptulose) et de leurs osazones (ribulose et xylulose) (*).

   3. Quant au 3-cétofructose, il a été identifié sur la base de l'expérimentation suivante. Dans un premier temps, le glucide 4 des figures 1 et 3 (glucide 5 de la figure 2) a été réduit par le borohydrure de potassium et les deux hexitols épimères formés ont été oxydés, dans un second temps, par l'eau de brome à 100°C (oxydation spécifique des fonctions alcooliques secondaires en position α des fonctions alcooliques primaires). L'identification, parmi les produits d'oxydation bromique, de l'allulose, du tagatose et du fructose (Fig. 4), nous conduit à la conclusion que les deux hexitols étaient l'altritol et le mannitol qui ne peuvent s'être formés, par réduction, qu'à partir du 3-cétofructose.

   4. Le glucide 1 des figures 1 et 3 (glucide 3 de la figure 2) est un désoxycétose dont la structure reste à déterminer. En effet, il donne toutes les réactions des cétoses et aussi celles des 2- et 3-désoses (réactivité avec la

---

(*) Les osazones du ribulose et du xylulose ont été identifiées en comparant leur comportement chromatographique en couche mince de Silicagel (système-solvant : chloroforme/dioxanne/tétrahydrofuranne/borate de sodium 0,1M (40 : 20 : 20 : 1,5)) (Rink et Herrmann) (14) avec celui de la ribosazone et de la xylosazone pures.

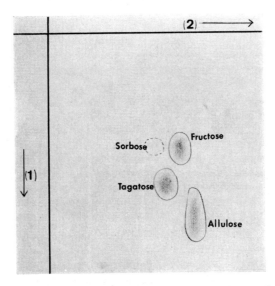

Fig. 4. — Chromatographie bidimensionnelle sur papier des cétoses obtenus par oxyda-tion bromique de l'altritol et du mannitol formés par la réduction du 3-cétofructose par le borohydrure de potassium. Papier Whatman n° 3. Système-solvant (1) : acétate d'éthyle/pyridine/eau (8 : 2 : 1) ; système-solvant (2) : phénol saturé d'eau (atmosphère ammoniacale). Révélation par l'urée chlorhydrique. Le cercle tracé en pointillé marque l'emplacement virtuel du sorbose, absent du milieu réactionnel.

Bi-dimensional chromatography on paper of the ketoses obtained by bromic oxidation of the altritol and the mannitol formed by the deduction of 3-ketofructose by borohydride of potassium. N° 3 Whatman paper. System-solvent (1) : acetate of ethyl/pyridine/water (8 : 2 : 1) ; system-solvent (2) : water saturated phenol (ammoniacal atmosphere). Revelation by hydrochloric urea. The dotted circle marks virtual position of the sorbose, absent from the reactive environment.

vanilline-acide perchlorique, avec de p-diméthylaminobenzaldéhyde-acide tri-chloracétique et avec l'acide périodique-acide thiobarbiturique).

5. En comparant la morphologie et les propriétés physiques des cris-taux des phénylosazones des cétoses urinaires et des cétoses purs de réfé-rence, nous confirmons la présence, dans les urines humaines, du D-fructose, du D-allulose, du D-*manno*-heptulose (Fig. 5) et du D-*altro*-heptulose.

6. Les urines de sujets sains ne renferment pas d'acide N-acétylneura-minique (*).

7. Le dosage des cétoses urinaires permet de fixer l'élimination journalière aux taux suivants : xylulose : 1,5 à 8 mg ; ribulose : 1,5 à 8 mg ; allulose : 15 à 30 mg ; 3-cétofructose : moins de 10 mg ; fructose : 7 à 30 mg ; *allo*-heptulose : moins de 5 mg ; *altro*-heptulose : 7 à 15 mg ; *manno*-heptulose : 5 à 10 mg.

_____

(*)   Jusqu'à présent, un seul cas de sialurie a été décrit (8).

Fig. 5. — Microphotographie de cristaux de la phénylosazone du *manno*-heptulose commercial (A) et urinaire (B).

Photomicrograph of crystals of the commercial (A) and urinary (B) *manno*-heptulose phenylosazone.

CONCLUSIONS

Les conclusions que nous pouvons tirer de notre étude sur les cétoses urinaires sont les suivantes :

1. Nous avons mis au point une série de procédés chromatographiques qui, associés à la détermination des propriétés physico-chimiques des cétoses natifs ou de leurs phénylosazones, nous a permis d'identifier dans les urines humaines normales, en proportions variant de 1 à 20 mg p. litre, les cétoses suivants : le ribulose, le xylulose, l'allulose, le 3-cétofructose, le D-fructose, les *allo-, altro* et *manno*-heptuloses et un désoxycétose inconnu.

2. Nos résultats confirment ceux de Montreuil *et al.* pour le fructose (7, 9) et l'allulose (10, 15) ; ceux de Futterman et Roe (2) pour le xylulose et le ribulose et ceux de White et Hess (16) pour les *manno-* et sédoheptuloses.

3. L'allulose, le 3-cétofructose, le *manno*-heptulose, l'*allo*-heptulose et la majeure partie du fructose sont d'origine « alimentaire ». En effet, la concentration de ces sucres dans les urines est étroitement liée au régime de l'individu et ils disparaissent complètement, sauf le fructose, après 48 h de suppression des aliments d'origine végétale. Au contraire, les autres cétoses continuent à être éliminés au même taux et sont donc d'origine « métabolique ».

Bibliographie

(1) DEDONDER, R., *C.R. Acad. Sci.,* **230** : 997, 1950.

(2) FUTTERMAN, S. et ROE, J.H., *J. Biol. Chem.,* **215** : 257, 1955.

(3) GODIN, P., *Nature,* **174** : 134, 1954.

(4) JERMYN, M.A. et ISHERWOOD, F.A., *Biochem. J.,* **44** : 402, 1949.

(5) KLEVSTRAND, R. et NORDAL, A., *Acta Chim. Scand.,* **4** : 1320, 1950.

(6) MAC COMB, E.A. et RENDING, V.V., *Arch. Biochem. Biophys.,* **95** : 316, 1961.

(7) MONTREUIL, J., 3e *Coll. Hôp. Saint-Jean, Bruges,* 209, 1955.

(9) MONTREUIL, J. et BOULANGER, P., *C.R. Acad. Sci.,* **236** : 337, 1963.

(8) MONTREUIL, J., BISERTE, G., STRECKER, G., SPIK, G., FONTAINE, G. et FARRIAUX, J.P., *C.R. Acad. Sci.,* **265 D** 97 : 1967, *Clin. Chim. Acta,* **21** : 61, 1968.

(10) MONTREUIL, J. et GOUBET, B., Chromatographie Symposium II, Bruxelles, *Société Belge des Sciences pharmaceutiques,* éd., 299, 1963.

(11) NOGGLE, G.R., *Arch. Biochem. Biophys.,* **43** : 238, 1953.

(12) PARTRIDGE, S.M., *Biochem. J.,* **42** ; 238, 1948.

(13) PARTRIDGE, S.M., *Biochem. Soc. Symp.,* **3** : 52, 1950.

(14) RINK, M. et HERRMANN, S., *J. Chromatog.,* **12** : 415. 1963.

(15) STRECKER, G., GOUBET, B. et MONTREUIL, J. *C. R. Acad. Sci.,* **260** : 999, 1965.

(16) SVENNERHOLM, E. et SVENNERHOLM, L., *Nature,* **181** : 1154, 1958.

(17) WHITE, A.A. et HESS, W.C., *Arch. Biochem. Biophys.,* **64** : 57, 1956.

Dosage par chromatographie automatique sur colonne des osamines présentes dans les hydrolysats chlorhydriques des glycoprotéines

par

Michel MONSIGNY, Geneviève SPIK et Jean MONTREUIL (*) (**).

L'identification et le dosage colorimétrique ou chromatographique des osamines les plus répandues dans les glycoprotéines : la glucosamine et la galactosamine, posent deux problèmes.

Le premier concerne la libération quantitative des osamines par hydrolyse acide. En effet, les liaisons « glycosaminidyl » sont très stables et nécessitent, pour être coupées, l'application de procédés drastiques qui risquent d'entraîner une destruction importante des osamines. L'hydrolyse de ces liaisons est donc un compromis entre la libération des osamines et leur destruction.

Le second problème est soulevé par l'identification et le dosage chromatographique des osamines généralement accompagnées de proportions importantes d'acides aminés.

L'étude que nous avons réalisée nous permet de proposer des solutions à ces deux problèmes.

I. Le problème de l'hydrolyse des liaisons « glycosaminidyl »

Le dosage colorimétrique des osamines par les méthodes dérivées du procédé original d'Elson et Morgan (5), d'une part, l'identification et le dosage des osamines par chromatographie ou par électrophorèse, d'autre part, nécessitent l'hydrolyse préalable des liaisons « glycosaminidyl ». Les principales causes d'erreur dans le dosage des osamines seront donc provoquées, soit par une destruction partielle de celles-ci, soit par une hydrolyse incomplète de leurs liaisons de conjugaison.

(*)   Faculté des Sciences ; Cité Scientifique de Lille-Annappes ; Laboratoire de Chimie Biologique ; BP 36 ; 59-Lille-Distribution (France).

(**) Avec la collaboration de Francis Décamps et de Renée Vandersyppe.

A. *Stabilité des osamines dans différentes conditions d'hydrolyse chlorhydrique*

1. Dans une première série d'expériences, nous avons étudié la stabilité de la glucosamine vis-à-vis de l'acide chlorhydrique 4 N et 5,6 N, concentrations qui sont classiquement utilisées pour hydrolyser les glycoprotéines en vue de doser, respectivement, les osamines par colorimétrie et les acides aminés par chromatographie sur colonnes.

TABLEAU I.

Influence des sels de fer sur la stabilité de la glucosamine (exprimée en p. 100 de récupération de la glucosamine) vis-à-vis de l'acide chlorhydrique 4 N à 100°C et 5,6 N à 105°C pendant des temps variables (a).

HCl	Nature du sel de fer (b)	Condi-tions (c)	Durée de l'action			
			1 h	2 h	3 h	4 h
4 N	Néant	V	102	101,5	100	103
		A	99,5	100	100	101
	Fe^{++}	V	100	99	93	92
		A	100	97	91	86
	Fe^{+++}	V	100	97	96,5	94
		A	99	97	96	96
			12 h	24 h	48 h	72 h
5,6 N	Néant	V	97,5	84	73	65
		A	92	81	70	63
	Fe^{+++}	V	90	83	68	53

(a) 12 mg de chlorhydrate de glucosamine sont dissous dans 20 ml d'acide chlorhydrique redistillé 4 N ou 5,6 N. Des fractions de 1 ml sont introduites dans des tubes qui sont ensuite scellés, les uns sous vide, les autres en présence d'air. La glucosamine a été ensuite dosée, à la fois, par la méthode d'Elson et Morgan, modifiée par Belcher, Nutten et Sambrook (1) et par chromatographie à l'Autoanalyseur Technicon. Nous avons, au préalable, vérifié (voir aussi Hartree (8)) que l'acide chlorhydrique ne détruisait pas l'osamine au cours de l'évaporation à siccité sous vide.

(b) 50 μg d'ions Fe^{++} ou Fe^{+++} p. ml d'acide chlorhydrique 4 N ; 2,5 μg d'ions Fe^{+++} p. ml d'acide chlorhydrique 5,6 N.

(c) V : sous vide ; A : en présence d'air.

Résultats.

Les résultats que nous avons obtenus sont rassemblés dans le tableau I. On voit que l'*acide chlorhydrique 4 N* n'altère pas la glucosamine, au bout de 4 h à 100°C, à la condition qu'il soit exempt de sels de fer et que « l'hydrolyse » soit effectuée à l'abri de l'oxygène de l'air. Au contraire, la présence de sels de fer, — principalement de sels ferreux —, provoque, même sous vide, la destruction d'une partie importante de l'osamine. L'*acide chlorhydrique 5,6 N* est beaucoup plus agressif, principalement quand il contient des sels de fer même à très faible concentration. En outre, l'action néfaste de l'oxygène est nettement plus marquée en présence de sels de fer.

Conclusions.

Les résultats que nous avons obtenus montrent la nécessité d'employer, comme l'avait déjà préconisé Neuberger (13), de l'acide chlorhydrique redistillé et rigoureusement exempt de sels de fer qui catalysent la destruction des osamines (voir aussi Hartree) (8). Celle-ci est, en outre, favorisée par la présence d'oxygène (voir aussi Ludowieg et Benmaman (10) ; Ward, Adams-Mayne et Wade (20) ; Walborg et Ward (19)), que l'on éliminera, avant de sceller les tubes à hydrolyse, soit par le vide, soit par un barbottage d'azote (Ludowieg et Benmaman (10) ; Wolfrom, Weisblat, Karabinos, Mc Neeley et Mc Lean (22) ; Ogston (14)) ou d'anhydride sulfureux (Pedersen et Baker) (15).

2. Dans une seconde expérience, nous avons appliqué le même mode opératoire pour étudier la stabilité de la glucosamine, de la galactosamine et de la mannosamine vis-à-vis de l'acide chlorhydrique 4 N et 5,6 N exempt de fer et d'oxygène.
Résultats.

a) Les résultats rassemblés dans le tableau II montrent que, dans ces conditions expérimentales, les trois osamines sont stables dans l'acide chlorhydique 4 N, à 100°C, pendant au moins 4 h. Ces résultats confirment ceux d'autres auteurs (3, 7, 9). Ils s'écartent de ceux d'Ogston (14) qui évaluait la destruction de la glucosamine dans l'acide chlorhydrique 4 N, à 100°C, à 1 p. 100 par heure.

b) Au contraire, l'acide chlorhydrique 5,6 N, dans les conditions d'hydrolyse des protéines, est plus agressif et détruit des proportions d'osamines variables avec la durée de l'hydrolyse et avec la nature de l'osamine, l'ordre décroissant des stabilités étant le suivant : galactosamine, glucosamine, mannosamine. Nos résultats concernant la destruction des osamines par l'acide chlorhydrique 5,6 N confirment ceux des autres auteurs (voir par exemple, 2, 8, 10, 12, 17, 21 et 23).

Conclusions.

a) L'hydrolyse d'osaminoglycannes, destinée à l'identification et au dosage des osamines, peut parfaitement être réalisée par l'acide chlorhydrique 4 N, à 100°C, pendant des temps variant de 1 à 4 h. En effet, dans ces conditions, les osamines ne subissent aucune destruction.

TABLEAU II.

vis-à-vis de l'acide chlorhydrique 4 N et 5,6 N, exempts de fer,
Stabilité des osamines (exprimée en p. 100 de récupération des osamines)
à 100°C, sous vide, pendant des temps variables.

HC1	Osamine	Durée de l'action de l'acide			
		1 h	2 h	3 h	4 h
4 N	GlcNH$_2$	102	101	100	103
	GalNH$_2$	100	99	100	102
	ManNH$_2$	100	102	100	100
		12 h	24 h	48 h	72 h
5,6 N	GlcNH$_2$	97	84	73	65
	GalNH$_2$	100	98	94	90
	ManNH$_2$	98	82	80	69

b) Au contraire, dans les conditions d'hydrolyse des protéines par l'acide chlorhydrique 5,6 N, à 105°C, des proportions importantes d'osamines sont détruites qui varient avec la nature de l'osamine. Toutefois, l'erreur de dosage pourra être partiellement compensée en extrapolant les courbes d'hydrolyse au temps zéro par un procédé identique à celui que l'on applique aux acides aminés acido-labiles comme la sérine et la thréonine. L'application de cette méthode d'extrapolation a déjà été préconisée par d'autres auteurs (16, 21, 23).

B. *Stabilité des liaisons osaminidyl.*

De nombreux procédés ont été décrits pour hydrolyser les liaisons osaminidyl et, devant la disparité des méthodes d'hydrolyse, nous avons effectué, au laboratoire, une étude critique systématique dans le but de préciser les meilleures conditions à appliquer aux glycoprotides pour libérer quantitativement les osamines et nous sommes parvenus à la conclusion que les conditions optimales étaient les suivantes : HC1 4 N, exempt de sels de fer, à 100°C, en tube scellé sous vide, pendant 4 h. Cette conclusion a été tirée sur la base des résultats expérimentaux précisés dans le tableau III. On voit, en effet, que les valeurs maximales sont données par l'hydrolyse effectuée avec l'acide chlorhydrique 4 N, dès après 2 h. Toutefois, par sécurité, nous prolongeons l'hydrolyse pendant 4 à 6 h.

TABLEAU III.

Libération de la glucosamine (exprimée en g p. 100 g de glycoprotéine et dosée par la méthode d'Elson et Morgan) par hydrolyse de l'ovomucoïde dans différentes conditions.

Durée de l'hydrolyse	Concentration de l'acide chlorhydrique		
	1 N	2 N	4 N
1 h	10,70	13,74	14,25
2 h	12,80	14,04	14,7
3 h	13,80	14,40	14,7
4 h	13,80	14,25	14,7

C. *Conclusions générales.*

1. L'oxygène et les sels de fer catalysent la destruction des osamines. Les hydrolyses seront donc toujours effectuées en tubes scellés sous vide ou sous atmosphère d'azote, avec de l'acide chlorhydrique redistillé, exempt de sels de fer.

2. L'ordre décroissant de stabilité des osamines vis-à-vis de l'acide chlorhydrique est le suivant : galactosamine, glucosamine et mannosamine.

3. L'acide chlorhydrique 4 N respecte les osamines pendant au moins 4 h, temps suffisant pour libérer quantitativement les osamines conjuguées. On peut donc l'utiliser pour hydrolyser des glyco-amino-acides et des glyco-oligopeptides. Toutefois, on effectuera systématiquement des cinétiques d'hydrolyse pendant des temps échelonnés entre 2 et 6 h, par exemple, et on retiendra les valeurs maximales.

4. Dans le cas de glycoprotéines, on emploiera l'acide chlorhydrique 5,6 N et on éliminera la cause d'erreur due à la destruction inévitable des osamines en effectuant, comme dans le dosage des acides aminés, des hydrolyses à 3 temps différents et en extrapolant au temps zéro.

II. Identification et dosage chromatographique des osamines

Les nombreux procédés de dosage, par chromatographie automatique sur échangeurs d'ions, de la glucosamine et (ou) de la galactosamine présentes dans les hydrolysats de glycoprotéines qui ont été décrits jusqu'à ce jour ne permettent pas d'effectuer simultanément le dosage des amino-acides. Par exemple, dans les conditions classiques d'utilisation des colonnes de l'Auto-analyseur Technicon, on obtient de mauvaises séparations de la glucosamine et de la valine (fig. 1). En outre, la galactosamine est éluée avec

TABLEAU IV.

Composition (en ml) des solutions introduites dans les 9 compartiments de l'« Autograd » Technicon en vue de la formation des gradients 1 (a), 2 (b) et 3 (c).

Compartiments	Tampon n° 1 pH 2,75 (d)	Tampon n° 2 pH 2,875 (d)	Tampon n° 4 pH 3,80 (d)	Tampon n° 5 pH 5,00 (d)	Tampon n° 6 pH 6,10 (d)	Nacl 2,5 M	Méthanol	Méthanol + n-butanol (1:1)
1	38 (38) [70]						2 (2)	5
2	12 (12) [75]	28 (28)						
3	[75]	40 (32)	(5)			(3)		
4	[60]	25 (25)	12 (10)	5		3 (5) [10]		
5		10 (10)	27 (25) [60]	5		3 (5) [10]		
6			[30]	40	40 (40)	[5]		
7				75	40 (40)			
8				75	40 (40)			
9				75	40 (40)			

(a) Colonnes de *Chromobeads C-2* de 0,6 x 65 cm ; (b) colonnes de *Chromobeads C-2* de 0,6 x 65 cm ; (c) colonnes de *Chromobeads A* de 0,6 x 130 cm. **Les volumes des solutions utilisées sont indiqués entre parenthèses ; Les volumes des solutions utilisées sont indiqués entre crochets ;** (d) *tampon n° 1* : solution de citrate de pH 2,75, contenant 10 % (v:v) de thiodiglycol ; *tampons n° 2, 3, 4* : solutions de citrate contenant 0,5 % (v:v) de thiodiglycol et ajustées respectivement à pH 2,875 ; 3,1 ; 3,80 ; *tampon n° 5* : solution de citrate, 0,6 M en chlorure de sodium, ajustée à pH 5,0 ; *tampon n° 6* : solution de citrate, 1 M en chlorure de sodium, ajustée à pH 6,10.

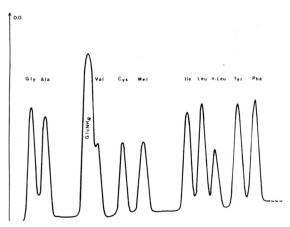

Fig. 1. — « Fraction osamines » d'un diagramme de chromatographie d'un hydrolysat de glycoprotéine effectué à l'Auto-analyseur Technicon avec le système-tampon préconisé par la firme.

« Osamine fraction » of a chromatograph diagram of a glycoprotein hydrolysate effected by the Technicon Auto-analyser with the plug system recommended by the firm.

la cystine. Nous avons mis au point une série de procédés qui éliminent cette cause d'erreur grâce à des modifications du gradient de pH et de force ionique réalisé dans un « Autograd » à 9 chambres.

A. *Matériel et méthodes.*

Les colonnes utilisées sont celles que fournit la firme Technicon : colonnes de « *Chromobeads A* » (0,6 x 130 cm) et de « *Chromobeads C-2* » (0,6 x 65 cm) thermorégulées à 60°C. Elles sont « stabilisées » par le passage, respectivement, du tampon n° 2 de pH 2,875 et du tampon n° 3 de pH 3,1 décrits dans le Tableau IV. En outre, leur pouvoir séparateur est maintenu par un traitement hebdomadaire avec une solution aqueuse d'EDTA à 1 g p. 100 ml, ajustée à pH 3 avec une solution de soude.

Les différents tampons utilisés dans la formation des gradients renferment tous, pour 1 litre, 17,85 g de citrate trisodique à 5 H_2O et 5 g de *Brij* 35. Le pH est ajusté par addition d'acide chlorhydrique 5,6 N redistillé. Leur composition est précisée dans le tableau IV. Le dosage des acides aminés et des osamines a été effectué par le réactif à la ninhydrine dans les conditions proposées par la firme Technicon. Simultanément, le dosage spécifique des osamines par la réaction d'Elson et Morgan a été réalisé à l'aide d'un dispositif automatique que nous avons mis au point (*) et qui est décrit dans la figure 2.

(*) La réaction d'Elson et Morgan, effectuée avec le dispositif automatique de Swann et Balazs (18) ne peut être utilisée pour le dosage des osamines présentes dans les éluats. En effet, d'une part, la coloration ne se développe pas lorsque les osamines sont dissoutes dans une solution tamponnée et, d'autre part, le réactif à l'acétyl-lacétone n'est pas suffisamment stable.

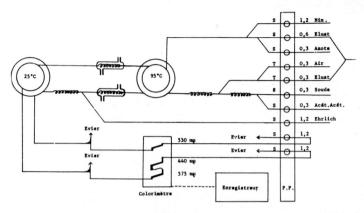

Fig. 2. — Schéma du dispositif de colorations simultanées. *Nin.* : réactif à la ninhydrine. *Soude* : solution aqueuse de soude 1,5 M. *Acét. Acét.* : acétylacétone (7 ml) dans une solution de phosphate monosodique 0,66 M (100 ml). *Ehrlich* : *p*-diméthylaminobenzaldéhyde (3,2 g) dans un mélange d'éthanol absolu (210 ml) et d'acide chlorhydrique (30 ml). (Ces deux derniers réactifs sont stables pendant plus de 24 h.). *P.P.* : pompe péristaltique. *S* : tubes en « Solvaflex ». *T* : tubes en « Tygon ». Les nombres placés à droite de la pompe péristaltique correspondent aux débits des fluides exprimés en ml/mn.

Diagram of the simultaneous dyeing layout. *Nin.* : reaction to ninhydrine. *Soda (Soude)* : Aqueous solution of soda 1,5 M. *Acet. Acet.* : Acetylacetone (7 ml) in a solution of monosodic phosphate 0.66 M (100 ml). *Ehrlich* : *p*-dimethylaminobenzaldehyde (3.2 g) in a mixture of absolute ethanol (210 ml) and hydrochloric acid (30 ml). (The two latter reagents are stable for more than 24 hours). *P.P.* : Peristaltic pump. *S* : « Solvaflex » tubes. *T* : « Tygon » tubes. The numbers indicated on the right of the peristaltic pump correspond to the flow of the fluids expressed in ml/mn.

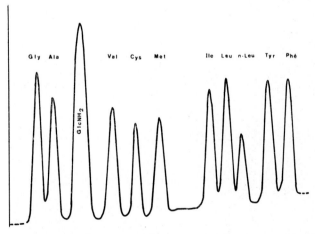

Fig. 3. — « Fraction osamines » d'un diagramme de chromatographie d'un hydrolysat de glycoprotéine effectué à l'Auto-analyseur Technicon avec le gradient n° 1 (voir texte et tableau IV).

« Osamine fraction » of a chromatograph diagram of a glycoprotein hydrolysate effected by the Technicon Auto-analyser with gradient N° 1 (see text and table IV).

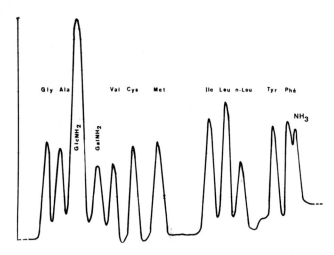

Fig. 4. — « Fraction osamines » d'un diagramme de chromatographie d'un hydrolysat de glycoprotéine effectué à l'Auto-analyseur Technicon avec le gradient n° 2 (voir texte et tableau IV).

« Osamine fraction » of a chromatograph diagram of a glycoprotein hydrolysate effected by the Technicon Auto-analyser with gradient N° 2 (see text and table IV).

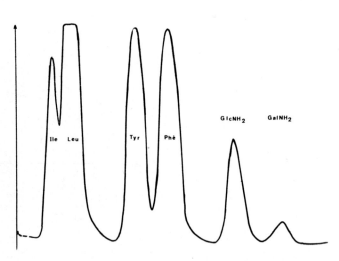

Fig. 5. — « Fraction osamines » d'un diagramme de chromatographie d'un hydrolysat de glycoprotéine effectué à l'Auto-analyseur Technicon avec le tampon n° 4 (voir texte et tableau IV).

« Osamine fraction » of a chromatograph diagram of a glycoprotein hydrolysate effected by the Technicon Auto-analyser with gradient N° 4 (see text and table IV).

B. *Résultats.*

La figure 3 montre que le gradient d'élution n° 1 permet de séparer, sur des colonnes de « *Chromobeads C-2* » (0,6 x 65 cm), la *glucosamine* de tous les acides aminés. Un résultat identique est obtenu avec le gradient n° 3 sur des colonnes de « *Chromobeads A* » (0,6 x 130 cm). La figure 4 illustre le pouvoir séparateur du gradient d'élution n° 2 qui fournit d'excellentes résolutions de la *glucosamine* et de la *galactosamine*. Toutefois, les pics de la phénylalanine et de l'ammoniac se confondent partiellement.

Le tampon n° 4 de pH 3,8 permet de doser, en 3 h, la glucosamine et (ou) la galactosamine quand celles-ci sont présentes en très faibles quantités dans des hydrolysats de glycoprotéines (fig. 5). En effet, les deux osamines se séparent très nettement des acides aminés qui sont élués en bloc, à l'exception de l'isoleucine, de la leucine, de la tyrosine et de la phénylalanine. Il est donc nécessaire de doser, sur une aliquote, les acides aminés séparés avec le gradient n° 3 pour effectuer, en associant les résultats fournis par les deux chromatographies, le dosage des osamines et de tous les acides aminés.

C. *Conclusions.*

Les modifications que nous avons apportées à la composition des solutions tamponnées, classiquement utilisées dans l'analyse automatique des acides aminés, permet de doser simultanément les acides aminés, la glucosamine et la galactosamine présents dans les hydrolysats de glycoprotides. Les procédés que nous proposons ont été éprouvés au Laboratoire par de nombreux essais et ont déjà fait l'objet de plusieurs applications (voir, par exemple, les références 4, 6 et 11).

Bibliographie

(1) BELCHER, R., NUTTEN, A.J. et SAMBROOK, C.M., *Analyst,* **79** : 201, 1954.

(2) BOAS, N.F., *J. Biol. Chem.,* **204** : 553, 1953.

(3) BOURRILLON, R. et MICHON, J., *Bull. Soc. Chim. Biol.,* **41** : 267, 1959.

(4) DESCAMPS, J., MONSIGNY, M. et MONTREUIL, J., *C.R. Acad. Sci.,* **266** : 1775, 1968.

(5) ELSON, L.A. et MORGAN, W.T.J., *Biochem. J.,* **27** : 1824, 1933.

(6) GRIMMONPREZ, L., TAKERKART, G., MONSIGNY, M. et MONTREUIL, J., *C.R. Acad. Sci.,* **265** : 2124, 1967.

(7) HAAB, W. et ANASTASSIADIS, P.A., *Can. J. Biochem. Physiol.,* **39** : 671, 1961.

(8) HARTREE, E.F., *Anal. Biochem.,* **7** : 103, 1964.

(9) JOHANSEN, P.G., MARSHALL, R.D. et NEUBERGER, A., *Biochem. J.,* **77** : 239, 1960.

(10) LUDOWIEG, J. et BENMAMAN, J.D., *Anal. Biochem.,* **19** : 80, 1967.

(11) MONSIGNY, M., ADAM-CHOSSON, A. et MONTREUIL, J., *Bull. Soc. Chim. Biol.,* **50** : 857, 1968.

(12) MURACHI, T., *Biochemistry,* **3** : 933, 1964.

(13) NEUBERGER, A., cité par HARTREE, E.F., *Anal. Biochem.,* **7** : 103, 1964.

(14) OGSTON, A.G., *Anal. Biochem.,* **8** : 337, 1964.

(15) PEDERSEN, J.W. et BAKER, B.E., *J. Sci. Food Agr.,* **5** : 549, 1954.

(16) SANDERSON, A.R., STROMINGER, J.L. et NATHENSON, S.G., *J. Biol. Chem.,* **237** : 3603, 1962.

(17) SOWDEN, J.C., *Soil Sci.,* **88** : 138, 1959.

(18) SWANN, D.A. et BALAZS, E.A., *Anal. Biochem.,* **12** : 565, 1965.

(19) WALBORG, E.F. et WARD, D.N., *Biochim. Biophys. Acta,* **78** : 304, 1963.

(20) WARD, D.N.. ADAMS-MAYNE, M. et WADE, J., *Acta Endocrinol,* **36** : 73, 1961.

(21) WARTH, A.J., *Biochim. Biophys. Acta,* **101** : 315, 1965.

(22) WOLFROM, M.L., WEISBLAT, D.I., KARABINOS, J.V., Mc NEELEY, W.H. et Mc LEAN, J., *J. Amer. Chem. Soc.,* **65** : 2077, 1943.

(23) YOUNG, F.E., SPIZIZEN, J. et CRAWFORD, I.P., *J. Biol. Chem.,* **238** : 319, 1963.

Solution du problème du dosage chromatographique des oses « neutres » constituant les glycoprotéines [*]

Geneviève SPIK, Gérard STRECKER et Jean MONTREUIL [**]

Laboratoire de Chimie Biologique de la Faculté des Sciences de Lille [***]
et Service de Biochimie Cellulaire de l'Institut de Recherches sur le Cancer
de Lille [****]

L'exploration de la structure des groupements glycanniques des glyco-protéines nécessite la connaissance de leur composition en oses à une mole près. Or, pour une glycoprotéine déterminée, les résultats varient souvent dans des proportions très importantes suivant les auteurs. Par exemple, dans le cas de l'orosomucoïde le rapport mannose/galactose oscille entre 0,5 et 1. En outre, le dosage chromatographique des oses « neutres » (aldohexoses, aldopentoses, 6-désoxyhexoses) présents dans les hydrolysats acides des glycoprotéines révèle souvent un déficit qui atteint parfois 50 p. 100 par rapport aux dosages colorimétriques des oses neutres totaux effectués préa-lablement sur les glycoprotéines natives (voir, à ce sujet, réf. 4). Dans un mémoire antérieur (4), nous avions décrit une série de procédés qui sem-blaient avoir résolu cette énigme. Toutefois, l'observation répétée, au cours de nos recherches, de ce « déficit chromatographique » nous a incités à remettre en question nos protocoles expérimentaux et à les vérifier soigneu-sement.

La détermination de la composition en oses neutres des glycoprotéines comporte les étapes suivantes : 1) hydrolyse acide des glycoprotéines ; 2) pu-rification des hydrolysats, généralement à l'aide d'échangeurs d'ions ; 3) chro-matographie quantitative des oses neutres sur papier, sur colonne ou en phase gazeuse. Les raisons qui pouvaient être invoquées pour expliquer le « déficit chromatographique » étaient donc les suivantes :

[*]   Ce travail a bénéficié d'une subvention de l'INSERM (Convention CR-66-228).
[**]   Avec la collaboration technique de Mlles Annick Poiteau et Renée Van-dersyppe.
[***]   Cité Scientifique d'Annappes ; BP 36 ; 59-Lille-Distribution (France).
[****]   Sac Postal 30 ; 59-Lille R.P. (France).

A. Imperfection des méthodes de dosages colorimétriques des oses « neutres ».

B. Imperfection des procédés de dosages chromatographiques.

C. Pertes au cours de l'hydrolyse acide.

D. Pertes au cours de la purification des hydrolysats.

Dans le présent mémoire, nous résumerons les résultats que nous avons obtenus en abordant les quatre points précédents.

I. Imperfection des méthodes de dosages colorimétriques des oses neutres

L'étude critique des procédés de dosages colorimétriques des oses neutres que nous avons effectuée a fait l'objet d'un mémoire (10) et d'une monographie (5). La principale cause d'erreur que nous avons mise en évidence provient du fait que les absorbances molaires des colorations données par les différents réactifs utilisés varient dans de larges proportions avec la nature de l'ose. Comme Winzler (11), nous éliminons cette cause d'erreur en introduisant, dans les séries de dosages, des solutions-témoins « internes » dont la composition molaire en oses est identique à celle des solutions de glucides ou de glycoprotéines étudiées. Par exemple, la composition centésimale d'un glycoprotide isolé du lait de Femme et riche en fucose, déterminée par la méthode à l'anthrone sulfurique, est de 62,1 p. 100 ou de 44,3 p. 100 suivant que le témoin interne est constitué par la solution, classiquement utilisée, de galactose + mannose, dans les proportions 1:1 ou par une solution de galactose + fucose, dans les proportions 3:2 telles qu'elles existent dans le glycoprotide natif (10).

II. Imperfection des procédés des dosages chromatographiques

Nous avons effectué une étude bibliographique et technique de nombreuses méthodes de dosages des oses « neutres » par chromatographie sur papier, sur couche mince, sur colonnes et en phase gazeuse. Elle nous a permis de conclure que ces procédés n'étaient pas responsables du « déficit chromatographique » que nous observions. En effet, quelle que soit la technique appliquée à l'analyse de solutions titrées d'oses purs, l'erreur n'excède jamais 15 p. 100 avec les procédés les plus grossiers (dosages par appréciation visuelle, planimétrie ou densitométrie) et la précision est de l'ordre de $\pm$ 1 à 2 p. 100 avec les méthodes les plus fines (6,8).

Au Laboratoire, nous appliquons toujours, dans les cas où une grande précision est recherchée, plusieurs procédés de conceptions différentes du point de vue de la chromatographie (chromatographie sur papier dans différents systèmes-solvants (*), sur colonnes et en phase gazeuse, par exem-

---

(*) Nous utilisons les deux systèmes-solvants suivants : *n*-butanol/acide acétique/eau (4:1:5) de Partridge (9) (papier Whatman n° 1 pendant 3 jours ou Whatman n° 3 pendant 2 jours) ; pyridine/acétate d'éthyle/eau (2:1:2) de Jermyn et Isherwood (3) (papier Whatman n° 1 pendant 18 à 24 h).

ple) et du dosage (dosage reductimétrique de Montreuil et *al.* (7) et dosage par élution des taches obtenues après révélation préalable des chromatogrammes, par exemple). Le tableau ci-contre montre que la récupération des oses est quantitative.

III. Pertes au cours de l'hydrolyse

A.   *Pertes par destruction des oses au cours de l'hydrolyse*

Dans un mémoire antérieur (4), nous avions montré que les aldohexoses et les 6-désoxy-hexoses étaient stables en présence de l'acide chlorhydrique 1 à 2 N à 100°C, pendant 2 à 4 h. (*). Toutefois, une cause d'erreur pouvait se glisser dans l'interprétation de nos résultats. En effet, la stabilité des oses avait été étudiée à l'aide de dosages colorimétriques, comme la méthode à l'orcinol sulfurique. Or, l'acide chlorhydrique pouvait parfaitement avoir détruit une certaine proportion des oses en donnant des chromogènes qui réagissaient avec l'orcinol. Pour éliminer cette cause d'erreur, nous avons soumis des solutions titrées d'oses à l'action de l'acide chlorhydrique et, après purification sur échangeurs d'ions, nous avons dosé les oses par chromatographie sur papier.

L'examen du tableau ci-contre montre que les oses sont stables dans les conditions d'hydrolyse des osides.

B.   *Pertes par hydrolyse incomplète des liaisons osidiques.*

Les expériences que nous avons effectuées sur différents composés ont montré que, à 100°C, l'acide chlorhydrique 1 N pendant 2 h ou 2 N pendant 1 h, libérait quantitativement les oses « neutres ». Toutefois, pour éviter toute cause d'erreur, nous préconisons le protocole expérimental suivant :

1) Les hydrolyses doivent être effectuées avec des concentrations en acide chlorhydrique différentes et pendant des temps variables (par exemple, HCl 1 N - 1,5 et 2 N pendant 1 et 2 h) et on retient les valeurs maximales.

2) On recherchera les osides, témoins d'une éventuelle hydrolyse incomplète, en examinant à la lumière de Wood, les chromatogrammes révélés avec les réactifs à l'oxalate ou au citrate d'aniline, à l'exclusion du réactif au phtalate qui réagit faiblement avec les oligosides.

3) On comparera systématiquement la quantité d'oses neutres, dosés par chromatographie, aux résultats fournis par les dosages colorimétriques effectués sur la glycoprotéine native.

C.   *Pertes dues à des réactions de transosidation*

1) *Réaction de Maillard.* Les oses et les osides sont susceptibles de se condenser avec les protides et cette réaction peut donc conduire à des causes d'erreur par défaut dans la détermination de la composition en oses des glycoprotéines. Toutefois, de l'ensemble des résultats parfois contradictoires

---

(*) Le xylose est stable en présence de HCl 1 N, pendant 4 h, mais il est sensible à l'action de HCl 2 N.

Récupération du galactose et du mannose et détermination du rapport mannose/galactose après chromatographie quantitative sur papier de solutions titrées des oses purs.

Traitement subi par la solution avant le passage sur échangeurs d'ions (a)	Quantités théoriques de chacun des oses	Dosage chromatographique du galactose et du mannose				Mannose/galactose (b)	
		Procédé au citrate d'aniline (c)		Procédé au ferricyanure (d)		Procédé au citrate d'aniline (c)	Procédé au ferricyanure (d)
		Galactose	Mannose	Galactose	Mannose		
Solution aqueuse non traitée	100 µg	100,2	98,7	101,8	97,6	0,98	0,97
Solution dans HCl 1,5 N non chauffée	100 µg	97,2	100,3	100	96,8	1,03	0,97
Solution dans HCl 1,5 N maintenue à 100°C pendant 1,5 h	100 µg	95,1	95,5	101,8	97,6	1	0,96

(a) Passage successif sur des colonnes (2 x 40 cm) de *Dowex-50 x 8* (*mesh 25-50* ; forme acétate) de 10 mg de galactose et de mannose. (b) Valeur théorique : 1. (c) Procédé de MONTREUIL et coll. (7).

obtenus à la suite des nombreux travaux suscités par la réaction de Maillard, nous pouvons tirer les enseignements suivants dont l'application élimine la réaction de condensation :

a) La concentration des hydrolysats acides en glucides et en protides doit être faible et de l'ordre de 0,5 à 1 p. 100 de glycoprotéide.

b) Les hydrolyses doivent être effectuées à l'obscurité et avec des produits exempts de sels de fer.

c) La réaction de Maillard étant maximale entre pH 7 et 10, les solutions de glucides et de protides seront maintenues acides par l'addition d'un peu d'acide acétique ou formique à toutes les étapes d'un mode opératoire.

d) Le dernier stade de la concentration d'une solution contenant des glucides et des protides est le plus favorable aux condensations de Maillard. Les conditions sont, en effet, optimales : concentrations élevées en glucides et en protides ; faible taux d'humidité ; pH tendant vers la neutralité. Les réactions seront inhibées ou au moins réduites en prenant la précaution d'acidifier de temps en temps le milieu en cours d'évaporation et d'effectuer cette dernière à l'obscurité.

e) Les solutions de glucides et de protides doivent être conservées au congélateur, *en milieu acide* et à l'obscurité quelques jours au plus.

2) *Réaction de condensation glucides-glucides.* Depuis les travaux de Fischer (2), on sait que les oses sont capables de se condenser par un mécanisme de transosylation qui s'effectue en milieu acide et qui est favorisé par les températures élevées et par de fortes concentrations en oses.

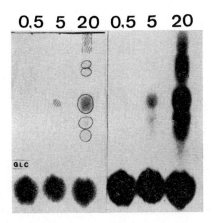

Fig. 1. — Transosylation de glucose « froid » et de ^{14}C-glucose (1 h à 100°C) en solution à différentes concentrations (0,5-5 et 20 p. 100) dans HCl 1 N. *A gauche :* chromatogramme révélé par le réactif à l'oxalate d'aniline. *A droite :* autoradiogramme.

Transosylation of « cold » glucose and ^{14}C-glucose (1 h at 100° C) in solution at different concentrations (0.5-5 and 20 p. 100) in HCl 1 N. On the left : chromatogram revealed by the reaction to oxalate of aniline. On the right : autoradiogram.

Nous avons vérifié que, pour une température de 100°C, les réactions de transosylation étaient nulles ou négligeables, même après 3 h de chauffage, dans les acides chlorhydriques 1 et 2 N, pour des concentrations finales en glucides n'excédant pas 0,5 à 1 p. 100 et qu'elles sont, au contraire, importantes pour des concentrations plus élevées. Pour sensibiliser la méthode et pour éviter des causes d'erreur dues à la formation d'osides non révélés par les réactifs classiquement utilisés en chromatographie sur papier, nous avons effectué les expériences en introduisant, dans les solutions d'oses étudiées, du ^{14}C-glucose et nous avons recherché, par autoradiographie, les osides formés (fig. 1). Le tableau ci-dessus montre que, dans les conditions que nous avons utilisées, aucune réaction de transosidation ne s'est produite.

IV. Pertes dues à l'emploi des échangeurs d'ions dans la purification

Nous employons au Laboratoire, depuis de nombreuses années, les échangeurs d'ions pour débarrasser les hydrolysats de glycoprotéines des composés « basiques » (osamines et protides) et « acides » (acide chlorhydrique, acides uroniques) qu'ils contiennent. Nous obtenons, de cette manière, une solution effluente « neutre » qui contient les oses « neutres ».

A. Dans une première série d'expérience, nous avons vérifié que le passage d'une solution titrée d'oses neutres sur le système combiné de colonnes d'échangeurs de cations et d'anions ne provoquait aucune perte

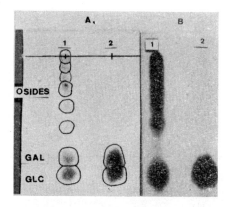

Fig. 2. — Chromatographie dans le système-solvant *n*-butanol/acide acétique/eau (4:1:5) d'hydrolysats chlorhydrique (HCl 1,5 N ; pendant 1,5 h ; 100°C) de lactose en présence de ^{14}C-glucose. A : révélation par le réactif à l'oxalate d'aniline. B : autoradiographie. En 1 subsistaient des traces d'acide chlorhydrique. L'hydrolysat 2 en était dépourvu.

Chromatograph in the system-solvent *n*-butanol/acetic acid/water (4 : 1 : 5) of hydrochloric hydrolysates (HCl 1.5 N ; for 1.5 h ; 100° C) of lactose in the presence of ^{14}C-glucose. A : Revelation by the reaction to oxalate of aniline. B : autoradiograph. In 1 traces of hydrochloric acid remained. The hydrolysate 2 had none.

en oses (voir le tableau ci-dessus). Les échangeurs d'ions ne retiennent donc pas et ne détruisent pas les glucides.

B. Dans une seconde série d'expériences, nous avons observé que les déficits chromatographiques s'accompagnaient toujours de l'apparition d'osides sur les chromatogrammes. La figure 2 est, à cet égard, démonstrative. Il s'agit de la chromatographie d'hydrolysats de lactose effectués en présence de ^{14}C-glucose. Dans un cas, la récupération chromatographique des oses était effective à 100 p. 100 ; dans l'autre cas existait un « déficit chromatographique » de près de 25 p. 100. La figure 2 montre que de nombreux osides sont apparus sur le chromatogramme. Or, l'hydrolyse du lactose avait été effectuée dans des conditions qui ne provoquaient pas de transosylation. Cette dernière ne pouvait donc s'effectuer qu'au cours de la concentration des effluents de colonnes.

C. Dans une dernière série d'expériences, nous avons vérifié cette hypothèse et décelé l'ultime cause d'erreur dont l'élimination supprime le « déficit chromatographique » des oses « neutres ». En effet, nous avons observé que les formes *acétate, carbonate* et *formiate* des échangeurs d'anions quels qu'ils soient *(Duolite A-102-D ; Dowex 1 x 2)*, laissaient souvent passer dans les solutions effluentes de faibles traces d'acide chlorhydrique (0,3 à 0,8 mg dans les conditions expérimentales suivantes : 25 ml d'acide chlorhydrique 1,5 N sur une colonne (2 x 40 cm) de *Duolite A-102-D* (forme *formiate*) ; lavage avec 500 ml d'eau distillée). Ces faibles quantités sont cependant suffisantes pour provoquer des transosylations au cours de la

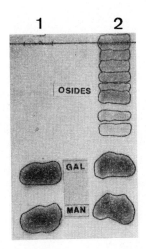

Fig. 3. — Chromatographie d'une solution aqueuse (10 ml) de galactose et de mannose (15 mg) évaporée en exsiccateur, additionnée (2) ou non (1) de 0,7 mg d'acide chlorhydrique. Chromatographie dans le système-solvant : *n*-butanol/acide acétique/eau (4:1:5).

Chromatograph of an aqueous solution (10 ml) of galactose and mannose (15 mg) evaporated in exsiccator, with (2) or without (1) the addition of 0.7 mg of hydrochloric acid. Chromatograph in the system-solvent : *n*-butanol/acetic acid/water (4 : 1 : 5).

concentration à siccité des solutions « neutres » effluentes comme nous avons pu le vérifier en évaporant en exsiccateur 10 ml d'une solution de galactose (15 mg) et de mannose (15 mg) contenant 0,7 mg d'acide chlorhydrique (fig. 3).

Ainsi s'explique, suivant que les solutions étaient totalement privées d'acide chlorhydrique ou en renfermaient encore des traces, le fait que les dosages chromatographiques des oses « neutres » étaient tantôt quantitatifs et tantôt déficitaires. Comme il est pratiquement impossible de déceler la présence de traces d'acide chlorhydrique dans les effluents de colonne, nous préconisons :

1) de concentrer ces derniers sous pression réduite en évitant de les évaporer à siccité et en ajoutant à plusieurs reprises du méthanol quand le volume des solutions atteint quelques millilitres ;

2) d'hydrolyser les glycoprotéines en présence de ^{14}C-glucose (quelques microcuries) et d'effectuer les autoradiographies des solutions utilisées pour réaliser les chromatogrammes quantitatifs.

V. Conclusions

1. Nous avons effectué, depuis plusieurs années, une étude critique systématique de la plupart des procédés de détermination de la composition centésimale et molaire des oses constituant les glycoprotéines. L'historique de chacune des questions que nous avons abordées et les résultats que nous avons obtenus ont été rassemblés dans les mémoires et dans les monographies référencés 4 à 8 et 10.

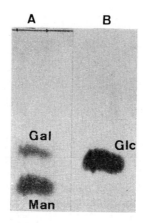

Fig. 4. — Chromatographie, dans le système-solvant : *n*-butanol/acide acétique/eau (4:1:5), d'un hydrolysat d'ovomucoïde effectué en présence de ^{14}C-glucose. A : révélation par le réactif à l'oxalate d'aniline ; B : autoradiographie.

Chromatograph, in the system-solvent : *n*-butanol/acetic acid/water (4 : 1 : 5), of an ovomucoid hydrolysate effected with ^{14}C-glucose present. A : revelation by the reaction to the oxalate of aniline ; B : autoradiograph.

2. L'expérience que nous avons acquise montre que les réactions de transosidation représentent la principale cause d'erreur dans la détermination de la composition molaire en oses « neutres » d'un oside ou d'un glycoprotide et qu'elles expliquent le « déficit chromatographique » que l'on observe fréquemment en comparant systématiquement les dosages effectués par colorimétrie sur le composé natif et par chromatographie sur le même composé hydrolysé. On évitera cette cause d'erreur en réalisant l'autoradiographie des chromatogrammes d'hydrolysats effectués en présence de ^{14}C-glucose. La figure 4 est, à cet égard, démonstrative car elle montre, dans le cas d'un hydrolysat d'ovomucoïde effectué dans ces conditions, l'absence de produits de transosidation.

Bibliographie

(1) DATE, J.W., *Scand. J. Clin. Lab. Invest.*, **10** : 149 et 444, 1958.

(2) FISCHER, E., *Ber.*, **23** : 3687, 1890.

(3) JERMYN, M.A. et ISHERWOOD, F.A., *Biochem. J.*, **44** : 402, 1949.

(4) MONTREUIL, J. et SCHEPPLER, N., *Bull. Soc. Chim. Biol.*, **41** ; 13, 1959.

(5) MONTREUIL, J. et SPIK, G., *Méthodes colorimétriques de dosage des glucides totaux*, Monographie n° 1, Lab. Chim. Biol. Fac. Lille éd., 1963.

(6) MONTREUIL, J. et SPIK, G., *Méthodes chromatographiques et électrophorétiques de dosage des glucides constituant les glycoprotéines*, Monographie n° 2, Lab. Chim. Biol. Fac. Sci. Lille éd., 1968.

(7) MONTREUIL, J., SPIK, G., DUMAISNIL, J. et MONSIGNY, M., *Bull. Soc. Chim. Fr.*, p. 239, 1965.

(8) MONTREUIL, J., SPIK, G. et KONARSKA, A., *Méthodes chromatographiques de dosage des oses « neutres »*, Monographie n° 3, Lab. Chim. Biol. Fac. Sci. Lille éd., 1967.

(9) PARTRIDGE, S.M., *Biochem. J.*, **42** ; 238, 1948.

(10) SPIK, G. et MONTREUIL, J., *Bull. Soc. Chim. Biol.*, **46** : 739, 1964.

(11) WINZLER, R.J., *Methods for determination of serum glycoproteins*, in GLICK, D., *Meth. Biochem. Anal.*, **2** : 290, Interscience éd., New-York, 1955.

Corrélation entre les structures moléculaires de quelques stéroïdes en C_{18} et leurs rétentions en chromatographie gaz-liquide

par

L.A. DEHENNIN et R. SCHOLLER

Fondation de Recherche en Hormonologie, 26, boulevard Brune, Paris 14e.

Introduction

Une des applications très importante de la chromatographie en phase gazeuse (C.P.G.) est l'identification des corps injectés à partir de leurs caractéristiques de rétention. En C.P.G. des stéroïdes quatre valeurs de rétention sont utilisées, à savoir :

1) le temps de rétention relatif (TRR)

$$TRR = \frac{t_{R\ (x)}}{t_{R\ (\text{réf})}}$$

$t_{R\ (x)}$: le temps de rétention d'un stéroïde (x)
$t_{R\ (\text{réf.})}$: le temps de rétention du corps de référence
(en général le 5α-cholestane).

2) la valeur ΔRmg, définie par Knights et Thomas (1)
$\log t_{R\ (x)} = \Sigma \Delta Rmg + \log t_{R\ (N)}$

Le logarithme du temps de rétention d'un stéroïde (x) est égal au logarithme du temps de rétention du stéroïde de base ($t_{R\ (N)}$) auquel on ajoute les incréments ΔRmg, dûs à la présence d'un ou plusieurs groupements fonctionnels sur le noyau stéroïdique.

3) le nombre stéroïdique (steroid number) défini par VandenHeuvel et Horning (2) (3).
$SN_{(x)} = S_{(N)} + F_1 + \dots\dots F_n$
$SN_{(x)}$: nombre stéroïdique d'un stéroïde (x)
$S_{(N)}$: nombre stéroïdique du stéroïde base, qui est par définition égal au nombre d'atomes de carbone présents dans le noyau stéroïdique (par exemple SN = 18 pour l'œstrane).
$F_1 \dots F_n$: facteurs dûs à l'introduction de groupements fonctionnels sur le stéroïde de base.

4) l'indice de rétention (I) défini par Kovats (4)

$$I^{ph.st.}_{T(x)} = 100 \; \frac{\log V_{g\,(x)} - \log V_{g\,(Pz)}}{\log V_{g\,(Pz+1)} - \log V_{g\,(Pz)}} + 100\,z$$

$V_{g\,(x)}$: volume de rétention spécifique du stéroïde (x) ;
$V_{g\,(Pz)}$: volume de rétention spécifique du n-alcane de la série C$_z$H$_{2z+2}$
$V_{g\,(Pz)} \leqslant V_{g\,(x)} \leqslant V_{g\,(Pz+1)}$

A débit de gaz porteur constant, on peut remplacer les (V_g) par les distances de rétention corrigées pour les volumes morts et mesurées sur l'enregistrement du chromatogramme ou par les temps de rétention corrigés.

L'unité méthylène (methylene unit) introduite par VandenHeuvel et Horning (5) est équivalente à I/100.

L'indice de rétention est la valeur de rétention la moins sensible aux variations de température et donc la plus appropriée aux échanges d'information inter-laboratoires, parce que sa détermination est basée sur la linéarité qui existe entre log V_g et le nombre d'atomes de carbone dans une série homologue de n-alcanes.

Cette communication a pour but de montrer les corrélations entre les valeurs de rétention et la structure de quelques œstrogènes pour lesquels le stéroïde de base est l'œstrane. Tous les œstrogènes naturels et la très grande majorité des œstrogènes de synthèse contiennent cependant un noyau A benzénique (fig. 1).

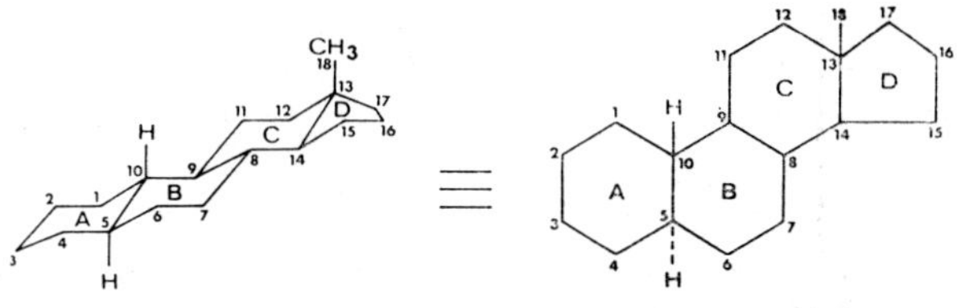

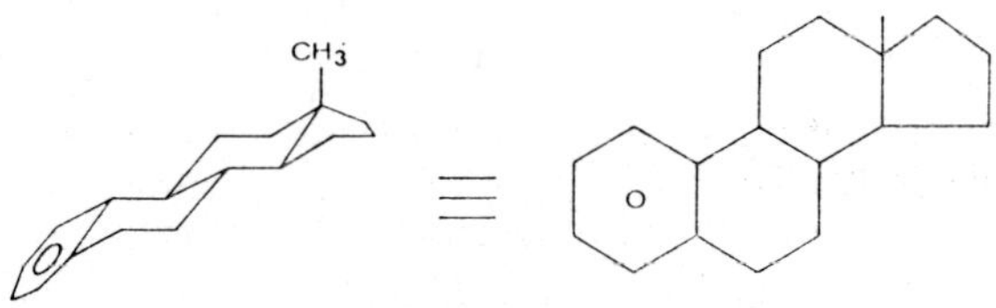

Fig. 1. — α-œstr.-1, 3, 5 (10)-triène.

Partie expérimentale

Les indices de rétention ont été déterminés graphiquement par interpolation linéaire sur une droite en portant les TRR par rapport à l'octacosane ($C_{28}H_{58}$) en ordonnée et les valeurs de I en abscisse (fig. 2)

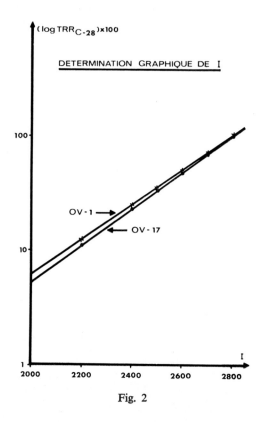

Fig. 2

Les différences entre les valeurs de I ainsi déterminées et celles calculées selon la formule de Kovats n'ont jamais dépassé l'erreur expérimentale qui est comprise entre 0,2 % et 0,3 %.

La CPG a été faite sur 10 phases stationnaires différentes : OV-1, OV-17, SE-30, F-50 Versilube, SE-52, XE-60, QF-1, NPGS et HIEFF-8B. Les conditions chromatographiques étaient les suivantes : colonnes en verre de $150\times$ 0,3 cm et remplies de Gas chrom Q 100-120 mesh imprégné à 1 % avec les différentes phases stationnaires ; — température de la colonne 200 °C ; — température de l'injecteur 220 °C ; — température du détecteur à ionisation de flamme 240 °C ; — débit du gaz porteur (azote) = 60 ml/mn à 30 psi. Les échantillons étaient injectés avec une seringue Hamilton de 10μl.

Résultats et discussion

Par analogie avec une série homologue de n-alcanes, on peut considérer une relation linéaire entre log t_R et le nombre d'atomes de carbone pour les hydrocarbures stéroïdiques. Les t_R des hydrocarbures stéroïdiques sont légèrement supérieurs aux t_R des n-alcanes correspondants sur une phase stationnaire non polaire telle que OV-1 (fig. 3).

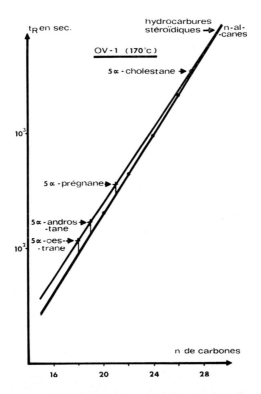

Fig. 3. — Variation du log t_R en fonction du nombre d'atomes de carbone pour les n-alcanes et les hydrocarbures stéroïdiques sur une phase stationnaire non polaire à 170 °C.

Variation of the log t_R versus the number of carbon atoms for the n-alcanes and the steroidic hydrocarbons on a non polar stationary phase at 170° C.

L'accroissement de la différence entre les t_R des hydrocarbures stéroïdiques et ceux des n-alcanes à nombres d'atomes de carbone identiques sur une phase stationnaire polaire, telle que OV-17, peut être expliqué par une certaine polarité des hydrocarbures stéroïdiques due à la fusion des cycles cyclohexaniques et cyclopentanique (6) (fig. 4).

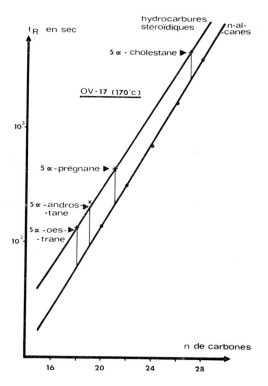

Fig. 4. — Variation du log t_R en fonction du nombre d'atomes de carbone pour les n-alcanes et les hydrocarbures stéroïdiques sur une phase stationnaire polaire à 170 °C.

Variation of the log t_R versus the number of carbon atoms for the n-alcanes and the steroidic hydrocarbons on a polar stationary phase at 170° C.

Les figures 5 et 6 font apparaître une relation linéaire entre les log t_R des hydrocarbures stéroïdiques, de l'œstra- 1, 3, 5 (10) -triène et l'inverse de la température absolue.

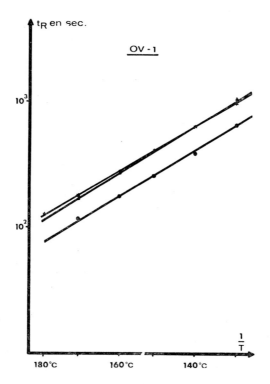

Fig. 5. — Variation du log t_R en fonction de l'inverse de la température absolue pour le 5α-œstrane (.——.), le 5α-androstane (Δ——Δ) et l'œstra-1, 3, 5 (10) -triène ($_x$——$_x$) sur OV-1.

Variation of the log t_R versus the inverse of the absolute temperature for the 5 α-estrane (.——.), the 5α-androstane (Δ——Δ) and the estra-1, 3, 5 (10)-triene (x——x) on OV-1.

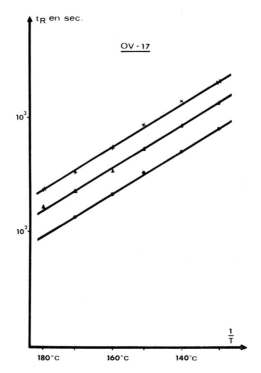

Fig. 6. — Variation du log t_R en fonction de l'inverse de la température absolue pour le 5α-œstrane (.———.), le 5α-androstane (Δ———Δ) et l'œstra-1, 3, 5 (10) -triène ($_x$———$_x$) sur OV-17.

Variation of the log t_R versus the inverse of the absolute temperature for the 5α-estrane (.———.) the 5α-androstane (Δ———Δ) and the estra-1, 3, 5, (10)-triene (x———x) on OV-17.

De la comparaison des figures 5 et 6 on peut déduire les conclusions suivantes :

— les différences entre les t_R obtenus pour le 5α-œstrane et pour le 5α-androstane en utilisant les phases stationnaires OV-1 et OV-17 sont pratiquement identiques ; ceci est dû à un même effet de la différence de volatilité et de masse moléculaire ;

— la présence d'un noyau benzénique planaire dans l'œstra-1, 3, 5 (10)-triène et sa volatilité plus faible par rapport au 5α-œstrane expliquant la différence des t_R de ces deux œstrogènes sur OV-1 ;

— la différence entre les t_R du 5α-œstrane et de l'œstra-1, 3, 5 (10)-triène sur la phase stationnaire OV-17 est le double de celle observée sur OV-1, ce qui veut dire que 50 % de l'augmentation de t_R est due à la polarité accrue de l'œstra-1, 3, 5 (10)-triène qui contient le sextet aromatique d'électrons π.

Dans le tableau A sont rassemblés les TRR par rapport au 5α-cholestane et les I correspondants. Ces valeurs de I sont d'une interprétation plus aisée et plus exacte que les TRR puisque chaque corps est encadré au plus près par deux n-alcanes. Par exemple : sur F-50 Versilube, le 3-hydroxy-œstra-1, 3, 5 (10)-triène a un I de 2375, c'est-à-dire que ce corps se trouve entre les I du triacosane ($C_{23}H_{48}$) et du tétracosane ($C_{24}H_{50}$) et plus précisément aux 3/4 de la distance de rétention qui sépare ces deux n-alcanes.

TABLEAU   A

STEROÏDE	OV-1		OV-17		SE-30		SE-52		SE-54		F-50		XE-60		QF-1		NPGS		HIEFF 8B	
	TRR	I	TRR	I	TRR	I	TRR	I	TRR	I	TRR	I	TRR	I	TRR	I	TRR	I	TRR	I
A₁ 5α-androstane	0,08	2040	0,08	2194	0,08	2020	0,08	2044	0,10	2100	0,08	2063	0,10	2205	0,09	2040	0,10	2175	0,11	2260
A₂ 5α-oestrane	0,05	1880	0,04	2010	0,05	1875	0,05	1904	0,06	1980	0,05	1948	0,08	2105	0,07	1935	0,07	2070	0,08	2155
A₃ oestra-1,3,5,(10)-triène	0,08	2020	0,10	2255	0,09	2075	0,09	2082	0,10	2100	0,09	2090	0,20	2415	0,22	2385	0,22	2500	0,25	2580
A₄ 3-hydroxy-1,3,5,(10)-triène	0,20	2286	0,42	2636	0,25	2363	0,28	2397	0,28	2400	0,24	2375	1,70	3155	0,58	2767	1,24	3064	2,25	3152
A₅ 17α-hydroxy-oestra-1,3,5,(10)-triène	0,15	2190	0,28	2528	0,19	2265	0,18	2282	0,18	2275	0,15	2215	0,75	2865	0,44	2640	0,40	3062	1,06	2975
A₆ 17β-hydroxy-oestra-1,3,5,(10)-triène	0,17	2218	0,33	2568	0,21	2300	0,20	2316	0,20	2308	0,17	2241	0,91	2932	0,52	2700	0,44	3129	1,35	3050
A₇ 17-oxo-oestra-1,3,5,(10)-triène	0,17	2215	0,34	2583	0,20	2270	0,19	2292	0,22	2335	0,18	2255	1,02	2975	0,78	2885	1,26	3068	0,90	2955
A₈ 3-méthoxy-oestra-1,3,5,(10)-triène-17-one	0,36	2456	1,09	2893	0,56	2495	0,48	2550	0,48	2570	0,44	2517	2,84	3303	1,87	3225	3,45	3385	2,90	3285
T.R. 5α-cholestane (mn)	12,5		17,0		15,7		21,3		18,4		18,4		7,1		7,4		5,5		5,4	
T.R. n-octacosane (mn)	14,8		12,5		18,0		23,7		18,8		20,5		4,8		6,5		4,0		3,6	

TABLEAU   B

groupement fonctionnel	OV-1	OV-17	SE-30	SE-52	SE-54	F-50	XE-60	QF-1	NPGS	HIEFF 8B
	ΔIg	ΔIg	ΔIg	ΔIg	ΔIg	ΔIg	ΔIg	ΔIg	ΔIg	ΔIg
19-méthyle	160	184	145	140	120	115	100	105	105	105
aromatisation du cycle A	140	245	200	178	120	142	310	450	430	325
3-hydroxy (phénolique)	266	381	288	315	300	285	740	382	564	572
17-oxo	195	332	195	210	235	165	560	500	568	375
17α-hydroxy	170	263	190	200	175	125	450	255	562	395
17β-hydroxy	198	313	225	234	208	151	517	315	629	465
3-méthoxy (phénylique)	232	320	225	258	235	262	328	340	317	330
17α-hydroxy ⟶ 17β-hydroxy	28	50	35	34	33	26	67	60	67	70
17-oxo ⟶ 17α-hydroxy	-25	-69	-5	-10	-60	-40	-110	-245	-6	20
17-oxo ⟶ 17β-hydroxy	3	-19	30	24	-25	-14	-43	-185	61	90
3-hydroxy ⟶ 3-méthoxy	-25	-71	-63	-57	-65	-23	-412	-42	-247	-242

Le tableau B donne les incréments ΔI_g dus, soit à l'introduction dans le stéroïde de base d'un groupement fonctionnel, soit à l'épimérisation ou soit à la réduction de quelques-unes des fonctions. La position particulière du groupement 19-méthyle angulaire dans la molécule stéroïdique peut expliquer qu'à l'introduction de ce groupement corresponde un I supérieur à l'I du méthane (I = 100). Les valeurs ΔI_g pour l'aromatisation du cycle A et pour l'introduction de groupements fonctionnels sont nettement supérieures sur les phases stationnaires polaires (XE-60, QF-1, NPGS et HIEFF-8B) d'où leur intérêt pour la séparation des mélanges complexes et stéroïdes.

Les valeurs ΔI_g pour les groupements 3-hydroxy et 17β-hydroxy ont permis de vérifier la concordance entre les valeurs calculées et mesurées de I pour l'œstradiol-17β sur OV-1 et XE-60.

— *sur OV*-1

œstra — 1,3,5 (10)-triène	:	I = 2020
3-hydroxy	:	ΔI_g = 266
17β-hydroxy	:	ΔI_g = 198
œstradiol-17β (calculé)	:	I = 2484
œstradiol-17β (mesuré)	:	I = 2480

— *sur XE*-60

œstra - 1,3,5 (10)-triène	:	I = 2415
3-hydroxy	:	ΔI_g = 740
17β-hydroxy	:	ΔI_g = 517
œstradiol-17β (calculé)	:	I = 3672
œstradiol-17β (mesuré)	:	I = 3666

Des écarts importants entre valeurs de I, calculées et mesurées, peuvent être observés dans les cas où il y a entre les groupements fonctionnels une interaction favorisée par la configuration stérique de la molécule.

Conclusion

L'utilisation des I pour les stéroïdes dans le but d'établir des corrélations entre la structure et les propriétés en CPG est préférable à toute autre valeur de rétention, puisque de toutes ces valeurs c'est I qui varie le moins en fonction de la température. La détermination graphique de I, telle qu'utilisée ici, rend l'obtention de ces valeurs moins fastidieuse. Connaissant I pour le stéroïde de base, il est facile de prévoir I pour un stéroïde de la même famille à partir des ΔI_g correspondants aux différents groupements fonctionnels, à condition que l'interaction entre ces groupements soit faible ou nulle. L'opération inverse qui consiste à obtenir des renseignements structuraux à partir des ΔI_g est également possible, surtout si l'on emploie des phases stationnaires sélectives où les ΔI_g sont bien distincts.

Références

(1) KNIGHTS, B.A. and THOMAS, G.H. *Nature,* **194** : 833, 1962. *Anal. Chem.,* **34** ; 1046, 1962. *J. Chem. Soc.* 3477, 1963.

(2) VANDENHEUVEL, W.J.A. and HORNING, E.C. *Biochim. et Biophys. Acta,* **64** : 416, 1962.

(3) HAMILTON, R.J., VANDENHEUVEL, W.J.A. and HORNING, E.C. *Biochim. et Biophys. Acta* **70** ; 679, 1963.

(4) KOVATS, E. *Helv. Chim. Acta,* **41** : 1915, 1958. *Helv. Chim. Acta,* **46** : 2705, 1963. *Z. Anal. Chem.,* **181** : 351, 1961.

(5) VANDENHEUVEL, W.J.A., GARDINER, W.C. and HORNING, E.C. *Anal. Chem.,* **36** ; 1550, 1964. *J. Chrom.,* **19** ; 263, 1965,

(6) HORNING, E.C. and VANDENHEUVEL, W.J.A. *Advances in chromatography,* Vol. 1, p. 153. Eds. Giddings, J.C. and Keller, R.A.

Identification, par vision directe, des 17-cétostéroïdes isolés par chromatographie sur colonne

par

Dr. François ESTEVE FORRIOL, pharmacien.

Valencia (Espagne)

Entre les multiples problèmes que pose un processus chromatographique, celui de l'identification des substances isolées n'est pas le moindre.

Dans certains cas les différentes fractions d'un chromatogramme peuvent être reconnues avec une certaine facilité (substances colorées, réactions avec les réactifs indiqués, lampe de Wood, rayons ultra-violets). Mais parfois la détection « a priori » et l'analyse postérieure présentent de grandes difficultés. Tel est le cas des substances incolores sensibles à l'emploi de réactifs, ou corps très semblables par leur composition ou plus simplement des stéréoïsomères, comme certains stéroïdes, dont les différences se fondent simplement sur la différente disposition d'un même groupe fonctionnel dans l'espace (isomérie « alfa » — « beta »).

Cette communication résume les résultats expérimentaux obtenus à travers de deux cents chromatographies sur colonne d'alumine des 17-cétostéroïdes urinaires, d'après la méthode de Dingemanse et collaborateurs. Elle a pour objet l'exposer un procédé simple pour identifier dans le chromatogramme différents 17-cétosteroïdes sans avoir recours à des vérifications ultérieures. Bien que cette identification ne soit pas absolument exacte, elle peut servir pratiquement dans tous les cas appliqués à la clinique.

Bien entendu pour des travaux de haute recherche scientifique, nécessitant une exactitude à toute épreuve, seule l'obtention du spectre infra-rouge des stéroïdes représente la méthode d'élection, mais elle implique l'utilisation d'un appareil coûteux et d'un personnel spécialisé ainsi que de délicates manipulations.

La méthode du fractionnement chromatographique des 17-cétostéroïdes sur colonne d'alumine, d'après Dingemanse, comporte schématiquement les opérations suivantes :

a) Obtention des 17-cétostéroïdes et détermination quantitative.
b) Adsortion.

c) Elution.

d) Captation des éluats en 48 fractions.

e) Réaction colorée (Méthode Zimmermann, Callow, etc.) dans chaque fraction.

f) Tracé de la courbe chromatographique, en situant en abscisses les différentes fractions et en ordonnées les coefficients d'extinction correspondants.

g) Identification des différentes ondes.

Chez les sujets normaux la courbe chromatographique présente huit ondes qui englobent les stéroïdes suivants :

Onde I.	— Produits artificiels de l'hydrolyse.
Onde II.	— i-Androstanolone, Isoandrostérone.
Onde III.	— Déhydroisoandrostérone.
Onde IV.	— Androstérone.
Onde V.	— Etiocholanolone.
Onde VI.	— 11.-Hydroxy-androstérone.
Onde VII.	— 11-Hydroxy-étiocholanolone.
Onde VIII.	— 17-cetostéroïdes non identifiés.

Les stéroïdes sont élués successivement dans cet ordre et si la position des ondes dans le chromatogramme ne subissait pas de déviations à droite ou gauche, c'est-à-dire, si les éluats des différents stéroïdes ne souffraient pas de variations et s'obtenaient toujours dans les mêmes fractions, l'analyse quantitative serait terminée. Mais il n'en est pas ainsi parce que différentes conditions ambiantes, des impuretés dans l'extrait chromatographique ou dans les liquides d'élution, des petites différences dans la force d'adsortion de l'oxyde d'alumine et d'autres impondérables provoquent des altérations. Bien que l'ordre de succession ne soit pas modifié il se produit des déplacements indésirables qui rendent nécessaire une estimation qualitative.

Celle-ci se réalise selon divers procédés :

1. - On trace des courbes témoins avec des stéroïdes purs. Chacun donne une onde en cloche qui le situera entre certaines fractions. En comparant ces ondes « témoins » avec celles obtenues dans le chromatogramme problème on identifie celles-ci. Il faut opérer dans les mêmes conditions. Ce procédé représente une détermination qualitative par l'emploi de valeurs Rf appliquées à la chromatographie en colonne.

2. - On ajoute à l'extrait original un stéroïde pur ; on réalise une nouvelle chromatographie et on compare avec l'extrait original « non marqué ». L'augmentation de l'onde correspondante l'identifiera sans erreur.

3. - Par spectrographie infra-rouge, après isolement et purification du stéroïde problème.

Ayant utilisé d'une manière exhaustive les deux premières méthodes dans nos travaux, nous pouvons en faire la critique. En effet la première méthode donne fréquemment des résultats inexacts ; on trouve de notables différences en comparant les courbes témoins avec le chromatogramme problème. On a même observé des différences en obtenant des courbes témoins d'un même stéroïde d'origine commerciale différente.

La deuxième méthode est plus exacte mais non infaillible, même en travaillant dans les mêmes circonstances expérimentales. Il est souvent difficile de distinguer des ondes très proches. La quantité de stéroïdes purs ajouté à l'extrait original masque fréquemment les ondes contiguës. Cette méthode oblige, même pour la localisation d'un seul stéroïde qui servira de référence quand aux autres, à réaliser deux chromatographies parallèles ou successives. Il convient de rappeler que la réalisation d'une seule chromatographie représente un travail de plusieurs jours.

L'emploi de la spectrographie infra-rouge, méthode la plus exacte, nécessite, comme nous l'avons déjà indiqué, l'emploi de matériel de grande valeur et un personnel spécialisé.

Dans nos travaux nous avons adopté le deuxième procédé, en marquant l'extrait original avec des stéroïdes purs, ce qui nous oblige à répéter plusieurs fois la même chromatographie. Mais en même temps, nous observons avec attention la colonne d'alumine, en notant avec soin les changements visibles qui se produisent pendant l'adsortion et l'élution. Ceci est possible parce que les extraits obtenus présentent toujours une coloration plus ou moins forte qui varie depuis le jaune jusqu'au rouge obscur ; ceci est dû à la présence d'urochromes ou d'autres pigments se formant peut-être au cours de l'hydrolyse précédente.

Durant l'adsortion, les stéroïdes et les substances colorées sont retenus dans la partie supérieure de la colonne, teignant une bande plus ou moins large selon le contenu de l'extrait en pigments.

Au début de l'élution, les substances adsorbées commencent un déplacement du haut vers le bas, une frange grise obscure avance rapidement, elle se recueille généralement sur la première et deuxième fraction, teignant les éluats en gris-vert sale, fréquemment avec fluorescence. Ces premières fractions recueillent les produits de l'onde I.

Ensuite, plus lentement, se déplace une zone de couleur rougeâtre, d'après Dingemanse, elle correspond à l'« uroroséine », et selon ses affirmations elle s'élue presque toujours avec la Déhidroisoandrostérone Nous avons réalisé cette vérification dans tous les cas, et en effet, les fractions qui recueillent cet urochrome contiennent ce stéroïde (onde III) avec de très légères différences de déplacement.

Par la suite il se détache de la zone supérieure de la colonne, un pigment gris cendré qui se déplace lentement et qui coïncide avec l'élution de l'androstérone (onde IV).

Ensuite une bande étroite jaune commence à émigrer qui se recueille conjointement avec l'étiocholanolone (onde V).

Nous avons réalisé les vérifications pertinentes en « marquant » les extraits avec la Déhidroïsoandrostérone, l'Androstérone et l'Etiocholanolone, en utilisant des extraits de diverses pigmentations et les différences n'ont été que d'intensité de couleurs. Nous avons toujours obtenu les mêmes colorations, coïncidant, avec une marge d'erreur minime, avec l'élution des trois stéroïdes cités. Dans notre large expérience nous n'avons jamais obtenu des extraits avec une pigmentation tellement faible qu'elle rende impossible l'identification des ondes III et IV au moins.

Avec ces observations nous pouvons affirmer que les variations chromatiques qui se produisent sur la colonne d'alumine pendant l'élution sont en relation avec l'élution de stéroïdes déterminés. Ceci constitue un procédé rapide et suffisamment exact pour les déterminations de routine en ce qui concerne l'identification de trois pics principaux qui apparaissent normalement dans ce type de chromatographie.

Il n'est pas nécessaire d'insister sur le fait que pour éliminer quelques doutes qui peuvent surgir pour la vérification, non seulement des ondes II, VI et VII, mais encore pour les ondes III, IV et V, on doit réaliser de nouvelles chromatographies avec l'extrait primitif « marqué » avec les stéroïdes correspondants.

Il est recommandé de connaître approximativement l'intensité du flux chromatique qui est recueilli dans chaque fraction, par les signes conventionnels habituels (+, ++, +++, etc.). Généralement l'intensité de couleur observée dans chaque fraction est en raison directe de la quantité de stéroïde recueilli, de telle sorte que la fraction la plus colorée correspond au point maximum de l'onde.

Dans un autre ordre d'idées et pour terminer nous devons insister sur l'importance qu'ont, en clinique, les trois ondes (III, IV et V) dont la signification est importante dans le diagnostic de certaines maladies endocrines.

CONCLUSIONS

Nous avons réalisé deux cents chromatographies de 17-cétostéroïdes sur colonne d'alumine. Dans toutes nous avons observé avec attention les variations chromatiques qui se produisent pendant l'élution. Ces observations, macroscopiques et subjectives, donc avec une certaine marge d'erreur, se concrétisent de la façon suivante :

1. — Dans les extraits pigmentés on confirme l'observation de Dingemanse qui affirme que la Déhydroïsoandrostérone est éluée avec l'urochrome de couleur rouge « uroroséine ».

2. — Nos propres observations nous permettent d'affirmer que l'androstérone s'élue conjointement avec un autre pigment gris plus ou moins obscur.

3. — Ensuite l'Etiocholanolone est éluée avec une substance jaune.

4. — Ces observations, applicables aux déterminations cliniques, permettent l'identification assez exacte, de trois ondes fondamentales du chromatogramme, sans avoir à recourir, sauf éventuellement, à d'autres procédés de détection qualitatifs.

The use of gas-chromatography in the detection of some industrial poisons in blood

S. FATI,
Institute of industrial medicine, University of Naples
(Director, Professor S. Caccuri)

The detection and recovery of some aromatic hydrocarbons such as benzene, toluene and xylene in the blood of workers and animals poisoned by these toxics have always been a problem of industrial medicine.

Several methods have been carried out for dosing the amount of benzene that can be recovered from the blood of non-poisoned and of poisoned subjects, but the results are still doubtful.

Simonin (1) states that benzene is not present in the blood of non-poisoned workers, while according to Fabre (2) the amount of this poison commonly present in non-poisoned men and animals is in the proportion of 70 - 80 micrograms for a hundred milliliters of blood. Schrenk and co-workers (3) state, on the other hand, that the normal amount of benzene in the blood of non-poisoned animals is 170 micrograms per cent.

For these reasons, today the determination of benzene and homologous in blood as a test of benzene poisoning is not of practical clinical application (Morelli and Mazzella di Bosco (4)).

We have carried out gas-chromatographic researches with N_2-flame ionization detector chromatograph for the identification and determination of benzene, toluene and xylene concentration in the blood of animals exposed to these substances.

The chromatograph was equipped with two columns, two meters long, containing 60 - 80 mesh chromasorb washed supports coated with 10 % silicone gum rubber SE 30 stationary phase.

The standards have been made with physiological solution containing known amounts of benzene, and known amounts of benzene, toluene and xylene. This was to determine the retention time of the substances and the amount that could be recovered by the chromatographic analysis.

The animals employed in our researches were rabbits, five of which were exposed for thirty minutes to benzene (concentration of 5000 p.p.m.), and five other animals exposed to a mixture of benzene (7000 p.p.m.), toluene (3500 p.p.m.) and xylene (3500 p.p.m.).

Samples of 1 μl of blood were injected directly into the vaporizer and the concentration of benzene, toluene and xylene absorbed into the blood was determined immediately after exposure, after thirty minutes, after one hour, and after two hours.

In the first picture is to be seen the chromatogram of the blood of a rabbit before benzene poisoning. Benzene is not present in the sample. The second is the chromatogram of the blood of the same animal immediately after poisoning. The amount of benzene present in the sample can be calculated as 300 micrograms per cent.

The next picture (n° 3) shows the blood of the same animal one hour after poisoning. Benzene is still present but in a small amount. Also after two hours (picture n° 4) benzene can be detected in blood but not calculated.

In the fifth picture is the blood of another rabbit before poisoning with benzene, toluene and xylene mixture. These three aromatic hydrocarbons are not present. Immediately after poisoning (picture n° 6) we can see all these three substances. In the seventh chromatogram is the blood of the same animal thirty minutes after poisoning, in the eighth one hour, and in the last (picture n° 9) the blood two hours after poisoning.

On the basis of our researches we can state that in the blood of non-poisoned animals no aromatic hydrocarbons are present.

After exposure the amount of benzene, toluene and xylene that can be recovered depends both on the concentration of toxics employed and the time that has elapsed between poisoning and withdrawal of the blood specimen. We can also state that, according to Schrenk and co-workers, the saturation of the circulating blood, either for benzene or for homologous, occurs rapidly and is reached within thirty minutes ; that the rate of elimination of these aromatic hydrocarbons from the blood after exposure is very rapid, so that although two hours after the poisoning they can still be detected it is in a small, not dosable amount.

Finally, our researches demonstrate that benzene, toluene and xylene can be easily identified in the blood of poisoned subjects by means of gas-chromatography through a simple method that can be employed as routine in laboratories of industrial medicine.

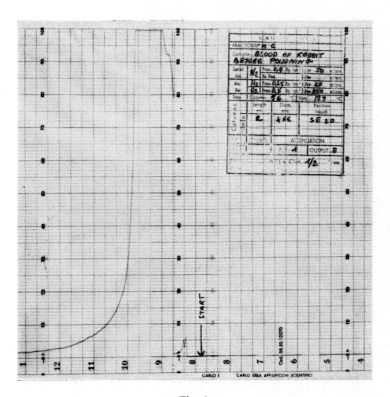

Fig. 1.

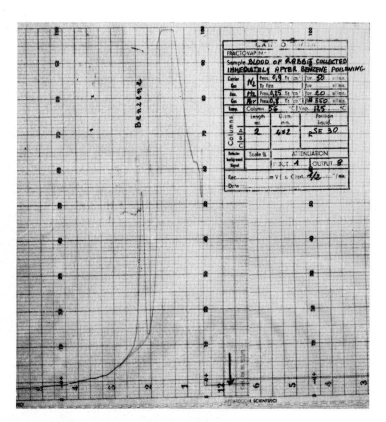

Fig. 2.

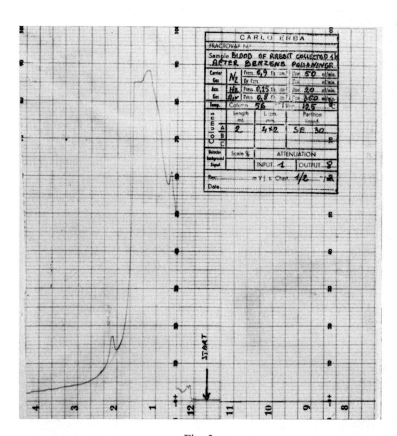

Fig. 3.

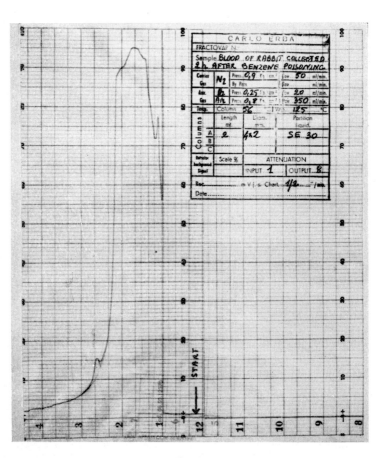

Fig. 4.

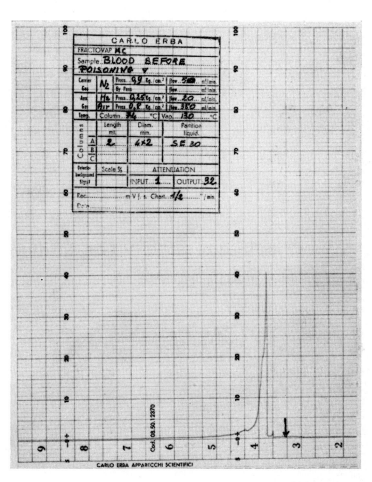

Fig. 5.

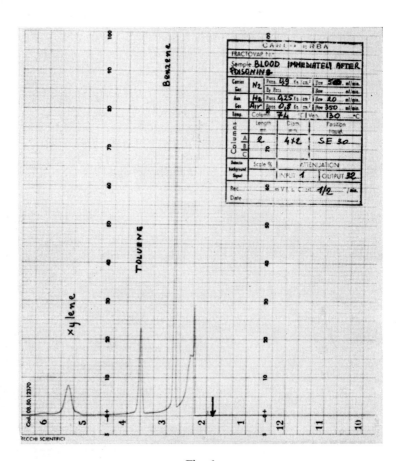

Fig. 6.

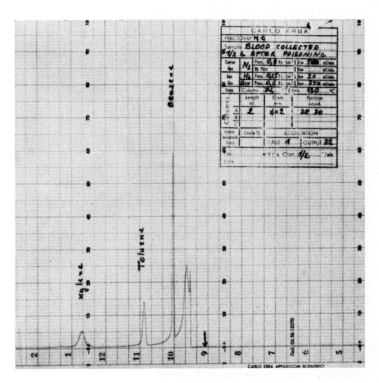

Fig. 7.

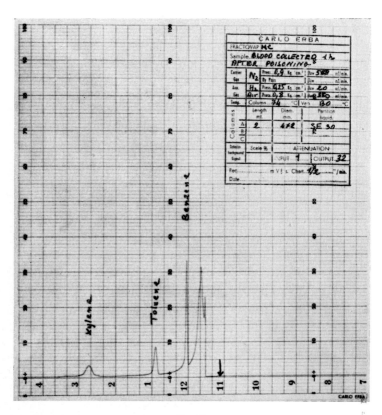

Fig. 8.

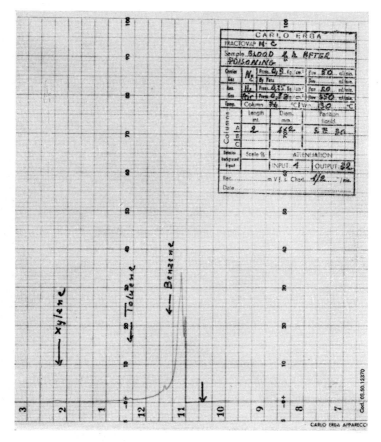

Fig. 9.

Bibliography

(1) SIMONIN, N. La Médecine du Travail, Paris. Ed. Masson et Cie, 1957.

(2) FABRE, A. Métabolisme du benzène et du toluène dans l'organisme, Paris. Herman et Cie. Ed., 1948.

(3) SCHRENK, H.H., YANT, W.P., PEARCE, S.J., PATTY, F.A. and SAYERS, R.R. Absorption, distribution and elimination of benzene by body tissues and fluids of dogs exposed to benzene vapor. J. Ind. Hyg. and Toxic, 23 ; 21, 1941.

(4) MORELLI, A. and MAZZELLA DI BOSCO, M. Malattie da benzolo. Caccuri S. La Medicina del laroro, Milano. Ed. Wassermann. 1963.

Nouveaux aspects de la chromatographie fonctionnelle des stéroïdes en couche mince

Belisario P. LISBOA, Alba I. PALOMINO* et Giulio ZUCCONI**

Hormonlaboratoriet, Kvinnokliniken, Karolinska sjukhuset et
Wenner-Grens Institut, Stockholms Universitet, Suède.

Résumé.

L'application de chromatographie en couche mince fonctionnelle à la séparation des groupes de stéroïdes possédant une polarité voisine est étudiée dans ce travail.

La différence de vitesse de condensation des hydrazines avec des groupes carbonyles, selon leur situation par rapport au noyau stérolique est utilisée pour la séparation des cétostéroïdes par élatographie.

Des couches imprégnées à l'acide borique se sont montrées utiles pour la séparation des stéroïdes isomères possédant une structure α,β-glycol attaché au noyau stérolique ou liée à la chaîne latérale.

Introduction.

La chromatographie en couche mince s'est révélée très utile dans la résolution et caractérisation d'un très grand nombre de substances stéroïdiques non séparables par d'autres procédés de fractionnement chromatographique.

Pour permettre la résolution des stéroïdes de polarité assez voisine, différentes techniques chromatographiques ont été développées (1, 2), telles que la chromatographie uni-dimensionnelle et bi-dimensionnelle multiple du type ascendant, descendant et horizontal ; la chromatographie de partage, en utilisant entre autres comme phase stationnaire la formamide, le pro-

---

* Boursier du Gouvernement suédois (Styrelsen för Internationell Utveckling SIDA = Swedish International Development Authority) ; adresse actuelle : Univérsidad de Cauca, Facultad de Medicina, Popayan, Cauca, Colombie.

** Adresse actuelle : Clinica Ostetrica e Ginecologica, Università di Firenze, Italie.

pylèneglycol, l'undecane ; la chromatographie continue — « over-run chromatography » — développée suivant les techniques précédemment indiquées : le développement chromatographique effectué à basse température ; l'usage des adsorbants spéciaux ou de mélange d'adsorbants ; et finalement, la formation des dérivés avant ou pendant le développement chromatographique.

Ces différentes techniques ont été précieuses pour la résolution d'un nombre élevé de stéroïdes en C_{18}, C_{19} et C_{21}, des stérines et d'autres substances à noyau stérolique. Cependant, quelques paires de substances possédant une polarité très proche sont restées inséparables. C'est le cas de quelques stéroïdes isomères et des stéroïdes qui diffèrent l'un de l'autre par la présence ou par la position de doubles liaisons isolées ou conjuguées.

Becker a développé pour la chromatographie sur papier une technique connue sous le nom d'élatographie (3), selon laquelle un mélange de substances de polarité voisine doit migrer, au cours du développement chromatographique, à travers, une zone du papier imprégné du réactif adéquat. Ce réactif doit être susceptible de former un dérivé avec une des substances du mélange, cependant que l'autre restait inchangé. Cette technique a été employée en couche mince pour la séparation des stéroïdes alcooliques et cétoniques en utilisant comme hydrazine le réactif T de Girard (4).

Les travaux de Foster (5), Frahn et Mills (6) ont montré l'intérêt de l'emploi d'agents capables de former des complexes (complex-forming agents) pour la séparation des substances polyhydroxylées par électrophorèse. Après la formation des chélates avec ions borates, un très grand nombre d'hydrates de carbone ont été séparés par la chromatographie sur papier et en couche mince (7). Ce procédé a été employé aussi pour la séparation électrophorétique des dihydroxystéroïdes présentant une structure glycolique (8).

Nous présentons ici l'application des couches minces imprégnées partiellement à la ligne du départ, ou totalement, pour la séparation de quelques « groupes critiques » de stéroïdes.

Matériel et méthodes.

1. Réactifs.

Tous les solvants utilisés pour le développement chromatographique ont été redistillés avant usage. A l'exception de l'anisaldéhyde (99 % ; S-5209, Kebo AB, Suède) tous les autres réactifs ont été d'une pureté optimale (pro analyse). L'hydrazine de l'acide isonicotinique (INH) et l'acide borique ont été obtenus de AB Bofors Nobelkrut, Bofors, Suède, et de Merck AG, Darmstadt, Allemagne, respectivement.

2. Stéroïdes.

5α-Androsta-1-ene-3,17-dione a été obtenu par oxydation de 17β-hydroxy-androsta-1-ène -3-one avec 17β-hydroxystéroïd-oxydoreductase d'origine bactérienne (P. testosteroni) ; 5β-prégnane-3β, 17α, 20β-triol a été obtenu par réduction de 3β, 17α-dihydroxy-5β-prégnane-20-one au moyen de borohydrure de potassium (9) et le prégna-5-ène-3β, 17α, 20α-triol par réduction de 3β, 17α-dihydroxy-prégna-5-ène-20-one avec aluminiumhydrure de lithium en tetrahydrofurane (10). La plupart des stéroïdes * utilisés dans cette investigation ont été très aimablement mis à notre disposition par le Dr. S. Bernstein (16α-hydroprégna-4-ène-3, 20-dione), le Prof. E. Diczfalusy (16-epi-œstriol, 17-epi-œstriol, 6α, 7α-dihydroxy-œstradiol, 20β, 21-dihydroxy-prégna-4-ène-3-one, prégna-5-ène-3β, 20α, 21-triol, prégna-5-ène-3β, 20β, 21-triol), le Dr. J. Fishman (15α-hydroxy-œstriol), le Dr. D. Fukushima (prégna-5-ène-3α, 20β-triol, prégna-5-ène-3β, 16α-20α-triol, prégna-5-ène-3β, 16α, 20β-triol), le Dr. T.F. Gallagher (prégna-5-ène-3α, 16α, 20β-triol), le Prof. W. Klyne (16-epi-œstriol-3-méthyléther, 17-epi-œstriol-3-méthyléther, 5β-prégnane-3-one, 5β-prégnane-3β, 16α, 20β-triol, 3α-hydroxy-5α-prégnane-20-one, 5β-prégnane-3β, 16β, 20α-triol, 5β-prégnane-3α, 17α, 20β-triol, prégna-5-ène-3β, 17α, 20β-triol, 5β-prégnane-3α, 11β, 17α, 20β-trétol, 5β-prégnane-3α, 11β, 17α, 20α-tétrol), le Dr. R. Neher (3β, 16α-dihydroxy-prégna-5-ène-20-one), le Dr. G. Snatzke (prégna-4-ène-3β, 15α, 20β-triol), le Dr. G. H. Thomas (3α, 6α-dihydroxy-5α-prégnane-20-one) et le Dr. J. Ufer (œstrone, œstradiol, œstriol, 17β-hydroxy-5α-androsta-1-ène-3-one, 5β-prégnane-3α, 17α, 20α-triol).

Les stéroïdes, dont l'origine n'est pas mentionnée ci-dessus ont été fournis par Steraloids Inc. (Pauling, New York, U.S.A.).

3. Méthodes.

La chromatographie en couche mince a été développée suivant la méthode ascendante dans des cuves standard en utilisant du gel de silice (Silica gel G, Merck AG, Darmstadt, Allemagne) comme adsorbant. Les conditions chromatographiques ont été décrites précédemment (11, 12). Les systèmes de solvants suivants ont été employés : (I) acétate d'éthyle-cyclohexane-éthanol 45 : 45 : 10 ; (II) acétate d'éthyle-cyclohexane 50 : 50 ; (III) chloroforme-éthanol 90 : 10 ; (IV) chloroforme-éthanol 95 : 5 ; (V) benzène-éthanol 90 : 10 ; (VI) acétate d'éthyle-n-hexane-acide acétique-éthanol 72 - 13,5 : 10 : 4,5 ; (VII) acétate d'éthyle-n-hexane-acide-acétique 75 : 20 : 5 ; (VIII) phase organique, après séparation, d'un mélange de n-butanol-tert-butanol-eau 1 : 1 : 1 ; (IX) benzène-méthanol 75 : 25.

L'imprégnation des couches en gel de silice a été faite différemment selon le dérivé à obtenir. Dans le cas de couches imprégnées à l'acide borique, les couches ont été préparées en mélangeant 50 g de gel de silice G avec 100 ml d'une solution d'acide borique à 10 % ; les plaques ont été laissées pendant la nuit à la température ambiante et chauffées à 100° C pendant 30 minutes peu avant l'usage. L'imprégnation des plaques à l'INH a été faite seulement dans la région de la ligne de départ, sur une largeur de 1,5 à 2,0 cm, avec une solution à 0,5 % d'hydrazine dans 10 % d'acide acétique ; les stéroïdes doivent être déposés sur la ligne de départ avant

---

*) Nomenclature systématique des stéroïdes dont la dénomination courante a été employée au cours de ce travail : testostérone = 17β-hydroxy-androsta-4-ène-3-one ; œstrone = 3-hydroxy-œstra-1,3,5(10)-triène-17-one ; œstradiol = œstra-1,3,5(10)-triène-3, 17β-diol ; œstriol = œstra-1,3,5(10)-triène-3, 16α, 17β-triol ; 16-epi-œstriol = œstra-1,3,5(10)-triène-3, 16β, 17β-triol ; 17-epi-œstriol = œstra-1,3,5(10)-triène-3, 16α, 17α-triol ; 16, 17-epi-œstriol = œstra-1,3,5(10)-triène-3, 16α, 17α-triol ; 15α-hydroxy-œstriol = œstra-1,3,5(10)-triène-3, 15α, 16α, 17β-tétrol ; 6α, 7α-dihydroxy-œstradiol = œstra-1,3,5(10)-triène-3, 6α, 7α, 17β-tétrol.

que la solution de l'INH soit complètement évaporé, et une heure avant le développement chromatographique. Pour la détection des stéroïdes dans les plaques, on a utilisé la réaction avec l'anisaldehyde-acide sulfurique (13).

Résultats et discussion

Deux méthodes chromatographiques fonctionnelles ont été utilisées pour la résolution des stéroïdes dont la séparation est impossible ou seulement partielle par les méthodes classiques de fractionnement en couche mince : la formation des hydrazones avec l'acide isonicotinique et de chromatographie en gel de silice imprégné à l'acide borique. Dans le premier cas le dérivé est formé par le traitement de la ligne de départ avec le réactif utilisé avant la déposition du stéroïde à chromatographier (élatographie). Le tableau 1 résume les résultats élatographiques obtenus dans le système benzène-méthanol 75 : 25 pour les stéroïdes cétoniques de la série du prégnane. Des cétones en position 3, 6, 17 et 20 dans le noyau stéroïdique montrent une réactivité vis-à-vis de l'hydrazine de l'acide isonicotinique (INH), cependant

TABLEAU I.

Migration chromatographique (en centimètres) des stéroïdes cétoniques et leurs isonicotinylhydrazones formés par la technique élatographique. Adsorbant : gel de silice .G (Merck AG, Allemagne). Système de solvants : benzène-méthanol 75 : 25. Pour d'autres détails, voir le texte.

	Stéroïde libre	Hydrazones		
		mono-	di-	tri-
5α-androstane-17-one	12,4	8,9		
5β-prégnane-3-one	12,2	8,7		
prégna-4-ène-3,20-dione	11,0	6,8	4,2	
5α-prégnane-3,20-dione	11,5	6,8	4,2	
5β-prégnane-3,6,20-trione	9,4	6,1	3,6	1,1
5β-prégnane-3,11,20-trione	9,8	6,0	3,4	
5α-prégnane-3,11,20-trione	9,8	6,5	4,1	
5β-prégnane-3,12,20-trione	10,5	6,3	3,9	
20β-hydroxy-5α-prégnane-3-one	10,5	6,0		
20β-hydroxy-prégna-4-ène-3-one	10,0	5,7		
3α-hydroxy-5β-prégnane-20-one	10,3	6,4		
3α-hydroxy-5α-prégnane-20-one	10,0	6,2		
3α,6α-dihydroxy- 5α-prégnane-20-one	6,3	4,2		
3β,16α-dihydroxy-prégna-5-ène-20-one	7,2	4,5		
12α-hydroxy-5β-prégnane-3,20-dione	7,8	4,8	3,4	
16α-hydroxy-prégna-4-ène-3,20-dione	9,0	5,5		
17α-hydroxy-5β-prégnane-3,20-dione	8,5	5,8		
3α-hydroxy-5β-prégnane-11,20-dione	8,5	5,9		
3β-hydroxy-5α-prégnane-11,20-dione	8,5	5,9		
3β,17α-dihydroxy-5α-prégnane-11,20-dione	7,2	4,5		

que les groupes cétoniques en position 11 et 12 ne forment pas d'hydrazones dans les conditions expérimentales choisies. La vitesse de formation des hydrazones diffère nettement suivant le groupe cétonique considéré. Smith et Foell (14) ont remarqué la formation immédiate des hydrazones avec des groupes cétoniques en C-3 non saturés en Δ^4, cependant que les hydrazones en C-20 non-saturées en Δ^{16} réagissent très lentement ; seulement quelques minutes sont nécessaires pour la formation des isonicotinoylhydrazones avec des cétones en C-3 et C-20 non conjugués. Ce fait nous a amené à séparer la progestérone de la 16-déhydroprogestérone (15), la prégnanolone de la 16-déhydraprégnanolone (16) et la prégnénolone de la 16-déhydroprégnéno-lone (10) par élatographie en utilisant la formation des hydrazones avec l'INH.

Il faut remarquer qu'en général cette technique permet plus souvent l'identification des stéroïdes en question que leur analyse quantitative. En effet, la vitesse de combinaison des divers groupes carbonyles avec des

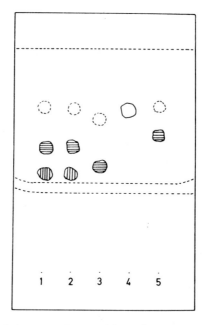

Fig. 1. — Développement élatographique dans une couche en gel de silice traitée par l'hydrazine de l'acide isonicotinique des différents stéroïdes en C_{19}. Système de solvants : la phase supérieure d'un mélange de n-butanol, *tert*-butanol et l'eau (1:1:1). Stéroïdes : (1) andostra-4-ène-3,17-dione, (2) 5α-androsta-1-ène-3,17-dione, (3) androsta-1,4,6-triène-3,17-dione, (4) 17β-hydroxy-androsta-1,4-diène-3-one et (5) 17β-hydroxy-androsta-4-ène-3-one.

Elatographic development in a silica gel layer treated by the hydrazine of isonicotinic acid of the different steroids in C_{19}. System-solvents : the higher phase of a mixture of n-butanol, *tert*-butanol and water (1:1:1). Steroids : (1) androsta-4-ene 3,17-dione, (2) 5α-androsta-1-ene-3,17-dione, (3) androsta-1,4,6-triene-3,17-dione, (4) 17β-hydroxy-androsta-1,4-diene-3-one and (5) 17β-hydroxy-androsta-4-ene-3-one.

hydrazines étant extrêmement variable selon leur type et position dans le noyau stérolique, il s'en suit la formation de plusieurs dérivés hydrazones pour un même stéroïde polycétonique. Par conséquent l'application quantitative de la méthode concerne essentiellement les stéroïdes mono-cétoniques et devient beaucoup plus problématique dans l'évaluation des stéroïdes polycétoniques.

Les études tenant compte de la réactivité des groupes cétoniques vis-à-vis du réactif de Girard (17) montrent que des diénone homoannulaires ne forment pas d'hydrazones ; ce fait nous a permis de séparer en couche mince des stéroïdes en C_{19} d'une polarité très proche après la formation des Girard-hydrazones (11). L'hydrazine de l'acide isonicotinique peut substituer le réactif T de Girard lorsqu'il s'agit de séparer ou identifier des stéroïdes cétoniques non saturés possédant des structures diènone homoannulaire (androstane-1,4-diène-3,17-dione), diènone hétéroannulaire (androstane-4,6-diène-3,17-dione), triènone hétéroannulaire (androstane-1,4,6-triène-3,17-dione) ou simplement énone (5α-androstane-1-ène-3,17-dione ; androstane-4-ène-3,17-dione).

La figure 1 montre un exemple élatographique ou la formation des hydrazones avec l'hydrazine de l'acide isonicotinique peut amener à la séparation de la paire testostérone (17β-hydroxy-androsta-4-ène-3-one)/1-déhydro-testostérone (17β-hydroxy-androsta-1,4-diène-3-one) ou à l'identification de l'androsta-1,4,6-triène-3,17-dione en présence de l'androsta-4-ène-3,17-dione ou du 5α-androsta-1-ène-3,17-dione.

Les couches en gel de silice imprégné à l'acide borique se sont révélées également très satisfaisante pour obtenir la résolution des paires critiques de stéroïdes, c'est-à-dire, des stéroïdes possédant une polarité aussi voisine pour empêcher leur séparation chromatographique dans des couches non imprégnées. Deux types de « paires critiques » ont pu être résolus dans des couches imprégnées à l'acide borique, à savoir, des α, β-glycols nucléaires et des glycols de la chaîne latérale dans les stéroïdes en C_{21}.

Le Tableau 2 nous montre les valeurs Rf obtenues pour plusieurs stéroïdes alcooliques de la série du prégnane dans des couches en gel de silice d'une part, et dans des couches en gel de silice/acide borique dans l'autre. Il se trouve que, dans des couches imprégnées, les stéroïdes possédant une structure glycolique en $16\alpha,20\alpha$-, $16\beta,20\alpha$-, $17\alpha,20\alpha$- et $17\alpha,20\beta$-migrent beaucoup plus que dans des couches non imprégnées. Au contraire, notons que les stéroïdes présentant une structure en $15\alpha,20\beta$- et $16\alpha,20\beta$-diol ne changent pas de polarité dans les deux types de couches employées. En observant la composition des systèmes de solvants employés, il apparaît que le pH exerce une influence très marquante sur la variation de polarité d'un stéroïde lorsqu'on passe d'une couche à l'autre ; en effet, dans des systèmes acides, la modification de migration de certains stéroïdes reste peu importante par exemple pour le 20,21-glycol-isomères étudiés.

En utilisant du gel de silice imprégné à l'acide borique, il est possible d'obtenir une séparation complète des stéroïdes acétaldéhydogéniques et formaldéhydogéniques avec une structure glycol en 17,20 ou 20,21. Dans la figure 2 est illustrée l'application des couches en gel de silice/acide borique pour la séparation des isomères $20\beta,21$-dihydroxy-prégna-4-ène-3-one, $17\alpha,20\alpha$-dihydroxy-prégna-4-ène-3-one et $17\alpha,20\beta$-dihydroxy-prégna-4-ène-3-one

TABLEAU II.

Valeurs Rf × 100 obtenues en couche mince pour des stéroïdes alcooliques en $C_{21}O_3$ après chromatographie ascendante en gel de silice (a) ou en gel de silice imprégné d'acide borique (b). Pour la composition des systèmes de solvants, voir le texte.

Stéroïdes [*]	SYSTEMES CHROMATOGRAPHIQUES												
	I		II		III		IV		V		VI		VII
	(a)	(b)	(a)	(b)	(a)	(b)	(a)	(b)	(a)	(b)	(a)	(b)	(a)	(b)
P⁵-3β,15α,20β-triol	30	33	6	4	28	31	12	15	16	14	48	49	29	28
5βP-3β-16 β,20α-triol	37	52			38	62	18	50			50	61		
5αP-3β,16α,20β-triol	27	26			24	27	10	11	14	13	44	42	24	22
P⁵-3α,16α,20β-triol	27	30	2	5	22	24	8	12	16	17	41	40	19	16
P⁵-3β,16α,20α-triol	29	40	14	15	38	47	15	28	22	31	46	49	27	29
P⁵-3β,16α,20β-triol	25	28	8	11	26	27	10	15	14	15	43	43	24	22
5βP-3α,17α,20α-triol	26	45	5	15	30	55	10	30	20	38	52	58		
5βP-3α,17α,20β-triol	32	50			35	56	15	38	19	41	54	58	37	39
5βP-3β,17α,20 β-triol	39	52			41	62	22	53	26	48	61	60	45	48
P⁵-3β,17α,20α-triol	38	51	11	26	36	59	14	50	23	45	57	62	40	44
P⁵-3β,17α,20β-triol	41	56	16	31	40	57	22	51	30	46	61	65	45	48
P⁵-3β,20α,21-triol	26	43	5	15			10	30	15	36	46	54	28	35
P⁵-3β,20β,21-triol	29	46	7	17			10	26	18	40	48	58	31	37

[*] P = prégnane, P⁵ = prégna-5-ène.

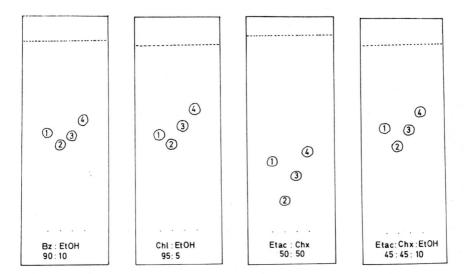

Fig. 2. — Chromatogrammes en gel de silice imprégné à l'acide borique de la testostérone (1), 20β,21-dihydroxy-prégna-4-ène-3-one (2), 17α,20α-dihydroxy-prégna-4-ène-3-one (3) et 17α,20β-dihydroxy-prégna-4-ène-3-one (4), développés en 4 systèmes de solvants ; Bz = benzène, Chl = chloroforme, Chx = cyclohexane, Etac = acétate d'éthyle et EtOH = éthanol.

Chromatograms with silica gel impregnated with boric acid of testosterone (1), 20β,21-dihydroxy-pregna-4-ene-3-one (2), 17α, 20α-dihydroxy-pregna-4-ene-3-one (3) and 17α, 20β-dihydroxy-pregna-4-ene-3-one (4), developed in 4 solvent systems, Bz = benzene, Chl = chloroform, Chx = cyclohexane, Etac = ethyl acetate and EtOH = ethanol.

dans 4 systèmes de solvants. Ces deux derniers glycols n'ont pas pu être séparés auparavant dans des couches en gel de silice en employant les mêmes systèmes de solvants utilisés ici (17).

Dans la figure 3 sont reproduits les chromatogrammes en gel de silice (a) et en gel de silice imprégné à l'acide borique (b) des 5β-prégnanetétrols avec épimérisme au carbone 20. Les chromatogrammes montrent que dans des couches minces traitées à l'acide borique, une séparation nette a été obtenue pour cette paire de stéroïdes non séparable dans des couches non traitées (18).

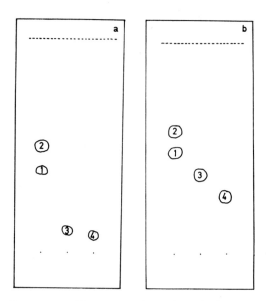

Fig. 3. — Chromatographie en couche mince de la testostérone (1), androsta-4-ène-3,17-dione (2), 5β-prégnane-3α,11β,17α,20β-tétrol (3) et 5β-prégnane-3α,11β,17α,20α-tétrol (4), développée en gel de silice (a) et en gel de silice imprégné à l'acide borique (b). Système de solvants : benzène-éthanol 90 : 10.

Thin layer chromatography of testosterone (1), androsta-4-ene-3, 17-dione (2), 5β-pregnane-3α, 11β, 17α, 20β-tetrol (3) and 5β-pregnane-3α, 11β, 17α, 20α-tetrol (4), developed with silica gel (a) and with silica gel impregnated with boric acid (b). System-solvents : benzene-ethanol (90 : 10).

La méthode de développement chromatographique en couches minces de gel de silice traité à l'acide borique a été appliquée avec succès pour la séparation des α,β-glycols nucléaires, en permettant une résolution remarquable des isomères cisoïdes d'une part, et transoïdes d'autre part. La figure 4 donne, d'une part, une image de la résolution de l'œstriol et d'autre part celle des 16- ou 17-épi-œstriol. Ces deux derniers stéroïdes présentent une structure glycol-cis au moyen D, forment des complexes cycliques avec du bore, cependant que ni l'œstriol, ni le 16-17-épi-œstriol ne le forment. Des complexes similaires sont formés avec d'autres stéroïdes glycoliques, telles que le 15α-hydroxy-œstriol et le 6α,7α-dihydroxy-œstradiol, dont l'ordre de polarité vis-à-vis de l'œstriol change lorsqu'on utilise des couches en gel de silice traité à l'acide borique.

Au délai de l'application de cette méthode pour la séparation et l'identification des stéroïdes isomères présentant une structure glycolique cis/trans l'emploi des couches imprégnées peut encore rendre de grands services lorsqu'on se propose de purifier des stéroïdes glycoliques cisoïdes des contaminants à plus forte polarité.

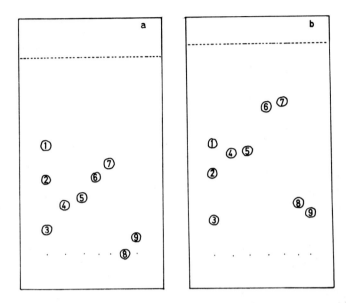

Fig. 4. — Chromatographie en couche mince de gel de silice (a) ou de gel de silice imprégné à l'acide borique (b), développée en benzène-éthanol 90 : 10 des stéroïdes œstrogènes : (1) œstrone, (2) œstradiol, (3) œstriol, (4) 16-épi-œstriol, (5) 17-épi-œstriol, (6) 16-épi-œstriol-3-méthyl-éther, (7) 17-épi-œstriol-3-méthyléther, (3) 15α-hydroxy-œstriol, (9) 6α, 7α-dihydroxy-œstradiol.

Thin layer chromatography with silica gel (a) or silica gel impregnated with boric acid (b) developed in benzene-ethanol 90 : 10 of the estrogenic steroids (1) estrone, (2) estradiol, (3) estriol, (4) 16-epi-estriol, (5) 17-epi-estriol, (6) 16-epi-estriol-3-methyl-ether, (7) 17-epi-estriol-3-methyl-ether, (8) 15α-hydroxy-estriol, (9) 6α, 7α-dihydroxy-estradiol.

Remerciements

Les stéroïdes nécessaires à la conduite de ce travail, nous ont été aimablement fournis par MM. le Dr. S. Bernstein (USA), le Prof. E. Diczfalusy (Suède), le Dr. J. Fishman (USA), le Dr. D. Fukushima (USA), le Dr. T.F. Gallagher (USA), le Dr. G. Snatzke (Allemagne), le Dr. G.H. Thomas (Grande-Bretagne) et le Dr. J. Ufer (Allemagne). Nous leur exprimons toute notre gratitude.

Une bourse du Gouvernement suédois (S.I.D.A.) a permis à l'un des auteurs (A.I.P.) de collaborer à ce travail. Nous tenons à remercier Mlle Tove T. Jahnsson pour sa collaboration technique dans nos expériences et le Dr. Horace Micheli pour son assistance pendant la préparation du manuscrit.

Bibliographie

1. KIRCHNER, J.G. Thin-Layer Chromatography, dans *Technique of Organic Chemistry* (E.S. Perry et A. Weissberger, Edts), Vol. XII, New York, Intercsience Publishers, 1967.
2. LISBOA, B.P. Thin-layer Chromatography of Steroids, Sterols and Related Compounds, dans *Steroids and Terpenoids,* Clayton R.B. Edt. (Methods in Enzymology, vol. 15, Colowick, S.P. et Kaplan, N.O., Serie - Editeurs) New York, Academic Press, 1968.
3. BECKER, A. *Z. anal. Chemie,* **174** : 161, 1960.
4. LISBOA, B.P. *J. Chromatogr.,* **24** : 475, 1966.
5. FOSTER, A.B. *Advanc. Carbohydr. Chem.* **12** : 81, 1957.
6. FRAHN, J.L. et MILLS, J.A. *Austral. J. Chem.,* **12** : 65, 1959.
7. MORRIS, L.J. Specific Separations by Chromatography on Impregnated Adsorbents, dans New Biochemical Separations, (James A.T. et Morris, L.J. Edts), London, D. van Nostrand Co. Ltd., p. 296, 1964.
8. BLANK, R.H., HAUSMANN, W.K., HOLMLUND, C.E. et BOHONOS, N. *J. Chromatogr.,* **17** : 528, 1965.
9. LISBOA, B.P. *Steroids,* **6** : 605, 1965.
10. LISBOA, B.P. *J. Chromatogr.* (1968) in press.
11. LISBOA, B.P. *J. Chromatogr.,* **19** : 81, 1965.
12. LISBOA, B.P. *Steroids,* **7** : 41, 1966.
13. LISBOA, B.P. *J. Chromatogr.,* **13** : 391, 1964.
14. SMITH, L.L. et FOELL, T. *J. Chromatogr.,* **9** : 339, 1962.
15. LISBOA, B.P. *Steroids,* **8** : 319, 1966.
16. LISBOA, B.P. International Symposium IV Chromatographie Electrophorese, Bruxelles, 1966, Presses Académiques Européennes, Bruxelles, p. 88, 1968.
17. LISBOA, B.P. *Acta endocrinol.,* **43** : 47, 1963.
18. WHEELER, O.H. *Chem. Rev.,* **62** : 205, 1962.
19. LISBOA, B.P. Separation and Characterisation of Sterols and Steroids by Thin-layer Chromatography, dans Chromatographic Analysis of Lipids (G.V. Marinetti, edt.). Volume 2, p. 57, New York, Marcel Dekker Inc., 1968.

Progestatifs associés aux œstrogènes

Détermination chromatographique quantitative sur silicagel

par

J.J. THOMAS et L. DRYON

Laboratoire de Chimie pharmaceutique organique.
Université Libre de Bruxelles.

Samenvatting.

Ondanks de aanzienlijke veranderingen in de struktuur van de voornaam-ste progestatieve derivaten die « per os » bruikbaar zijn, is het mogelijk ze te identificeren door een algemene dunne laag chromatografische methode. Deze kenmerking kan nog meer volledig gemaakt worden door het meten, na elutie, van het U.V. of het I.R. spectrum.

De kwantitatieve bepaling van het progestatieve derivaat, alsook van de œstrogeen die aanwezig is in een hoeveelheid die soms 50 tot 70 maal verminderd wordt, is door chromatografische afscheiding uit-gevoerd. Vervolgens wordt een kwantitatieve elutie en een spektrofoto-metrische of kolorimetrische of nog een fluorimetrische techniek ver-wezenlijkt, in vergelijking met een inwendige standaard die op een gelijke wijze gechromatografieerd en behandeld wordt.

$$* \overset{*}{} *$$

Les recherches effectuées en vue de produire des agents progestatifs bien actifs par voie buccale ont conduit à l'obtention de structures parfois assez éloignées de celle de l'hormone naturelle, la progestérone. En effet, si l'introduction du groupement éthinyl $-C \equiv CH$ qui vient remplacer la chaîne $-CO-CH^3$ en position 17 favorise la résorption par voie buccale, elle conduit cependant à des dérivés dont l'activité est masculinisante et celle-ci doit être atténuée par différents remaniements de la molécule initiale.

Ces importantes modifications peuvent être appréciées par les exemples qui suivent, dont les principales caractéristiques de structure permettent de distinguer trois catégories :

1° *Groupe 19-méthyl,* dont les dérivés conservent la fonction 3-céto et le méthyl en 19 : Médroxyprogestérone acétate ;
Chlormadinone acétate.

2° *Groupe 19-nor* : possède la fonction 3-céto, mais pas de fonction céto-nique en 20 (présence du radical éthinyl) ; les dérivés sont tous déméthylés en 19 et en double liaison est parfois déplacée ou répétée dans d'autres cycles : 19-Noréthistérone (Norlutine) ; Noréthinodrel ; Norgestrel ; Norgestriènone.

3° *Groupe 3-désoxy* : dont la fonction 3-céto est généralement supprimée (ou réduite en fonction alcool) : lynestrénol ; allylestrénol ; 17-éthinyl-estrènediol diacétate (Ethynodiol) ; 3-désoxy-7-acétoxy-6-méthyl-prégnè-none (Anagestone acétate).

Tableau I.

1er groupe.

17α hydroxy-6-méthyl progestérone acétate — Médroxyprogestérone acétate

17α acétoxy-6-chloro-6-déhydro-progestérone — Chlormadinone acétate

2me groupe.

17α éthynyl-4-estrène 17β ol-3 one — 19-noréthistérone

17α éthynyl-5(10)-estrène 17d-3 one — Noréthinodrel

dl-13-éthyl-17α éthynyl-17β hydroxy-4-gonène-3 one — Norgestrel

17α éthynyl-17 hydroxy.estra-tiène-3 one — Norgestriènone (Planor)

3me groupe.

17α éthinyl-estrénol (Lynestrénol)

17α allyl-estrénol (Gestanon)

17α éthinyl-estrène-diol diacétate. (Ovulène) — Ethynodiol diacétate

3-Désoxy-17α acétoxy-6-méthyl prégnènone (Anagestone acétate)

Les modifications de structure font apparaître de notables différences dans le comportement chimique de ces dérivés, soit entre eux, soit par rapport à la progestérone ; aussi, la caractérisation et le dosage des différents types de progestatifs posent-ils un problème analytique qui présente différents aspects. Des publications relativement récentes (1, 2, 3, 4, 5) ont traité cette détermination qualitative et quantitative, mais le plus souvent en se plaçant du point de vue de telle ou telle substance gestagène prise en particulier.

Nous nous proposons d'aborder ce problème de façon aussi générale que possible, et d'établir des points de comparaison entre les dérivés. Par une méthode simple de chromatographie sur couche mince de silicagel, nous rechercherons des conditions opératoires qui, du moins pour la détermination qualitative, seront applicables à la plupart des progestatifs utilisés en gestagénothérapie par voie orale ou comme anti-ovulatoires, c'est-à-dire en association avec une faible proportion d'un œstrogène tel que l'éthinylœstradiol ou son 3-méthyl-éther, le mestranol.

I. Détermination qualitative.

L'opération chromatographique d'identification, pour être de portée aussi générale que possible, peut s'effectuer sur les progestatifs appartenant aux trois groupes, en utilisant le même solvant mobile, le mélange éther-cyclohexane (80-20 ml). La quantité déposée par capillaire est de 20 à 25 µg. Après migration, puis séchage de la plaque, la révélation des spots est obtenue par pulvérisation d'un mélange à volumes égaux d'acide sulfurique concentré et d'éthanol; on porte ensuite à 100° pendant 5 minutes.

Les principaux progestatifs des deux premiers groupes sont aisément reconnus par les données caractéristiques de Rf, de coloration et de fluorescence sous rayonnement à 350 mµ qui sont rassemblés à la figure 1.

Les valeurs de Rf sont respectivement :

Chlormadinone acétate :	0,34	Norgestriènone :	0,49
Médroxyprogestérone acétate :	0,46	Lynestrénol :	0,92
Noréthistérone acétate :	0,55	Allylestrénol :	0,91
Noréthinodrel :	0,73	Ethynodiol diacétate :	0,88
Norgestrel :	0,66		

A noter que le Norgestriènone présente déjà, avant tout traitement, un spot de fluorescence bleue. Celui-ci, après pulvérisation sulfurique et chauffage, prend une coloration jaune, avec forte fluorscence verte.

L'éthynodiol donne par pluvérisation et avant chauffage une coloration rose-mauve qui passe au brun-clair par chauffage à 100° C.

Les gestagènes du groupe 3 (Lynestrénol ; Allylestrénol ; Ethynodiol) présentent, par ce solvant mobile, des Rf, assez voisins ; il est donc nécessaire pour les différencier de recourir à un autre mélange de solvant, notamment le mélange benzène-éthanol (95-5 ml).

Les Rf obtenus sont alors de :

Lynestrénol : 0,55   Allylestrénol : 0,69   Ethynodiol diacétate : 0,77

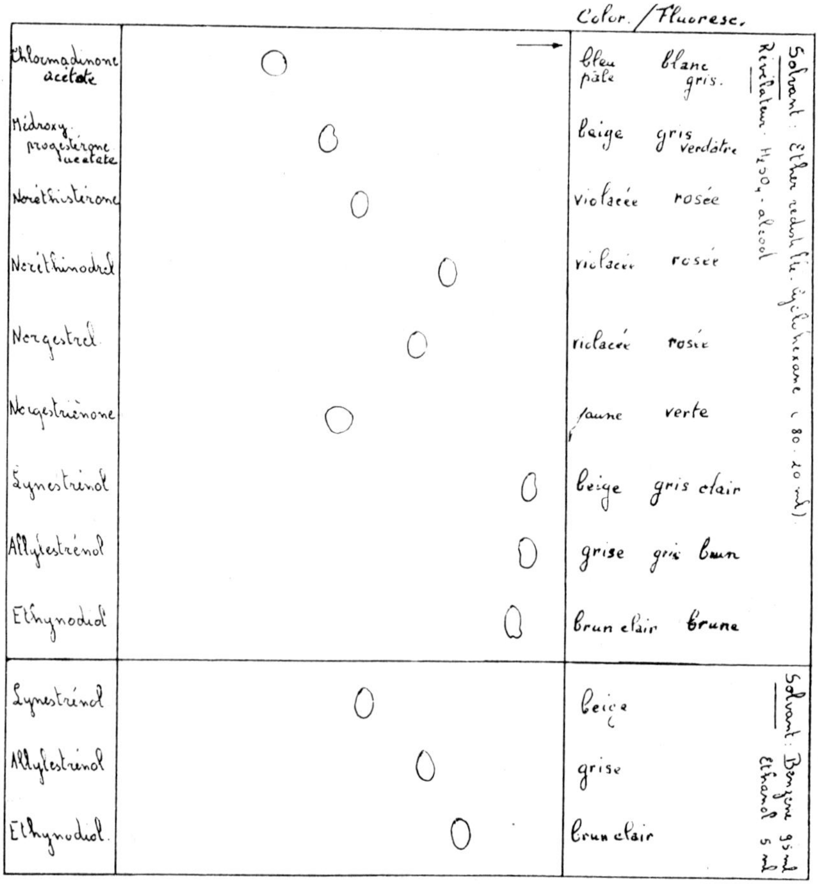

Figure 1.

Spectres U.V.

La caractérisation chromatographique peut être utilement complétée par la détermination des spectres U.V. à la concentration de 5 à 10 μg par ml d'éthanol (voir opération quantitative).

Ces spectres sont caractéristiques pour le *chlormadinone acétate* avec une bande intense à 284 mμ ; pour la *norgestriènone,* bande très large vers 342 mμ ; ainsi que pour le *noréthinodrel* qui tel quel ne présente qu'une très faible absorption mais qui, après chauffage à l'ébullition en milieu chlorhydrique, s'isomérise en noréthistérone et possède alors une forte bande à 240 mμ (voir opération quantitative ci-après et figure 6).

Spectres I.R.

Pour les autres dérivés, il est préférable de recourir à la détermination du spectre I.R., en pellet de KBr (20 μg pour 6 mg de KBr), et notamment pour les gestagènes du *3ᵐᵉ groupe* qui ne présentent qu'une très faible absorp-

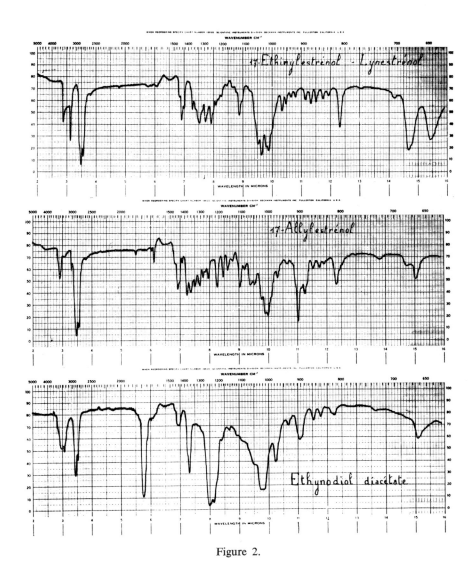

Figure 2.

tion dans l'U.V., étant dépourvu de fonction cétonique en position 3. (Lynestrénol ; Allylestrénol ; Ethynodiol diacétate.) Voir figure 2.

Les spectres infra-rouges sont également importants pour permettre de distinguer *entre-eux* les progestatifs qui possèdent la fonction 3-céto Δ 4 et qui, de ce fait, présentent tous *la même bande* dans l'ultra-violet à 240 mμ ; c'est le cas notamment pour la Médroxyprogestérone, la 19-Noréthistérone et son acétate, et le Norgestrel, dont les spectres I.R. sont donnés ci-après (figure 3).

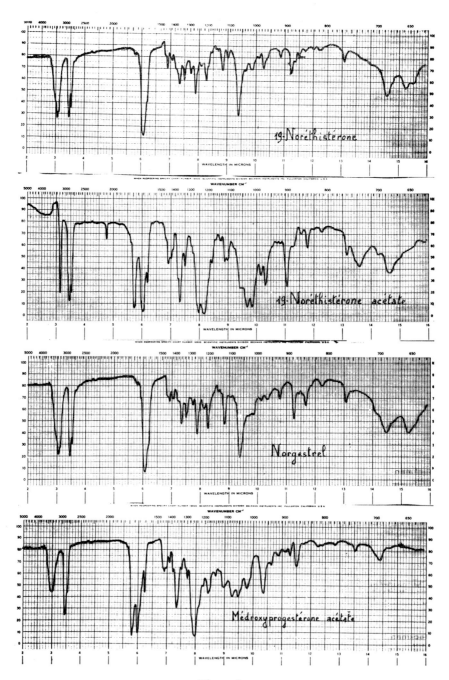

Figure 3.

II. Détermination quantitative.

Méthode générale.

Les préparations antiovulatoires utilisent des proportions d'œstrogène généralement très faibles : elles sont 20 fois moindres que celle du progestatif dans quelques cas, mais peuvent encore être réduites à 50 ou 70 fois moindre dans d'autres types d'associations.

Pour cette raison, la séparation chromatographique ne semble pas s'effectuer favorablement par l'utilisation de « couches épaisses » selon la technique de chromatographie préparative (silicagel, PF 254) car la nécessité d'opérer sur des proportions élevées de progestatif par rapport à l'œstrogène fait le plus souvent apparaître des bandes chromatographiques dont le tracé est sinueux et irrégulier.

Dans cette éventualité, la technique qui semble la plus adéquate consiste à utiliser de préférence la couche mince de 0,25 mm de silicagel PF 254 et à accroître la surface de dépôt par l'emploi de plaques plus larges, ou mieux de plusieurs plaques de 20 × 20 cm. L'extrait de la préparation contenant le mélange hormonal est distribué à l'aide d'un dispositif du genre « Chromatocharger » de Camag ou d'un appareil analogue, à raison de 0,25 à 0,50 ml de solution par plaque de 20 × 20 cm, ce qui correspond au dépôt de 25 μg d'œstrogène, et de 0,50 à 2,0 mg de progestatif. On utilise ainsi 1 à 2 plaques suivant les quantités nécessaires. De part et d'autre de cette ligne de dépôt, on ménage des spots de contrôle qui serviront par révélation à l'aide d'un réactif à repérer la position de migration de la ligne d'œstrogène et de la ligne de progestatif, en protégeant toute la région centrale par un écran.

On a dû recourir à cette technique de repérage par réactifs, malgré l'emploi de silicagel fluorescent, notamment lorsqu'il s'agit de progestatifs du 3me groupe peu absorbants dans l'U.V.

Pour accroître la précision, des standards internes de l'œstrogène et du progestatif sont également introduits par la même technique de distribution.

Après développement chromatographique, on procède au séchage et à la délimitation des plages de migration (repérées soit par la suppression de fluorescence, soit par la révélation par réactifs des spots-repères latéraux), et on prélève ces plages par raclage de la couche de silicagel. On reprend ainsi :

a) les échantillons à doser ;

b) les standards internes traités de même façon ;

c) le « blanc silicagel » constitué par la *même surface de plage* de silicagel, prélevé sur une plaque simplement traitée par le solvant mobile, et séchée dans les mêmes conditions que pour les autres plaques.

Ces différents prélèvements sont extraits par le méthanol à l'aide d'un extracteur du type Kumagawa (on peut aussi utiliser l'extracteur de Soxhlet, à la condition de procéder à un lavage préalable et très prolongé des cartouches à utiliser et après contrôle de l'absence dans le méthanol évaporé d'impuretés réagissant à l'acide sulfurique concentré).

Les différents échantillons de liquides d'extraction sont évaporés à sec à température assez douce sous courant d'air ou d'azote et les résidus d'éluats sont repris par 10 ml de cyclohexane. On filtre et on reprend un volume approprié pour les différents dosages. Ce volume est évaporé à sec et le résidu est repris pour le dosage spectrophotométrique, colorimétrique ou fluorimétrique suivant les cas envisagés. Les mesures s'effectuent toujours par rapport au standard interne et vis-à-vis du blanc silicagel indiqué ci-dessus.

1° Dosage de l'œstrogène.

Il s'agit le plus souvent de l'éthinyloestradiol ou de mestranol (le 3-méthyléther).

La quantité à prélever sur le chromatogramme est de l'ordre de 50 mg. On peut utiliser les méthodes décrites par Tsilifonis et Chafetz (6), par Heusser (7), par Shroff et Huettemann (8) ou Urbanyi et Rehm (1) ou encore par Comer, Hartsaw et Stevenson (2), et par Khoury (10). Nous utilisons un mode opératoire inspiré de la technique de Heusser, mais qui en diffère par le fait que la réaction s'effectue sur l'extrait de chromatographie et non sur le silicagel directement, ainsi que par la composition du réactif sulfurique.

Cette opération se résume comme suit :

Après évaporation de la prise du cyclohexane, le résidu sec (50 à 100 μg) est dissous dans 2 ml d'acide acétique concentré. On ajoute progressivement 8 ml de H_2SO_4 conc. Il se développe une coloration rose. Après 10 minutes, comptées après la fin de l'addition de H_2SO_4, on effectue la mesure à 530 mμ par rapport au blanc réactif.

Les résultats sont donnés dans les exemples qui suivent.

2° Dosage du progestatif.

La méthode adoptée devra varier suivant le type de structure rencontré et déterminé par l'opération qualitative intiale. Lorque le progestatif appartient aux groupes 1 et 2 et possède la fonction 3-céto Δ 4 qui correspond à une importante bande d'absorption à 241 mμ, la détermination quantitative peut s'effectuer par spectrophotométrie directe et mesure de l'absorption à 241 mμ.

Ces mêmes dérivés des groupes 1 et 2 peuvent également être dosés par la méthode à l'hydrazide isonicotinique de E.J. Umberger (11) ; la coloration obtenue est mesurée à 380 mμ. Cette détermination présente l'avantage de ne pas être perturbée par la présence de l'œstrogène, et permettrait même d'éviter la séparation chromatographique ; toutefois, celle-ci reste nécessaire pour doser l'œstrogène.

Pour les progestatifs du groupe 3, les méthodes précédentes sont inapplicables ; dans les exemples qui sont donnés ci-après, notamment pour le 17-Ethinylestrénol (Lynestrénol), la détermination a été effectuée par fluorimétrie. Le résidu sec de l'éluat chromatographique, correspondant à une dose de 100 à 200 μg de lynestrénol, est traité par 10 ml de H_2SO_4 conc. à température de 20° C pendant 20 minutes.

La détermination de la fluorescence s'effectue par excitation à 360·

380 mμ. et mesure à 555 μ., par rapport à un standard interne traité de façon identique. Les mesures ont été obtenues à l'aide du dispositif fluorimétrique de Beckman ou de l'appareil de Turner.

Les résultats sont donnés ci-après.

Exemples d'opérations faites sur des préparations pharmaceutiques

1er GROUPE.

I. Médroxyprogestérone acétate.

a) Médroxyprogestérone acétate 5 mg, utilisé seul (*).

b) Médroxyprogestérone acétate 5 mg, associé à l'Ethinylœstradiol 0.150 mg.

Extraction des hormones par le chloroforme.

La séparation chromatographique est obtenue au moyen du solvant constitué d'un mélange : éther - cyclohexane (80 - 20 ml).

Le Rf du Médroxyprogestérone acétate est de 0,46.

Le Rf de l'Ethinylœstradiol acétate est de 0,76.

La figure 4 donne également les colorations et fluorescence observées par révélation après pulvérisation d'un mélange à parties égales de H_2SO_4 et éthanol, puis chauffage à 100° C pendant 5 minutes.

— *Dosage de l'Ethinylœstradiol* (voir méthode générale).

Ces déterminations obtenues après séparation chromatographique par la méthode colorimétrique, par rapport à un standard interne, donnent une valeur moyenne de *102,6* % de la quantité introduite dans la préparation.

— *Dosage du Médroxyprogestérone acétate* (voir méthode générale).

1. Par spectrophotométrie après chromatographie.
   Ces déterminations donnent de bons résultats (lorsqu'il s'agit de la préparation sans œstrogène [a]) avec une moyenne d'erreur de 3 % par rapport au standard interne (figure n° 4).

2. La méthode à l'isoniazide permet d'obtenir des résultats satisfaisants en présence de l'œstrogène, avec une moyenne d'erreur également de 3 % par comparaison à la quantité de Médroxyprogestérone introduite dans la préparation.

Le coefficient d'absorption obtenu par action du réactif isoniazide-alcool absolu-HCl présente une valeur de E 1 % à 380 mμ. = 574.

(*) Nom de la spécialité : « Provera ».

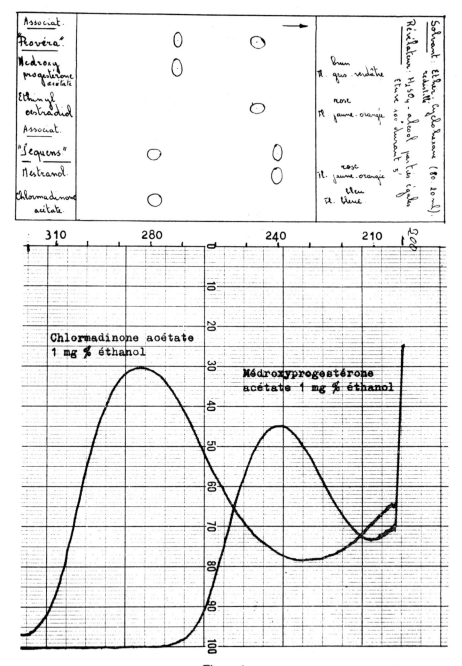

Figure 4.

II. Chlormadinone acétate.

Association : Chlormadinone acétate 2 mg (*).
Mestranol 0,08 mg.

Les principes actifs sont extraits par le chloroforme, puis chromatographiés comme dans l'exemple précédent à l'aide du solvant mobile : Ether - cyclohexane (80 - 20 ml).

Le chromatogramme de la figure 4 montre les Rf de 0,34 pour la Chlormadinone acétate et de 0,83 pour le Mestranol, ainsi que les réactions colorées et fluorescences caractéristiques :
Spot bleuté et fluorescence bleu-grise pour le Chlormadinone acétate.
Spot rose et fluorescence jaune orangé pour le Mestranol.

— *Dosage du Mestranol.*

On applique également la méthode générale et la détermination colorimétrique par le mélange acide acétique - acide sulfurique. Les résultats obtenus pour ces préparations ont été légèrement moins favorables, avec une moyenne d'erreur de 5 % par rapport à la quantité introduite.

— *Dosage du Chlormadinone acétate.*

Le Chlormadinone acétate, par la présence d'un groupement chromophore supplémentaire en position 6, présente au lieu de la bande habituelle à 240 mμ, un renforcement et un déplacement de celle-ci vers 285 mμ. Cette bande intense permet la détermination spectrophotométrique directe, après séparation chromatographique, élution et comparaison à un standard interne, avec une erreur de l'ordre de 1 à 2 %.

L'application de technique à l'isoniazide, référée ci-dessus (page 412) à un mélange de chlormadinone et de Mestranol sans séparation chromatographique préalable permet de retrouver la quantité de progestatif présente avec une erreur de 4,5 % en moyenne.

Le coefficient d'absorption obtenu par action du réactif isoniazide-alcool-HCl est de E 1 % à 380 mμ. = 606.

2me GROUPE.

I. Noréthistérone.

Association : Noréthistérone acétate 4 mg (**).
Ethinylœstradiol 0,05 mg.

Même mode opératoire pour l'extraction des principes actifs par le chloroforme, puis chromatographie avec le même solvant mobile : Ether - cyclohexane (80 - 20 ml).

Le chromatogramme de la figure 5 montre les Rf de 0,53 pour le Noréthistérone acétate et 0,76 pour l'Ethinylœstradiol, ainsi que leurs réactions de coloration et de fluorescence respectives.

(*) Nom de la spécialité « Provera ».

(**) En spécialité pharmaceutique sous le nom de « Anovlar ».

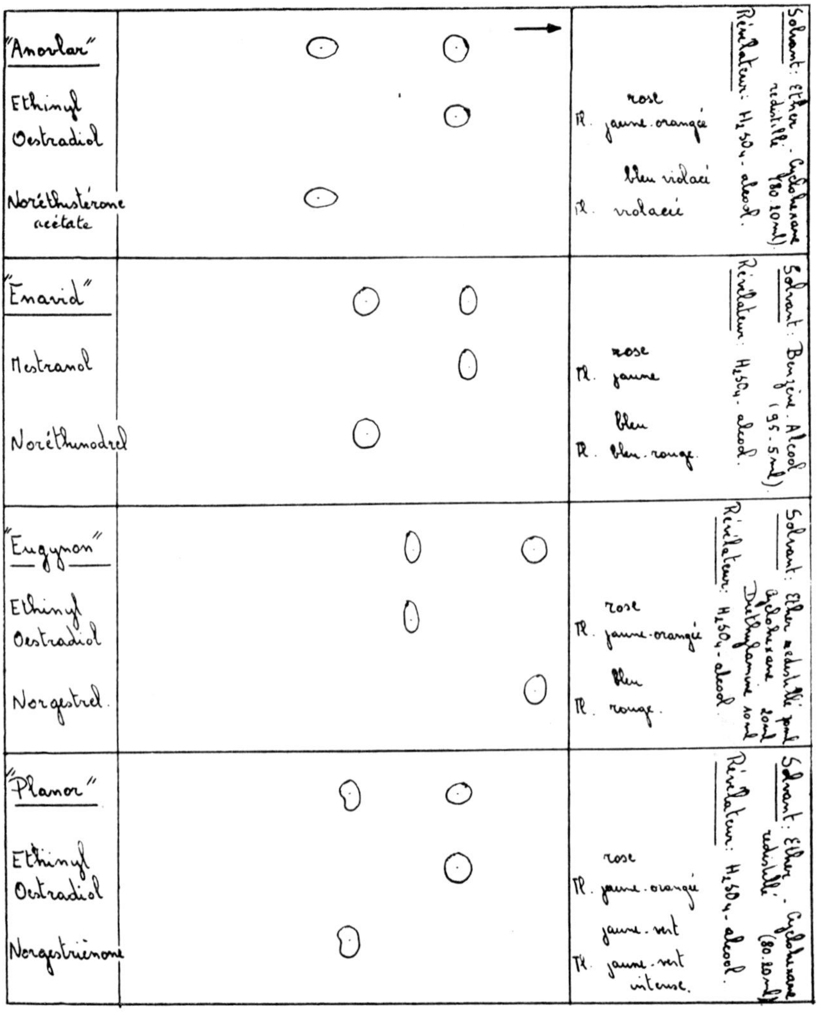

Figure 5.

— *Dosage de l'Ethinylœstradiol.*

Voir l'opération décrite pour le 1er Groupe : elle est suivie sans modification et donne des résultats analogues avec une erreur moyenne qui ne s'écarte pas de plus ou moins 3 % de la quantité introduite.

— *Dosage du Noréthistérone acétate.*

L'acétate de Noréthistérone est déterminé par spectrophotométrie à 240 mμ. On effectue la séparation par chromatographie quantitative suivant la technique générale, élution et comparaison à un standard interne traité de façon identique. On retrouve ainsi la teneur en Noréthistérone acétate avec une précision de ± 3 % de la quantité introduite dans le mélange.

II. Noréthinodrel.

Association : Noréthinodrel 9,85 mg (*).

Mestranol 0,15 mg.

Extraction des principes actifs par le chloroforme et chromatographie en utilisant pour solvant mobile le mélange *benzène - éthanol* (95 - 5 ml). Le chromatogramme de la figure 4 donne la valeur du Rf du Noréthinodrel = 0,56 et pour le Mestranol = 0,78 pour ce type de solvant. Les réactions de coloration et de fluorescence, obtenues après pulvérisation du mélange éthanol - acide sulfurique et chauffage à 100° C, complètent la caractérisation de ces deux constituants.

— *Dosage du Mestranol.*

L'œstrogène est déterminé par la méthode générale rappelée pour le groupe 1. Pour ce genre de préparation, il n'y a pas de perte sensible et on retrouve avec une erreur de ± 1 à 2 % la teneur introduite.

— *Dosage de Noréthinodrel.*

Ce progestatif présente la particularité de posséder une double liaison dans le 1er cycle du noyau stéroïde, mais située en position Δ 5-10. Or, cette structure peut aisément s'isomériser par ébullition en milieu acide et donner le déplacement de la double liaison en position plus stable en Δ 4-5, ce qui reconstitue la formule de la 19-Noréthistérone.

Voir ces détails de structure au tableau I (2me Groupe).

La technique d'isomérisation a été décrite par Bastow (12).

Nous avons appliqué cette technique pour le dosage du Noréthinodrel, mais en introduisant une importante modification, car après l'isomérisation, il est apparu qu'il était plus favorable d'effectuer également la séparation chromatographique.

Notre mode opératoire est le suivant :

A 5 ml de la solution méthanolique extractive (contenant une dose de 5 à 10 mg % de Noréthinodrel) on ajoute 5 ml d'HCl N et on place au B.M. à l'ébullition pendant 5 minutes. Après refroidissement, on neutralise par 5 ml de NaOH N. On évapore à sec, on reprend le résidu par de l'éthanol absolu et on filtre. Cette solution est utilisée pour effectuer les dépôts chromatographiques suivant la technique générale décrite, page 416, étant donné qu'à ce stade, on opère en fait sur la Noréthistérone. On utilise donc le mélange éther - cyclohexane (80 - 20 ml) exactement comme pour cette détermination. Le standard interne sera constitué par le Noréthinodrel pur, qui subira un mode opératoire identique de façon à compenser les légères pertes de ces manipulations. On obtient ainsi des résultats dont les limites d'erreur ne dépassent pas 3 à 4 % de la quantité mise en œuvre. Les mesures du spectre d'absorption U.V. avant et après l'isomérisation apportent un élément de confirmation pour l'identification du Noréthinodrel (voir figure 6).

(*) En spécialité pharmaceutique sous le nom de « Enavid ».

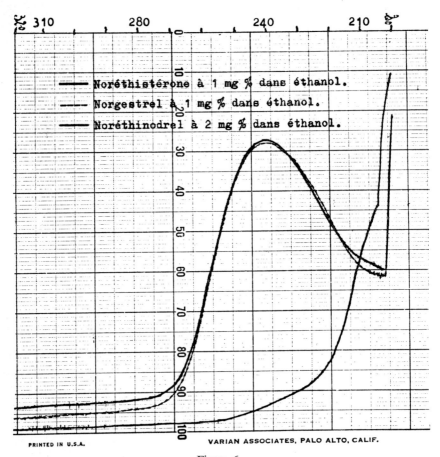

Figure 6.

III. Norgestrel.

Association : Norgestrel 0,5 mg (*).
(DL-13 éthyl-17 α-éthinyl-17 β hydroxy-4-gonène-3 one)
Ethinylœstradiol 0,05 mg.

L'extraction des deux constituants actifs se fait par le chloroforme.

Pour la détermination chromatographique, les mélanges habituels de solvants non polaires ne permettent pas une bonne séparation des deux constituants de cette association. Nous avons utilisé dans ce cas, pour solvant mobile, le mélange éther-cyclohexane-diéthylamine (70 - 20 - 10 ml).

Une séparation nette peut ainsi être obtenue (voir figure 5).

Le Rf du Norgestrel = 0,92 et pour l'éthinylœstradiol = 0,65.

Les réactions de coloration et de fluorescence complètent la caractérisation.

(*) En spécialité pharmaceutique sous les noms de « Eugynon » ou de « Stédiril »

— *Dosage de l'Ethinylœstradiol.*

Il n'y a pas de remarque au sujet de ce dosage colorimétrique effectué par la même technique que pour les cas précédents. La teneur obtenue comporte une erreur de plus ou moins 5 % par rapport au taux introduit.

— *Dosage du Norgestrel.*

Le comportement du Norgestrel est assez semblable à celui de la Noréthistérone dont il est l'homologue supérieur (groupe éthyl en 13, au lieu d'un méthyl). Voir spectre U.V. à la figure 6. Le procédé de dosage peut être calqué sur celui de la Noréthistérone, page 416.

Après chromatographie et élution, détermination spectrophotométrique à 240 mμ par rapport au standard interne.

La précision des valeurs obtenues est de $\pm$ 2 à 3 %.

La méthode à l'isoniazide, avec pour coefficient d'absorption par action du réactif E 1 % à 380 mμ = 754, permet d'atteindre le même degré de précision.

IV. Norgestriènone.

Association : Norgestriènone 2 mg (*).
                (17α éthinyl-17 hydroxy-estratriène-3 one)
                Ethinylœstradiol 0,05 mg.

On applique la même technique opératoire que pour la Noréthistérone avec le solvant éther - cyclohexane (80 - 20 ml), page 416.

Rf de l'Ethinylœstradiol = 0,76.

Ce gestagène est facilement identifié par la variation de fluorescence et la coloration jaune après pulvérisation du réactif éthanol - sulfurique et chauffage à 100° C (voir figure 5 et page 407).

Rf de l'Ethinylœstradiol = 0,76.

— *Dosage de l'Ethinylœstradiol.*

Même technique que précédemment. Résultats analogues avec un coefficient d'erreur qui ne dépasse pas $\pm$ 5 %.

— *Dosage de la Norgestriènone.*

Ce progestatif possède une structure à trois doubles liaisons conjuguées ∙ il y correspond une large bande d'absorption U.V. située vers 338-342 mμ: cette bande est déjà très intense pour une concentration de 5 μg/ml (voir figure 7).

Par la technique chromatographique décrite pour la Noréthistérone (p. 416), la détermination spectrophotométrique à 340 mμ par rapport à un standard interne, permet d'atteindre une précision de $\pm$ 2 à 3 % par rapport à la teneur théorique du mélange.

(*) En spécialité pharmaceutique sous le nom de « Planor ».

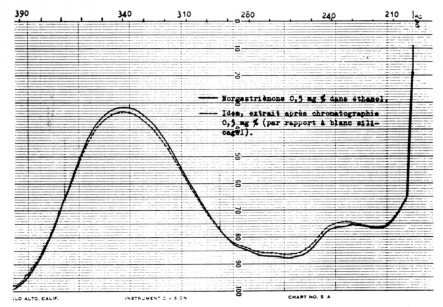

Figure 7.

La détermination par la méthode à l'isoniazide, effectuée sans séparation chromatographique, conduit à des résultats moins précis, avec une marge d'erreur de ± 8 %.

Le coefficient d'absorption à 380 mμ obtenu par l'action de ce réactif à l'isoniazide est de E 1 % = 1296.

3me GROUPE.

La caractéristique de ce groupe est l'absence de fonction cétonique en position 3. Pour l'Ethynodiol, cependant, ce carbone est porteur d'une fonction alcool acétylée. Ces gestagènes, nous l'avons vu, sont dépourvus de forte bande d'absorption U.V. Ils ne réagissent pas avec le réactif à l'isoniazide - éthanol absolu - HCl.

Par contre, ils présentent par action de l'acide sulfurique une intense coloration orangée, accompagnée parfois de fluorescence. Ce caractère sera mis à profit pour la détermination quantitative après chromatographie. soit du progestatif utilisé seul et accompagné d'α-tocophérol, soit d'une association du progestatif et d'un œstrogène.

I. 17-ethinyl-estrénol ou Lynestrénol.

Association : Lynestrénol 5 mg (*).
                Mestranol 0,15 mg (voir figure 8).

(*) En spécialité pharmaceutique sous le nom de « Lyndiol ».

Figure 8.

— *Dosage du Mestranol.*

Après séparation chromatographique utilisant pour solvant mobile le mélange éther - cyclohexane (35 - 65 ml), le Mestranol est élué par extraction, et dosé suivant les conditions opératoires antérieures. La même précision de ± 2 à 3 % est obtenue.

— *Dosage du Lynestrénol.*

La détermination par fluorimétrie s'effectue sur le résidu d'évaporation du liquide d'élution (méthanol) contenant 50 à 150 μg du gestagène. On traite par 10 ml d'acide sulfurique pour analyse en agitant fréquemment et après un contact de 20 minutes, on mesure la fluorescence à 555 mμ et excitation à 380 mμ à l'aide du dispositif de Beckman adapté au spectrophotomètre.

L'échelle d'étalonnage est établie en utilisant le standard interne de Lynestrénol qui a subi le même traitement et dont on prépare plusieurs résidus d'évaporation pour constituer l'échelle de concentrations voisines de la concentration de l'échantillon à mesurer.

La précision obtenue par cette méthode est de ± 3 à 4 %.

II. **Allylestrénol.**

Cet Allylestrénol (5 mg) est utilisé en tant que gestagène associé à l'α-tocophérol (*).

La séparation chromatographique est obtenue dans de bonnes conditions par le solvant mobile benzène - éthanol (95 - 5 ml).

Voir résultats donnés à la figure 8.

— *Dosage de l'Allylestrénol.*

Après cette séparation, puis élution au méthanol, le dosage colorimétrique est obtenu en traitant une quaitité de l'ordre de 50 à 100 µg (résidu d'évaporation de la solution méthanolique) par 2 ml d'acide acétique conc.. puis, après mise en solution, par 8 ml d'acide sulfurique pour analyse. On mélange et après 10 minutes de contact, on mesure l'absorption au spectrophotomètre pour le sommet principal situé à 356 mµ. sous 10 mm (voir fig. 9).

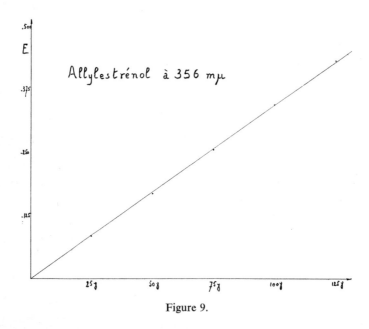

Figure 9.

La coloration est stable pendant 10 à 15 minutes.

Par rapport au standard interne, la précision est de ± 3 à 4 %.

— La même technique est applicable à la séparation d'une association d'Allylestrénol et de Mestranol avec pour solvant mobile un mélange éther - cyclohexane dans une autre proportion : 20 ml d'éther et 80 ml de cyclohexane.

La précision obtenue est du même ordre.

(*) En spécialité pharmaceutique sous le nom de « Gestanon ».

III. **Ethynodiol diacétate.**

Association : Ethynodiol diacétate 1 mg (*).
Mestranol 0,1 mg (voir figure 8).

— *Dosage de l'Ethynodiol diacétate.*

L'opération de chromatographie par le mélange benzène - éthanol (95 - 5 ml), de même que l'élution par le méthanol, ne présentent aucune différence avec la technique générale.

La détermination quantitative s'obtient par colorimétrie : on traite le résidu d'évaporation du méthanol (prise de 50 à 150 µg) par 10 ml d'acide sulfurique pour analyse, en agitant pour favoriser la mise en solution.

La coloration est développée en 25 minutes. On procède aux mesures photométriques à 445 mµ sous 10 mm. L'éluat du standard interne est traité de même façon et permet de contrôler le diagramme d'absorption qui suit la loi de Lambert-Beer.

Exemple : Pour 50 µg lecture = 0,145
    »   100 »   »   = 0,292
    »   150 »   »   = 0,426

Comme pour les opérations précédentes, la précision est ± 3 à 4 %.

En résumé, il apparaît que, malgré une certaine diversité dans les chaînes latérales et fonctions chimiques caractéristiques des principaux progestatifs de synthèse utilisables par voie buccale, il est cependant possible de les identifier par une méthode générale de chromatographie sur couche mince, en complétant par la mesure du spectre U.V. et surtout I.R.

La détermination quantitative du progestatif et de l'œstrogène qui lui est associé (quoique en quantité 50 à 70 fois moindre), s'effectue, après séparation chromatographique et élution quantitative, par mesure spectrophotométrique ou colorimétrique, ou encore par fluorimétrie, par rapport à un standard interne, chomatographié et traité de façon identique.

Bibliographie

1. URBANYI et REHM. *Journ. of Pharm. Sciences,* **55** : 501, 1966.
2. COMER, HARTSAW et STEVENSON. *Journ. of Pharm. Sciences,* **57** : 147, 1968.
3. BICAN-FISTER, T. *Journ. of Pharm. Sciences,* **57** : 169, 1968.
4. VELLUZ. *Annales Pharmac. Franç.,* **25** : **69**, n° 1, 1967.
5. SHROFF A.P. et HUETTEMANN. *Journ. of Pharm. Sciences,* **57** : 882, n° 5, 1968.
6. TSILIFONIS, A.C. et CHAFETZ, L. *Journ. of Pharm. Sciences,* **56** : 625, 1967.
   TSILIFONIS, A.C. et CHAFETZ, L. *Journ. of Pharm. Sciences,* **57** : 1000, 1968.
7. HEUSSER, D. *Dtsch. Apotheker Ztg.,* **106** : 411, 1966.
8. SHROFF, A.P. et HUETTEMANN. *Journ. of Pharm. Sciences,* **56** : 654, n° 5, 1967.

(*) Préparation pharmaceutique spécialisée sous le nom de « Ovulène ».

9. BOUGHTON, O.D. et coll. *Journ. of Pharm. Sciences,* **55** : 951, 1966.
10. KHOURY, A.J. et CALI, L.J. *Journ. of Pharm. Sciences,* **56** : 1485, 1967.
11. UMBERGER, E.J. *Analytical Chemistry,* **27** : 768, 1955.
12. BASTOW. *Journ. of Pharmacy and Pharmacol.,* **19** : 41, 1967.

***

Nous désirons exprimer nos vifs remerciements aux firmes : Searle-Pharbil, Schering-Coles, Roussel-Labunis, et Organon qui nous ont offert gracieusement les progestatifs utilisés pour l'élaboration de ce travail.

Dünnschichtchromatographische Bestimmung
von Xylonest und o-Toluidin

H. STRUCK und W. PAKUSA,
Aus der Biochemischen Abteilung (Leiter : Priv.-Doz. Dr. H. Struck)
der II. Chirurgischen Universitätsklinik (Direktor : Prof. Dr. W. Schink)
Köln - Merheim.

Wir befassen uns seit geraumer Zeit mit Stoffwechsel und Verteilung verschiedener Lokalanaesthetika. Eines der neueren Präparate ist das Xylonest, welches auch Citanest oder Prilocain genannt wird. Chemisch ist es α-Propylamino-2-methylpropionanilid. Als ein wesentliches Stoffwechselprodukt des Xylonest wird das o-Toluidin angesehen, welches wir sowohl in Organextrakten als auch im Harn nachweisen konnten (Abb. 1).

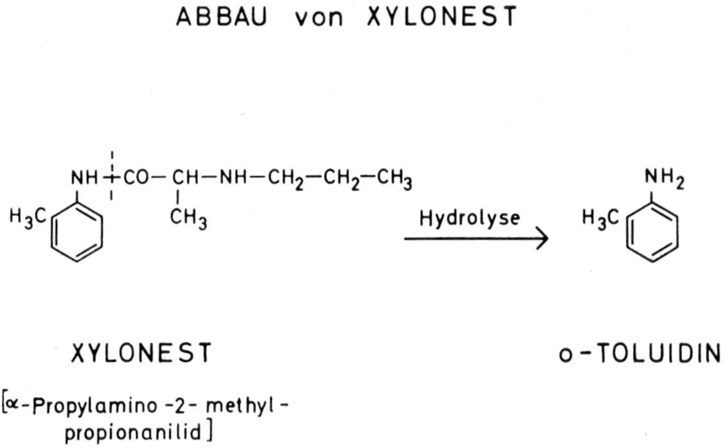

ABBAU von XYLONEST

XYLONEST   o-TOLUIDIN

[α-Propylamino -2- methyl -
propionanilid]

Abb. 1. — Decomposition of Xylonest.

Sie sehen links das Molekül des Xylonest, welches an der Carbonimidbindung sehr leicht spaltbar ist und, wie das Formelbild zeigt, dann in das o-Toluidin übergeht. Um das Ausmaß dieses Überganges im Körper verfolgen zu können, wurde zunächst eine quantitative Methode zur Bestimmung des o-Toluidin entwickelt. Wenn es gelang, für das Xylonest eine geeignete und reproduzierbare Hydrolysemethode zu finden, mußte das o-Toluidinverfahren auch auf das Xylonest anwendbar sein.

Über die von uns entwickelte und inzwischen auch an zahlreichen Beispielen erprobte Methode möchten wir Ihnen berichten.

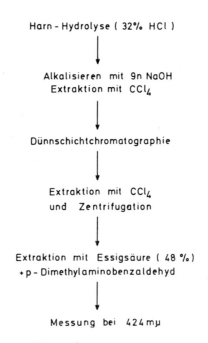

Abb. 2. — Schema of the urine analysis of Xylonest.

Im obigen Bild (Abb. 2) sehen Sie schematisch den Analysengang dargestellt. Die Körperflüssigkeit oder das Organhomogenat, in unserem Fall Harn, werden mit HCl für 16 Stunden bei 105° - 110° C behandelt, um, wie im ersten Bild demonstriert, aus dem Xylonest das o-Toluidin zu erhalten. Anschließend wird mit einem Überschuß an 9 n NaOH alkalisiert und mit dem gleichen Volumen Tetrachlorkohlenstoff extrahiert. Die organische Phase wird für die Dünnschichtchromatographie verwendet. Nach der Chromatographie wird das Kieselgel im Bereich des o-Toluidin abgekratzt und wiederum mit Tetrachlorkohlenstoff (5 ml) extrahiert. Der abzukratzende Kieselgelbereich wird durch eine angefärbte Leitsubstanz ermittelt. Die Extraktion mit Tetrachlorkohlenstoff (5 ml) ist nach 5 Minuten Schütteln beendet. Jetzt wird das Kieselgel abzentrifugiert und zwar 15 Minuten bei 8 000 U/Min. 4 ml der Tetrachlorkohlenstoffphase werden vom Niederschlag vorsichtig agbenommen und mit 4 ml 48%iger Essigsäure geschüttelt (5 Minuten). Die wäßrige Phase wird abgetrennt (3 ml) und mit 3 ml Farbreagens versetzt. Dieses besteht aus einer 5%igen Lösung von p-Dimethylaminobenzaldehyd in Methylglykoll. Es bildet sich sofort eine intensive gelbe Farbe. Das Maximum des Farbkomplexes liegt bei 424 mµ. Die Extinktion wird bei dieser Wellenlänge gemessen. An Hand einer Eichkurve kann dann die Konzentration ermittelt werden. Der Farbkomplex ist über Tage unter Luftabschluß stabil. Der Xylonestgehalt ergibt sich aus der Extinktionsdifferenz, welche bei der Messung des Harns vor und nach Hydrolyse entsteht.

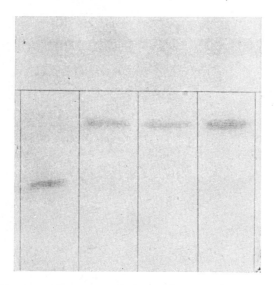

Abb. 3. — Trennung zwischen Xylonest und o-Toluidin.
Separation between Xylonest and o-Toluidin.

In der Abbildung sehen Sie ein Chromatogramm, welches die Trennung zwischen Xylonest und o-Toluidin zeigt. Die Anfärbung erfolgte in diesem Fall mit Ninhydrin. Das Laufmittel ist Methylacetat. Das nächste Bild (Abb. 4) bringt tabellarisch einige Daten, welche bei der Dünnschichtchromatographie von Bedeutung sind, wie Laufmittel, Rf-Wert, Sprühreagenz, Anfärbung und Nachweisbarkeitsgrenze.

Sie sehen, daß der Nachweis des o-Toluidin mit p-Dimethylaminobenzaldehyd recht empfindlich ist, es können noch weniger als 1 µg deutlich sichtbar gemacht werden, wie die Abbildung 5 demonstriert.

Im Bild erkennen wir die Eichkurve für das o-Toluidin in essigsaurer Lösung nach Reaktion mit p-Dimethylaminobelzaldehyd. Die Messung des Farbkomplexes erfolgt bei 424 mµ. Die Reaktion ist außerordentlich empfindlich. Oberhalb 15 µg ist das Lambert Beer'sche Gesetz schon nicht mehr eindeutig gültig. Bei höheren Konzentrationen muß die zu messende Lösung dann verdünnt werden.

Dünnschichtchromatographie von Xylonest und o-Toluidin

Substanz	Laufmittel	R$_F$ - Wert	Sprühreagenz	Farbe	Nachweisbarkeit in µg
Xylonest	Methylacetat	0.38 ± 0.03	p - Dimethylamino - benzaldehyd	—	—
o - Toluidin		0.73 ± 0.04		gelb	< 1
Xylonest	Cyclohexan/ Methanol ges.	—	Ninhydrin	hellbraun	< 20
o - Toluidin		0.30 ± 0.05	Ninhydrin	gelbbraun	< 5

Abb. 4. — Thin layer chromatography of Xylonest and o-Toluidin.

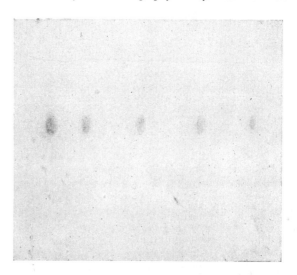

Abb. 5. — Nachweis von o-Toluidin mit p- Dimethylaminobenzaldehyd.
Proof of o-Toluidin with p-Dimethylaminobenzaldehyde.

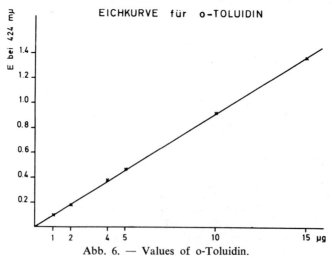

Abb. 6. — Values of o-Toluidin.

Das folgende Bild (Abb. 7) stellt ein Chromatogramm dar, auf welchem Xylonest vor und nach Hydrolyse aufgetragen wurde. Als Vergleich ist das o-Toluidin ebenfalls auf der Platte. Man kann an diesem Chromatogramm sehr schön zumindest qualitativ sehen, wie das Xylonest nach der Hydrolyse in das o-Toluidin übergeht.

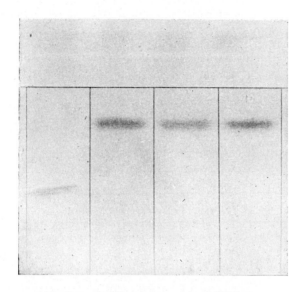

Abb. 7. — Xylonest vor und nach Hydrolyse.
Xylonest before and after hydrolysis.

Die jetzt folgende Tabelle (Abb. 8) zeigt die Wiederfindungsraten in Prozent nach einzelnen Analysenschritten sowohl für das Xylonest als auch für das o-Toluidin. In der ersten Spalte sehen Sie die Zahlen über die Extraktionszeit in Abhängigkeit von der Konzentration an o-Toluidin dargestellt. Man erkennt deutlich, daß schon nach 5 Minuten die Gesamtmenge an o-Toluidin in die Tetrachlorkohlenstoff-Phase übergegangen ist.

Extraktion von o-Toluidin					Hydrolyse und Extraktion von Xylonest		Harn + o-Toluidin		Harn + Xylonest	
Konz µg	Zeit min				Konz µg	Wiederfindung %	Konz µg	Wiederfind %	Konz µg	Wiederfind %
	5	10	15	20						
10	101	106	105	100	10	90	5	87 ± 8	5	89 ± 9
20	97	99	108	97	20	93	10	91 ± 10	10	93 ± 9
30	97	95	97	99	30	89				
					40	91				

Abb. 8. — Verlustberechnung im Analysengang (Angaben in % Wiederfindung).
Calculation of loss during analysis (quantities found expressed in %).

In der nächsten Kolumne sind die Zahlen über die Hydrolyseausbeuten in Abhängigkeit der Xylonestkonzentration wiedergegeben. Man kann auch hier deutlich sehen, daß über den gesamten Konzentrationsbereich eine praktisch 100%ige Hydrolyse des Xylonest zu o-Toluidin erfolgt ist.

Die letzte Spalte zeigt nun den Prozentsatz an Xylonest bzw. o-Toluidin, der wiedergefunden wurde, wenn 5 oder 10 μg Reinsubstanz der zu untersuchenden Ausgangslösung, in diesem Fall Harn, zugesetzt worden waren. Die Wiederfindungsrate von 85 - 90 % ist zufriedenstellend. Auch die Schwankungsbreite, die unter ± 10 % liegt, ist bei diesen geringen Konzentrationen durchaus zu vertreten.

Einige Werte, welche wir mit der eben geschilderten Methode erhalten haben, sind in der folgenden Tabelle zusammengestellt (Abb. 9). Es handelt sich dabei um Ratten bzw. Patienten, welche das Lokalanaesthetikum Xy-

Xylonestbestimmungen im Harn

Rattenharn				
Anzahl n	injizierte Menge mg / kg	Art der Injektion	mg Xylonest / 24 h	% des appliziert. Xylonest
3	100	i. m.	0.916	4.0
2	85	i. m.	1.400	4.6
1	100	i. v.	0.390	1.8
2	10	i.v.	0.022	0.9
Humanharn				
6	6.2	i.m.	18.572	3.4
5	6.3	i.m.	15.710	4.0

Abb. 9. — Determination of Xylonest in urine.

lonest erhalten hatten und von denen Harn gesammelt worden war. Es sind sowohl die absoluten Werte im 24-Stunden-Harn als auch die Prozentzahlen bezogen auf die applizierte Menge an Lokalanaesthetikum angegeben. Im 24-Stunden-Harn werden im Durchschnitt nur 5 % der Ausgangsmenge wiedergefunden.

Die Methode wurde von uns auch an Organen geprüft und gibt auch dort zufriedenstellende Ergebnisse. Die Organe wurden mit 32 % HCl homogenisiert, das ausgefällte Eiweiß abzentrifugiert (3 000 U/Min) und anschließend dem eingangs skizzierten Analysengang unterworfen.

Die Reaktion zwischen o-Toluidin und p-Dimethylaminobenzaldehyd, die unserer Methode zu Grunde liegt, wurde von uns nicht näher untersucht. Es dürfte sich aber aller Wahrscheinlichkeit nach eine Schiff'sche Base bilden, wie es das letzte Bild (Abb. 10) zeigt. Wir konnten uns im

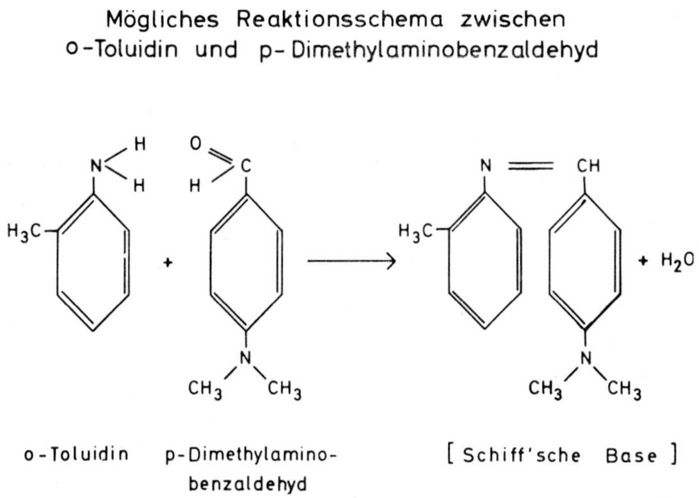

Mögliches Reaktionsschema zwischen
o-Toluidin und p-Dimethylaminobenzaldehyd

o-Toluidin   p-Dimethylamino-   [Schiff'sche Base]
             benzaldehyd

Abb. 10. — Schema of possible reaction between o-Toluidin and
p-Dimethylaminobenzaldehyde.

Reagenzglasversuch überzeugen, daß zahlreiche andere aromatische Amine ebenfalls eine positive Reaktion ergaben. Bei aliphatischen Aminen, die auch nicht zu einem System konjugierter Doppelbindungen führen würden, bleibt die Reaktion aus.

Durch die dünnschichtchromatographische Isolierung von o-Toluidin bzw. Xylonest ist die von uns beschriebene Methode als spezifisch anzusehen.

Vapor-programmed thin-layer chromatography, a new technique for improved separations [*]

R. A. de ZEEUW

Laboratory for Pharmaceutical and Analytical Chemistry, State University,
Antonius Deusinglaan 2, Groningen, The Netherlands.

Recent investigations in our laboratory (1, 2) on the influence of solvent vapor in thin-layer chromatography have shown that if development is done in unsaturated chambers then separations with multicomponent solvents are more efficient than in saturated chambers. The improvements are caused by a concentration gradient of adsorbed vapor of the more polar solvent components on the dry part of the plate, the steepness of the gradient being mainly dependent on the rate of evaporation of the solvent components and on their affinity to the adsorbent used.

It will be clear that little or no control can be exercised on the extent to which the gradient develops on the plate during development. With highly polar solvent components the gradient may become too steep on the lower part of the plate, with the upper parts showing no gradient at all, because maximum vapor adsorption will take place there. With less polar solvent components the gradient may become too flat or will not reach its maximum within the time given due to fast development. Furthermore, differences in the evaporation velocity between the solvent components may cause failures in the desired gradient. Thus, although unsaturated chambers may well give improved separations, the vapor conditions are not necessarily optimum for every case.

We therefore have searched for a new TLC development technique, providing a more efficient control of the vapor processes during development. This has led to a new apparatus, the Vapor-Programming chamber, which allows full vapor control all over the plate, thus making it possible to affect the migration rate of each individual spot which is, of course, one of the main conditions for successful separations.

---

[*] Full details can be found in : Rokus A. de Zeeuw. *Anal. Chem.*, **40** : 2134, 1968.

Apparatus. The Vapor-Programming chamber consists of three parts, all of chromium plated brass (fig. 1). The ground plate, (A), $20 \times 20 \times 1$ cm, which has an internal water circulating system for cooling purposes, the inlet and outlet being visible at F, the solvent reservoir, (B), ($20 \times 1 \times 1$ cm) and the trough chamber (C), with 21 troughs, inner diameter 6 mm, depth 12 mm, and partition walls of 2 mm. The side walls are 5 mm. Figure 2 shows the assembled Vapor-Programming chamber. The troughs are to be filled with a series of liquid mixtures of appropriate compositions to give suitable vapors. Most often the liquids show increasing polarity from bottom to top. This can be done for example by using mixtures of two solvent components, one polar, one non-polar, with the mixtures having an increased proportion of the polar component. Development of the plate takes place horizontally, the adsorbent downward, facing the troughs and can be continued for unlimited time as excess of solvent can evaporate at the top of the plate. The latter process can be accelerated by passing warm water through the warm water tube (D), at the top, which is insulated from the ground plate by asbestos.

Fig. 1. — The Vapor-Programming chamber in parts. A : ground plate, B : solvent reservoir, C : trough chamber. The ground plate is equipped with a warm water tube (D), fixation clamps (E), and an internal water-circulating system for cooling purposes, with the inlet and outlet visible at F.

Fig. 2. — The assembled Vapor-Programming chamber.

Fig. 3. — The Vapor-Programming chamber at work. The solvent makes the adsorbent transparent with the underlying troughs and the filter paper becoming visible.

The solvent is brought upon the plate by a strip of filter paper, and two springy metal strips underneath the solvent reservoir provide a suitable contact between the strip and the plate.

The plates for Vapor-Programmed TLC are normal 20×20 cm ones with a usual layer thickness of 0.25 mm, but any flat layer between 0.20-2 mm can also be used. The side edges and the top edge are stripped 5 mm wide.

The plate rests on two small Teflon strips with a thickness of about 0.5 mm, the layers being 0.25 mm, on both sides of the plate. Thus, the space between the plate and the troughs is very small because otherwise vapor currents will interfere. On the other hand, the layers must not touch the trough walls. If so, development stops immediately.

Four clamps (E), fix the plate and the trough chamber.

So, the use of the VP-chamber permits vapor adsorption by the adsorbent from the underlying troughs and by filling these troughs with suitable mixtures the vapor conditions can be controlled all over the plate. Thus, optimal vapor gradients can be obtained since every desired polarity can be applied to the various parts of the plate via the vapor phase.

There is one difficulty, however. Solute spots are of finite size and if we should subject these spots to a stepwise gradient, migration of the upper parts of the spots will be speeded sooner than the lower parts and tailing will consequently result. Fortunately, this can be prevented by interspersing troughs with solvents of low polarity between the troughs containing the more polar mixtures, thus having a decelerating effect on the migration of the spots, particularly on the upper parts. If, for example a separation is required of closely related sulfonamides, ether can be used as running solvent. The first two troughs, the latter underneath the starting points, are also filled with ether. Trough 3 is then filled with ether-methanol 80 : 20, in which the methanol has an accelerating effect. Troughs 4 and 5 are filled with ether as decelerator, trough 6 is filled with ether-methanol 50 : 50 troughs 7 and 8 with ether, trough 9 with ether-methanol 20 : 80 and so on. With these combinations of accelerating and decelerating forces compact spots can be obtained indeed.

There remain a few points which should have some more attention.

The thickness of the strips plays an important role in the applicability of a certain gradient. It will be clear that due to the differences in liquid compositions in the troughs strong vapor diffusion will occur if the space between the trough walls and the adsorbent is too large. These diffusions may be well cause anomalies in the direction of solvent flow and worthless separations will result. On the other hand it must be avoided the layers touching the walls. If so, development stops immediately, and the spots readily diffuse along the trough walls. Furthermore, it should be noted that most adsorbents swell to some extent when wetted by the solvent, the rate of swelling being dependent on the polarity of the solvent. So, the thickness of the strips should be adapted to this phenomenon and, of course, to the dimensions of the adsorbent layer. In our experiments with layers of 0.25 mm good results were obtained with strips of 0.3-0.5 mm for solvents like hexane, benzene, chloroform and ether, 0.5-0.8 mm for solvents containing acetone, ethylacetate and lower alcohols, whereas ammonia containing solvents required strips of 1 mm.

Saturation. To be sure that reproducible quantities of vapor are available for adsorption it proved to be necessary to acclimatize the plate over the filled troughs for 10 minutes after it was fixed into position. After this saturation period, in which the small volumes over the troughs become almost saturated, the solvent reservoir is filled and development is started. If high volatile liquid mixtures are used the saturation time may even be reduced to 5 minutes.

Temperature. It will be clear that constancy of the room temperature within $\pm$ 1° C is required to obtain reproducible separations. It was found however, that with Vapor-Programmed TLC at room temperatures higher than 22° C poor separations were obtained in comparison with the results at lower room temperatures. Over 22° C the migration rates of the spots markedly decreased. Presumably this is caused by too much vapor adsorption on the plate, which is then followed by condensation. Thus, solvent transport from the solvent reservoir will be diminished and consequently the migration rates of the spots will decrease.

So, cooling of the Vapor-Programming chamber is necessary and we therefore have equipped the ground plate with an internal tube system which can be connected to a cooling water bath thermostat. The temperature of the cooling system, if kept constant, thus enables a good control of the vapor adsorption rate. In our investigations at average room temperatures of $\pm$ 21° C optimal separations were obtained with cooling temperatures of 18-20° C.

Some applications of Vapor-Programmed TLC will be shown in a second paper.

The Vapor-Programming chamber will soon be available from C. Desaga, Heidelberg.

References

1. DE ZEEUW, R.A. *J. Chromatog.,* **32** : 43, 1968.
2. DE ZEEUW, R.A. *Anal. Chem.,* **40** . 915, 1968.

Some applications of vapour-programmed TLC in difficult separations of closely resembling substances

by

Rokus A. de ZEEUW

Laboratory for Pharmaceutical and Analytical Chemistry, State University,
Antonius Deusinglaan 2, Groningen, The Netherlands.

Regarding the application possibilities of Vapour-Programmed TLC (1) it will be obvious that this technique will be very suitable in separations of closely related compounds which are inadequately separated by the classical technique. During development in the VP-chamber the faster running spots of a mixture will pass into plate areas enriched in the more polar vapour components, thus producing an acceleration of their migration rates. At the same time the slower running spots pass through areas with a lower concentration of the polar vapour components. The migration rate of these will thus be affected to a smaller extent and hence better separations will result. We further presume that the separation improvements are not only caused by the acceleration forces but also by a selective deceleration of the interspersed decelerating troughs, the slower running substances being decelerated to a higher extent than the faster running substances.

In the following we will show some results of Vapour-Programmed TLC, whereas saturated normal tank chambers are used as controls. Experimental details are described earlier (1).

The separation of butter yellow, sudan red G, indophenol and 4-nitroaniline in N-chambers using benzene as solvent is far from optimal as can be seen in Fig. 1A. The spread of the spots over the plate is poor because only one-third of the plate is utilized and furthermore, sudan red G and 4-nitroaniline do not separate. Changing the solvent to benzene-chloroform (80 + 20) does not really improve the separation or the spreading (Fig. 1B). The spots cover higher parts of the plate but now 4-nitroaniline and indophenol are not separated, while the spread remains the same. In this case the use of unsaturated chambers could not provide better results, the gradient being too flat if benzene-chloroform mixtures are used and becoming too steep with more polar components than chloroform.

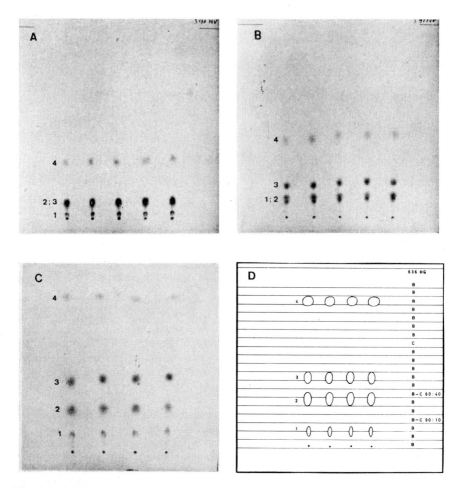

Fig. 1. — Improved separation of colour dyes in the Vapour-Programming chamber as compared to classical development in saturated N-chambers.

A : N-chamber, solvent benzene. Temp. 21.0°C, rel. humidity 26 %, development 36 min., saturation 45 min.

B : N-chamber, solvent benzene-chloroform (80 + 20). Temp. 22.0° C, rel. humidity 30 %, development 39 min., saturation 45 min.

C : VP-chamber, solvent benzene. Temp. 21,7°C, rel. humidity 29 %, development 110 min., saturation 10 min., strips 0,3 mm, cooling 19° C.

D : Position of the troughs during development in the VP-chamber and the liquid compositions therein.

1 = indophenol, 2 = 4-nitroaniline, 3 = sudan red G, 4 = butter yellow, B = benzene, C = chloroform. Adsorbent = silica gel GF 254, load = 20 μg of each substance.

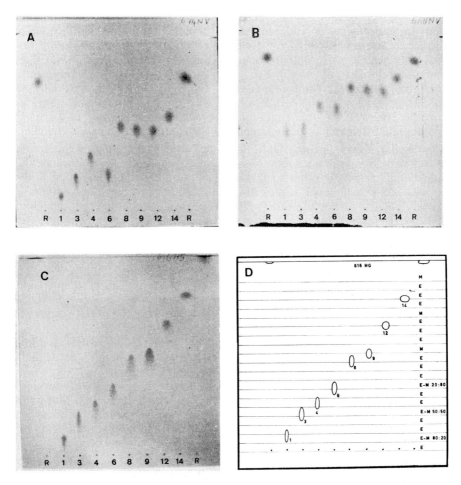

Fig. 2. — Improved separation of sulfonamides in the Vapour-Programming chamber as compared to classical development in saturated N-chambers.

A : N-chamber, solvent ether-methanol (90 + 10). Temp. 21.5º C, rel. humidity 35 %, development 38 min., saturation 45 min.

B : N-chamber, solvent ether-methanol (80 + 20). Temp. 21.4º C, rel. humidity 33 %, development 39 min., saturation 45 min.

C : VP-chamber, solvent ether-methanol (95 + 5). Temp. 21.5º C, rel. humidity 35 %, development 63 min., saturation 10 min., strips 0.5 mm, cooling 19ºC.

D : Position of the troughs during development in the VP-chamber and the liquid compositions therein. 1 = sulfaguanidine, 3 = sulfisomidine, 4 = sulfathiazole, 6 = sulfacetamide, 8 = sulfadimidine, 9 = sulfapyridine, 12 = sulfisoxazole, 14 = sulfanilamide, R = reference 4-nitroaniline, E = ether, M = methanol. Adsorbent = silica gel GF 254, load = 3 μg of each substance.

Development in the Vapour-Programming chamber however gives a much better separation as can be seen in Fig. 1C. Benzene is used as solvent, together with a simple gradient of benzene-chloroform mixtures and benzene as decelerating liquid. The four substances are clearly separated and the entire plate is utilized for the spreading. If necessary the mutual distances between the spots can be changed by application of slightly different gradients. So, if a fifth component was present between sudan red G and 4-nitroaniline, the distance between these spots could be sufficiently enlarged, so that the fifth component would easily fit in. The overall-time for the separation in the VP-chamber is somewhat longer than in the N-chamber because of the continuous development, but this is partially compensated by the less time needed for the saturation of the VP-chamber.

Similar improvements could be obtained in the separation of a selection of sulfonamides : sulfaguanidine, sulfisomidine, sulfathiazole, sulfacetamide, sulfadimidine, sulfapyridine, sulfisoxazole and sulfanilamide. These substances are poorly separated in N-chambers using ether-methanol (90 + 10) as solvent (Fig. 2A). Changing the solvent ratio to ether-methanol (80 + 20) only results in higher Rf values but the separation becomes worse (Fig. 2B).

With the VP-chamber all substances can be clearly separated however, with a spread of nearly 80 % (Fig. 2C). The separation is now suitable for identification purposes. Ether-methanol (95 + 5) is used as solvent, in combination with an ether-methanol vapour gradient and ether as decelerator. The overall-time for this separation in the VP-chamber is less than in the N-chamber, due to the short saturation period of the former.

The reproducibility of the separations in the Vapour-Programming chamber is good and at least equal to that obtained in the classical technique. It will be obvious, however, that the factors involved in the reproducibility such as room temperature, cooling temperature, relative humidity, saturation, preparation of the layers, thickness of the spacers etc. must be kept constant. Small variations in these parameters will have a great influence on the vapour-gradient, and, subsequently, on the separation.

Thus, it becomes clear that Vapour-Programmed TLC offers a great many new possibilities in the analysis of closely related substances which are often inadequately separated by the classical techniques. It allows full control on the migration of each individual spot and this is one of the main conditions for a successful separation.

Furthermore, Vapour-Programmed development can also be valuable in preparative TLC, the better separations allowing more efficient isolations.

References

(1) DE ZEEUW, R.A. *Anal. Chem.*, **40** : 2134, 1968.

La chromatographie des arylamines et des arylaminométhylphtalimides

par

H. AMÂL, E. GURSU et S. DEMIR.

Résumé

Les dérivés phtalimidométhyliques de 23 amines sont préparé en chauffant les solutions hydroalcooliques des amines avec une solution également hydroalcoolique de l'hydroxyméthylphtalimide. Ils ont chromatographié sur plaques couvertes de gel de silice G ou HF 254 - 366 (Merck).

Les dérivés ont en général des valeurs de Rf, plus grandes que des amines eux-mêmes.

En utilisant cette méthode on peut facilement séparer d'une part un mélange qui contient les sulfonamides tels que la sulfanilamide, sulfamérazine, sulfaméthazine, sulfathiazole, sulfadiazine et sulfapyridine, d'autre part l'acide p-aminobenzoïque et ses dérivés, tels que benzocaïne, orthocaïne, procaïnamide, procaïne, butocaïne et tetracaïne.

* * *

La chromatographie directe des amines, surtout des amines primaires et secondaires, en couche mince, cause quelquefois des difficultés provenant de l'adsorption des amines par les adsorbants usuels. Diverses méthodes ont été proposées pour la séparation et l'identification des amines.

Plusieurs auteurs ont directement caractérisé sur papier ou en couche mince, quelques amines aromatiques, révélants les chromatogrammes par pulvérisation des réactifs différents. Smyth et Mckeown (1) ont utilisé le nitrate d'argent ammoniacal, le p-diméthylaminobenzaldéhyde et le naphtoquinone sulfonate de sodium pour différents amines aromatiques comme les aminophénols, les phénylènes diamines, les aminotoluènes, les anisidines isomer o- m- et p-, sur papier. Gillio-Tos (2) et collaborateurs ont fait la chromatographie en couche mince des isomers o-, m-, p-, des aminophénols, des toluidines, des acides aminobenzoïques, des anisidines, des anilines halogènes, etc. et ont utilisé l'irradiation UV ou bien des réactifs différents.

Plus d'auteurs encore ont caractérisé des amines à l'état de leurs dérivés benzoylés, 3,5-dinitrobenzoylés (3, 4, 5) toluènsulfonylés (5, 8), diméthyla-minonaphtalène sulfonylés (6) etc. Jart et Bilger (7) ont utilisé le 4-(Phény-lazo) benzène sulfonylchlorure pour préparer les dérivés des amines et ont constaté que cette technique convient très bien, surtout pour les six premiers amines saturés, aliphatiques et qu'il a plusieurs avantages par rapport aux autres réactifs ; mais pour les groupes aminés qui sont attachés directement aux noyaux du benzène il est impossible de préparer du dérivé phénylazo-benzène sulfonylé. Gasparic et Borecky (3) ont caractérisé les amines à l'état 3,5-dinitrobenzamide par chromatographie sur papier et ont réussi à séparer les dérivés des amines primaires et secondaires, aliphatiques. Ils ont constaté que, avec cette technique, on peut identifier les dérivés beau-coup plus facilement que les amines à l'état libre ou même que les sels des amines.

Teichert et coll. (4) ont utilisé le même dérivé pour la chromatographie en couche mince de quelques amines aliphatiques. Parihar et coll. (8) ont réussi à l'identifier et séparer les amines ou les mélanges des amines en les transformant à l'état de tosylate. Le résultat le plus remarquable de ce travail est de pouvoir préparer les tosylates directement sur le chromato-gramme et de pouvoir séparer les mélanges des isomers des amines aroma-tiques, tels que les isomers o-, m-, p- ou α- et β-.

Gerlach et Senf (5), en même temps que Parihar et coll. ont chromato-graphié quelques amines à l'état de la p-toluène sulfonamide, de la benza-mide et de la 3,5-dinitrobenzamide, et ils ont obtenu des résultats repro-ductibles avec des p-toluène sulfonamides de 7 amines aromatiques tandis que, avec les benzamides de même 7 amines, les résultats n'étaient pas satisfaisants.

Tous ces auteurs ont conclu que la chromatographie des dérivés des amines donne des résultats supérieurs que les amines eux-mêmes.

Nous envisageons ici le problème de caractérisation des dérivés phta-limidométhyliques de 23 amines aromatiques et de l'identification de quel-ques-uns dans leurs mélanges. La chromatographie en couche mince de dérivés phtalimidométhyliques des amines n'a pas été étudiée jusqu'à main-tenant. Les dérivés phtalimidométhyliques de l'acide p-aminobenzoïque, de p-aminobenzoate, d'éthyle (9, 10), de m-amino-p-hydroxy-benzoate de mé-thyle, de p-aminobenzoate de diéthylaminoéthyle de la sulfanilamide (10) de p-aminophénole de p-phénylène diamine, de l'α-naftylamine (11) et de diphénylamine (9) ont été préparés pour l'identification des amines aroma-tiques à l'aide de leurs points de fusion.

Le temps de transformation des amines à l'état de leurs dérivés, avec un rendement suffisant pour isoler et cristalliser la substance formée, est assez long et change de 30 min. à plusieurs heures. D'autre part la formation de dérivé phtalimidométhylique commence dès que l'amine est mélangé avec l'hydroxyméthyle phtalimide. Nous avons profité de ce phénomène pour caractériser les amines par la chromatographie. Nous avons première-ment préparé les dérivés phtalimidométhyliques des amines suivants : l'acide p-aminobenzoïque, benzocaïne, orthocaïne, procaïne, procaïnamide, buta-caïne, tetracaïne, sulfanilamide, sulfamérazine, sulfaméthazine, sulfathiazole, sulfadiazine, sulfapyridine, α- et β-naphtylamine, o-, m-, et p-aminophénols, m- et p-phénylène diamine, 2,4-diaminophénole, diphénylamine et éthoxy-diaminoacridine. Nous avons déterminé les points de fusion des produits préparés et nous les avons utilisés, comme substance de référence pour nos recherches.

PREPARATION DES DERIVES PHTALIMIDOMETHYLIQUES DES AMINES

On peut préparer ces dérivés suivant la technique donnée par Winstead et Heine (9) et pour cela il faut chauffer 0.3 g (0.002 mol) de phtalimide avec 0.2 ml de l'aldéhyde formique dans une solution éthanolique à 80 % et ajouter à cette solution une petite quantité (quelques mg) d'amine ou de mélange des amines à examiner, puis chauffer pendant 5 - 10 min. La solution ainsi obtenue contient de N-arylaminométhylphtalimide à côté de l'amine, de la phtalimide, de l'hydroxyméthylphtalimide qui n'ont pas été entrés en réaction et de dérivé N-bisméthylphtalimidique. Nous avons utilisé cette solution directement pour la chromatographie en couche mince.

LA CHROMATOGRAPHIE

Les couches minces ont été préparées avec du gel de silice G et HF 254-366 (Merck) suivant la technique usuelle, sur les plaques de verre de 10 x 20 et 20 x 20 cm ; épaisseur de couche étant 0.25 mm, ils ont été activé à 105 - 110° pendant 30 min.

L'hydroxyméthylphtalimide et la phtalimide ont été utilisées comme substances de comparaison (les deux ont les mêmes valeurs de Rf dans les éluants 1, 2, 3, 4 et 5). Chaque dérivé a été comparé par l'amine lui-même en appliquant sur les plaques, les amines et son dérivé l'un à côté de l'autre. La quantité nécessaire est environ 2-5 µl et cette quantité est suffisante pour produire une tache bien visible. Les éluants employés sont les suivants :

1-Chloroforme-méthanol (80 : 20)
2-Chloroforme-méthanol (80 : 30)
3-Ethanol-ammoniaque 25 % (80 : 20)
4-n-Butanol-acide acétique-eau (4 : 1 : 5)
5-Chloroforme.

Les chromatogrammes ont été examinés sous la lumière UV 254 mµ où elles sont révélées par pulvérisation premièrement d'une solution de nitrite de sodium à 5 % dans l'alcool qui contient 2 ml HCl, puis chauffés à 50° pendant 5 min. et après par pulvérisation d'une solution dans l'alcool de l'α-naphtylamine à 5 %.

Les valeurs de Rf des amines et celles des dérivés sont rassemblées dans les tableaux I, II et III (les lettres A et D expriment Amines et Dérivés respectivement).

TABLEAU I

Valeurs Rf des amines et des dérivés phtalimidométhyliques
(en 100 Rf)

Substance	Eluants							
	1		2		3		5	
Phtalimide	70		80		65		40	
	A	D	A	D	A	D	A	D
l'Acide p-aminobenzoïque	9	31	15	38	55	61	0	35
Benzocaïne	84	94	70	84	70	80	0	21
Orthocaïne	57	73	69	84	61	70	0	20
Procaïne	41	79	40	56	75	57	0	14
Procaïnamide	6	24	12	41	63	69	46	27
Butacaïne	64	6			77	83	15	25
Tetracaïne	47	58			73	59	0	19

TABLEAU II

Valeurs Rf des amines et des dérivés phtalimidométhyliques
(en 100 Rf)

Substance	Eluants					
	1		2		3	
Phtalimides	70		80		65	
	A	D	A	D	A	D
Sulfanilamide	34	49	45	49		
Sulfamérazine	43	56	61	72	60	69
Sulfaméthazine	46	59	71	85	62	66
Sulfathiazole	36	63	50	64		
Sulfadiazine	46	57	55	67	61	71
Sulfapyridine	35	59	58	68	68	76

TABLEAU III

Valeurs Rf des amines et des dérivés phtalimidométhyliques
(en 100 Rf)

Substance	Eluants							
	1		2		3		5	
Phtalimide	70		80		65		40	
	A	D	A	D	A	D	A	D
α-Naphtylamine	75	87					68	28
β-Naphtylamine	75	84					88	27
o-Aminophénol	55	81	75	92			} 10 24	40
m-Aminophénol	40	67	58	74			5	29
p-Aminophénol	48	20	69	55			0	27
m-Phenylénediamine	47	0	47	0				
p-Phenylénediamine	43	52	41	61	72	79		
2,4-Diaminophénol	17	0					0	20
Ethoxydiaminoacridine	13	32					0	26
Diphénylamine	81	94					0	24

La température est de 28° et la durée de migration du solvant (1) étant 47 min. A la même température la durée de migration du solvant (2) étant 45 min.

L'examen des tableaux I, II et III montre qu'avec les éluants qui contiennent chloroforme et méthanol, on peut avoir des valeurs Rf reproductible pour presque tous les dérivés des amines essayés, mais avec les solvants qui contiennent ammoniaque ou l'acide acétique, la séparation des amines et des dérivés n'est pas bonne. Le chloroforme seul, convient assez bien pour le dérivé de l'acide p-aminobenzoïque tel que PABA, la benzocaïne, orthocaïne, procaïne, procaïnamide, butacaïne et tetracaïne, mais ne convient pas pour des sulfonamides.

RESULTATS ET CONCLUSIONS

Les résultats de cette investigation démontre que :

Pour la chromatographie en couche mince de 23 amines examinés il est mieux de les transformer en leurs dérivés phtalimidométhyliques ayant la plupart des valeurs Rf beaucoup plus grands que celle d'amine elle-même.

La préparation des dérivés phtalimidométhyliques est facile et on peut préparer les solutions à chromatographier en chauffant pour quelques minutes les ingrédients dans un tube à essais. Il est possible de chromatographier les produits de la réaction, sans avoir besoin de les isoler.

Avec la méthode donnée on peut arriver à séparer le mélange de dérivés phtalimidométhyliques de PABA, benzocaïne, procaïne, orthocaïne, procaïnamide, butacaïne et tetracaïne ; qui à l'état d'amine libre leurs mélanges donnent trois taches allongées avec l'éluant 1. Le mélange de dérivés des sulfonamides mentionnés au tableau II donnent six taches bien séparées qui, elles aussi, donnent trois taches allongées, quand on les chromatographie à l'état libre avec éluants 2.

Donc les dérivés phtalimidométhyliques des amines ont plusieurs avantages sur les amines aussi bien que sur les dérivés utilisés avant (comme 3,5-dinitrobenzamides, p-toluène sulfonamides, 4-(phénylazo) benzène sulfonamides) puisque la réaction avec l'hydroxyméthyle phtalimide peut être appliqué aussi aux amines aromatiques, la préparation peut être conduite facilement soit par l'amine soit par les solutions d'amines ou même les sels d'amines.

Bibliographie

(1) SMYTH, R.B. et McKEOWN, G.G. *J.Chromatog.*, **16** : 454, 1964.
(2) GILLIO-TOS, M., PREVITERA, S.A. et VIMERCATI, A. *J. Chromatog.*, **13** : 571, 1964.
(3) GASPARIC, J. et BORECKY, J. *J. Chromatog.*, **5** : 466, 1961.
(4) TEICHERT, K., MUTSCHLER, E. et ROCHELMEYER, H. *Deuts. Apoth. Ztg.*, **100** ; 283, 1960.
(5) GERLACH, H. et SENF, J.H. *Pharm. Zentralhalle*, **105** : 93, 1966.
(6) SEILER, N. et WIECHMANN, M. *Experientia*, **21** : 203, 1965.
(7) JART, A. et BIGLER, A. J., *J. Chromatog.*, **29** : 255, 1967.
(8) PARIHAR, D.B., SHARMA, S.P. et TEWARI, K.C. *J. Chromatog.*, **24** : 443, 1966.
(9) WINSTEAD, M.B. et HEINE, H.W. *J. Am. Chem. Soc.*, **77** : 1913, 1955.
(10) AMAL, H. et DEMIR, S. *J. Fac. Pharm. Istanbul*, **3** : 1, 1967.
(11) HEINE, H.W., WINSTEAD, M.B. et BLAIR, R.P. *J. Am. Chem. Soc.*, **78** : 672, 1956.

Determination of estrogens in oily solutions by means of gas chromatography

by

G. CAVINA, G. MORETTI and P. SINISCALCHI.

The determination of estrogens in oily solutions for pharmaceutical use requires a preliminary separation in order to be able to apply successive analytical procedures.

Also in regard to the relatively small amounts of the active compounds used (few mg/ml), the difficulties arising in this determination have not left much possibility of choice and therefore the methods described in literature, USP XVI (1), B.P. 1963 (2), for estradiol-3-benzoate, estradiol-3,17-dipropionate and estrone, N.F. XII (3) for estradiol-17-cyclopentylpropionate are based nearly exclusively on the use of the Kober-reagent (4), after modification, known as the Iron-Kobër-reagent (5).

Critical views as to these methods have been described in papers by Snair and Schwinghammer (1961) (6) and Urbanyi and Rehm (1966) (7).

Recently gas chromatographic methods have been described by Talmage, Penner and Geller (1965) (8), by Boughton et al. (1966) (9) and by Schulz (1965) (10) for the ethinyl estradiol and its methyl ether, which are estrogens used mostly in tablets ; for these estrogens, applications of colorimetric methods based on a convenient modification (Brown, 1955 (11)) of the Kober method (4) have also been described (Longecker (1961) (12) and Ercoli, Vitali and Gardi (1964) (13)).

It is evident how the methods of separation of estrogens from oil, complicated and in some cases not quantitative, influence the choice of the successive method of determination.

We carried out a series of experiments in order to verify the possibilities of using gas chromatographic methods for the analysis of oily solutions of estrogens at low concentrations (2 mg/ml) adopting for the preliminary separation of estrogens in oil a procedure of thin-layer chromatography with continuous elution as described by us already for the quantitative analysis of other steroids in oily solutions (14 and 15) or the procedure of separation between hexane and 85 % ethanol according to NF XII (3) with reciprocally saturated solvents (Cavina, Cingolani and Giraldez) (16).

RESULTS

I. - Analysis of diesters of estradiol and monoesters of estrone.

As an example of the analysis of the diesters of estradiol, the determination of its dipropionate in purified and neutralized olive oil solution for pharmaceutical purposes, at a concentration of 2 mg/ml, was studied ; similarly for the monoesters of estrone, the determination of estrone-3-benzoate at the same concentration and in the same solvent as indicated above was studied. As indicated in A of the experimental part, the separation by TLC was performed with 50 μl of a diluted oily solution with heptane (2 to 5).

Parallely with the gas chromatographic analysis, we studied the application of the colorimetric reaction to TLC eluates, using the sulfuric acid-hydroquinone reagent according to Brown (11) with the Allen's correction formula (17) in order to eliminate the small absorbance due to traces of contaminating material.

The results are shown in Table I : values obtained with both methods are satisfactory and in good agreement, as to precision and accuracy. In these cases the gas chromatographic analysis is particularly convenient. The compounds are stable under the conditions of analysis, and they can be determined directly without the need of preparing derivatives.

In figures 1 and 2 the gas chromatograms of estrogens isolated from oily solutions are shown. The separation of the steroid band, representing

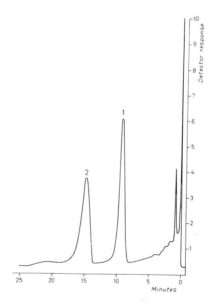

Fig. 1. — Gas chromatogram of the eluate after chromatography of the oily solution of estradiol dipropionate. Peak 1: 5 α-cholestane ; peak 2: estradiol dipropionate. Column : 2 % SE-30 on sil. 80-100 mesh Chromosorb G, 2.20 m length, 240° C.

TABLE I

Recovery obtained by gas chromatographic and colorimetric analysis.

Steroid	Quantity of steroid used (in 50 µl) for the analysis.	Recovery by gas chromatographic method.		Recovery by colorimetric method (Absorbancies are corrected with Allen's formula).	
	µg	µg ± s.d.	%	µg ± s.d.	%
Estradiol dipropionate in alcoholic solution	40.0	40.3 ± 0.8	100.7 ± 2.0 (4)	38.0 ± 0.9	95.0 ± 2.2 (7)
Estradiol dipropionate in oily solution	40.0	39.7 ± 0.8	99.2 ± 2.0 (5)	38.1 ± 1.3	95.2 ± 3.2 (5)
Estrone benzoate in alcoholic solution	40.0	39.6 ± 1.0	99.0 ± 2.5 (5)	38.9 ± 0.4	97.2 ± 1.0 (5)
Estrone benzoate in oily solution	40.0	38.3 ± 0.3	95.7 ± 0.7 (5)	38.8 ± 1.6	97.0 ± 4.0 (5)

The number of determinations is indicated in brackets.

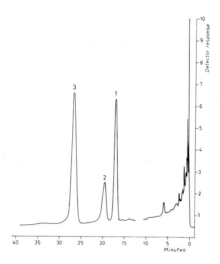

Fig. 2. — Gas chromatogram of the eluate after chromatography of the oily solution of estrone-3-benzoate. Peak 1 : 5α-cholestane-3β-ol acetate. Peak 2 : minor component of the oil (not identified). Peak 3 : estrone-3-benzoate. Column : 3 % JXR on sil. 100-120 mesh Gas Chrom P, 2.20 m length, 230° C.

approximately 0.2 % of the oily solution, can be defined as good in the two cases studied. In fact, only with estrone benzoate a small peak (n. 2) due to an unknown component of the oil can be observed on the gas chromatogram, which however does not interfere with the determination of the estrogen. As a rule, in other cases studied here, the gas chromatogram of the blanks does not show any interfering peak in the portion corresponding to the steroid and internal standard peaks.

In Table I the recovery data obtained with samples containing only the steroid in an alcohol solution are reported as well. It can be seen that between these data and those obtained for oily solutions, using both the gas chromatographic and colorimetric methods, there are no appreciable differences as to recovery and precision. This demonstrates the convenience of using the thin-layer chromatographic method proposed for the isolation of estrogens, and also the accuracy of analytical procedures employed successively.

II. - Analysis of 3- or 17-monoesters of estradiol.

In this case it is necessary to protect the hydroxyls. We performed this by transforming them into the respective trimethylsilyl ethers (TMSE). In the examples presented, different methods of steroid separation from oily solutions have been applied and described. For the analysis of estradiol-3-benzoate, the technique described under A in the experimental part is applied : a thin-layer chromatography with continuous elution for the separation of the estrogen from oil. Also in this case parallel analysis of the eluate from TLC by the gas chromatographic and the colorimetric methods were carried out. Results are reported in Table II. In figure 3 a gas chromatogram of the analysed estrogen is shown.

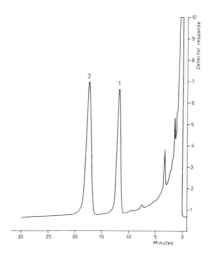

Fig. 3. — Gas chromatogram of the eluate after chromatography of the oily solution of estradiol-3-benzoate. Peak 1 : cholesteryl propionate. Peak 2 : estradiol-3-benzoate-17-TMSE. Column : 2 % SE-30 on 80-100 mesh Gas Chrom Q, 2.20 m length, 235º C.

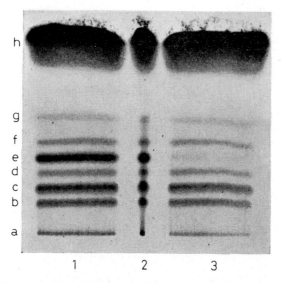

Fig. 4. — Thin-layer chromatography with solvent HEAA 80 : 20 : 1, continuous development for 3 hours. 1) estradiol-17-cyclopentylpropionate in oily solution ; 2) idem reference lane ; 3) reference oil blank. a) monoglycerides (traces) ; b) 1,2-diglycerides ; c) 1,3-diglycerides ; d) and f) minor components (not identified) ; e) estradiol-17-cyclopentylpropionate ; g) free fatty acids ; h) triglycerides. Detection : 50 % H_2SO_4 and heating.

TABLE II

Recovery obtained by gas chromatographic and colorimetric analysis.

Steroid	Quantity of steroid used for the analysis	Analytical procedure for separation and gas chromatography	Recovery by gas chromatographic method		Recovery by colorimetric method (Absorbancies are corrected with Allen's formula)	
			mean ± s.d.	%	mean ± s.d.	%
Estradiol-17β-cyclopentyl propionate in oily solution (2 mg/ml)	μg 40.0	B	μg 39.7 ± 1.1	99.2 ± 2.7 (6)	μg 38.9 ± 1.1	97.2 ± 2.7 (6)
Estradiol-3-benzoate in oily solution (2 mg/ml)	μg 40.0	A	μg 40.0 ± 1.3	100.0 ± 3.2 (5)	μg 40.5 ± 1.0	101.2 ± 2.5 (5)
Estradiol-17β-valerate in oily solution (10 mg/ml)	mg 10.0	C	mg 9.54 ± 0.06	95.4 ± 0.6 (5)	mg 9.46 ± 0.23	94.6 ± 2.3 (5)

The number of determinations is indicated in brackets.

For the analysis of estradiol-17-cyclopentylpropionate a thin-layer chromatography has also been used in order to separate it from oil. In figure 4 a clear example of the necessity of this separation by the continuous flow technique is shown. We used a method of saponification in the presence of the gel for elution of the steroid, followed by the preparation of the ditrimethylsilyl ether of estradiol obtained by hydrolysis of the ester (the technique is described under B in the experimental part). The colorimetric method mentioned above was also applied to the hydrolysis product. The results are shown in Table II. In figure 5 a gas chromatogram of the analysed estrogen is shown.

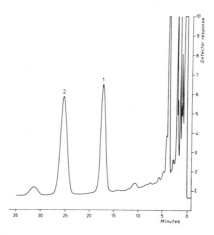

Fig. 5. — Gas chromatogram of the eluate after chromatography of the oily solution of estradiol-17-cyclopentylpropionate after saponification to estradiol and TMSE preparation. Peak 1: 5 α-cholestane. Peak 2: estradiol diTMSE. Column: 2% SE-30 on 80-100 mesh Gas Chrom Q, 2.20 m length, 190° C.

We should like to stress the contemporary elution and saponification which enable the analysis, by gas chromatography, of all estradiol derivatives. Our results are comparable with and are, for this estrogen, better than those obtained by Bylliar and Eikness (1965 (18)), who saponified various steroid acetates, including estradiol, with cholinesterase and sodium carbonate. Better results obtained by us with alkali in the presence of silica gel are probably due to the buffering effect of the latter which is confirmed by observations of Hornstein, Crowe and Ruck (1967) (19) on the saponification of lipids under conditions similar to those of our experiments.

The contemporary elution and saponification can be advantageous also for the application of the colorimetric analysis: in fact in the example of estradiol-17-cyclopentylpropionate the direct determination of estrogen in the chloroform eluate from the gel was disturbed by interfering substances which were eliminated by saponification.

For the analysis of estradiol-17-valerate in oily solution, the method of simple extraction with hexane and ethanol (85 %) was used. Higher concentrations of this steroid are used and therefore a concentration of 10 mg/ml was considered, while the extraction was performed from 1 ml of the sample (technique described under C of the experimental part). Before applying the gas chromatographic analysis, the steroid extracted into ethanol was subjected to trimethylsilylation. To the alcohol solution also the colorimetric method mentioned above was applied. The results are presented in Table II. : the data demonstrate the convenience of using the two analytical procedures proposed. Values obtained for estradiol-17-valerate by the extraction procedure C indicate that the recovery is only a little less than that obtained in other cases by separation with TLC. In figure 6 a gas chromatogram of the analysed estrogen is shown.

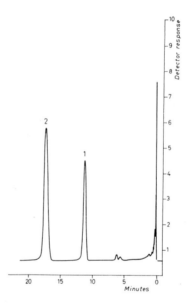

Fig. 6. — Gas chromatogram of estradiol-17-valerate separated from the oily solution by partition between hexane and 85 % ethanol. Peak 1 : estradiol-17-valerate-3-TMSE. Peak 2 : 5 α-cholestane-3β-ol acetate. Column : 3 % JXR on sil. 100-120 mesh Gas Chrom P, 1.80 m length, 230° C.

III. - **Analysis of steroïd mixtures.**

The possibility of using procedures of separation of estrogens from oily solution in hexane and 85 % ethanol for successive gas chromatography allows to foresee ulterior possibilities of analysis of estrogen mixtures with other steroids of hormonal character, particularly androgens and progestogens. In order to evaluate these possibilities we studied the conditions of separation of a certain number of steroids on 4 different stationary phases. The data are reported in Table III., showing various possibilities of separation as follows :

TABLE III

Relative retention times on different stationary phases.

Steroid	3 % JXR on sil. 100-120 mesh Gas Chrom P; length of column 2.20 m ; 240°.	3 % QF-1 on 100-120 mesh Gas Chrom Q; length of column 1.80 m ; 240°.	2 % SE-52 on 80-100 mesh Gas Chrom Q; length of column 2.20 m ; 230°.	3 % XE-60 on 90-100 mesh Anakrom AS ; length of column 2.20 m ; 240°.
Estradiol dipropionate	0.61	1.39	0.68	1.79
Estradiol-3-benzoate-17-TMSE	1.90	2.26	2.20	3.46
Estradiol-17-cyclopentyl propionate-3-TMSE	1.95	1.97	2.30	2.62
Estradiol-17-valerate-3-TMSE	0.62	—	—	—
Progesterone	0.32	1.90	0.32	2.07
Methyltestosterone	0.23	0.90	0.22	—
Testosterone propionate	0.38	1.64	0.38	1.71
Testosterone cyclopentyl propionate	1.96	6.61	2.21	8.03
Cholestane	0.36	—	0.46	0.21
Cholestanol acetate	1.00	1.00	1.00	1.00

a) On 3 % JXR, testosterone propionate from estradiol-3-benzoate, or 17-cyclopentylpropionate ; progesterone from estradiol dipropionate and testosterone cyclopentylpropionate.

b) On 3 % QF-1, estradiol dipropionate from progesterone and testosterone cyclopentylpropionate ; estradiol-3-benzoate or 17-cyclopentyl-propionate from testosterone cyclopentylpropionate.

c) On 2 % SE-52 progesterone from estradiol-3-benzoate or 17-cy-clopentylpropionate.

d) On 3 % XE-60 testosterone propionate from progesterone and estradiol-3-benzoate or 17-cyclopentylpropionate.

In practice the analytical possibilities are restricted, before all for compounds at low concentrations because some minor components of the oil diglycerides, free fatty acids and sterols), which remain in the 85 % ethanol phase, are present in sufficient quantity to interfere with the steroids to be analysed or with internal standards. Other complications may arise in the presence of $\Delta_4 - 3$ CO steroids, as they may, under certain conditions of the preparation of the TMSE, give rise to the respective 3-enol ethers, the peaks of which may cause interference.

An example illustrating these difficulties is given for the analysis of a mixture of testosterone cyclopentylpropionate and estradiol-17-cy-clopentylpropionate (50 mg and 2.5 mg respectively per 1 ml of oil). According to procedure C, the extraction technique allows, with some

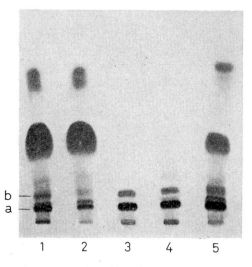

Fig. 7. — Thin-layer chromatography with solvent HEAA 70 : 30 : 1, normal development. 1) Oily solution of 50 mg testosterone cyclopentylpropionate (TCPP) and 2.5 mg estradiol-17-cyclopentylpropionate (ECPP) per ml. 4 μl. 2) Hexane phase after extraction (quantity corresponding to 4 μl of oily solution). 3) Pure steroids in amounts corresponding to 4 μl of oily solution, a) TCPP, 200 μg ; b) ECPP, 10 μg. 4) Ethanol phase after extraction (quantity corresponding to 4 μl of oily solution). 5) Hexane washings (quantity corresponding to 10 times the ethanol phases chromatographed under 4). Detection : 50 % H_2SO_4 and heating.

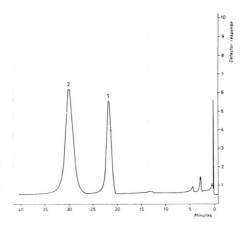

Fig. 8. — Gas chromatogram of testosterone-17-cyclopentylpropionate in mixture
with estradiol-17-cyclopentylpropionate separated from oily solution by partition
between hexane and 85 % ethanol. Peak I: estrone-3-benzoate (internal standard).
Peak 2: testosterone-17-cyclopentylpropionate. Column: 3 % QF-1 on 100-120 mesh
Gas Chrom Q, 1,80 m length, 240° C.

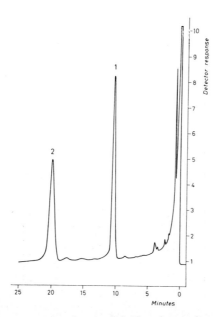

Fig. 9. — Gas chromatogram of estradiol-17-cyclopentylpropionate separated by
thin-layer chromatography from testosterone-17-cyclopentylpropionate in the 85 %
ethanol phase. Peak 1: 5 α-cholestane-3β-ol acetate. Peak 2: estradiol-17-cyclopentyl-
propionate-3-TMSE.

TABLE IV

Recovery obtained by gas chromatographic analysis.

Steroid	Quantity of steroid used for the analysis mg	Analytical procedure for separation and gas chromatography	Recovery mg ± s.d.	Recovery %
1. Testosterone cyclopentylpropionate in oily solution, 50 mg/ml associated with (2).	50.0	C	47.0 ± 1.8	94.0 ± 3.6 (6)
2. Estradiol cyclopentylpropionate in oily solution, 2.5 mg/ml, associated with (1)	2.50	C + A	2.19 ± 0.04	87.6 ± 1.6 (5) (*)

The number of determinations is indicated in brackets. (*) In each extraction two thin layer chromatographies and two gas liquid chromatographic analyses were performed and the mean is considered.

losses, the separation of both steroids sufficiently free of impurities (Fig. 7). The gas chromatographic analysis of testosterone cyclopentylpropionate can be directly applied to the extract in 85 % ethanol as this steroid is stable under the conditions of analysis (Fig. 8).

The determination of estradiol-17-cyclopentylpropionate in the 85 % ethanol extract after the TMSE preparation has not as yet given good results and this determination is actually the object of our further critical studies. As an alternative we verified the quantitative separation of estradiol -17-cyclopentylpropionate from testosterone cyclopentylpropionate in the 85 % ethanol extract by means of thin-layer chromatography with HEAA 50 : 50 : 1 and elution with chloroform.

The gas chromatographic analysis as performed with trimethylsilylation in position 3 (Fig. 9) is satisfactory although with a not very good recovery, probably due to losses in the extraction procedure (see Fig. 7). The results of the analysis of this mixture are shown in Table IV.

It can be said in conclusion, that the application of the gas chromatographic technique to the determination of steroids in pharmaceutical preparations seems to open a field of research promising interesting results. This is particularly valid for those preparations which are difficult to analyse by traditional methods owing to the vehicle and the complex formulation.

EXPERIMENTAL PART

The solutions of compounds to be analysed were prepared as follows : a) in ethanol at a concentration of 0.8 mg/ml for estradiol dipropionate and estrone benzoate ; b) in oil at a concentration of 2 mg/ml for estradiol dipropionate, estrone benzoate, estradiol-3-benzoate, estradiol-17-cyclopentylpropionate ; c) in oil at a concentration of 10 mg/ml for estradiol-17-valerate ; the mixture of testosterone cyclopentylpropionate and estradiol-17-cyclopentylpropionate contained 50 mg/ml and 2.5 mg/ml respectively. The oily solutions (b) were diluted by 2 to 5 with heptane (concentration of steroid 0.8 mg/ml, of oil 0.4 mg/ml). The oily solutions (c) were diluted by 1 to 5 with heptane.

For the separation technique according to (A), our method of thin-layer chromatography with continuous elution was reported in a previous paper (Cavina and Moretti 1966 (14)). The plates were coated with Merck Silica Gel G, in some cases mixed with 0.3 % Dupont Luminescent Chemical 609 ; the thickness was 0,5 mm. For quantitative work 50 µl of the oily solution diluted with heptane (40 µg steroid and approx. 20 mg oil) were applied in a line of 6 cm length. On the same plate, if necessary, a blank performed with equally diluted oil was applied. The conditions of the development were as follows :

1) estradiol dipropionate, solvent hexane-ethyl ether-acetic acid (HEAA) 70 : 30 : 1 with continuous elution for 1 hour ;

2) estrone benzoate, solvent HEAA 90 : 10 : 1 with continuous elution for 7 hours ;

3) estradiol-3-benzoate, solvent HEAA 50 : 50 : 1 with continuous elution for 1,5 hours ;

Operating conditions for gas

Steroid	Column	Derivative	Column temper. C^o	Injector temper. C^o
Estradiol dipropionate	2 % SE-30 on sil. 80-100 mesh Chromosorb G ; 2.20 m length	—	240	270
Estrone-3-benzoate	3 % JXR on sil. 100-120 mesh Gas Chrom P ; 2.20 m length	—	230	265
Estradiol-17-cyclo pentylpropionate	2 % SE-30 on 80-100 mesh Gas Chrom Q ; 2.20 m length	diTMSE	190	260
Estradiol-3-benzoate	2 % SE-30 on 80-100 mesh Gas Chrom Q ; 2.20 m length	mono TMSE	235	260
Estradiol-17-valerate	3 % JXR on sil. 100-120 mesh Gas Chrom P ; 2.20 m length	mono TMSE	230	270
Testosterone-17-cyclopentylpropionate	3 % QF-1 on 100-120 mesh Gas Chrom Q ; 1.80 m length	—	240	270
Estradiol-17-cyclopentylpropionate	3 % JXR on sil. 100-120 mesh Gas Chrom P ; 2.20 m length	mono TMSE	240	270

(*) : (A) indicates calculations made by area of the peaks ; (H) indicates calculations made by height of the peaks.

V

chromatographic analysis.

Detector temper. C^o	Nitrogen flow rate ml/min.	Attenuation	Steroid calibration range in µg (*)	Retention time of the steroid in min.	Internal standard	Retention time of the standard in min.
240	40	× 20	5 — 15 (A)	15	5α-cholestane (5 µg)	9
230	60	× 10	4 — 12 (H)	27	5α-cholestane —3β ol acetate (4 µg)	17
190	40	× 5	2.5 — 7.5 (H)	25	5α-cholestane (8 µg)	12
235	40	× 5	4 — 12 (H)	17	cholesteryl propionate (5 µg)	12
250	40	× 50	50 — 150 (H)	11	5α-cholestane-3β-ol-acetate (200 µg)	17
250	40	× 50	250 — 750 (A)	30	estrone-3-benzoate (250 µg)	21
250	40	× 20	15 — 35 (A)	20	5α-cholestane-3β-ol-acetate (20 µg)	10

4) estradiol-17-cyclopentylpropionate, solvent HEAA 80 : 20 : 1 with continuous elution for 3 hours. In the case of the mixture (c) the separation from testosterone-17-cyclopentylpropionate and minor components of oil was carried out in the solvent HEAA 50 : 50 : 1 with normal elution after a previous partition between hexane and 85 % ethanol.

For the separation technique according to (A), the elution was carried out with 25 ml chloroform transferring the removed layer on a small column of 0.8 cm diameter and collecting the eluate in a tared 25 ml vessel (solution Cp).

Two 6.5 ml aliquots were used for the colorimetric and two 5 ml aliquots for the gas chromatographic analysis.

For the separation technique according to (B), the elution conditions were as follows : after the identification of the steroid band (only the lane of the reference sample was sprayed with the Folin-Ciocalteau reagent diluted 1 : 5 with water (Mitchell and Davies, 1954 (20)) and then with 4 N NaOH, estrogens giving a blue colour), the silica gel corresponding to this band (approx. 2 x 7.5 cm) was scraped off and transferred into a 30 ml glass stoppered test tube. The saponification of the estrogen ester was carried out with 4 ml 10 % KOH (w/v) in methanol refluxing in a boiling water bath for 30 minutes. After cooling, 3 ml of 2 N HCl, 3 ml of water and 6 ml of ethyl ether (peroxide free and freshly distilled) were added. After shaking, the ether phase was sucked off with a Pasteur pipet and transferred into a 25 ml separator equipped with a teflon stopcock. Four successive extractions each with 2 ml of ether were carried out. The extracts were then collected and washed with 2 ml of 2.5 % (w/v) $NaHCO_3$ and twice with 2 ml of water. The ether phase was dehydrated with anhydrous sodium sulfate and filtered into a 25 ml volumetric flask. The sample was dried by evaporation of the ether under a nitrogen current, 0.5 ml of methanol was added and filled up with chloroform (Solution Cp).

The separation technique according to (C) was carried out as follows : in a 100 ml separator (with a teflon stopcock) 50 ml of 85 % ethanol (previously saturated with hexane) and 5 ml of the sample diluted by 1 to 5 with heptane were introduced and shaken. Then 35 ml hexane (previously saturated with 85 % ethanol) were added and the mixture was thoroughly shaken, the phases were allowed to separate until clear, and the alcohol phase was transferred into another (250 ml) separator, in which 25 ml of hexane were first introduced. The hexane phase in the first separator was extracted successively three times with 25 ml of 85 % ethanol each, and all alcohol phases were collected in the second separator. This was then shaken and both phases were carefully separated. The alcohol phase together with two 10 ml 85 % ethanol washings of the hexane layer was transferred into a tared 200 ml flask and filled up with 95 % ethanol (solution Cp).

For the gas chromatographic analysis a calibration curve was first done, referring the steroïd to an internal standard. The calibration ranges are reported in Table V. For the calibration the value of $R = A/B \times C/D$ (21) (A = sample area or height, B = standard area or height, C = standard amount, D = sample amount) was determined. The preparation of the TMSE was performed as described by Boughton et al. (9). Analyses were carried out on a Perkin — Elmer model 801 gas chromatograph with a hydrogen flame ionization detector and glass columns. The columns and the operating conditions were different for the various steroids and are described in Table V.

Concerning the colorimetric analysis, the reaction with sulfuric acid and hydroquinone as described by Brown (1955) (11) for estrogens and their methyl ethers was applied. For estrone derivatives the reagent was prepared with 66 % H_2SO_4 and for estradiol and its derivatives with 60 % H_2SO_4. Two methods were used to compensate the small absorption caused by traces of minor components of the oil and impurities coming from the gel : by subtracting the blank with only oil or only gel, or by introducing the correction formula according to Allen (1950, (17)) with measurements at three wavelengths.

References

(1) Pharmacopea of the United States XVI, 277, 280, 282, 285.

(2) British Pharmacopoeia, 540, 1964.

(3) National Formulary XII, 156.

(4) KOBER, S. *Biochem. Z.,* **239** : 209, 1931.

(5) HAENNI, E.O., *J. Amer. Pharm. Ass., Sci. Ed.,* **39** : 544, 1950.

(6) SNAIR, D. W. and SCHWINGHAMMER, L.A. *J. Pharm. & Pharmacol.,* **13** : 148, 1961.

(7) URBANYI, T. and REHM, C.R., *J. Pharm. Sci.,* **55** : 501, 1966.

(8) TALMAGE, J.M., PENNER, M.H. and GELLER, M. *J. Pharm. Sci.,* **54** : 1194, 1965.

(9) BOUGHTON, O.D., BRYANT, R., LUDWIG, W.J. and TIMMA, D.L., *J. Pharm. Sci.,* **55** : 961, 1966.

(10) SCHULZ, E.P. *J. Pharm. Sci.,* **54** : 146, 1965.

(11) BROWN, J.B., *Bioch. J.,* **60** : 185, 1955.

(12) LONGECKER, H., *Acta Endocrinol,* **37** ; 14, 1961.

(13) ERCOLI, A., VITALI, R. and GARDI, R. *Steroïds,* **3** : 497, 1964.

(14) CAVINA, G. and MORETTI, G., *J. Chromatog,* **22** : 41, 1966.

(15) CAVINA, G. MORETTI, G. and GIOCOLI, G. *Boll. Soc. Ital. Biol. Sper.,* **42** : 116, 1966.

(16) CAVINA, G., CINGOLANI, E. and GIRALDEZ, A. *Il Farmaco Ed. Prat.,* **17** : 149, 1962.

(17) ALLEN, W. M. *J. Clin. Endocrinol. & Metab.,* **10** : 71, 1950.

(18) BILLIAR, R. B. and EIK-NES, K.B. *Anal. Biochem.,* **13** : 11, 1965.

(19) HORNSTEIN, I., CROWE, P. and RUCK, J.B. *J. Chromatog.,* **27** : 485, 1967.

(20) MITCHELL, F.L. and DAVIES, R.E., *Bioch. J.,* **56** : 690. 1954.

(21) CELESTE, A. and TURCZAN, J. *J. Ass. Offic. Agr. Chemists,* **46** : 1055, 1963.

Ein einfaches analytisches Verfahren zur schnellen Bestimmung von Ectoparasiticiden in Viehbädern

von

G. SZÉKELY, D. EBERLE

Zentrale Forschung der Fa. J.R. Geigy A.G., Basel.

In bedeutenden Vieh- und Schafzuchtländern wie Argentinien, Australien, Neuseeland entstehen auch heute noch durch Räude und Hautmyiasis (blow-fly strike) bei Wollschafen, sowie durch Zecken bei Rindern grosse wirtschaftliche Verluste. Der Bekämpfung und Verminderung der Schadinsekten widmen die Regierungen dieser Länder daher grösste Aufmerksamkeit.

Die Fa. J.R. Geigy A.G. in Basel hat zur Bekämpfung von Ectoparasiten verschiedene Phosphorsäureester entwickelt :

GS 13'006

0,0-Diaethyl-S-[2-methoxy-1,3,4-thiadiazol-5-(4H)-onyl-(4)-methyl]-dithiophosphat.

Diazinon®

0,0-Diaethyl-0-[2-isopropyl-4-methyl-pyrimidin-(6)-yl]-thiophosphat.

Diazinon wird bevorzugt zur Bekämpfung der Räudemilben und gegen Myiase verursachende Fliegen eingesetzt. Der neue Phosphorsäureester GS 13'006 hat in der Zeckenbekämpfung bisher interessante Ergebnisse geliefert.

Die Bekämpfung der Ectoparasiten erfolgt in Argentinien fast ausschliesslich in sogenannten Viehbädern. Hierbei handelt es sich um grosse Wasserbecken, in denen Hunderte von Rindern, Schafen periodisch gebadet werden.

Beim Bad-Durchgang werden die Köpfe der Rinder mit langen Stangen untergetaucht, um auch die besonders an den Ohren sitzenden Zecken zu erfassen.

Die von den Tieren abtropfende Flüssigkeit gelangt wieder in die Bäder zurück. Die Wirksubstanz-Konzentration in den Viehbädern muss auch nach dem Baden einiger Hundert Tiere noch so hoch sein, dass während der kurzen Tauchzeit eine möglichst quantitative Abtötung der Ectoparasiten gewährleistet wird. Sie darf jedoch nicht so hoch werden, dass für die Tiere, falls sie Badflüssigkeit trinken, die Gefahr einer P-Ester-Vergiftung besteht. Daher ist eine fortlaufende analytische Kontrolle der Viehbad-Konzentration an Ort und Stelle unerlässlich. Im Falle von GS 13'006 soll die Bad-Konzentration zwischen 0.005 und 0.015 % liegen. Wir mussten deshalb eine analytische Schnellmethode entwickeln, die so einfach war, dass sie auch von analytisch völlig ungeschultem Personal, wie die Gauchos, durchgeführt werden konnte, die aber noch genau genug sein musste, um Bad-Konzentrationen, die bedeutend ausserhalb des genannten optimalen Bereiches lagen, zu erkennen und durch Wasser- bzw. AS-Zugabe zu korrigieren.

Die Methode der Wahl war ein schicht-chromatographisches Verfahren.

Die Probenahme aus den Bädern erfolgt sehr einfach. An einem Stock sind 3 geschlossene Stöpselflaschen befestigt, die senkrecht in das Bad eingetaucht werden. Nach dem Eintauchen der Stange werden die Stöpsel, die jeweils an einer Schnur befestigt sind, herausgezogen, die Flaschen füllen sich mit der Viehbadlösung, die Stange wird herausgezogen und der Inhalt der 3 Flaschen repräsentiert eine Badprobe vom Grunde, von der Mitte und von der Oberfläche des Badtroges.

Die Wirksubstanz wird anschliessend mit Benzol aus der Viehbad-Lösung extrahiert. Ein Aliquot des Benzol-Extraktes wird dünnschichtchromatographisch getrennt und semiquantitativ ausgewertet. Anstelle der konventionellen Dünnschicht-Platten oder -Folien verwendeten wir die Desaga Chromatotubes®. Es handelt sich dabei um Glasröhren, die auf ihrer Innenseite mit Kieselgel beschichtet sind (Hersteller : C. Desaga GmbH, Heidelberg). Das « Tube » selbst ist die Trennkammer und die Dünnschicht auf der Innenseite des Tubes ist gegen Verletzung geschützt. Elastische Verschlusskappen schliessen die aktivierte Schicht ab und erübrigen eine Reaktivierung vor der Trennung.

Experimentelles : 100 ml Viehbad werden im Scheidetrichter mit 10 ml Benzol geschüttelt. Nach kurzem Stehen lässt man die wässerige Phase ab und trägt mit Hilfe einer kalibrierten Glaskapillare 2 µl der benzolischen Phase auf den Startpunkt der Chromatotube AT auf (AT = aktive Tubes). Daneben trägt man als Vergleich 5 verschiedene Konzentrationen der Wirksubstanz auf, entsprechend 90, 70, 50, 30 und 10 % der Anfangskonzentration des Viehbades. Bei hoher relativer Luftfeuchtigkeit sollen Extrakt und Vergleichslösungen rasch aufgetragen werden. Es ist sinnvoll, zuerst alle Substanzlösungen in Mikrokapillaren aufzuziehen und in den Kapillarhaltern griffbereit zu halten. Vor dem Auftragen entfernt man nur eine der beiden Verschlusskappen, damit keine nennenswerte Konvektion im « Tube » entsteht. Nach dem Auftragen der Substanzlösungen wird die zweite Verschlusskappe abgenommen und dafür eine Entlüftungskappe mit Kanüle aufgesetzt. Dann muss das « Tube » unverzüglich in den Fliessmittelbehälter geschoben werden. Dieser wurde schon vorher mit dem Fliessmittel Benzol gefüllt (10 ml). Die Laufstrecke beträgt 10 cm. Die Flecken werden mit Joddampf nachgewiesen. Dazu stellt man das « Tube » in einen leeren Glasbehälter, in welchem sich einige vorgewärmte Jod-Körner befinden. Der R_f-Wert beträgt 0,36 für GS 13'006 und 0,21 für Diazinon. Die Auswertung erfolgt visuell mit Hilfe der oben erwähnten Eichreihe. Die relative Standardabweichung der Methode beträgt ca. 50 %. Die übrigen im Viehbad vorhandenen Verunreinigungen — meist Tierexkremente — haben bis jetzt die Auswertung nicht gestört.

Diese einfache Analysenmethode wird seit 2 Jahren in den argentinischen Pampas von analytisch ungeschultem Personal erfolgreich angewendet. Zur Erleichterung der Bestimmungen wurde von der Fa. Geigy eine leicht verständliche Analysenvorschrift sowie ein normierter Koffer, welcher alle notwendigen Geräte, Lösungsmittel, Eichreihen enthält, entwickelt.

Mit diesem kurzen Referat möchten wir nur zeigen, welch unerwartete Arbeits- und Einsatz-Gebiete chromatographischen Methoden noch offen stehen. Für die Ueberlassung der Farb-Dias sowie für die Einführung der Methode in der Praxis danken wir Herrn Dr. V. Flück von der Forschung Agro-Chemikalien der Fa. J.R. Geigy A.G., Basel.

Sur la présence de dérivés flavoniques dans les boutons floraux de l'Eugenia Caryophyllata (L) Thunbg

par

Mme M. HOTON-DORGE,
chef de travaux aux Laboratoires de Pharmacognosie et de Pharmacie pratique
de l'Université libre de Bruxelles

ISOLEMENT DE QUATRE FLAVONOIDES

Jusqu'à présent, aucun travail ne signale la présence de flavonoïdes dans les boutons floraux d'Eugenia Caryophyllata bien que plusieurs auteurs (1) aient mis en évidence des dérivés structure « Chromone » et que, d'autre part, H. Schmid (2, 3, 4, 5) a montré qu'il existe entre les chromones de l'Eugenia Caryophyllata (eugénine, eugénitine, isoeugénitol et isoeugénitine) et certains flavones du bois de pin (techtochrysine, strobochrysine, strobopinine) une identité de structure du noyau chromone de leur molécule (Tableau I).

Il nous a donc paru intéressant de rechercher la présence de flavonoïdes dans les boutons floraux et de voir si l'on pouvait rattacher leur structure à celles des chromones identifiées et citées ci-dessus.

Isolement des flavonoïdes.

La présence d'une part d'huile essentielle et d'acide oléanique en quantité importante, et d'autre part, la présence des chromones nous a tout naturellement amené à éliminer ces composés de la poudre par extraction par l'éther au Soxhlet. La poudre est ensuite épuisée par percolation avec de l'éthanol à 94c. La solution éthanolique, brun foncé, est évaporée à sec et le résidu repris par l'éther acétique, solvant de choix pour dissoudre aussi bien les hétérosides flavoniques que leurs aglycons.

Cette solution éthéro-acétique donne nettement la réaction de la cyanidine et sera purifiée par agitation :

1. avec une solution saturée de bicarbonate de sodium pour éliminer les composés acides présents ;

TABLEAU I

EUGENINE ⟷ TECTOCHRYSINE

EUGENITINE ⟷ STROBOCHRYSINE

ISOEUGENITOL

ISOEUGENITINE ⟷ STROBOPININE OU CRYPTOSTROBINE

2. avec une solution de borax à 2 % afin de séparer les flavonoïdes qui possèdent deux fonctions -OH voisines dans le noyau latéral C et 1 OH situé au carbone 5 (6) ;

3. avec de l'eau jusqu'à neutralité de la solution aqueuse.

Ainsi débarrassée de la plupart des constituants annexes, cette solution éthéro-acétique est desséchée sur sulfate de sodium anhydre (Sol. éthéro-acétique B) et évaporée à sec. Son résidu représente 0,2 % du poids de la poudre préalablement épuisée à l'éther.

Nous n'envisagerons dans ce travail que les dérivés flavoniques présents dans cette solution éthéro-acétique B, bien que nous puissions dès à présent signaler la présence de quercétine dans la solution aqueuse boratée.

TABLEAU II

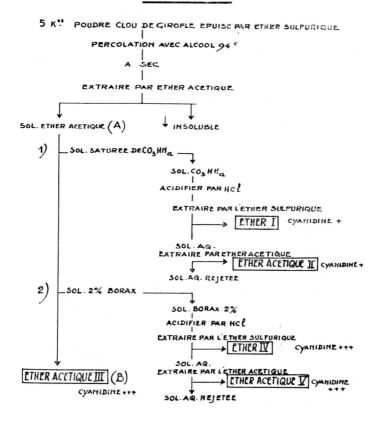

Traitement de la solution éthéro-acétique B.

Le résidu résultant de l'évaporation de cette solution est dissous dans le méthanol. L'insoluble est filtré. La solution méthanolique présente une intense réaction de la cyanidine.

Dans le but de déterminer le nombre de flavonoïdes présents et avant de tenter leur séparation sur colonne de polyamide, nous avons effectué une chromatographie sur couche mince de cette solution en appliquant la technique suivante :

Nous utilisons des plaques de 7 cm de côté.

2 g de poudre de polyamide Merck pour chromatographie sur couche mince sont mis en suspension dans 8 ml de méthanol et agités très vivement pendant 1 minute. Cette suspension est versée sur les plaques et répartie de façon aussi homogène que possible en couche mince.

Le solvant utilisé est un mélange de chloroforme/méthanol/méthyl éthyl cétone/acide acétique (7 : 2 : 0,5 : 0 ; 1).

L'ascension du solvant s'effectue dans un récipient cylindrique de 7,5 cm de diamètre et 7,5 cm de hauteur, fermé par une plaque de verre.

Après développement, le chromatogramme est séché à l'air libre pendant 5 minutes. L'examen du chromatogramme en U.V. à 350 mμ nous laisse voir quatre bandes fluorescentes en orange et une large bande fluorescente en jaune-clair au front de solvant.

Chacune des bandes fluorescentes oranges est grattée. La poudre est recueillie dans un tube à essai contenant du méthanol (5 ml). La suspension est agitée cinq minutes et filtrée. Les solutions méthanoliques sont concentrées à 1 ml. Chacune de ces solutions donne la réaction de la cyanidine. Nous avons donc à faire à des dérivés flavoniques et nous nous proposons de les séparer.

Après pulvérisation d'un chromatogramme avec une solution alcoolique à 1 % de chlorure d'aluminium, les quatre bandes fluorescentes en orange aux U.V. à 350 mμ apparaissent fluorescentes en jaune intense. La bande jaune qui migre avec le solvant reste inchangée.

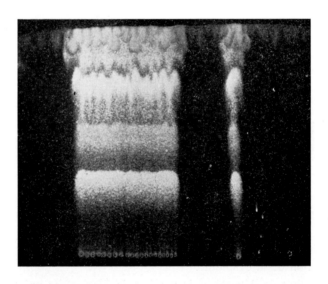

Séparation par chromatographie sur colonne d'Ultramid B.M.K. 228.

(Badische Anilin und Soda Fabrik — Ludwigstrafen am Rhein)

L'Ultramid est d'abord lavé par du CH_3OH bouillant pour éliminer la plus grande partie des particules de polyamide solubles dans ce solvant.

L'Ultramid est introduit dans une colonne de 50 cm de hauteur et 4,5 cm de diamètre contenant du méthanol.

Après tassement, la colonne d'Ultramid qui a une hauteur de 33 cm est lavée au benzène jusqu'au moment où l'évaporation de celui-ci ne laisse plus de résidu de polyamide.

La solution méthanolique à chromatographier est concentrée à 20 ml et ensuite versée sur la colonne. L'élution de la colonne est alors réalisée successivement par le benzène, par des mélanges benzène - méthanol (9 : 1) ; benzène - méthanol (8 : 2) et finalement par du méthanol pur.

Lors de la première élution au benzène, on récolte trois fois 500 ml, deux fois 500 ml pour le deuxième éluant et 500 ml pour le troisième. L'élution finale avec le méthanol pur se poursuit jusqu'à absence de coloration de la solution.

Les différentes fractions de 500 ml récoltées au cours de l'élution de la colonne par les divers éluants sont évaporées à sec et les résidus sont repris par l'éthanol à 94^c. Ces solutions alcooliques contiennent une forte proportion de polyamide soluble.

L'élution benzénique de la colonne fournit ainsi les solutions alcooliques a, b, c ; l'élution par le mélange benzène - méthanol (9 : 1) donne les solutions d et e tandis que l'élution par le mélange benzène - méthanol (8 : 2) fournit la solution alcoolique f et l'élution finale par le méthanol donne la solution g.

Ces différentes fractions sont chromatographiées sur couche mince afin de nous permettre de réunir les solutions de composition identique. Quatre fractions sont ainsi différenciées.

La fraction I, correspondant à la réunion des solutions a, b, c provenant de l'élution de la colonne par le benzène.

La fraction II, correspondant aux solutions d et e réunies et provenant de l'élution par le mélange benzène - méthanol (9 : 1).

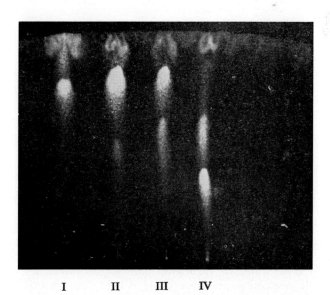

I   II   III   IV

La fraction III est identique à la fraction f de l'élution par le mélange benzène - méthanol (8 : 2).

La fraction IV correspond à la fraction g de l'élution méthanolique finale de la colonne.

Ces quatre fractions sont chromatographiées sur couche mince et examinées en U.V. à 350 mμ après pulvérisation du chromatogramme avec la solution alcoolique à 1 % de chlorure d'aluminium.

Cet examen nous montre que la séparation des flavonoïdes obtenue au cours de l'élution de la colonne par les différents solvants utilisés est encore grossière et qu'une nouvelle chromatographie de chacune d'elles est nécessaire. Nous avons décidé de purifier chacune des quatre fractions précédentes par chromatographie sur colonne de 15 g d'Ultramid et avons recueilli, à l'aide d'un collecteur de fractions, des volumes successifs de 5 ml.

Comme nous l'avons fait précédemment, la chromatographie sur couche mince de chacune des solutions récoltées nous a permis de réunir celles qui présentaient des spots de Rf identiques.

Ces dernières solutions sont alors évaporées à sec et le résidu est repris par l'éther, ce solvant ayant l'avantage sur l'alcool de dissoudre peu de polyamide.

La fraction I ainsi traitée nous a permis d'isoler deux des flavonoïdes présents à l'état pur. Nous les appellerons momentanément X$_1$ et X$_2$. Ils se retrouvent mélangés et accompagnés d'une autre substance dans les fractions du liquide d'élution de la colonne.

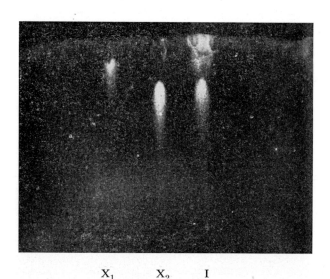

X$_1$ X$_2$ I

Cette méthode de séparation est satisfaisante mais son inconvénient principal est la solubilité de l'Ultramid dans la succession des éluants utilisés, malgré le traitement préalable de la colonne par chacun d'eux.

Séparation par chromatographie préparative sur couche mince de polyamide Merck des fractions III et IV.

10 mg des fractions III et IV en solution alcoolique sont chromatographiées sur plaque de polyamide Merck de 40/20 cm. La solution alcoolique est déposée à 1 cm du bord inférieur de la plaque sous forme d'un trait continu.

Le solvant utilisé est le même que celui utilisé dans toutes les chromatographies en couche mince citées précédemment.

Après développement du chromatogramme, la plaque est séchée cinq minutes à l'air libre et examinée en lumière U.V.

Toutes les bandes fluorescentes en jaune-orange à l'U.V. sont grattées et les fractions de polyamide ainsi récoltées sont extraites séparément au soxhlet par du méthanol. Les solutions méthanoliques évaporées abandonnent un résidu que l'on reprend par l'éther. Les solutions éthérées sont filtrées et examinées à nouveau par chromatographie sur couche mince.

De la fraction III, il se sépare ainsi un flavonoïde à peu près pur X_3 et de la fraction IV, deux flavonoïdes, l'un correspondant à X_3 et le second que nous désignons momentanément par X_4.

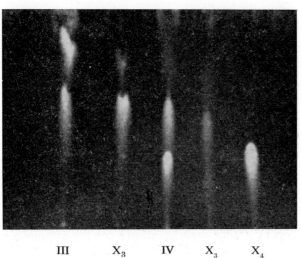

III X_3 IV X_3 X_4

La purification des fractions III et IV sur couche mince préparative de polyamide Merck ne nous ayant pas donné des produits purs, nous avons tenté d'effectuer la séparation des flavonoïdes de ces diverses fractions sur colonne de polyamide MN SC_6 de Macherey-Nagel (grosseur des grains 0,16 mm) qui présente l'avantage d'être quasi insoluble dans les solvants utilisés pour la chromatographie.

Technique : 15 g de polyamide MN SC$_6$ sont mis en suspension dans une colonne à chromatographier (H = 85 mm ; ϕ = 30 mm) contenant du méthanol. La colonne est lavée préalablement au méthanol et au benzène.

La fraction III (150 mg) est dissoute dans 5 ml de méthanol et la solution est versée au sommet de la colonne.

On opère l'élution successivement par du benzène, par des mélanges benzène - méthanol (9 : 1) ; benzène - méthanol (8 : 2) et finalement par du méthanol pur.

Des fractions de 5 ml sont recueillies au cours de l'élution, fractions dont on vérifie la composition par chromatographie sur couche mince. Les fractions qui montrent des spots identiques sont réunies. L'examen du chromatogramme nous indique que la substance X$_3$ à l'état pur passe en solution dans l'éluant benzène - méthanol (8 : 2).

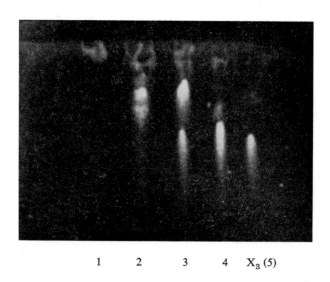

1   2   3   4   X$_3$ (5)

La fraction IV traitée dans les mêmes conditions nous a permis d'obtenir à l'état pur le flavonoïde X$_3$ identifié dans les fractions II et III ainsi qu'un autre flavonoïde (X$_4$).

Nous avons donc isolé quatre flavonoïdes à l'état pur.

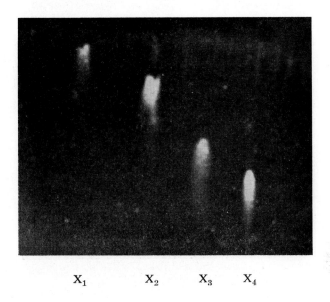

$$X_1 \qquad X_2 \qquad X_3 \qquad X_4$$

Ces flavonoïdes, provisoirement appelés X_1, X_2, X_3 et X_4, sont des aglycons. En effet, ils ne sont pas attaqués par hydrolyse. Leur Rf respectif obtenu par chromatographie en couche mince de polyamide est mentionné dans le tableau suivant.

TABLEAU IV

CHROMATOGRAPHIE SUR COUCHE MINCE DES DERIVES FLAVONIQUES DES BOUTONS FLORAUX D'EUGENIA CARYOPHYLLATA (L) THUNBG

- SUR POLYAMIDE MERCK
- SYSTEME DE SOLVANT : CHLOROFORME / METHANOL/METHYL-ETHYL-CETONE/ ACIDE ACETIQUE $(7:2:0,5:0,7)$

SUBSTANCE	RF	FLUORESCENCE EN U.V. A 350 $m\mu$ APRES PULVERISATION AVEC SOL. ALC. 1 % CL_3 AL.
X_1	0,90	JAUNE INTENSE
X_2	0,76	ID.
X_3 : RHAMNETINE	0,50	ID.
X_4 : KAMPFEROL	0,32	ID.

Après examen de ce tableau, nous remarquons que X_3 présente le même Rf que la rhamnétine et X_4 le même Rf que le kampferol.

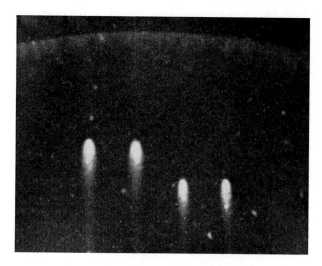

X_3 Rhamnetine X_4 Kampferol

La fusion alcaline oxydante (technique de D. MOLHO (7)) d'un dérivé flavonique permet d'obtenir avec de bons rendements l'acide correspondant au noyau latéral C par dégradation du flavonoïde par l'eau oxygénée en milieu alcalin.

Dans le cas de X_1 et de X_3, nous obtenons l'acide protocatéchique.

Dans le cas de X_2 et de X_4, nous obtenons l'acide hydroxybenzoïque.

TABLEAU III

RHAMNETINE = X_3 FUSION ALCALINE OXYDANTE →   ACIDE PROTOCATÉCHIQUE

KAMPFEROL = X_4 FUSION ALCALINE OXYDANTE →   ACIDE P. HYDROXYBENZOÏQUE

L'identification se fait par chromatographie sur couche mince de Kieselgel G. Merck.

Solvant : Chloroforme/éther acétique/ acide formique (5 : 4 : 1) (8).

Révélateur : solution aq. 2 % chlorure ferrique.

Le point de fusion de X_3 ne subit pas d'inflexion en mélange avec la rhamnétine. De même, le point de fusion de X_4 reste inchangé en mélange avec le kampférol.

X_1 et X_2 sont en cours d'identification.

En conclusion dans la solution éthéro-acétique B examinée, les flavonoïdes X_1 et X_3 (rhamnétine) se trouvent en faible proportion par rapport aux flavonoïdes X_2 et X_4 (kampférol).

X_3 (rhamnétine) est présent surtout dans la solution boratée qui entraîne les flavonoïdes possédant 2 OH voisins dans le noyau latéral C et 1 OH au carbone 5.

Bibliographie

(1) SCHMID, H. und BOLLETER, A. Über die Inhaltstoffe von Eugenia Caryophyllata (L) Thunbg. V. Isolierung des Isoeugenitins. *Helv. Chem. Acta,* **33** : 1170, 1950.

(2) SCHMID, H., Natürlich vorkommende Chromone. Progrès dans la Chimie des Substances organiques naturelles. Vol. XI, p. 151 - 153.

(3) LINSTEDT, G. Constituents of Pine Haertwood XXVI. A general discussion. *Acta Chem. Scand.,* **5** : 129, 1951.

(4) LINSTEDT, G. and MISIORNY. Constituents of Pine Haertwood XXIX. A Synthesis of Strobochrysin diméthylether (5-7 dimethoxy-6 Methylflavone) *Acta Chem. Scand.,* **6** : 1212, 1952.

(5) ERDTMAN, H. Über einige Inhaltsstoffe des Kernholzes der Koniferenordnung Pinales. *Holz, als Roh- und Werkstoff.,* **11** ; 245, 1953.

(6) WACHTMEISTER, C.A. Paper Chromatography on Borate impregnated Paper. *Acta Chem. Scand.,* **5** ; 976, 1951.

(7) MOLHO, D. *Bull. Soc. Chim. Fr.,* p. 39, 1956.

(8) STAHL, E. und SCHORN, J. *Z. Physiol. Chem. Hoppe Seyler's,* **325** : 263, 1961.

Quantitative thin-layer chromatography on liquid anion exchangers

Part I : An investigation into some of the parameters involved in the direct densitometric determination of zinc

by

R.J.T. GRAHAM, L.S. BARK and D.A. TINSLEY
Department of Chemistry and Applied Chemistry
The University of Salford, Lancashire.

Summary

A study has been made of some of the parameters involved in the direct densitometric determination of zinc on cellulose layers impregnated with the liquid ion exchanger Primene JM-T-hydro-chloride. The chromogenic reagent PAN was employed to locate the zinc spots.

Of the factors associated with the instrument (the Joyce-Loebl Chromoscan) the most important has been shown to be the setting of the baseline control. The most important factor concerned with the chromatographic system is shown to be the difficulty of distributing the chromogenic reagent uniformly over the layer.

In spite of these problems it has been shown that an accuracy of about 5 % can be achieved at the 1 microgram level.

INTRODUCTION

Inorganic ions, previously separated by thin-layer chromatography, have been quantitatively determined in a number of ways. These are 'in situ' methods (1 - 4), including spot area measurements (1, 2, 3) radiochemical methods (3, 4) and by spectroscopy involving transmitted light (3), as well as by methods involving the removal of the inorganic species from the layers prior to their quantitative determination (5 - 7).

The layers used for these investigations have been limited to laboratory prepared silica gel layers (1, 2, 3) precoated silica gel layers (4) and cellulose layers (5 - 7), so that the investigations have been limited to normal adsorption and normal partition techniques.

We have already reported the results of our investigations into the qualitative separations of metal ions by reversed phase thin-layer chromatography on a neutral organo-phosphorus substrate (8 - 10) and on layers impregnated with long chain amines in the form of their hydrochlorides (11 - 13). Because of this, and also because of the absence of direct quantitative data obtained from such reversed phase thin-layer chromatographic systems we decided to investigate the possible applications of quantitative investigations in one of our reversed phase systems, namely the system Primene JM-T-hydrochloride/hydrochloric acid (12 - 13).

Early ' in situ ' methods of qualitative inorganic T.L.C. analysis (1 - 4) were carried out on systems which gave Rf values of low reproducibility whereas the system investigated here has been shown to yield highly reproducible Rf values. This fact is of considerable importance because our quantitative studies were carried out using a direct densitometric method. It has been shown by Blank and co-workers (14), and Thomas et al. (15) that the precision of quantitative results obtained by this method can be affected by the Rf values of the sample scanned. This phenomenon has also been investigated by Dallas (16) who investigated a number of other parameters which are of importance in the precision direct densitometry of coloured substances on silica gel thin layers.

In addition to the problems considered by Dallas (16) further problems arise in the direct densitometry of colourless species by reversed phase systems. In order to assess these additional effects, we have examined the parameters affecting the densitometry of a standard red dye on cellulose layers (the support medium for the impregnants in our reversed phase systems) as well as those affecting the precision of the results obtained for the element zinc in the chosen system.

EXPERIMENTAL

Chromatography

(a) *Chromatography of the Standard Dye*

Cellulose (15g MN 300 HR) was slurried with water (90mls) and the coated plates were allowed to air dry overnight.

A solution of the standard red food dye (Red 10B) was applied to the layers and eluted with n-butanol ; water : glacial acetic acid (17) (20 : 12 : 5 v/v/v) by our sandwich chamber technique (17). When the solvent front had reached the appropriate point (14.0cms $\pm$ 0.5cms) the layers were removed from the chamber, dried and scanned.

(b) *Reversed Phase System*

Cellulose (15gMN 300 HR) was slurried with a solution of the amine hydrochloride in chloroform (70ml of 0.3M) which had been prepared under the standard conditions previously described (12, 13). Zinc (11) chloride solution (1μl of a 1mg/ml solution) was applied to the layers using a Hamilton syringe (1μl capacity).

The elution of the chromatograms with the hydrochloric acid eluent, in our sandwich chamber (17), the drying of the plates and the visualisation of the zinc spot with the chromogenic reagent PAN have already been described.

Direct Densitometry

The instrument used throughout the work was the « Chromoscan » recording and integrating densitometer with a thin layer attachment *. The instrument may be used for both reflectance and transmission densitometry, but in the work reported here we were concerned only with its use as a reflectance densitometer. The main features of the instrument have been discussed by Dallas (16).

In scanning the standard spots a 10 mm x 1 mm slit aperture was used throughout. Unless otherwise stated each standard spot was scanned twenty five times and the integrator counts quoted represent the mean counts.

RESULTS AND DISCUSSION

In his study, Dallas (16) preferred not to use the integrator with which the instrument is provided. We have orientated our study to a consideration of the factors which affect the efficiency of the integrator. These are :
  i)   instrument drift ;
 ii)   the linearity of the integrator ;
iii)   the speed of scanning the spot relative to the chart speed ;
 iv)   the use of cams to adjust the pen deflection ;
  v)   base line adjustment ;
 vi)   the centering of the spot under the light beam.

Each of these parameters was investigated using the standard red dye on cellulose layers.

i) Instrument Drift

This was noticeable in readings taken immediately after the instrument was switched on, but became negligible if a period of half an hour was allowed to elapse between first switching on the instrument and scanning the first spot.

ii) The Linearity of the Integrator

The scale deflection, produced by adjustment of the base line control, and the numbers of integral counts per minute produced by this deflection is given in Table 1.

---

* Available from Joyce Loebl and Co., Gateshead on Tyne, England.

TABLE I

Linearity of the integrator attachment

Scale deflection (cm)	1	2	3	4	5	6	7	8	9	10
Counts/min	147	338	524	698	862	1047	1275	1460	1658	1849

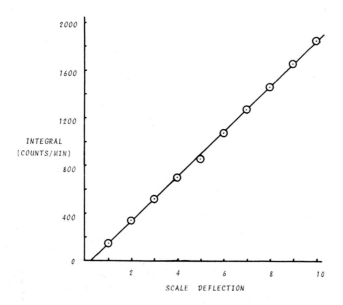

Figure 1 shows that a plot of scale deflection vs counts per minute is linear, though the line does not quite pass through the origin. This failure of the line to pass through the origin is not unexpected because the instrument possesses a low background count which may be adjusted by the base line control to 1 count/ 5 seconds.

iii) The Speed of Scanning the Spot relative to the Chart Speed.

The standard spot was scanned using each of the drive ratios 1 : 2 and 1 : 4 (specimen speed : chart speed) the chart speed being maintained constant. These results are given in Table II.

TABLE II

The effect of the specimen/chart drive ratio on reproducibility

Drive Ratio	Mean Count	Standard Deviation
1 : 2	207.7	0.48 %
1 : 4	413	0.66 %

From the results, it can be seen that the difference in reproducibility is insignificant. However, the 1 : 4 ratio scans the spot more slowly and hence allows the instrument more time to respond to changes in the light intensity. Thus for subsequent work we used the 1 : 4 ratio.

iv) The Use of Cams to Adjust the Pen Deflection

Three cams, A, B and C are provided with the instrument. These produce linear scale deflections in the ratios 1 : 1 . 3 : 2. 4. The standard spot was scanned with each cam in position in turn, and average counts of 170, 215, 413 were obtained for cams A, B and C respectively giving a ratio of 1 : 1.27 : 2.42 which confirms the accuracy of the system. For maximum sensitivity in subsequent work a drive ratio of 1 : 4 and cam C were used.

v) The Base Line Adjustment

The standard spot was scanned, resetting the base line control after each scan. A standard deviation of 1.63 % was obtained from the mean. This compares unfavourably with a standard deviation of 0.66 % for a similar test in which the base line control was not adjusted after its initial setting.

vi) The Centering of the Spot under the Light Beam

One of the problems associated with the successive scanning of either several spots on the same plate, or several different spots is the question of the correct positioning of the spot under the light beam, particularly as this has to be done visually. This was checked by removing the plate from the carriage after each scan and recentering the spot before commencing the next scan. A standard deviation of 0.83 % was obtained compared with 0.66 % for rescanning without altering the position of the plate relative to the light beam.

It is interesting to note that our standard deviation for repeat scanning based on integrator read out is smaller than that quoted by Dallas (16) for his preferred method of planimetry for determining the quantity of sample in the spot. This may be due to our use of a 10 mm x 1 mm slip aperture compared with the 1 mm diameter circular aperture used by this worker resulting in the positioning beam being more critical in the latter case.

The foregoing results confirm and compliment the findings of Dallas that a good degree of precision can be achieved in the densitometry of coloured compounds using the Chromoscan instrument. The greatest source of error was found to be introduced by attempting to adjust the base-line between each scan. For this reason we did not attempt to adjust the base line in the work in our attempts to appraise the instrument for work on colourless substances chromatographed by reversed phase systems.

Colourless Substance Chromatographed by Reversed Phase Thin-Layer Chromatography

The following additional factors must also be considered for colourless substances chromatographed by reversed phase thin-layer chromatography namely :

(a) the effect of the stationary phase ;

(b) the treatment of the layers after elution and before spraying ;

(c) the method of application of the chromogenic reagent to the layers.

(d) the stability of the coloured complex.

(a) The Effect of the Stationary Phase

The most probable effect contributing to lack of precision from this source is the non-uniform distribution of the stationary phase over the cellulose support. However, it has been shown by Duncan (18) that the slurry method of incorporating an impregnant into the support gives a reasonably uniform distribution of the impregnant over the layer.

Even so, we scanned a number of impregnated layers and it was observed that the base line, once initially adjusted remained constant when the same plate was scanned in several different positions across its width. From this we concluded that errors due to the method of impregnation will be minimal and that the presence of the impregnant is unlikely to affect any results obtained from impregnated layers.

(b) The Treatment of the Layer after Elution and before Spraying

The need to heat the plates in order to remove the hydrochloric acid eluent before spraying with the chromogenic reagent, PAN, has been discussed by us (8 - 13). Such treatment caused a darkening of the cellulose layer, and in particular a very dark zone immediately behind the solvent front. The lower edge of this zone was uneven and extended to a maximum distance of 0.5 cm behind the front. The scanning of eluted, dried chromatograms resulted in marked variation in the integral count in the region of this zone. However, scans between the lower end of this zone and the point of application of the samples yielded a fairly uniform base line.

From this, it can be seen that the technique will lead to imprecise results when attempts are made to scan spots immediately behind the solvent front, even when the upper, as well as the lower borders, of the spots are clearly demarkated. When the upper end of the spot mergers with the solvent front, further imprecision is introduced consequent upon the lateral zone spreading of the spot.

At this juncture we would like to add a further note concerning the phenomenon of lateral zone spreading. It is our opinion that this can lead to marked imprecision in all densitometric determinations, whether of coloured or colourless species, in normal or in reversed phase systems when the sample migrates with the solvent front, or when demixion occurs during the run. When this latter happens, those samples which run with the β etc. fronts, i.e. in polyzonal thin layer chromatography, also show lateral migration of the spots and hence attempts at a quantitative analysis

of such samples by direct methods will yield imprecise results. This parameter, as a contributory factor to imprecision in densitometric determinations, was not commented on by Dallas (16), probably because his use
of a single component mobile phase obviated the need for him to consider
this factor, but it is a factor which must be constantly borne in mind by
anyone attempting to use the technique.

(c) **The Method of Application of the chromogenic Reagent to the Layer.**

Chromogenic reagents may be applied to thin layer chromatograms.

(i)   by dipping the layer, face downwards into the reagent.

(ii)   by drawing a filter paper through the reagent and then impres-
sing the reagent loaded paper against the layer.

(iii)   by direct spraying the reagent on to the surface of the layer.

Both (i) and (ii) were tried as a means of visualising zinc in our reversed
phase system. When the layers were dipped into the alcoholic solution of
PAN, the layers became separated from the glass plate. Blotting the layers
with filter paper similarly damaged the layers. Thus we were forced to rely
on method (iii) and the results obtained using this method are given in
Table III.

TABLE III

Imprecision caused by uneven spraying of the plates

Plate Number	Standard Deviation %
1	4.3
2	3.2
3	5.2
4	3.9

8 spots of Zn (II) standard solution were applied across each plate.

Thus we see that the greater source of error in the direct densitometry
of colourless substances will be associated with the application of the
chromogenic reagent to the layer. To some extent, however, errors from
this source can be mitigated by spotting a sample of known concentration
on either side of the unknown sample because we observed that the average
difference between any two adjacent spots is 2.5 % compared with 5.2 %
maximum error for a whole plate.

(d) **The Stability of the coloured Complex.**

The stability of the zinc/PAN complex was investigated by counting
a 1 μg spot at regular time intervals after removal of the plate from the
ammonia vapour. The results are shown in Table IV.

TABLE IV

Stability of a zinc/pan spot

Time (min)	0	10	20	30	40
Integral	266	264	262	264	260

As all the spots on a plate can be counted in less than five minutes, the results in Table IV show that fading of the spots will not have a significant effect on the accuracy of our results.

CONCLUSIONS

Direct densitometry affords a quick and reasonably accurate method of quantitatively assessing thin layer chromatograms. The major source of error in the evaluation of colourless compounds is seen to be the unevenness of the application of the chromogenic reagent. Other sources of error may be associated with the instability of the coloured complex formed and the effects of demixion on the shape of the spots.

References

(1) KÜNZI, P., BÄUMLER, J., and OBERSTEG, J. Im. *Deut. Zt. für gerlicht Medizin,* **52** : 605, 1962.

(2) PURDY, S.J. and TRUTER, E.V. *Analyst,* **87** : 802, 1962.

(3) SEILER, H. *Helv. Chim. Acta,* **46** : 2629, 1963.

(4) MUZZARELLI, R.A.A. *Talanta,* **13** : 1689, 1966.

(5) GAGLIARDI, E. and LIKUSSAR, W. *Mikrochim. Acta,* p. 1053, 1965.

(6) GAGLIARDI, E. and PORKORNY, G. *Mikrochim. Acta,* p. 577, 1966.

(7) GAGLIARDI, E. and PORKORNY, G. *Mikrochim. Acta,* 1966.

(8) BARK, L.S., DUNCAN, G. and GRAHAM, R.J.T. *Analyst,* **92** : 31, 1967.

(9) BARK, L.S., DUNCAN, G. and GRAHAM, R.J.T. *Analyst,* **92** : 347, 1967.

(10) BARK, L.S., DUNCAN, G. and GRAHAM, R.J.T. 4th International Symposium Chromatog. and Electrophoresis, 1966, Brussels, Belgian Society for Pharmaceutical Sciences, p. 207, 1968.

(11) BARK, L.S., GRAHAM, R.J.T. and McCORMICK, D. 4th Intern. Symp. Chromatog. and Electrophoresis, 1966, Brussels, Belgian Society for Pharmaceutical Sciences, p. 199, 1968.

(12) GRAHAM, R.J.T., BARK, L.S. and TINSLEY, D.A. *J. Chromatog.,* **35** ; 416, 1968.

(13) GRAHAM, R.J.T., BARK, L.S. and TINSLEY, D.A. *J. Chromatog.,* **39** : 1969 (in the press).

(14) BLANK, M.L., SCHMIT, J.A. and PRIVETT, O.S. *J. Am. Oil Chemists' Soc.,* **41** : 371, 1964.

(15) THOMNS, A.E., SCHAROUN, J.E. and RALSTON, H. *J. Am. Oil Chemists' Soc.,* **42** : 789, 1965.

(16) DALLAS, M.S.J. *J. Chromatog.,* **33** : 337, 1968.

(17) BARK, L.S., GRAHAM, R.J.T. and McCORMICK, D. *Talanta,* **12** ; 122, 1965.

(18) DUNCAN, G. M.Sc. Thesis, University of Salford, 1966.

The Thin-layer chromatography of synthetic food dyes

by

R. J. T. GRAHAM* and A. E. NYA

The Department of Chemistry and Applied Chemistry
The University of Salford, Salford 5, Lancashire.

Summary

The 28 British food colours have been chromatographed on laboratory prepared silica gel thin-layers and on pre-coated silica gel layers using n-butanol : ethyl methyl ketone : ammonia (0.88) : water (5 : 3 : 1 : 1 v/v) as an eluent under conditions of sandwich chamber elution and normal tank elution. The former elution technique gave more reproducible Rf values but improved resolution was obtained using the latter technique.

INTRODUCTION

The introduction of precoated thin layer chromatography plates and foils has enabled the analyst to combine the advantages of thin-layer chromatograph with the simplicity of paper chromatography.

Synthetic food dyes have been chromatographed on a variety of different substrates (1 — 7) but in all cases laboratory prepared layers have been used, there being no reported use of pre-coated foils or plates for their separation.

In this paper a comparison is made between the separations of the 28 permitted British food colours on laboratory prepared plates and pre-coated silica gel layers under conditions of both sandwich chamber development and normal tank development.

EXPERIMENTAL

Aqueous solutions of the dyestuffs (0.1 % w/v) were prepared.

Silica gel G (30 g) was slurried with water (60 ml) in a fast electric mixer and the homogeneous slurry was used to coat clean glass plates (5 x 20 cm x 20 cm). These were allowed to air dry for 2 ½ hrs.

The laboratory prepared plates and the pre-coated layers were activated for $1\frac{1}{2}$ hrs. at $160°$ C in an air oven after which they were allowed to cool in a desiccator. When cold, they were spotted with the dyestuff solutions using a multispotting technique (8).

The spotted plates were then eluted either in a sandwich chamber (8, 9) or in a tank. In the latter case, the tank atmosphere was equilibrated for 10 minutes before the plates were introduced.

The eluent system was the organic phase of n : butanol : ethyl methyl ketone : ammonia (0.88) : water (5 : 3 : 1 : 1 v/v).

When the eluent front had travelled $12\frac{1}{2}$ cm from the point of application of the dyes, the plates were removed from the chamber or tank and dried in an oven to remove the eluent. The Rf values were then calculated.

RESULTS

These are given in Table 1.

DISCUSSION

(a) The Effect of the Nature of the Layers

In general, the order of Rf values is independent of the type of layer, laboratory prepared or precoated, bound or unbound layers. This is of course to be expected since the dyes are interacting probably by hydrogen bonding, with the same chemical substrate, silica gel, in each case. Some differences in Rf values are reported for the different layers but these cannot be rationalised except to state that in saturation chamber development the precoated glass plates (Merck) consistently gave the lowest Rf values.

(b) The Effect of the Manner of Development

The Rf values obtained under sandwich chamber conditions were almost always lower than those obtained using normal tank development. The reproducibility of former sets of Rf values were consistently higher ($\pm$ 0.02 Rf units) than was the reproducibility of those obtained using the latter method of elution ($\pm$ 0.05 Rf units). These results are in agreement with those observed previously by Graham (10).

These results are to be expected because under the former set of conditions almost instantaneous saturation of the tank atmosphere will occur whereas under the latter set of conditions evaporation of the components of the eluent system into the large void volume of the tank results in (a) a greater amount of eluent flowing through the layers than is apparent from the position of the eluent front and (b) gradient elution resulting from the differential evaporation of the components of the quaternary eluent system used.

Even so, in a number of cases the advantages to be gained from the increased resolution of the dyes chromatographed under tank conditions outweighs the disadvantage of loss of reproducibility inherent in this method of elution.

(c) **The Separation of the Dyes**

For convenience the 28 dyes are divided into 4 groups of similarly coloured dyes - namely

a) Brown dyes ;

b) Blue dyes - including Black PN and Green S ;

c) Yellow dyes - including the two permitted ; orange dyes - Orange G and Orange RN ;

d) Red dyes.

Group 1 : Brown Dyes.

Multiple spotting occurred with Brown FK. All the spots were clearly separated from both Chocolate Brown FB and Chocolate Brown HF neither of which moved.

Group 2 : Blue Dyes.

Surprisingly the dyestuff Indigo Carmine faded during the chromatographic process and could not subsequently be located on the layers.

The general order of the R_f values of the remaining 4 dyes of this group for most systems is :

Green S < Black BN < Blue VRS = Violet Carmine.

However, in system f i.e. a system of tank elution Green S and Black BN can be resolved as can Blue VRS and Violet Carmine.

Group 3 : Yellow Dyes.

The general order of the Rf values for members of this group for most systems is :

Yellow RY < Tartrazine < Yellow 2G < Yellow RFS (1)

= Orange G = Sunset Yellow < Naphthol Yellow < Yellow RFS (2) < Orange RN.

Thus with the exception of the trio Yellow RFS (1) Orange G and Sunset Yellow, this group of dyes can be resolved. However, the formation of two spots by Yellow RFS, the second of which is separate from all other yellow dyes enables the presence of this dye in mixtures to be detected.

Group 4 : Red Dyes.

Reference to Table 1 shows that many good separations of the dyes of this the largest group of permitted dyes is possible. However, the separation of Ponceau 4R and Amaranth is possible only in Systems 1 c d and h. Carmoisene and Ponceau 3R are also difficult to separate, but can be resolved in systems 1 e f and h. The multiple spotting of Ponceau SX enables it to be distinguished from Carmoisene and Ponceau 3R.

TABLE I
Rf values of permitted British dyes

Dyes	Rf Values × 100							
	a	b	c	d	e	f	g	h
Chocolate Brown FB	0	0	0	0	0	0	0	0
Chocolate Brown HF	0	0	0	0	0	0	0	0
Brown FK	18	19	10	18				
	33	42	21	31	42	53	39	
	47	58	33	58	53	80	56	88
Black PN	16	15	4	11	7	10	9	25
Violet BNP	32	39	21	30	33	63	41	81
Indigo Carmine	—	—	—	—	—	—	—	—
Blue VRS	31	40	19	31	31	49	25	80
Green S	12	17	6	13	8	17	8	28
Yellow 2G	19	19	8	14	15	20	20	—
Naphthol Yellow	34	52	23	35	41	68	48	85
Tartrazine	15	15	5	11	5	8	10	21
Yellow RFS	27	44	19	27	22	38	36	70
	47	66	35	61			59	
Sunset Yellow	30	38	18	29	33	49	28	86
Yellow RY	3	4	—	0	6	7		
	1	1	—	2	12	17	15	8
Orange G	28	44	18	26	27	51	36	72
Orange RN	51	70	35	68	58	86	59	95
Erythrosine BS	53	67	35	69	56	84	58	94
Ponceau MX	25	33	14	26	31	41	27	
	53	65	36	68		87	64	88
Ponceau SX	25	37	15	23	27	38	35	67
	33	44	22	29				
	38	54	27	39				
Red 2G	30	33	16	25	30	43	30	73
Ponceau 4R	14	17	3	8	12	15	11	42
Fast Red E	36	44	23	36	41	63	37	83
Ponceau 3R	25	39	15	29	21	53	32	72
Amaranth	16	17	78	14	8	14	15	23
Red 10B	25	39	16	24	25	44	33	78
Red FB	36	53	23	28	32	48	52	80
Red 6B	20	32	14	22	16	29	23	47
Carmoisene	25	34	15	26	13	32	29	44

Key:
a) Laboratory prepared plates/sandwich chamber development
b) Laboratory prepared plates/tank development
c) Precoated glass plates (Merck)/sandwich chamber development
d) Precoated glass plates (Merck)/tank development
e) Precoated foils (MN) with starch binder/sandwich chamber development
f) Precoated foils (MN) with starch binder/tank development
g) Precoated foil (MN) without binder/sandwich chamber development
h) Precoated foil (MN) without binder/tank development.

CONCLUSIONS

The system proposed for the separation of the 28 dyes permitted as additives for foods under British legislation represents an improvement on the only other thin-layer chromatographic system proposed for these dyes particularly with regards the separation of red dyes.

References

(1) WOLLENWEBER, P. *J. Chromatog.,* **7** : 557, 1962.
(2) SYNDINOS, E., KOTAKIS, G. and KOKKATI-KOTAKIS, E. *Riv. Ital. Sostanze Grasse,* **40** : 674, 1963.
(3) BARRETT, J.F. and RYAN, A.J. *Nature,* **199** : 372, 1963.
(4) RAMAMURTLY, M.K. and BHALERAO, V.R. *Analyst,* **89** : 740, 1964.
(5) DICKES, G.J. *J. Assoc. Public Analyst,* **3** : 49, 1965.
(6) DAVIDEK, J. and DAVIDKOVA, E. *J. Chromatog.,* **26** : 529, 1967.
(7) WANG, K.T. *Nature,* **213** : 212, 1967.
(8) BARK, L.S., GRAHAM R.J.T. and McCORMICK, D. *Talanta,* **12** : 122, 1965.
(9) BARK, L.S., DUNCAN, G. and GRAHAM, R.J.T. *Analyst,* **92** : 31, 1967.
(10) GRAHAM, R.J.T. *J. Chromatog.,* **33** : 125, 1968.

Détermination du R_{Mo} par chromatographie sur couche mince en phases inversées, comme méthode d'évaluation du coefficient de partage de dérivés phénothiaziniques

par

M. MERCIER — P. DUMONT

Université Catholique de Louvain, Laboratoire de Chimie thérapeutique.
Institut de Pharmacie.

I. INTRODUCTION

Dans le cadre d'un travail ayant pour objet l'étude des relations entre la structure chimique et l'activité neuroleptique des aminoalkylphénothiazines, nous avons examiné, dans quelle mesure les variations d'effets au niveau du système nerveux central, qu'entraînent différentes modifications structurales peuvent être attribuées à des changements dans la valeur du cœfficient de partage entre phases hydrophobe et hydrophile.

En effet, l'activité de surface prononcée que manifestent ces substances, (13), leur grande liposolubilité, les premiers indices dont on dispose concernant leur mécanisme d'action, et en particulier leurs effets sur la perméabilité membranaire (3, 4, 5, 11, 12, 14, 15), ainsi que les propriétés que l'on reconnait à la barrière hémato-encéphalique, impénétrable à la plupart des substances, nous conduisent à penser que l'activité psychotrope des phénothiazines pourrait dépendre étroitement de leur aptitude à quitter une phase aqueuse pour un milieu moins polaire, et donc trouver son expression dans la mesure de leur cœfficient de partage entre ces deux phases.

La détermination directe du cœfficient de partage des aminoalkylphénothiazines entre une phase aqueuse et une phase lipophile présente de sérieuses difficultés pratiques, dues à la très faible solubilité dans l'eau de ces substances ; aussi avons-nous fait appel à une méthode d'évaluation de ce paramètre, basée sur la mesure du R_{Mo} par chromatographie de partage sur couche mince, en phases inversées.

Cette méthode repose sur la relation théorique, établie par MARTIN (10), entre le cœfficient de partage α et la valeur du Rf, mesurée par chromatographie de partage liquide — liquide

$$\alpha = \frac{1}{r} \left(\frac{1}{Rf} - 1 \right)$$

où r est une constante pour le système de solvants envisagé, et représente le rapport des volumes des phases stationnaire et mobile.

Si nous combinons cette équation avec l'expression :

$$R_M = \log \left(\frac{1}{Rf} - 1 \right)$$

proposée par BATE-SMITH et WESTALL (1), nous obtenons la relation :

$R_M = \log \alpha \, r$.

R_M serait une grandeur constitutive et additive, proportionnelle au potentiel chimique standard μo nécessaire au transport, à température constante, d'un molécule d'une phase dans l'autre ; pour une série de composés présentant le même noyau de base, toute modification dans la valeur de R_M (ΔR_M) amenée par un substituant donné reflète donc un changement du potentiel chimique standard $\Delta\mu$o et possède la même signification que celle du paramètre π de HANSCH et FUJITA (6, 7, 8) :

$$\pi = \log \frac{\alpha_X}{\alpha_H}$$

où α_H et α_X représentent respectivement les cœfficients de partage du composé parent et d'un dérivé de substitution.

Dans le système chromatographique décrit ci-après dont la phase stationnaire hydrophobe est représentée par de la paraffine, l'emploi d'eau pure comme phase mobile n'entraîne, pour la plupart des phénothiazines, aucun déplacement de la tache de départ, en raison de la très faible hydrosolubilité de celles-ci. Aussi avons-nous été amené à utiliser, comme phases mobiles, des mélanges en proportion variable eau-méthanol, la linéarité théorique et expérimentalement observée de la relation entre les valeurs de R_M et la composition volumétrique du solvant hydrophile nous permettant d'extrapoler les résultats et de calculer pour chaque substance une valeur de R_{M0}, correspondant à une phase mobile constituée d'eau pure :

$R_{M0} = \log \alpha_o r$

où α_o représente le cœfficient de partage entre la paraffine et l'eau.

II. PARTIE EXPÉRIMENTALE

A. Description de la technique

Vingt grammes de Kieselgur G « MERCK » sont introduits dans une solution constituée de 4 ml de paraffine liquide et d'un mélange de 40 ml de dioxanne et 6 ml d'acétone ; la suspension homogène de Kieselgur G est coulée, en couche de 0,25 mm, sur des plaques de verre de 20 × 20 cm ; le solvant est ensuite évaporé sous vide ; la phase stationnaire est donc constituée d'un film hydrophobe de paraffine sur un support de Kieselgur G.

Les composés, utilisés sous forme de bases, sont mis en solution dans le chloroforme et appliqués, à raison de 10 µg par tache, à des intervalles de 1,5 cm, le long d'une ligne située à 3 cm. du bord inférieur de la plaque et à 1 cm. environ au-dessus du niveau du solvant de développement.

Les phases mobiles sont constituées de mélanges, en proportions variables, d'eau et de méthanol, contenant 1 % $^V/_V$. d'ammoniaque et équilibrés avec la phase stationnaire. L'addition d'ammoniaque permet d'éviter la formation de traînées.

Après le développement qui s'effectue, dans les conditions d'équilibre et à température constante, sur une distance d'environ 10 cm, mesurée depuis la ligne de départ des taches, les plaques sont séchées et la position des taches est déterminée, soit par examen en lumière ultra-violette qui fait apparaître les phénothiazines fluorescentes, soit après révélation par le réactif FNP, préconisé par FORREST (17) et composé de :

> Chlorure ferrique, solution aqueuse à 5 % (p/v), 5 ml
> Acide perchlorique, 20 % (v/v), 45 ml
> Acide nitrique, 50 % (v/v), 50 ml.

B. Résultats expérimentaux

a) *Etude préliminaire de la relation entre la valeur de R_M et la composition volumétrique de la phase mobile.*

Dans le tableau I, se trouvent réunies les valeurs des R_f et des R_M, mesurées pour trois aminoalkylphénothiazines : la lévomépromazine (1), la thiomépromazine (2) et la butyrylpérazine (3), en fonction de la proportion de méthanol dans la phase mobile.

1. Lévomépromazine	$X = OCH_3$	$R_1 = CH_2\text{-}CH\text{-}CH_2\text{-}N\begin{smallmatrix}CH_3\\ \\CH_3\end{smallmatrix}$ avec CH$_3$
2. Thiomépromazine	$X = SO_2\text{-}N\begin{smallmatrix}CH_3\\ \\CH_3\end{smallmatrix}$	$R_1 = CH_2\text{-}CH\text{-}CH_2\text{-}N\begin{smallmatrix}CH_3\\ \\CH_3\end{smallmatrix}$ avec CH$_3$
3. Butyrylpérazine	$X = CO\text{-}CH_2\text{-}CH_2\text{-}CH_3$	$R_1 = CH_2\text{-}CH_2\text{-}CH_2\text{-}N\bigcirc N\text{-}CH_3$

Tableau I : Variation des valeurs des Rf(x 10²) et des R_M en fonction de la proportion de CH_3OH dans la phase aqueuse mobile.

%CH₃OH (v/v)	1		2		3	
	Rf* (x10²)	R_M	Rf* (x10²)	R_M	Rf* (x10²)	R_M
5			6;7	1,20;1,12	2;3	1,69;1,51
10,8						
12,9			8;9	1,06;1,00		
21,6			10;11	0,95;0,91	4;3,5	1,38;1,44
32,4	2,5;2,5	1,59;1,59	21,5;21,5	0,56;0,56	7;6	1,12;1,20
43,2	3;3	1,51;1,51	54;52	-0,07;-0,06	21;21	0,58;0,58
51,6	6;6	1,20;1,20	70;71	-0,37;-0,39		
54,1	8,5;8,5	1,03;1,03	79;80	-0,57;-0,60	50;51;52	0,00;-0,02 -0,04
64,5	23;23	0,52;0,52	87;88	-0,83;-0,85		
64,8	22;24	0,55;0,50			71;72;73	-0,39;-0,41 -0,43
74,2	41;46	0,16;0,07				
75,6	46;50	0,07;0,00	93;92	-1,15;-1,05	82;87	-0,66;-0,82
84	68;68	-0,33;-0,33				
86,4	72;70	-0,41;-0,41	96;96	-1,40;-1,40	89;88	-0,85;-0,92
95	86;86	-0,80;-0,80				

L'accroissement de la proportion de méthanol dans la phase mobile entraîne donc une augmentation progressive des valeurs de Rf, ce qui est l'expression, si l'on s'en réfère à l'équation de MARTIN :

$$\alpha = \frac{1}{r} \left(\frac{1}{Rf} - 1 \right) \text{ ou } Rf = \frac{1}{\alpha r + 1} \qquad \text{(Eq. 1),}$$

d'une diminution du coefficient de partage entre phases hydrophobe et hydrophile.

(*) Les deux ou trois valeurs de Rf données pour chaque composition de la phase mobile correspondent à des expériences conduites à des jours différents ; chacune de ces valeurs est la moyenne de 4 déterminations faites en plaçant la tache de départ en un point différent de 4 plaques chromatographiques.

Or, KEMULA et BUCHOWSKI (9) ont montré que lorsqu'une des phases liquides formant le système chromatographique est constituée d'un mélange de deux solvants 1 et 2, et que l'autre phase est représentée par un liquide pur, non ou très peu miscible aux deux solvants précédents, le coefficient de partage d'une substance entre ces deux phases est donné par la relation :

$$\log \alpha = \mu_1 \log \alpha_1 + \mu_2 \log \alpha_2 \qquad \text{(Eq. 2)}$$

dans laquelle : μ_1 et μ_2 représentent les fractions volumétriques des deux constituants de la phase binaire

: α_1 et α_2 désignent les coefficients de partage de la substance entre la phase unique et les solvants 1 et 2 pris isolément.

L'équation (2) peut également s'écrire :

$$\log \alpha r = \mu_1 \log \alpha_1 r + \mu_2 \log \alpha_2 r \qquad \text{(Eq. 3)}$$

ou, tenant compte de l'expression $R_M = \log \alpha r$

$$R_M = \mu_1 R_{M1} + \mu_2 R_{M2} \qquad \text{(Eq. 4)}$$

Graphiquement, la relation entre Rf et μ_1 et μ_2 se traduira, si l'on tient compte des équations (1) et (3), par une courbe sigmoïde.

La valeur de R_M, quant à elle, est une fonction linéaire de la composition volumétrique du solvant binaire, avec comme limites à l'ordonnée, les valeurs de R_{M1} et R_{M2}.

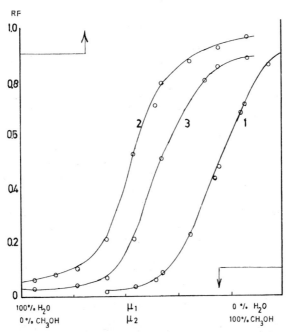

Fig. 1. — Variation des Rf en fonction de la composition volumétrique de la phase mobile.

Variation of the Rf's in function of the volumetric composition of the mobile phase.

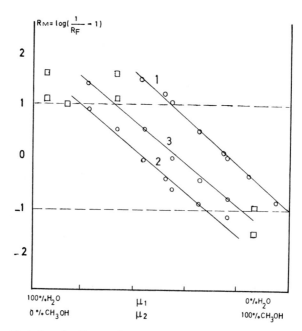

Fig. 2. — Variation des R_M en fonction de la composition volumétrique de la phase mobile.

Variation of the R_M's in function of the volumetric composition of the mobile phase.

Dans les figures 1 et 2, se trouvent représentées, pour les trois substances précédentes, les variations des Rf et des R_M en fonction de la composition volumétrique de la phase mobile eau-méthanol.

Nous pouvons vérifier, l'allure sigmoïdale de la relation entre Rf et la composition de la phase mobile (fig. 1), ainsi que la relation linéaire unissant les valeurs de R_M et la proportion de méthanol dans la phase hydrophile (fig. 2).

Une déviation à la linéarité est cependant observée pour les valeurs de R_M situées en dehors de l'intervalle compris entre — 1 et + 1, correspondant à des valeurs de Rf situées entre 0,1 et 0,9 environ ; les valeurs trop faibles (<0,1) ou trop élevées (>0,9) de Rf sont en effet à écarter : dans ces régions, on doit s'attendre à des déviations dues notamment à l'établissement de gradients entre le volume des phases stationnaire et mobile et donc à la disparition de la constance de r.

b)   *Calcul des valeurs de R_{M0} des aminoalkylphénothiazines*

Une fois vérifiée la linéarité de la relation entre R_M et la composition volumétrique de la phase mobile, nous pouvons calculer, pour chacune des substances expérimentées, la valeur de R_{M0} ainsi que les limites de confiance (t = 0,05) de celle-ci.

Tableau II : Influence de la nature de la chaîne aminée latérale sur les valeurs de R_{M0}

Substituant X du C_2	Chaîne aminée R_1	R_{M0}
H	- CH$_2$-CH$_2$-CH$_2$-N(CH$_3$)(CH$_3$)	$2,73 \pm 0,15$
	- CH$_2$-CH(CH$_3$)-CH$_2$-N(CH$_3$)(CH$_3$)	$3,35 \pm 0,19$
	- CH$_2$-CH(CH$_3$)-N(CH$_3$)(CH$_3$)	$2,54 \pm 0,01$
	- CH$_2$-CH(CH$_3$)-CH$_2$-N⟨ ⟩N-CH$_2$-CH$_2$-O-CH$_2$-CH$_2$OH	$1,29 \pm 0,16$
Cl	- CH$_2$-CH$_2$-CH$_2$-N(CH$_3$)(CH$_3$)	$3,40 \pm 0,08$
	- CH$_2$-CH$_2$-CH -N⟨ ⟩N-CH$_3$	$2,71 \pm 0,22$
	- CH$_2$-CH$_2$-CH -N⟨ ⟩N-CH$_2$-CH$_2$OH	$2,01 \pm 0,15$
CN	- CH$_2$-CH$_2$-CH$_2$-N(CH$_3$)(CH$_3$)	$2,47 \pm 0,31$
	- CH$_2$-CH(CH$_3$)-CH$_2$-N(CH$_3$)(CH$_3$)	$2,86 \pm 0,10$
	- CH$_2$-CH$_2$-CH$_2$-N⟨ ⟩-OH	$1,20 \pm 0,02$
	Chlorprotixène	$3,68 \pm 0,29$

Tableau III : Influence de la nature du substituant et de modifications de structure du noyau sur les valeurs de R_{M0}

Chaîne aminée R_1	Substituant X du C_2	R_{M0}
- CH_2-CH_2-CH_2-N$\big\langle$ CH_2 / CH_3	H	$2,73 \pm 0,15$
	CH_3	$2,97 \pm 0,09$
	Cl	$3,40 \pm 0,08$
	$COOCH_3$	$2,72 \pm 0,15$
	CN	$2,47 \pm 0,31$
	OCH_3	$2,72 \pm 0,41$
	CF_3	$3,98 \pm 0,52$
- CH_2-CH -CH_2-N$\big\langle$ CH_3 / CH_3 ; CH_3	H	$3,35 \pm 0,19$
	OCH_3	$3,29 \pm 0,31$
	CN	$2,86 \pm 0,10$
	SO_2-N$\big\langle$ CH_3 / CH_3	$2,24 \pm 0,06$
	oxochlorpromazine	$- 0,42$
	oxomémazine	$0,73 \pm 0,13$

Avant d'aborder la discussion des résultats repris dans les tableaux II et III, il nous semble essentiel d'attirer l'attention sur deux observations qui apportent une confirmation directe de la validité du procédé décrit :

1) Il paraissait indispensable de s'assurer que la nature du constituant non-aqueux de la phase mobile n'avait pas quelqu'incidence sur les valeurs de R_{M_0} ; dans ce but, nous avons déterminé ce paramètre pour plusieurs des substances étudiées, en remplaçant le méthanol par de l'acétone, toutes les autres conditions expérimentales restant identiques ; la bonne concordance des deux séries de valeurs (tableau IV) permet de conclure à une absence d'effets liés à la nature du solvant non-aqueux présent dans la phase mobile.

TABLEAU IV

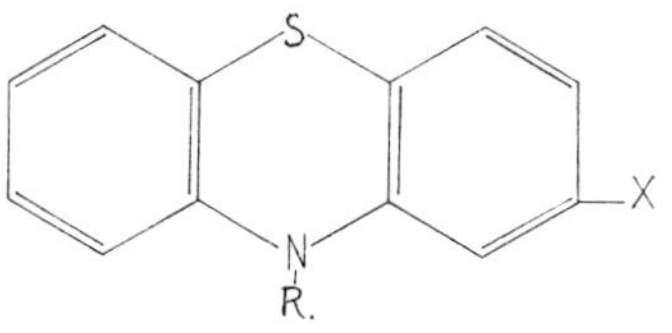

X	R_1	R_{M_0} solvant (eau-méthanol)	R_{M_0} solvant (eau-acétone)
H	$CH_2\text{-}CH_2\text{-}CH_2\text{-}N(CH_3)_2$	$2{,}73 \pm 0{,}15$	$2{,}59 \pm 0{,}37$
CH	$CH_2\text{-}CH_2\text{-}CH_2\text{-}N(CH_3)_2$	$2{,}47 \pm 0{,}31$	$2{,}21 \pm 0{,}22$
CN	$CH_2\text{-}CH(CH_3)\text{-}CH_2\text{-}N(CH_3)_2$	$2{,}86 \pm 0{,}10$	$2{,}88 \pm 0{,}12$
$COCH_3$	$CH_2\text{-}CH_2\text{-}CH_2\text{-}N\underset{}{\diagup\diagdown}N\text{-}CH_2\text{-}CH_2OH$	$0{,}50 \pm 0{,}06$	$0{,}52 \pm 0{,}17$
H	$CH_2\text{-}CH(CH_3)\text{-}CH_2\text{-}N\underset{}{\diagup\diagdown}N\text{-}CH_2\text{-}CH_2\text{-}O\text{-}CH_2\text{-}CH_2OH$	$1{,}29 \pm 0{,}16$	$1{,}21 \pm 0{,}23$

2) Si nous comparons (cfr. tableau V) d'une part les valeurs de log α mesu-
    rées dans le système octanol-eau (8), pour les dérivés de substitution du
    benzène avec celles de R_{Mo} des dérivés de la promazine, et par ailleurs,
    celles de π (= log $α_X$ — log $α_H$) et de $ΔR_{Mo}$, un parallélisme tout-à-fait
    remarquable se manifeste dans les 2 cas :

    — Les paramètres log α et R_{Mo} ont été mesurés à l'aide de systèmes
       utilisant des phases hydrophobes différentes ; ainsi la corrélation
       observée entre ces deux constantes vient à l'appui de l'équation de
       COLLANDER (2), qui établit une relation linéaire entre les log
       $α_I$ d'une substance dans un premier système de solvants et le log
       $α_{II}$ de celle-ci dans un second système.
    — Bien que calculées à partir de noyaux chimiquement très différents,
       les constantes π et $ΔR_{Mo}$ présentent une excellente concordance qui
       plaide en faveur de leur signification en tant que grandeurs consti-
       tutives.

TABLEAU V

Substituant	log α (octanol - eau)	π	R_{Mo} (paraffine - eau)	$ΔR_{Mo}$
CN	1,56	–0,57	2,47	–0,26
OCH_3	2,11	–0,02	2,72	–0,01
$COOCH_3$	2,12	–0,01	2,72	–0,01
H	2,13	0	2,73	0
CH_3	2,69	0,56	2,97	0,24
Cl	2,84	0,71	3,40	0,67
CF_3	3,53	1,40	3,98	1,25

III. DISCUSSION

1. Influence du substituant en C_2

A l'examen du tableau III, il est possible de distinguer trois groupes de
substituants, selon le sens de leur effet sur la balance lipophile hydrophile.

a) les substituants à caractère nettement hydrophile, tels

$$SO_2 - N\begin{matrix} CH_3 \\ \\ CH_3 \end{matrix} \quad et \quad C \equiv N.$$

b) les substituants à caractère lipophile, CH_3, Cl, CF_3

c) les substituants qui n'entraînent pas de modification importante :
   $COOCH_3$ et OCH_3.

Cette classification est en accord avec la répartition généralement admise des substituants, basée sur leur effet sur les caractères de solubilité des molécules organiques.

Il n'est pas sans intérêt d'observer que, parmi ces substituants, ce sont ceux dont le caractère, soit lipophile, soit hydrophile, est le plus accentué, qui confèrent à la molécule correspondante la meilleure efficacité neuro-leptique ; les substituants à caractère intermédiaire, tels que $COOCH_3$ et OCH_3, qui ne modifient pas ou peu le coefficient de partage de la promazine, n'en améliorent pas non plus l'activité thérapeutique.

2. Influence de la nature de la chaîne amino alkylée

Le remplacement de la chaîne diméthylaminopropylique par la chaîne méthyl-4 pipérazinopropylique s'accompagne d'une diminution sensible de R_{M_0} ; celle-ci est très nette lors de la substitution par une chaîne hydroxy-4 pipéridino ou hydroxyéthyl-4 pipérazinopropylique.

Cette diminution est parallèle à une amélioration très nette de l'activité neuroleptique, et traduit selon toute vraisemblance une possibilité accrue de formation de ponts hydrogène avec l'eau.

3. Influence de modifications du noyau de la molécule

a. Le remplacement du groupement $\overset{\diagdown}{N}$—CH_2— de la chlorpromazine par le groupement $\overset{\diagdown}{\underset{\diagup}{C}} = CH$— du chlorprothixène, dont l'activité neuro-leptique est à peu près équivalente à celle de la précédente, n'entraîne aucune modification sensible de R_{M_0}.

b. L'oxydation, au niveau du soufre, du noyau de la phénothiazine, qui donne naissance à deux substances dénuées d'activité neuroleptique, l'oxochlorpromazine et l'oxomémazine, s'accompagne d'une très nette diminution de la valeur de R_{M_0}, qui traduit l'augmentation de la polarité du noyau.

Bibliographie

(1) BATE - SMITH, E.C. and WESTALL, R.G. *Biochim. Biophys. Acta,* **4** : 427, 1950.
(2) COLLANDER, R. *Physiol. Plant.,* **7** : 420, 1954.
(3) GEY, K.F. and PLETSCHER, A. *Extrapyramidal system and neuroleptics,* Montréal, J.M. Bordeleau, éd. Ed. psychiatriques.
(4) GEY, K.F. and PLETSCHER, A. *J. Pharmacol. Exptl. Therap.,* **133** : 18, 1961.
(5) GUTH, P.S. *Feder. Proc.,* **21** : 1100, 1962.
(6) HANSH, C., MUIR, R.M., FUJITA, T., MALONEY, P.P., GEIGER, F. and STREICH, M., *J. Amer. Chem. Soc.,* **85** ; 2817, 1963.
(7) HANSCH, C. and FUJITA, T. *J. Amer. Chem. Soc.,* **86** : 8, 1964.
(8) IWASA, J., FUJITA, T. and HANSH, C. *J. Med. Chem.,* **8** : 2, 1965.
(9) KEMULA, W. and BUCHOWSKI, H. *Roczniki Chem.,* **29** : 718, 1955.
(10) MARTIN, A.J.P. *Biochem. Soc. Symp. Camb.,* **3** : 4, 1949.
(11) NATHAN, H.A. *Nature,* **192** : 471, 1961.
(12) QUASTEL, J.H. *Psychopharmacol. Service Center Bull.,* **2** : 55, 1962.
(13) LEERMAN, N.P. and BIALY, H.S. *Biochem. Pharmacol.,* **12** : 1181, 1963.
(14) SPIRTES, M.A. and GUTH, P.S. *Nature,* **190** : 274, 1961.
(15) SPIRTES, M.A. and GUTH, P.S. *Biochem. Pharmacol.,* **12** : 37, 1963.
(16) SU, C. and LEE, C.Y. *Brit. J. Pharmacol.,* **15** : 88, 1960.
(17) FORREST, I.S. and FORREST, F.M. *Clin. Chem.,* **6** : 11, 1960.

Thin-layer chromatography of naturally occurring hydroxyanthrones, hydroxydianthrones and hydroxyanthraquinones

by

R. P. LABADIE *

Summary

A thin layer chromatographic method for the separation of the naturally occurring substances chrysophanol, chrysophanolanthrone, chrysophanoldianthrone, emodin, emodinanthrone, emodindianthrone, physcion, physcionanthrone and physciondianthrone is described. For their separation on a micro scale and the identification of each compound Kieselgel/Kieselgur layers are used. A pyridin reagent is described by which the hydroxyanthrones and the hydroxydianthrones as a group may be distinguished from the hydroxyanthraquinones.

For the separation on a preparative scale a preliminary fractionation of the sample into two parts is made on Kieselgel G layers. The subsequent separation of each fraction into its constituents is performed on Kieselgel/Kieselgur layers.

INTRODUCTION

Investigation of anthraquinone extracts from plant material in many cases shows that anthraquinones are accompanied by corresponding reduced compounds as anthrones and dianthrones (1, 2, 3, 4, 5, 6, 7, 8, 9). Confining our attention to anthrones, anthraquinones and dianthrones with a 1,8 -dihydroxy- and 1,8, 1',8', -dihydroxy substitution pattern in common, we have selected a group of substances derived from three widely occurring natural anthraquinones, namely chrysophanol, emodin and physcion.

* Pharmaceutisch Laboratorium, afd. Pharmacognosie. Hugo de Grootstraat 32, Leiden, Nederland.

The present paper will be concerned with the thin layer chromatography of a sample containing about equal parts of compounds chrysophanol, emodin, physcion, chrysophanolanthrone, emodinanthrone, chrysophanoldianthrone, physcionanthrone, physciondianthrone and emodindianthrone.

By the micro scale separation the substances are identified by their localization, their colours in daylight, in U.V.-light and after having been sprayed with a 5 %-KOH solution in methylalcohol and a newly introduced pyridin/methylalcohol (1 : 1) reagent.

Since anthrones and dianthrones are easily oxidized under normal atmospheric conditions, precautions are required to be able to handle these compounds in preparative scale chromatography.

The method of thin layer chromatography described here, is therefore based on exclusion of daylight and air influences.

EXPERIMENTAL

Sample substances

Chrysophanol, emodin and physcion were isolated from chrysarobin by preparative thin layer chromatography (10).

Chrysophanolanthrone and emodinanthrone were prepared by reduction of the corresponding anthraquinone with stannochloride according to Auterhoff (11).

Physcionanthrone was isolated from Chrysarobin by means of preparative thin layer chromatography.

Chrysophanoldianthrone, emodindianthrone and physciondianthrone were prepared by oxidation of the corresponding anthrones with ferrichloride under nitrogen and in the dark according to Auterhoff (11).

The sample mixture contains about equal weight parts of these nine substances in an ethylacetate solution.

Purification of Kieselgur G (E. Merck)

A slurry was prepared from 200 g Kieselgur G (E. Merck) and 1 L. 36 %-hydrochloric acid and put away overnight. The adsorbent is then filtered off and washed thoroughly with water to remove the acid. After having been dried for three hours at 120-130° C the adsorbent may be used for the preparation of the thin layers.

Preparation of thin layer plates

1. A slurry was prepared from 35 g of a mixture of Kieselgel G (E. Merck), Kieselgur G (E. Merck) (1 : 6) and 70 ml. water and this was applied to glass plates 200 x 200 mm or 50 x 200 mm by means of a Unoplan Leveller (Shandon & Co) with the spreader set at 0.30 mm. The plates were dried at 110° C for one hour and stored over anhydrous silicagel.

2. A slurry was prepared from 100 g Kieselgel G (E. Merck) and 170 ml. water, and this was applied to glass plates 400 x 200 mm., by

means of a Unoplan Leveller (Shandon & Co) with the spreader set at 1.5 mm. The plates were first dried for 30 min. at room temperature, then for one hour at 40° C, and at last at 120° C for one hour and stored over anhydrous silicagel.

3. A slurry was prepared from 100 g of a mixture of Kieselgel G (E. Merck) and purified Kieselgur G (E. Merck) (1 : 6) and 160 ml. water and this was applied to glass plates 400 x 200 mm. by means of a Unoplan Leveller (Shandon & Co) with the spreader set at 1,5 mm. The plates were first dried for 30 min. at room temperature, then for one hour at 40° C and at last at 120° C for one hour, and stored over anhydrous silicagel.

Solvent systems

I. Ethylacetate.

II. Petroleumether (bp. 40-60° C)/ ethylformate/ formic acid (90 : 4 : 1, v/v/v).

III. Petroleumether (bp. 40-60° C)/ ethylformate/ 36 %-hydrochloric acid (85 : 15 : 0,5, v/v/v).

Application of samples

For the separation on a micro scale the solutions of the sample substances are applied in the usual way by means of a micropipet or a capillary tube.

For the separation on a preparative scale, using 1,5 mm thick layered plates, a solution of the sample mixture is applied as a thin band.

The application in band-form is performed by a special adapted apparatus *, by which three operational conditions are made possible.

1. A buret, containing the sample solution, is vertically movable so that its capillary outlet may be adjusted at a suitable hight (ca 1 mm) above the surface of the thin layer.

Furthermore the buret can move automatically back and forth over a desired length (max. 45 cm.).

Independent of this movement, a switching contact controls the mechanism which presses the sample solution out of the buret and prevents it from running.

2. The thin layer plate can be heated to a desired temperature up to about 80° C, in order to accelerate the evaporation of the solvent.

3. In the course of the application the apparatus is covered with a perspex cover and a constant stream of an inert gas (nitrogen) is passing through it.

---

* The author is indebted to Mr. J.v.d. Spek and Mr. B. Janssen who have constructed the apparatus.

Chromatography and elution

The chromatographic separations are carried out in the dark. The chromatographic chamber is saturated with the solvent to be used, by means of lining the walls with filter paper which dips into the solvent.

For the separation on a micro scale a 0.30 mm thick Kieselgel/Kieselgur (1 : 6) layered plate is used. The development takes place with solvent system II over a distance of 15 cm, then subsequently 10 cm with solvent system III. (Fig. 1)

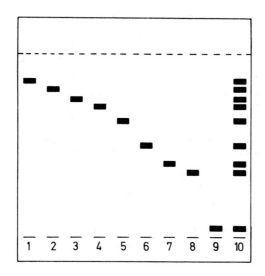

Fig. 1. — Separation of the sample on a 0.30 mm thick Kieselgel G/Kieselgur (1 : 6) layer. Development : 15 cm with solvent system II, and subsequently 10 cm with solvent system III. 1 = chrysophanolanthrone, 2 = chrysophanol, 3 = physcionanthrone, 4 = physcion, 5 = chrysophanoldianthrone, 6 = physciondianthrone, 7 = emodin, 8 = emodinanthrone, 9 = emodindianthrone, 10 = sample mixture.

For the separation on a preparative scale, the sample mixture applied to a 1.5 mm thick Kieselgel G layered 400 x 200 mm plate is developed 15 cm with solvent I. (Fig. 2a)

The bands A and B formed are eluted as quick as possible with ethylacetate in the dark and each elute is concentrated under reduced pressure, the temperature not exceeding 35° C.

Fraction A contains : chrysophanolanthrone, physcionanthrone, chrysophanoldianthrone, physciondianthrone and emodinanthrone.

Fraction B contains : chrysophanol, emodin, physcion and emodindianthrone.

The components of fraction A, applied to a 1.5 mm thick Kieselgel G/Kieselgur (1 : 6) layered 400 x 200 mm plate, are separated by developing with solvent system II over a distance of 15 cm. (Fig. 2b)

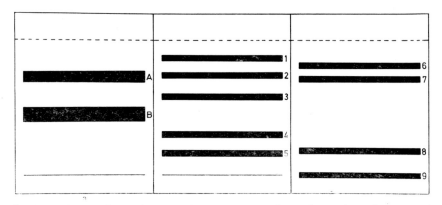

Fig. 2. — Preparative scale separation of the sample. a. Separation of the sample mixture in two bands, A and B, on a 1.5 mm thick Kieselgel G layer. Development: 15 cm with solvent system I. b. Separation of the components of band A, on a 1.5 mm thick Kieselgel G/Kieselgur (1 : 6) layer. Development: 15 cm with solvent system II. 1 = chrysophanolanthrone, 2 = physcionanthrone, 3 = chrysophanol-dianthrone, 4 = physciondianthrone, 5 = emodinanthrone. c. Separation of the components of band B, on a 1.5 mm thick Kieselgel G/Kieselgur (1 : 6) layer. Development: 15 cm with solvent system II. 6 = chrysophanol, 7 = physcion, 8 = emodin, 9 = emodindianthrone.

The components of fraction B, applied also to a 1.5 mm thick Kieselgel G/Kieselgur (1 : 6) layered 400 x 200 mm plate, are separated by developing over a distance of 15 cm with solvent system II. (Fig. 2c)

The components of all fractions are eluted with ethylacetate.

The anthrones and dianthrones are eluted in the dark and concentrated under reduced pressure.

The sample mixture and all fractions are applied as solutions in ethylacetate.

Detection and identification

In addition to their localization, the substances are characterized by their colours which appear on the developed chromatogram.

a. in daylight.

b. in U.V.-light (Philips 57236 E/70 HPW 125 III D6).

c. after having been sprayed with a 5 %-KOH solution in methylalcohol.

d. after having been sprayed with a pyridin/methylalcohol (1 : 1, v/v) mixture and the chromatogram having been heated for 15 min. at 120° C.

The bands on the preparative chromatograms are traced by observation of the still wet chromatogram in U.V.-light for a short time.

RESULTS AND DISCUSSION

For the separation of mixtures of natural occurring hydroxyanthrones, -dianthrones and their corresponding anthraquinones, thin layer chromatography on Kieselgel/Kieselgur layers proved to be a good method. In table 1 some characteristics of each of the substances are collected. In general the hydroxyanthrones and the hydroxydianthrones investigated may be distinguished as a group from the hydroxyanthraquinones. First on wet chromatograms developed with acidic solvent systems, observed in U.V.-light, the hydroxyanthrones and the hydroxydianthrones show a red colour,

TABLE I

	Colour before spraying		Colour after spraying	
Substance	*daylight*	*UV-light*	*KOH*	*Pyridin*
Chrysophanol	yellow-orange	brightly-orange	red-purper	yellow-orange
Physcion	»	»	red	»
Emodin	»	»	red	»
Chrysophanolanthrone	weak-yellow	reddish	yellow → red brown	violet
Physcionanthrone	»	»	»	»
Emodinanthrone	»	»	yellow → blue purper	»
Chrysophanoldianthrone	»	dark-red	yellow	»
Physciondianthrone	»	»	»	»
Emodindianthrone	»	»	»	»

whereas hydroxyanthraquinones are yellow. Secondly the pyridin reagent distinguishes clearly between a member of the reduced derivatives and the hydroxyanthraquinones, by producing a violet colour with the former and a yellow colour with the latter. The precautions taken in the course of the described preparative chromatographic method are necessary to obtain substances which are not partly oxidized and thus pure enough for structural analysis. This applies only to anthrones and dianthrones. With regard to the capacity for the preparative thin layer plates described, the following is of importance :

a) On a 40 x 20 cm, 1,5 mm thick Kieselgel G layered plate an amount of up to 25 mg of the sample mixture can be separated distinctly into the two fractions.

b) On a 40 x 20 cm plate, 1,5 mm thick Kieselgel/Kieselgur (1 : 6) layered plate an amount of up to 10 mg of each fraction can be separated into its constituents.

Literature

(1) STOLL, A., BECKER, B. and KUSSMAUL, W. *Helv. chim. Acta,* **32** : 1892 - 1903, 1949.
(2) STOLL, A., BECKER, B. and HELFENSTEIN, A. *Helv. chim. Acta,* **33** : 313 - 336, 1950.
(3) LEMLI, J. *Pharm. Tijdschr. Belg.,* **39** : 67 - 68, 1962.
(4) LEMLI, J., DEQUEKER, R. and CUVEELE, J. *Pharm. Weekblad,* **98** : 500 - 502, 1963.
(5) LEMLI, J., DEQUEKER, R. and CUVEELE, J. *Planta Med.,* **12** : 107 - 111, 1964.
(6) LEMLI, J. *Lloyda,* **28** ; 63 - 67, 1965.
(7) KINGET, R. *Planta Med.,* **14** : 460 - 464, 1966.
(8) EDER, R. and HAUSER, F. *Arch. Pharm.,* Berl. **263** : 321 - 347, 1925.
(9) EDER, R. and HAUSER, F. *Arch. Pharm.,* Berl. **263** : 436 - 451, 1925.
(10) LABADIE, R. and BAERHEIM SVENDSEN, A., *Pharm. Weekblad,* **102** : 615 - 617, 1967.
(11) AUTERHOFF, H. and SCHERFF, F.C. *Arch. Pharm.,* **293** ; 918 - 925, 1960.

Über die Verwendung von homogenen Azeotropengemischen in der Dünnschichtchromatographie

Dr. Erhard RÖDER

In der Dünnschichtchromatographie hängt die Reproduzierbarkeit der Rf-Werte von zahlreichen Parametern ab. Die wesentlichen Faktoren, die den Rf-Wert beeinflussen, sind das Adsorbens, der Aktivitätsgrad des Adsorbens, die Schichtdicke, der Sättigungsgrad der Kammer, die Arbeitstechnik, Laufstrecke, Substanzmenge, Temperatur und Laufmittel.

Durch systematische Untersuchungen gelang es, für die meisten dieser Parameter Standardbedingungen zu erarbeiten, wodurch die Ergebnisse der Rf-Werte verbessert werden konnten. Als unbefriedigend gelöst gilt nach wie vor das Problem der Laufmittel.

Benutzt man zum Chromatographieren reine Lösungsmittel, dann bleibt die Zusammensetzung der flüssigen Phase und die der Gasphase zwar gleich — wenn man von dem Vorhandensein der Luft absieht — aber leider sind reine Lösungsmittel nur beschränkt anwendbar. Der Grund hierfür liegt darin, dass die Polarität der reinen Lösungsmittel eine konstante physikalische Grösse bildet.

Nur in Lösungsmittelgemischen kann die Polarität und damit die Elutionswirkung variiert und den Bedürfnissen der zu chromatographierenden Substanzen angepasst werden. Die Lösungsmittelgemische sind meist aus Flüssigkeiten mit verschieden hohen Siedepunkten zusammengesetzt. Bei mehrmaligem Gebrauch ändert sich deren Zusammensetzung durch unterschiedlich schnelles Verdunsten der einzelnen Komponenten. Da hierdurch auch die Elutionswirkung des Laufmittels verändert wird, wirkt sich dies nachteilig auf die Konstanz der Rf-Werte aus. Trotz dieser Mängel finden Laufmittelgemische weitverbreitete Anwendung. Gegen die zu starke Veränderung des Laufmittelgemisches hilft man sich dadurch, dass eine Kammerfüllung mit einem Gemisch dieser Art höchstens bis zu drei mal zum Chromatographieren benutzt und der Rest verworfen wird. Die hierbei erzielten Rf-Werte bleiben ungenau und ausserdem ist dieses Verfahren unwirtschaftlich.

Sucht man nach Möglichkeiten, diese Schwierigkeiten zu umgehen, bieten sich homogene azeotrope Lösungsmittelgemische an, die alle Vorzüge der reinen Lösungsmittel in sich vereinen, ohne deren Nachteile zu besitzen. Da von den üblichen Lösungsmitteln zahlreiche Kombinationen als azeotrope Gemische bekannt sind, ist hierdurch eine gewisse Variation in der Polarität gewährleistet. Ihr Hauptvorzug besteht aber darin, dass sie sich in weiten Temperatur- und Druckbereichen wie reine Lösungsmittel verhalten. Dies sei an zwei Beispielen, dem Methanol/Aceton-Gemisch und dem Aethanol/Wasser-Gemisch erläutert (1).

Dort, wo die Gerade des Azeotrops die Siedekurven von Methanol, bzw. von Aceton schneidet, liegen die Grenzen des azeotropen Bereichs. Eine solche Begrenzung ist für das Wasser/Aethanol-Gemisch nach oben überhaupt nicht mehr gegeben. Diese beiden Beispiele geben das Verhalten der meisten azeotropen Gemische wieder und zeigen, dass die azeotropen Gemische in keinem der gezeigten Druck- und Temperaturbereiche ihre Azeotropie verlieren.

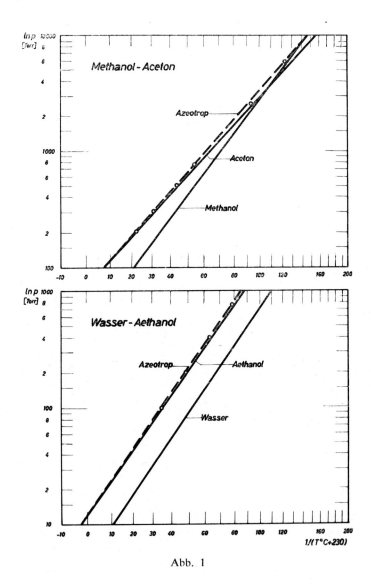

Abb. 1

Bei Anwendung geeigneter azeotroper Gemische als Laufmittel sollten beim Chromatographieren unter Berücksichtigung der schon eingangs erwähnten Paramentern gut reproduzierbare Rf-Werte erhalten werden. Diese Vermutung hat sich bestätigt und hat ihren Niederschlag gefunden bei der Auftrennung einiger Gestagene, Oestrogene, Alkaloide und zahlreicher Sulfonamide.

Hierüber wurde bereits an anderer Stelle berichtet (2), (3). In diesem Referat soll über die Dünnschichtchromatographie einiger Psychopharmaka, Betäubungsmittel und Mutterkornalkaloide berichtet werden.

Nr	Laufmittelgemisch (Gew. prozent)		$Kp \, {}^{760T}_{°C}$	$DK \, {}^{25°}_{E}$	Trenngemisch für :
1	Methanol	(12,0)	55,5	22,05	Psychopharmaka
	Aceton	(88,0)			
2	Äthanol	(96,0)	78,2	25,40	Psychopharmaka
	Wasser	(4,0)			
3	Isopropanol	(16,3)	66,2	5,75	Psychopharmaka
	Isopropyläther	(83,7)			
4	Methanol	(17,8)	50,8	8,35	Psychopharmaka
	Methylacetat	(48,6)			
	Cyclohexan	(33,6)			
5	Methanol	(17,7)	54.0	10,75	Psychopharmaka
	Methylacetat	(82,3)			
6	Chloroform	(87,4)	53,4	9,80	Betäubungsmittel
	Methanol	(12,6)			
7	Methanol	(39,1)	57,5	13,40	Betäubungsmittel
	Benzol	(60,9)			
8	Tetrachlormethan	(12,6)	56,0	19,30	Betäubungsmittel
	Aceton	(87,4)			
9	Äthanol	(31,7)	68,0	7,50	Betäubungsmittel
	Benzol	(68,3)			
10	Dichlormethan	(92,7)	37,8	10,50	Mutterkornalkaloide Betäubungsmittel
	Methanol	(7,3)			
11	Chloroform	(92,0)	59,4	6,05	Mutterkornalkaloide
	Äthanol	(8,0)			
12	Chloroform	(17,0)	79,9	17,30	Mutterkornalkaloide
	2-Butanon	(83,0)			
13	Aceton	(67,5)	53,0	13,75	Mutterkornalkaloide
	Cyclohexan	(32,5)			

Abb. 2

In Anlehnung an die in der Literatur beschriebenen Laufmittel wurden etwa 60 azeotrope Gemische, die eine gute Trennwirkung erwarten liessen, ausgewählt und geprüft. Die Herstellung erfolgte in der Weise, dass die in der Literatur der azeotropen Gemische genannten Anteile in Gewichtsprozenten zusammengegeben und anschliessend destilliert wurden, um eventuelle Verunreinigungen zu entfernen. Nur diejenige Fraktion, die den Siedepunkt des azeotropen Gemischs zeigte, wurde für die Dünnschichtchromatographie verwendet. Als Laufmittel zeigten sich 13 Gemische geeignet. In Abbildung zwei sind ihre Zusammensetzung und einige Eigenschaften zu ersehen. Die Dielektrizitätskonstante ist als Mass für die Polarität mit angegeben.

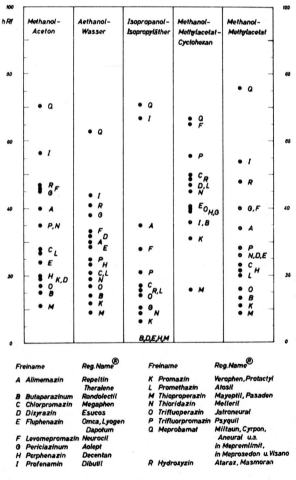

Abb. 3

Abb. 3. — The Rf values of some drugs used in psychopharmacology.

Wegen des Mangels an geeigneten basischen azeotropen Laufmittehn wurde mit natriumkarbonatalkalischen Platten gearbeitet, da es sich bei den vorliegenden Substanzgruppen um Stoffe mit mehr oder weniger basischem Charakter handelt. Bereitet wurden die Platten mit Kieselgel GF$_{254}$ « Merck » und 0,1 normaler Natriumkarbonatlösung auf Glasplatten der Grösse 20 x 20 cm. Die Schichtdicke betrug 0,3 mm. Nach dem Aktivieren bei 105° wurden die Platten bei Zimmertemperatur und einer relativen

Übersicht über die hRf-Werte der Betäubungsmittel

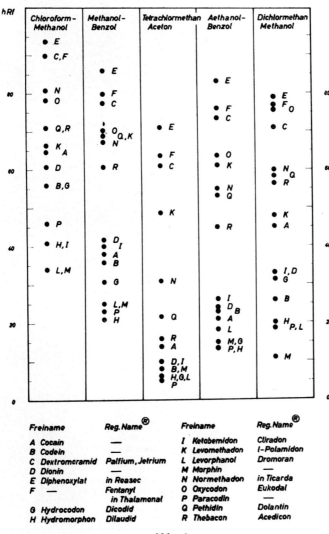

Abb. 4

Abb. 4. — Some examples of the Rf values of anesthetics.

Luftfeuchtigkeit von 40 bis 50 % aufbewahrt und unter diesen Bedingungen in der üblichen Weise zum Chromatographieren verwendet. Die Steighöhe betrug 15 cm. Als Detektionsmittel dienten neben UV-Licht die üblichen Anfärbereagentien. Von den ermittelten Rf-Werten sind in den folgenden Übersichtsbildern die Mittelwerte von 5 Einzelbestimmungen jeder Substanz erfasst und der mittlere Fehler der Einzelwerte errechnet. Er beträgt durchschnittlich 3 % für die angegebenen hRf-Werte. Zu erwähnen ist, dass die Ergebnisse mit nur einer Kammerfüllung erzielt wurden. Es konnten mehrere Platten nacheinander entwickelt werden, bis der Flüssigkeitsspiegel eine untere Markierung, die bei 75 ml lag, erreicht hatte. Dann wurde

Übersicht über die hRf-Werte einiger Mutterkornalkaloide

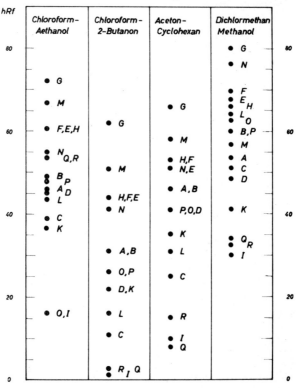

Lysergsäure-Alkaloide
A Dihydroergocornin
B Dihydroergocristin
C Dihydroergotamin
D Erginin
E Ergocornin
F Ergocristin
G Ergocristinin
H Ergocryptin

I Ergometrin
K Ergometrinin
L Ergotamin
M Ergotaminin
N Ergosin
O (LSD)
Clavin-Alkaloide
P Agroclavin
Q Elymoclavin
R Penniclavin

Abb. 5

Abb. 5. — The hRf values of some alkaloids from the ergot of rye.

mit demselben Laufmittel bis zur oberen Markierung, die bei 100 ml lag, aufgefüllt und weiterchromatographiert. Azeotrope Gemische können also beliebig oft benutzt werden und liefern trotzdem konstante Rf-Werte.

Im folgenden Bild (3) sind in einer schematischen Übersicht die Trennmöglichkeiten einiger Psychopharmaka, die bis auf zwei Ausnahmen der Gruppe der Phenothiazine angehören, in fünf Laufmitteln wiedergegeben. Bei systematischer Anwendung dieser Laufmittel ist eine eindeutige Indizierung möglich. Bemerkenswert ist der hohe hRf-Wert von Meprobamat. Er erlaubt eine sichere Bestimmung des Meprobamats in seinen Kombinationspräparaten.

Das nächste Bild (4) zeigt die entsprechenden Ergebnisse für die wichtigsten Betäubungsmittel, das letzte Bild (5) die der Mutterkornalkaloide.

Die Dünnschichtchromatographie mit homogenen azeotropen Gemischen ist überall dort zu empfehlen, wo besonders genaue Rf-Werte angestrebt werden und der Aspekt der Wirtschaftlichkeit berücksichtigt werden muss. Dies ist beispielsweise bei Reihenuntersuchungen der Fall. Da einige moderne Arzneibücher die Dünnschichtchromatographie als Untersuchungsmethode für die in ihnen beschriebenen Substanzen aufgenommen haben, erscheint hier die Dünnschichtchromatographie mit homogenen azeotropen Gemischen besonders geeignet.

Literatur

(1) COX, E., *Ind. Eng. Chem.,* **15** : 592, (1923) R.

(2) RÖDER, E., *Dtsch. Apotheker-Zeitung,* **107** : 1007, (1967). (Laufmittel 2 : Äthylacetat/Cyclohexan = 54/46).

(3) RÖDER, E., MUTSCHLER, E. und ROCHELMEYER, H., *Arch. Pharmaz.,* **301** : 624, (1968).

Caractérisation des dérivés de la phénothiazine après séparation chromatographique sur couche mince. Influence des substituants du noyau sur le Rf de quelques dérivés

par

Jean MEUNIER,
Chaire de Chimie appliquée à la Biologie et aux Expertises
dans l'Armée, Hôpital d'Instruction des Armées du Val de Grâce, Paris 5e, France.

I. Introduction

Les méthodes chromatographiques occupent aujourd'hui une place importante dans les moyens techniques dont dispose l'analyste pour séparer et identifier les molécules médicamenteuses et leurs métabolites. En particulier, la chromatographie sur couche mince d'adsorbant (CCM) a reçu des applications pratiques pour l'étude des grandes familles chimiques ou pharmacologiques. En pharmacie industrielle, la CCM permet d'identifier des molécules médicamenteuses et de contrôler leur pureté. En toxicologie, les avantages de la CCM soulignés récemment par Eliakis et Coutselinis (8) et par Truhaut et ses coll. (26) la font apparaître comme une méthode complémentaire rapide et efficace pour aider à l'identification d'une molécule d'une « série médicamenteuse » après qu'une ou plusieurs réactions chimiques caractérisant une structure fondamentale aient permis de déceler et de classer une molécule dans un grand groupe médicamenteux pour lequel l'homogénéité structurale n'est pas toujours synonyme d'actions pharmacologiques identiques. Il en est ainsi pour les dérivés de la phénothiazine.

Le but de ce travail est, d'une part, de préconiser une nouvelle technique spécifique et sensible de détection des dérivés de la phénothiazine (et de leurs métabolites) sur des chromatogrammes, d'autre part, d'apporter les résultats d'une étude concernant l'influence de divers substituants du noyau phéno-thiazinique sur les valeurs de Rf de quelques dérivés qui les possèdent.

II. Mise en évidence des dérivés de la phénothiazine
sur des Chromatogrammes

Les premières études de séparation des dérivés de la phénothiazine par CCM ont été effectuées par Sunshine (25) et par Noirfalise (21, 22, 23) tandis que Nadeau (20), Eisdorfer (7), Feigl (9), Demoras (6) et leurs coll. avaient précédemment utilisé la chromatographie sur papier pour séparer quelques phénothiazines médicamenteuses. En CCM l'adsorbant le plus fréquemment utilisé est le gel de silice préconisé en particulier par Kiger et Kiger (16) pour la diagnose des dérivés usuels à usage thérapeutique.

La révélation de ces molécules après séparation chromatographique peut s'effectuer par diverses méthodes. L'examen du chromatogramme en lumière ultra-violette, le passage dans une atmosphère de vapeurs d'iode, la pulvérisation du réactif iodoplatinique, du réactif iodobismuthique, d'une solution de chlorure de palladium, conduisent à la mise en évidence de spots sans que ces méthodes utilisent le caractère très oxydable des dérivés phénothiaziniques. D'autres réactifs qui mettent à profit le potentiel d'ionisation bas de la phénothiazine et de ses dérivés conduisent à des colorations spécifiques, sensibles et différentielles. En particulier, l'action d'un accepteur d'électron en milieu très acide engendre des formes radicalaires qui sont le support de colorations utilisables en analyse. C'est ainsi que deux espèces de réactifs sont, depuis plusieurs années, employées pour détecter des spots phénothiaziniques après séparation chromatographique :

— l'acide sulfurique à concentration plus ou moins grande (14, 15, 24) en présence d'un stabilisateur comme l'éthanol (1, 4, 5, 13, 27) ou d'une molécule oxydante comme un sel ferrique (17,26) ;

— les divers réactifs de Forrest et Forrest et notamment le réactif « universel » (12) encore appelé réactif FPN (acide nitrique à 50 % : 50 v, acide perchlorique à 20 % : 45 v, solution de chlorure ferrique à 5 % : 5 v) préconisés primitivement pour la recherche rapide des dérivés de la phénothiazine et de leurs métabolites dans les urines.

Dans le cas de l'acide sulfurique seul, certaines impuretés métalliques ou l'oxygène dissout dans le milieu réactionnel jouent le rôle d'accepteurs d'électron.

D'autres réactifs plus rarement utilisés reposent sur un principe identique : acide phosphorique et acide iodique (2), acide nitrique dilué et sel ferrique (10,24), acide nitreux (11) vapeurs de brome (3).

Ces réactifs oxydants produisent des colorations différentes essentiellement liées à la nature du substituant sur le carbone 3 du noyau phénothiazinique. De tels révélateurs apportent un élément supplémentaire — à côté de l'observation en lumière ultra-violette et de la mesure du Rf — pour identifier une molécule phénothiazinique après séparation chromatographique. Néanmoins, les réactifs précédents possèdent plusieurs inconvénients :

— pulvérisation dangereuse ;

— fugacité des réactions colorées avec certains réactifs ;

— difficulté de conservation de chromatogrammes imprégnés d'acide fixe concentré et en particulier d'acide sulfurique. De plus, l'utilisation du réactif F.P.N. rend les papiers chromatographiques friables après quelques jours selon Demoras et coll. (6) ;

— mesure des Rf délicate.

A ces inconvénients s'ajoutent les effets du pouvoir très oxydant ou de l'excès d'un réactif qui peuvent conduire au dépassement du premier stade coloré intéressant, en provoquant la succession — plus ou moins rapide — d'étapes d'oxydation ou d'oxygénation dont les colorations, lorsqu'elles existent, ne sont plus utilisables pour l'identification.

Pour éviter l'ensemble de ces inconvénients sans perdre en spécificité, en rapidité et en sensibilité, nous proposons l'utilisation d'un accepteur d'électron organique, la benzoparaquinone, et d'une atmosphère de vapeurs chlorhydrique pour mettre en évidence les spots phénothiaziniques. Les vapeurs chlorhydriques, se trouvant au-dessus d'une nappe d'acide chlorhydrique fumant et dans une enceinte close, assurent le milieu acide pour que la benzoparaquinone (que nous avons proposée pour la caractérisation et le dosage des dérivés de la phénothiazine, 18, 19,) engendre les formes radicalaires colorées. Comme nous l'avons montré (18), ces colorations possèdent un spectre d'absorption dans le visible stable, et, de ce fait, la mesure de la longueur d'onde du maximum d'absorption est une valeur caractéristique du dérivé étudié. La valeur de cette longueur d'onde dépend surtout du substituant sur le carbone 3 du noyau phénothiazinique (radical R_1). Si, du fait de l'existence de plusieurs molécules médicamenteuses possédant le même substituant R_1, la mesure de la longueur d'onde du maximum d'absorption de la coloration développée par la benzoparaquinone en milieu acide ne peut suffire à identifier un dérivé, elle permet de le classer dans un groupe λ_{max} PQ très voisines et de limiter le nombre de témoins à déposer sur le support de chromatographie.

Le Tableau I, qui rassemble les structures chimiques de 25 dérivés de la phénothiazine, donne la valeur de la λ_{max} PQ de chacun de ces dérivés.

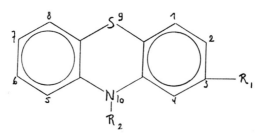

Structure générale d'un dérivé de la phénothiazine.
General structure of a derivate of phenothiazine.

III. Influence de divers substituants du noyau phénothiazinique sur le Rf de quelques dérivés

31-Les séparations chromatographiques ont été effectuées sur couche mince de gel de silice F 254 Merk dans les conditions suivantes :

TABLEAU I.

Structure chimique des dérivés de la phénothiazine utilisés. λ max PQ : valeur de la longueur d'onde du maximum d'absorption de la coloration obtenue pour chaque dérivé avec la benzoparaquinone en milieu acide phosphorique (18).

NOM COMMUN	REFERENCE	R_1	R_2	λ max PQ
méthoxy-3 phénothiazine	4672 RP	OCH_3	H	553 nm
chloro-3 phenothiazine	4688 RP	Cl	H	526
diméthylsulfamoyl-3 phénothiazine	8510 RP	$SO_2N(CH_3)_2$	H	513
cyano-3 phénothiazine	9270 RP	CN	H	514
lévomépromazine	7044 RP	OCH_3	$CH_2\text{-}CH(CH_3)\text{-}CH_2\text{-}N(CH_3)_2$	567
méthopromazine	4632 RP	OCH_3	$CH_2\text{-}CH_2\text{-}CH_2\text{-}N(CH_3)_2$	563
périméthazine	9159RP-1317AN	OCH_3	$CH_2\text{-}CH(CH_3)\text{-}CH_2\text{-}N{<}{>}N\text{-}OH$	568
chlorpromazine	4560 RP	Cl	$CH_2\text{-}CH_2\text{-}CH_2\text{-}N(CH_3)_2$	527
chlorproéthazine	4909 RP	Cl	$CH_2\text{-}CH_2\text{-}CH_2\text{-}N(C_2H_5)_2$	526
pipamazine	Lab. Grémy-Longuet	Cl	$CH_2\text{-}CH_2\text{-}CH_2\text{-}N{<}{>}\text{-}CO\text{-}NH_2$	527
prochlorpérazine	6140 RP	Cl	$CH_2\text{-}CH_2\text{-}CH_2\text{-}N{<}{>}N\text{-}CH_3$	528
perphénazine	Lab. Cétrane	Cl	$CH_2\text{-}CH_2\text{-}CH_2\text{-}N{<}{>}N\text{-}CH_2\text{-}CH_2OH$	527
thiopropazate	Lab. Searle (USA)	Cl	$CH_2\text{-}CH_2\text{-}CH_2\text{-}N{<}{>}N\text{-}CH_2\text{-}CH_2\text{-}O\text{-}CO\text{-}CH_3$	528
promazine	3276 RP	H	$CH_2\text{-}CH_2\text{-}CH_2\text{-}N(CH_3)_2$	512
alimémazine	6549 RP	H	$CH_2\text{-}CH(CH_3)\text{-}CH_2\text{-}N(CH_3)_2$	510
thiazinamium	3554 RP	H	$CH_2\text{-}CH(CH_3)\text{-}N\ (CH_3)_3$ +	516
diéthazine	2987 RP	H	$CH_2\text{-}CH_2\text{-}N(C_2H_5)_2$	513
triflupromazine	Vesprin(ND)USA	CF_3	$CH_2\text{-}CH_2\text{-}CH_2\text{-}N(CH_3)_2$	498
trifluopérazine	7623 RP	CF_3	$CH_2\text{-}CH_2\text{-}CH_2\text{-}N{<}{>}N\text{-}CH_3$	497
thiopropérazine	7843 RP	$SO_2N(CH_3)_2$	idem -	509
butapérazine	Lab. Bayer	$CO\text{-}CH_2\text{-}CH_2\text{-}CH_3$	idem -	517
propiomazine	Lab. Wyeth Byla	$O\text{-}CO\text{-}CH_2\text{-}CH_3$	$CH_2\text{-}CH(CH_3)\text{-}N(CH_3)_2$	517
thioridazine	Lab. Sandoz	$S\text{-}CH_3$	$CH_3\text{-}CH_2\!-\!{<}{>}N\text{-}CH_3$	635
acéprométazine	1664 CB	$CO\text{-}CH_3$	$CH_2\text{-}CH(CH_3)\text{-}N(CH_3)_2$	516
cyamépromazine	7204 RP	CN	$CH_2\text{-}CH(CH_3)\text{-}CH_2\text{-}N(CH_3)_2$	508

— Dépôt de 0,01 micromole de chacun des dérivés ;
— Activation des chromatoplaques 1 heure à 105°C-110°C ;
— Saturation des enceintes durant 12 heures (parois tapissées de papier Arches 304) ;
— Migration du solvant à partir de la ligne des dépôts : 10 centimètres ;
— Température : 20°C.
— Solvant A : Acétone 95, ammoniaque 20 % 5 ;
— Solvant B : Acétate d'éthyle 56, acide acétique 17, éthanol 17, eau 10 ;
— Solvant C : chloroforme 50, méthanol 50 ;
— Séchage des plaques à l'air libre 10 minutes puis à 50°C 30 minutes.
— Révélation : pulvérisation d'une solution de benzoparaquinone 10^{-3}M dans le dichloroéthane ; passage des plaques durant 3 à 4 minutes dans une atmosphère de vapeurs chlorhydriques (au-dessus d'une nappe d'acide chlorhydrique fumant d = 1,18). Sensibilité : 0,1 µg de chlorpromazine.

32-Les dérivés de la phénothiazine utilisés ont été choisis de telle manière à pouvoir étudier l'influence des divers substituants R_1 et R_2 sur les valeurs des R_f mesurées après chromatographie dans les trois systèmes solvants précédents.

a) *Chromatographie 1* : R_1 commun, R_2 différent.

$R_1 = Cl$
- Chloro-3 phénothiazine
- Chlorpromazine
- Chlorproéthazine
- Pipamazine
- Prochlorpérazine
- Perphénazine
- Thiopropazate

$R_1 = OCH_3$
- Lévomépromazine
- Méthopromazine
- Périméthazine
- Méthoxy-3 Phénothiazine

$R_1 = H$
- Promazine
- Alimémazine
- Diéthazine
- Thiazinamium

b) *Chromatographie 2* : R_1 différent, R_2 commun.

chlorpromazine	Cl	diméthylaminopropyl
méthopromazine	OCH_3	»
promazine	H	»
trifluopromazine	CF_3	»

trifluopérazine	CF_3	méthyl-4' pipérazino 1')-3 propyl
Thiopropérazine	$SO_2N(CH_3)_2$	»
Butapérazine	$CO-CH_2-CH_2-CH_3$	»
Prochlorpérazine	Cl	»

c) *Chromatographie 3* : R_1 différent, R_2 commun :

alimémazine	H	diméthylamino-3 méthyl- 2 propyl
lévomépromazine	OCH_3	»
cyamépromazine	CN	»

cyano -3 phénothiazine	CN	H
méthoxy-3 phénothiazine	OCH$_3$	H
chloro-3 phénothiazine	Cl	H
diméthylsulfamoyl -3 phénothiazine	SO$_2$ N(CH$_3$)$_2$	H

Le Tableau II rassemble les valeurs des Rf St de chacun des dérivés dans les trois systèmes solvants utilisés. Le calcul du Rf relatif au Rf d'un étalon (chlorpromazine) permet de neutraliser les effets non négligeables de petites variations dans les conditions de la chromatographie auxquelles les dérivés de la phénothiazine semblent très sensibles. Le calcul du Rf St a été préconisé antérieurement par Noirfalise (21) et Sunshine (25).

TABLEAU II.

Valeur des Rf St de 25 dérivés de la phénothiazine (cf. Tableau I) dans trois systèmes solvants (valeur moyenne de trois mesures).

Nature des systèmes solvants et conditions de chromatographie : cf. § 1 Chap. III).

$$\text{Rf St} = \frac{\text{Rf du dérivé}}{\text{Rf de la chlorpromazine}}$$

	Solvant A	Solvant C	Solvant B
ACEPROMETAZINE	1,07	1,30	0,88
ALIMEMAZINE	1,27	1,15	0,91
BUTAPERAZINE	0,55	1,22	0,77
CHLORPROETHAZINE	1,13	1,27	1,06
CHLORPROMAZINE	1,00	1,00	1,00
CYAMEPROMAZINE	1,50	1,31	0,97
DIETHAZINE	1,21	1,26	0,95
LEVOMEPROMAZINE	1,35	1,19	0,95
METHOPROMAZINE	0,83	0,91	0,88
PERIMETHAZINE	1,26	1,43	1,00
PIPAMAZINE	0,99	1,25	0,98
PERPHENAZINE	0,75	1,35	0,82
PROCHLORPERAZINE	0,45	1,10	0,75
PROMAZINE	0,77	0,76	0,85
PROPIOMAZINE	1,10	1,50	1,02
THIAZINAMIUM	0,04	0,15	0,35
THIOPROPAZATE	1,28	1,77	1,15
THIOPROPERAZINE	0,49	1,05	0,56
THIORIDAZINE	0,87	1,17	1,03
TRIFLUOPERAZINE	0,70	1,16	0,85
TRIFLUOPROMAZINE	1,12	1,06	1,08
Chloro-3 PHENOTHIAZINE	1,83	1,75	Front du solvant
Cyano-3 PHENOTHIAZINE	1,80	1,80	»
Méthoxy-3 PHENOTHIAZINE	1,79	1,75	»
Diméthylsulfamoyl-3 PHENOTHIAZINE	1,77	1,77	»

IV. Discussion et conclusions

Le Tableau II met en évidence le meilleur pouvoir de séparation du solvant A. L'utilisation de deux solvants ne permet pas dans tous les cas d'identifier avec certitude un dérivé phénothiazinique. Par exemple, les solvants A et C n'apportent pas une séparation suffisante des couples. [prochlorpérazine - thiopropérazine], [acéprométazine - chlorproéthazine], [alimémazine - lévopromazine], [trifluopérazine - perphénazine]. Mais dans les quatre cas, la mesure de la longueur d'onde du maximum d'absorption de la coloration développée par la benzoparaquinone en milieu acide phosphorique (18) permet de lever l'incertitude (lorsque cette réaction n'a pas été utilisée pour sélectionner les témoins déposés en même temps que le dérivé à identifier). Dans le cas d'un mélange de deux dérivés dont les λ_{max} PQ respectives seraient trop voisines pour fournir deux maximums nets dans le spectre d'absorption de la coloration, il est alors nécessaire de pratiquer une troisième chromatographie dans un autre système solvant et de confirmer la diagnose par la mesure de la λ_{max} PQ après élution des spots.

Les chromatographies réalisées et le Tableau II montrent :

— que les quatre dérivés substitués différemment en 3, mais sans chaîne latérale fixée sur l'azote 10 possèdent le même R_f dans les trois systèmes solvants utilisés ;

— que pour les dérivés substitués sur l'azote 10, la nature de cette chaîne latérale est responsable en très grande partie du R_f des dérivés qu'ils soient ou non porteurs d'un groupement R_1 sur le carbone 3 ;

— que pour les dérivés possédant une chaîne latérale commune fixée sur l'azote 10, la nature du substituant sur le carbone 3 module l'action de la chaîne liée à l'azote sur la valeur des R_f dans les trois systèmes solvants utilisés.

Remerciements :

Nous remercions vivement la Société des Usines Chimiques Rhône Poulenc, les Laboratoires Bayer, Roger Bellon, Cétrane, Grémy Longuet, Sandoz, Searle, Spécia pour les échantillons des dérivés de la phénothiazine qu'ils ont bien voulu mettre à notre disposition.

Références

(1) BEHN, W., FRAHM, M., FRETWURST, E. Über den diaplacentaren Übergang von Phenothiazinderivaten. *Klinische Wochenschrift*, **34** : 872, 1956.

(2) CALO, A., MARIANI, A., MARIANI-MARELLI, O. Separazione jonoforetica della cloropromazina e sua determinazione colorimetrica. *Estratto dai rendiconti dell'istituto superiore di sanita*, **20** : 802-810, 1957.

(3) CLARKE, V., COLE, E.R. A colour reaction for psychotropic drugs on thin layer chromatograms. *Journ. Chrom.*, **24** : 259-261, 1966.

(4) COCCIA, P.F., WESTERFELD, W.W. The metabolism of chlorpromazine by liver microsomal enzyme systems. *Journ. Pharm. Exp. Therap.*, **157** : 446-458 N° 2, 1967.

(5) COCHIN, J., DALY, J.W. The use of thin layer chromatography for the analysis of drugs. Identification and isolation of phenothiazine tranquilizers and of antihistaminics in body fluids and tissues. *Journ. Pharm. Exp. Therap.*, **139** : 160-165, 1963.

(6) DEMORAS, Y., PARADO, C., DREVON, B. Isolement et caractérisation de quelques phénothiazines médicamenteuses par chromatographie. *Bull. Trav. Soc. Pharm. Lyon,* **7** : 88-94, 1963.

(7) EISDORFER (I.B.) ELLENBOGEN (W.C.) A chromatographic procedure for separating chlorpromazine derivatives. *Journ. Chromat.,* **4** : 329-333, 1960.

(8) ELIAKIS, E.C., COUTSELINIS, A.S. La chromatographie sur couche mince en toxicologie médico-légale. *Ann. Pharm. Franç.,* **25** : 361-364, 1967.

(9) FEIGL, F., HAGUENAUER-CASTRO, D. Two new sensitive spot tests for the detection of thiodiphenylamine. *Anal. Chem.,* **33** : 1412-1413, 1961.

(10) FISHMAN, V., GOLDENBERG, H. Metabolism of chlorpromazine organic extractable fraction from human urine. *Proc. Soc. Exp. Biol. Med.,* **164** : 99-103, 1960.

(11) FLANAGAN, T.L., REYNOLDS, L.W., NOVICK, W., LIN, J., RONDISH, T.H., VAN LOON, E.J. Bilary and urinary excretion patterns of chlorpromazine in the dog. *Journ. Pharm. Sciences,* **51** : 833-836, 1962.

(12) FORREST, F.M., FORREST, I.S., MASON, A.S. Review of rapid urinary tests for phenothiazine and related compounds. *Amer. Journ. Psych.,* **118** : 300-307, 1961.

(13) FRAHM, M., FRETWURST, E., SOEHRING, K. Papierchromatographischer Nachweis einiger Phenothiazinderivate im Harn. *Klinische Wochenschrift,* **34** : 1256-1262, 1956.

(14) GILETTE, J.R., KAMM, J.J. The enzymatic formation of sulfoxydes : the oxydation of chlorpromazine and 4-4'-diaminodiphenylsulfide by guinea pig liver microsomes. *Journ. Pharm. Exp. Therap.,* **130** : 262-267, 1960.

(15) HUANG, C.L., RUSKIN, B.H. Determination of serum chlorpromazine metabolites in psychotic patients. *Journ. Nerv. Ment. Diseases,* **139** : 381-386, 1964.

(16) KIGER, J.L., KIGER, J.G. Procédés de diagnose différentielle rapide et en série des dérivés usuels de la phénothiazine. *Ann. Pharm. Franç.,* **23** : 489-500, N° 7-8, 1965.

(17) KLEINSORGE, H., THALMANN, K., ROSNER, K. Die Ausscheidung des N-Methyl -piperidyl (3) -methyl phenothiazins beim Menschen ; ein Beitrag zur Kolorimetrie der Phenothiazinderivate. *Arzneimittel Forschung,* **9** : 121-123, 1959.

(18) MEUNIER, J. Action de la benzoparaquinone sur les dérivés de la phénothiazine utilisés en thérapeutique. I. Applications qualitatives. *Ann. Pharm. Franç.,* **26** ; 25-33, N° 1, 1968.

(19) MEUNIER, J., VIOSSAT, B. Action de la benzoparaquinone sur les dérivés de la phénothiazine utilisés en thérapeutique. II. Applications quantitatives. *Ann. Pharm. Franç.,* **26** : 429-441, n° 6, 1968.

(20) NADEAU, G., SOBOLEWSKI, G. Identification et dosage des drogues de la famille des phénothiazines par chromatographie sur papier. *Journ. Chrom.,* **2** : 544-546, 1959.

(21) NOIRFALISE, A. Mise en évidence et différentiation des dérivés phénothiaziniques par chromatographie sur couche mince. *Acta Clinica Belgica,* **20** : 273-278, 1956.

(22) NOIRFALISE, A. Mise en évidence de quelques dérivés phénothiazines par chromatographie en couche mince de silicagel. *Journ. Chrom.,* **19** ; 68-74, 1965.

(23) NOIRFALISE, A., GROSJEAN, M.H. Mise en évidence de quelques dérivés de la phénothiazine par chromatographie en couche mince. *Journ. Chrom.,* **16** : 236-237, 1964.

(24) ROBINSON, A. Biotransformations in vitro undergone by phenothiazine derivatives in a liver preparation. *Journ. Pharm. Pharmacol.,* **18** : 19-32, 1968.

(25) SUNSHINE, I. Use of thin layer chromatography in the diagnosis of poisoning. *Amer. Journ. Clin. Pathol.,* **40** : 576-582, 1963.

(26) TRUHAUT, R., BOUDENE, C., CLAUDE, J.R. Aperçus sur les applications de la chromatographie en couche mince (CCM) en analyse toxicologique. *Ann. Biol. Clin.,* **26** : 93-111, 1968.

(27) WECHSLER, M.B., WHARTON, R.N., TANAKA, E., MALITZ, S. Chlorpromazine metabolite pattern in psychotic patients. *Journ. Psych. Research.,* **5** : 327-333, 1967.

Kolorimetrische Bestimmung von Kohlehydraten nach dünnschichtchromatographischer Trennung auf Calciumcarbonat

von

G. STEHLIK

aus dem Institut für Biologie und Landwirtschaft am Reaktorzentrum
Seibersdorf (Leiter : Prof. Dr. K. Kaindl)

Summary

Carbohydrates were separated on thin layers of calcium carbonate, impregnated with sodium acetate and after solution of the layers in hydrochloric acid quantitatively estimated by means of the triphenyl tetrazolium chloride reaction.

Zusammenfassung

Kohlehydrate wurden auf Calciumcarbonatschichten, die mit Natriumacetat imprägniert waren, aufgetrennt und nach Auflösen des Schichtmaterials in Salzsäure quantitativ bestimmt mit Hilfe der Triphenyltetrazoliumchloridreaktion.

* * *

Es wurde versucht, für die dünnschichtchromatographische Auftrennung von Kohlehydraten ein Schichtmaterial zu verwenden, das sich bei einer nachfolgenden quantitativen Auswertung im verwendeten Reagens auflösen läßt, wie dies etwa für Calciumcarbonat im sauren Milieu zutrifft. Dadurch kann der oft zeitraubende Elutionsschritt umgangen werden, besonders wenn dieser Schritt wegen einer nachfolgenden Radioaktivitätsbestimmung quantitativ erfolgen soll.

Qualitative Trennung von Kohlehydraten auf CaCO₃-Schichten :

60 g $CaCO_3$ (« Calciumcarbonat gefällt » von Merck) wurden mit 72 ml Wasser bzw. 0,05 M Natriumacetatlösung zu einem Brei verrührt und

die Dünnschichtplatten (200 x 200 mm) mit der Streichvorrichtung nach STAHL (1) beschichtet (Schichtdicke 0,4 mm). Anschließend wurden die Platten während 30 Minuten bei 100-110° C getrocknet. Mengen von 15-20 µg der einzelnen Kohlehydrate bzw. ihrer Mischungen wurden 2 cm vom unteren Rand entfernt aufgetragen und unter Verwendung folgender Lauf-mittelkombinationen aufgetrennt :

a. Äthylacetat - Aceton - Wasser   4 : 5 : 1
b. Äthylacetat - Dimethylformamid - Wasser   15 : 3 : 1
c. Äthylacetat - Isopropanol - Wasser   65 : 23 : 12
d. Äthylacetat - Pyridin - Wasser   20 : 13 : 5

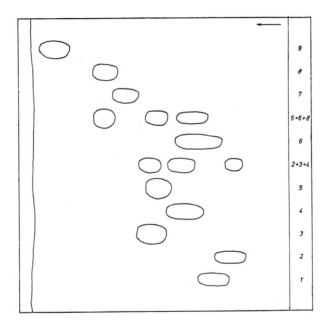

Dünnschichtchromatographische Auftrennung von Kohlehydraten

1. Saccharose   6. D-Galactose
2. Lactose   7. L-Arabinose
3. Fructose   8. D-Xylose
4. Glucose   9. D-Erythrose
5. L-Sorbose

Abb. 1. — Schichten aus reinem CaCO₃. Aufgetrennte Menge an Kohlehydraten 15 µg. Laufmittel : Äthylacetat-Aceton-Wasser (4 : 5 : 1).

Separation by thin layer chromatography of carbohydrates. Layers of pure $CaCO_3$. Quantity separated from carbohydrates 15 µg Solution : Ethyl acetate-acetone-water (4 : 5 : 1).

Die Sichtbarmachung der einzelnen spots erfolgte bei den nicht imprägnierten Schichten durch Besprühen mit einer 0,5 %igen sodaalkalischen Kaliumpermanganatlösung oder durch die Silbernitratreaktion nach TREVELYAN et al. (2), wobei die Platten zuerst mit einer Lösung von AgNO₃ in Aceton und anschließend mit methanolischer KOH besprüht werden. Zur Färbung auf den imprägnierten Schichten diente die Reaktion mit Triphenyltetrazoliumchlorid (3), die in einer modifizierten Form auch zur quantitativen Bestimmung herangezogen wurde.

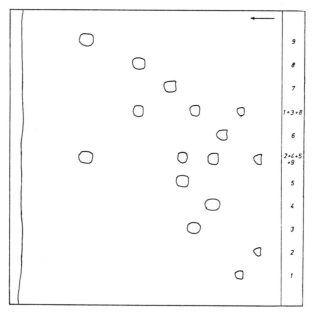

Abb. 2. — CaCO₃ imprägniert mit 0,05 M Natrium-acetat. Aufgetrennte Menge :
20 µg. Laufmittel : Äthylacetat-Pyridin-Wasser (20 : 13 : 5).

CaCO₃ impregnated with 0.05 M sodium acetate. Quaitity separated: 20 µg Solution :
ethyl acetate-pyridine-water (20 : 13 : 5).

Bei Verwendung von reinen CaCO₃-Schichten (Abb. 1) kam es bei der Auftrennung sogar geringer Substanzmengen bereits zu einer starken Vergrößerung der spots sowie zu unerwünschter Streifenbildung bei den Lösungsmittelkombinationen a., b. und c. Beim System d. wanderten die einzelnen Kohlehydrate mit der Lösungsmittelfront mit. Diese Nachteile konnten aber durch die Imprägnierung mit Natriumacetat unterdrückt werden (Abb. 2). Bei der Imprägnierung mit höheren Konzentrationen an Natriumacetat als 0,05 M wurden die Rf-Werte der Kohlehydrate sehr klein.

Die besten Trennergebnisse konnten an Schichten erzielt werden, die mit 0,05 M Natriumacetat imprägniert waren und mit dem Lösungsmittelgemisch Äthylacetat — Pyridin — Wasser (20 : 13 : 5) entwickelt wurden (siehe Abb.2).

Quantitative Bestimmung reduzierender Zucker mit Hilfe der TTC-Reaktion nach Trennung auf imprägnierten CaCO₃-Schichten :

Die quantitative Bestimmung kann grundsätzlich nach zwei Arten erfolgen. Entweder man lokalisiert die aufgetrennten Substanzen mit Hilfe eines Leitchromatogramms, kratzt dann an diesen Positionen die Schichte ab und löst diese im entsprechenden Reagens. Diese Methode besitzt jedoch den Nachteil, daß beim Abschaben Fehler entstehen können, da man ja die Lage der getrennten und zu bestimmenden Substanzen auf dem Chromatogramm nicht genau kennt. Besser ist es, die Platte mit einer geringen Menge Reagens zu besprühen, wodurch die getrennten Substanzen sichtbar werden. Die Schichte kann an dieser Stelle sehr genau abgeschabt und die Farbreaktion im Reagensglas zu Ende geführt werden. Anschließend wird das Trägermaterial durch tropfenweise Zugabe von Salzsäure aufgelöst.

Wir benützten ausschließlich diese zweite Methode. Um größere Quantitäten auftrennen zu können, wurden die Mischungen der Kohlehydrate bandförmig aufgetragen. Als Schichtmaterial diente Calciumcarbonat, das mit 0,05 M Natriumacetat imprägniert worden war. Als Lösungsmittelgemisch wurde Äthylacetat — Pyridin — Wasser (20 : 13 : 5) verwendet.

Nach dem Abdampfen der Lösungsmittel wurde das Chromatogramm zur Sichtbarmachung der Zuckerstreifen mit frisch bereitetem Triphenyltetrazoliumchloridreagens besprüht. Dieses wurde durch Mischen gleicher Volumina einer 1%igen äthanolischen Lösung von Triphenyltetrazoliumchlorid (TTC) und einer 0,25 N wässerigen Kalilauge hergestellt. Nach dem Besprühen wurden die Platten während 10 Minuten in einem Trockenschrank bei 65° C belassen und dann sofort in einer Kühltruhe abgekühlt. Die Zucker waren als rote Streifen sichtbar. Annähernd flächengleiche Stellen wurden abgeschabt, in graduierten Eprouvetten mit 3,5 ml Äthanol aufgeschlämmt und mit 0,2 ml 1%iger äthanolischer TTC-Lösung und 0,2 ml 1 N wässeriger KOH versetzt. Nach guter Durchmischung (Whirlimixer) wurden die Proben im Wasserbad 10 Minuten lang bei 65° C belassen und anschließend sofort mit Eiswasser abgekühlt. Nach Zugabe von 1,0 ml Wasser wurde bis zum Erreichen einer klaren Lösung bei starker Durchmischung (Whirlimixer) tropfenweise HCl conc. zugefügt. Bei einer abgeschabten Fläche von 6 cm² obiger Schichte wurden etwa 0,3 ml (9 Tropfen) benötigt. Ein Überschuß von Säure wurde vermieden. Die klaren Proben wurden dann in den graduierten Eprouvetten mit 96%igem Äthanol auf 10,0 ml aufgefüllt, ihre Extinktion am besten noch innerhalb von 15 Minuten in einem Zeiss PMQ II Spektralphotometer bei 485 nm gegen den Blindwert gemessen und für 1 cm Schichtdicke berechnet.

Da die Bildung des roten Reaktionsproduktes (Triphenylformazan) stark von den äußeren Bedingungen abhängt, ist es ratsam, bei jeder Bestimmung einen Standard mitlaufen zu lassen. Ebenso ist das Formazanbildungsvermögen für die einzelnen Zuckerarten verschieden, deshalb muß für jeden Zucker eine eigene Eichkurve erstellt werden.

Als Beispiel wird die quantitative Bestimmung von Glucose und Fructose nach dünnschichtchromatographischer Trennung auf diesen Schichten angeführt. Zur Erzielung einer guten Trennung soll der Glu-

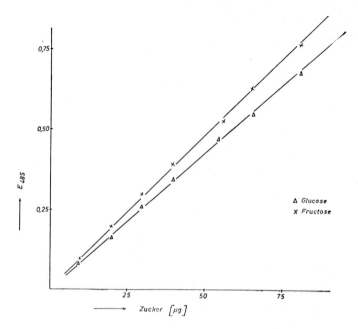

Abb. 3. — Abhängigkeit der Extinktion von der Zuckerkonzentration bei 485 nm
berechnet für 1 cm Schichtdicke (Gesamtvolumen 10,0 ml)

Dependence of the extinction versus the sugar concentration for 485 nm, calculated
for a layer thickness of 1 cm. (total volume 10.0 ml).

cosestreifen mindestens 10 cm Laufstrecke zurücklegen. Dies kann durch
Verwendung von Platten mit dem Ausmaß von 200 x 400 mm oder bei
200 x 200 mm Platten durch Anwendung der aufsteigenden Durchlaufchro-
matographie (4) erreicht werden. Die dabei erhaltenen Resultate ergeben
einen guten linearen Zusammenhang zwischen der Extinktion und der
angesetzten Zuckermenge (Abb. 3). Für eine Bestimmung der Radioaktivität
können aliquote Teile nach Zugabe von Dioxan und Szintillatoren (100 g
Naphthalin, 7 g PPO und 0,3 g POPOP auf 1000 ml Dioxan) in einem
Flüssigkeitsszintillationszähler gemessen werden.

Literatur

(1) STAHL, E.: Handbuch der Dünnschichtchromatographie, Berlin-Göttingen-Hei-
delberg. Springer-Verlag, 1962.
(2) TREVELYAN, E.W., PROCTER, D.P. und HARRISON, J.S. : *Nature*, **166** :
444, 1950.
(3) FISCHER, F.G. und DÖRFEL, H. : *Z. physiol. chem.*, **297** ; 164, 1954.
(4) BANCHER, E., SCHERZ, H. und KAINDL, K. : *Mikrochim. Acta*, 654, 1964.

Etude du baume du Pérou
de diverses origines par chromatographie en couche
mince et en phase gazeuse

A. MONARD et A. GRENIER

Laboratoire de Chimie du Service de Contrôle des Médicaments
de l'Association Pharmaceutique Belge.

INTRODUCTION.

Suivant le plan du recontrôle des spécialités poursuivi par notre laboratoire, nous nous sommes préoccupés de la détermination du baume du Pérou dans diverses formes pharmaceutiques. La chromatographie en phase gazeuse nous paraissait susceptible de résoudre la séparation et le dosage des deux principaux constituants : le benzoate et le cinnamate de benzyle.

Les normes en cinnaméine du baume du Pérou, données par la Pharmacopée Belge V ne pouvaient s'appliquer à cette méthode, la cinnaméine extraite par l'éther en milieu alcalin contenant d'autres substances que les deux esters précités. C'est dans ce but que nous avons comparé les résultats obtenus selon la P.B.V. à ceux fournis par la chromatographie en phase gazeuse, d'où nous avons tiré les normes adéquates pour cette dernière.

PROPRIETES PHYSIQUES ET COMPOSITION.

Le baume du Pérou est le produit de sécrétion du Myroxylon Pereirae, qui croît dans les forêts de San Salvador sur la côte du Pacifique. Il est principalement utilisé en usage externe comme cicatrisant, antiseptique et parasiticide. C'est un liquide sirupeux, brun noirâtre dont les caractéristiques physiques sont les suivantes :

PS 1,138 – 1,158
n_{D25} 1,5880 – 1,5952
I. acide 56 – 84[1]
I. saponification 241 – 287[2]

Certaines données physico-chimiques de la cinnaméine constituent des tests de pureté d'une grande importance :

n_{D25} 1,5695 – 1,5725 (un indice inférieur à 1,565 indique un baume artificiel)
I. saponification 230 – 240

Le baume du Pérou contient 20 à 28 % d'esters résineux dont le constituant principal est l'ester cinnamique du Pérurésinotannol et 50 à 62 % d'esters balsamiques : benzoate et cinnamate de benzyle. Il renferme également 8 à 10 % d'acides libres : acides cinnamique et benzoique provenant de l'hydrolyse des esters. Comme constituants mineurs, on peut noter l'alcool benzylique et le Péruviol, alcool sesquiterpénique à odeur de miel. On rencontre aussi des traces de vanilline et de coumarine ([3]).

ETUDE DU BAUME DU PEROU.

I. - Méthodes Chimiques.

Nous disposions de 9 échantillons de baume dont nous avons déterminé les constantes physico-chimiques et la teneur en cinnaméine selon la P.B.V.

Pour la détermination de l'indice d'acide, nous avons utilisé le Potentiographe E 436 Metrohm. Le mode opératoire est le suivant : 500 mg de baume sont dissous dans 25 ml de diméthyl formamide. La solution est titrée par le tétrabutylammonium hydroxyde N/10 dans le mélange isopropanol/méthanol (Merck).

Le tableau I donne les résultats obtenus. Les chiffres tombant en dehors des normes ont été soulignés.

En examinant les données du tableau, on constate que le lot n° 6 présente des anomalies pouvant indiquer une falsification.

Nous avons alors effectué sur ce lot quelques essais chimiques qui ont été tous négatifs quant à la présence de baume artificiel ou de constituants comme la térébenthine, le colophane, les huiles végétales.

TABLEAU I.

Caractéristiques physico-chimiques des 9 échantillons de Baume du Pérou.

Echantillons	1	2	3	4	5	6	7	8	9
$n_{D\,25}$ Baume du Pérou	1,5880	1,5895	1,5915	1,5895	1,5890	**1,578**	1,5915	1,5900	1,5900
$n_{D\,25}$ Cinnaméine	1,5695	1,5705	1,5695	1,571	1,5725	**1,5595**	1,5710	1,5720	1,5715
Indice d'acide	54	—	—	—	—	81	—	76	81
Teneur en cinnaméine	50,5%	57,5%	**47%**	60%	57%	61%	52,5%	52,5%	50,5%
Indice de saponification de la Cinnaméine	238	253	235	233	237	**225**	234	235	236

II. - Chromatographie en couche mince.

Nous avons poursuivi notre étude des 9 échantillons en effectuant la séparation de leurs constituants par chromatographie sur plaques de Kieselgel G.

Dimension des plaques : 20 cm × 20 cm.
Epaisseur du support : 0,6 mm.
Solvant : Benzène – Méthanol 95/5[4]
0,1 ml de chaque solution à 10 % dans le chloroforme est déposé à 2 cm d'intervalle.

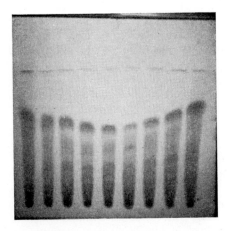

Fig 1. — Chromatogramme sur plaque de Kieselgel des 9 échantillons de baume du Pérou après pulvérisation du chlorure d'antimoine, observé à la lumière ordinaire.

Chromatogram on Kieselgel plate of the 9 samples of Peruvian balsam after pulverization of the chloride of antimony, observed by normal light.

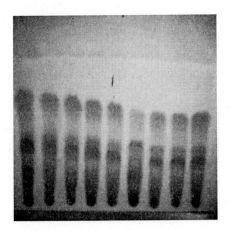

Fig. 2. -- Même chromatogramme observé à 350 mμ.

Same chromatogram observed at 350 mμ.

Après développement, la plaque est révélée par une solution de chlorure d'antimoine après un chauffage de 10 minutes à 100°. Les figures 1 et 2 montrent les chromatogrammes observés à la lumière ordinaire et à 350 mμ.

L'examen de la plaque ne décèle pas de différences notables entre la plupart des échantillons. Les lots 1 et 6 font exception. Sur le chromatogramme du premier, on observe quelques taches supplémentaires, tandis que le second se distingue plus nettement des autres échantillons.

III. - Chromatographie en phase gazeuse.

Le chromatographe Pye 104 modèle 26 est équipé de deux détecteurs à ionisation par flamme. L'enregistreur Sargent est muni d'un intégrateur Disc. Les colonnes en verre, de 1,5 m de long et 1/4″ de diamètre extérieur, sont remplies par du chromosorb G (AW-DMCS) 80–100 mesh, celui-ci est imprégné de 1,7 % de butanediol succinate.

Le débit du gaz vecteur (N_2) est de 38 ml/minute. Les injections sont effectuées avec une seringue Hamilton de 10 μl munie d'une aiguille de 11 cm de longueur pour déposer le soluté (0,4 μl) au sommet de la colonne.

A. Analyse qualitative.

Nous avons utilisé une température de 210° pour la séparation des différents constituants, qui ont été identifiés par la mesure de leur temps de rétention, à l'exception des pics 2 et 3 dont nous ne possédions pas d'échantillons.

Les chromatogrammes des 9 lots sont très semblables entre eux. On peut seulement noter des variations quantitatives ainsi que la présence d'un pic supplémentaire (pic n° 3) dans le chromatogramme du lot 1 (figures 3 et 4).

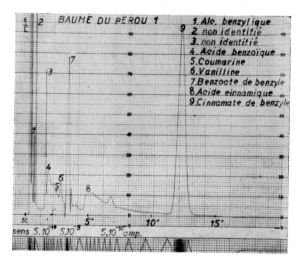

Fig. 3. — Chromatogramme du lot 1.

Chromatogram of batch 1.

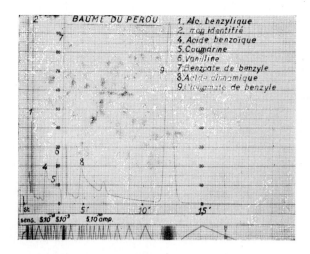

Fig. 4. — Chromatogramme du lot 2.

Chromatogram of batch 2.

TABLEAU II.

x_1	x_3	Doses calculées par :		x_2	x_4	Doses calculées par :	
Doses mises benzoate de benzyle (BB)	Sur-face en mm²	éq. 1	éq. 2	Doses mises cinnamate de benzyle (CB)	Sur-face en mm²	éq. 3	éq. 4
20,5 mg	284,–	20,65	20,47	7 mg	99,9	7,09	7,13
	281,5	20,47	20,28		98,5	6,97	7,03
	287,–	20,80	20,68		97,3	6,92	6,94
	282,–	20,48	20,32		97,37	6,91	6,95
30 mg	422,–	30,18	30,46	10 mg	137,7	9,95	9,87
	425,–	30,39	30,68		141,–	10,18	10,11
	416,5	29.93	30,06		143,–	10,29	10,26
	404,–	29,02	29,16		137,–	9,29	9,82
40 mg	560,–	40,39	40,45	15 mg	211,6	15,25	15,23
	552,7	39,87	39,92		209,–	15,06	15,05
	557,6	40,14	40,28		206,7	14,91	14,88
	553,–	39,80	39,95		204,–	14,72	14,68
45 mg	621,6	45,13	44,91	18 mg	255,–	18,32	18,38
	615,–	44,59	44,43		248,5	17,86	17,91
	620,–	44,89	44,80		247,7	17,82	17,85
	625,–	45,23	45,16		248,4	17,88	17,90

éq. 1; BB $= 0,0641\ x_3 + 0,0186\ x_4 + 0,57$ éq. 3 ; CB $= 0,0023\ x_3 + 0,0676\ x_4 + 0,32$
éq. 2; BB $= 0,0724\ x_3$ $\qquad — 0,0997$ éq. 4 ; CB $= \qquad 0,0726\ \ x_4 — 0,1189$

B. *Analyse quantitative.*

Une température de 220⁰ est appliquée à la colonne pour le dosage des deux esters balsamiques en présence de benzocaïne comme étalon interne.

Les temps de rétention du benzoate de benzyle, de la benzocaïne et du cinnamate de benzyle sont respectivement de 2,2′, 3′ et 7,8′.

Quatre mélanges de benzoate de benzyle et de cinnamate de benzyle à doses différentes ont été chromatographiés chacun quatre fois. La quantité de benzocaïne était la même dans chaque solution c.à.d. 60 mg. A partir des mesures de surfaces, deux formes d'équation ont été établies et les résultats obtenus sont repris dans le Tableau II. Pour le benzoate de benzyle, les coefficients de variation $\dfrac{(100s)}{x}$ sont respectivement

de 1,17 % pour la régression classique
et 1,09 % pour la régression multiple.

Pour le cinnamate de benzyle, ces coefficients sont respectivement de 1,52 % pour la régression classique et de 1,54 % pour la régression multiple. Les équations sont développées dans l'appendice qui suit cet article.

A 100 mg de chaque échantillon de baume du Pérou dissous dans 3 ml de chloroforme, 60 mg de benzocaïne ont été ajoutés.

La quantité de chaque ester ainsi que la somme des deux ont été calculées et sont exprimées en mg par gramme de baume (tableau III).

Si on compare le chiffre ainsi obtenu à la teneur en cinnaméine déterminée selon la P.B.V., on remarque que cette dernière contient environ 90 % d'esters balsamiques.

On peut donc conclure que la teneur en esters benzyliques doit atteindre au moins 45 % et au plus 55 % du poids total du baume.

Le rapport du benzoate au cinnamate de benzyle semble une valeur intéressante pour détecter une falsification par le benzoate de benzyle dans lequel cas il serait plus élevé. Ce rapport varie de 1,5 à 3,13 exception faite pour le lot 6 où il atteint 5,5.

Conclusions :

Si les tests physico-chimiques gardent toute leur valeur, seules les méthodes chromatographiques apportent des renseignements importants sur la composition normale du baume. La chromatographie en couche mince est la technique la plus appropriée pour déceler les falsifications, tandis que la chromatographie gazeuse permet une analyse quantitative précise des principaux constituants.

TABLEAU III.

Teneur en benzoate et cinnamate de benzyle des 9 échantillons,
exprimée en mg/gr de baume.

Echantillons	1	2	3	4	5	6	7	8	9
Benzoate de benzyle	324,7	421	294	318	320	435	322	351	342
Cinnamate de benzyle	134	136	134	200	198	80	157	112	114,5
Total	458,7	557	**428**	518	518	515	479	463	456,5
Benzoate benzyl / Cinnamate benzyl	2,40	3,10	2,20	1,59	1,62	**5,45**	2,05	3,13	3,00

DOSAGE DU BAUME DU PEROU DANS LES SPECIALITES PHARMACEUTIQUES.

La méthode chromatographique décrite ci-dessus a été appliquée au dosage des deux esters benzyliques du baume du Pérou dans diverses formes pharmaceutiques. Dans le but de contrôler l'extraction des principes actifs et rechercher l'interférence éventuelle d'autres constituants, nous avons reconstitué les préparations suivantes :

1. Comprimés :

Ac.gum. 6 mg — Ac.borique 186 mg — Amylum 193 mg — Saccharose 67 mg — Ess.thym 24,7 mg — Ess.romarin 34,4 mg — Ess. lavande 13 mg — Ess. bergamotte 2,7 mg — Baume du Pérou 53,4 mg — Bicarbonate Na 1,410 gr pour 1 comprimé.

2. Pommade I. :

Benzophénol 250 mg — Thymol 250 mg — Camphre 250 mg — Menthol 250 mg — Bithiol 500 mg — Baume du Pérou 725 mg — Oxyde Zn 5 gr — Amylum 5 gr — Vaseline q.s. pour 50 gr.

3. Pommade II. :

Oxyde Zn 24 gr — Baume du Pérou 549 mg — Cire blanche 500 mg — Huile de ricin 20,872 gr — Adeps lanae 5,500 gr.

Procédé 1.

Une quantité de poudre correspondant à 50 mg de baume du Pérou est agitée plusieurs fois avec 30 ml de chloroforme. Après filtration, on ajoute 30 mg de benzocaïne. La solution chloroformique est alors évaporée sous vide jusqu'à 3 ml.

Procédé 2.

Dans le cas des onguents, la meilleure méthode s'est révélée l'extraction par du méthanol en utilisant un agitateur magnétique chauffant. L'extraction est répétée 4 fois et la solution méthanolique est décantée après refroidissement sous un courant d'eau froide.

Les solutions d'extraction réunies, auxquelles on a ajouté préalablement la benzocaïne, sont refroidies dans la neige carbonique pour précipiter la majeure partie de la vaseline dissoute dans le méthanol. Après filtration, on évapore jusqu'au volume approprié.

La vaseline présente dans la solution injectée provoque une traînée sévère après le solvant, ce qui entraîne une erreur plus ou moins importante dans la mesure du pic du benzoate de benzyle et de la benzocaïne. La précipitation de la vaseline par la neige carbonique améliore la ligne de base.

Résultats :

Dans les trois préparations, nous avons retrouvé 97 % du benzoate de benzyle et 102 % du cinnamate de benzyle ajoutés.

Appendice sur le calcul des résultats

par

R. BONTEMPS et J. PARMENTIER

Quatre mélanges de benzoate de benzyle et de cinnamate de benzyle à doses différentes ont été chromatographiés chacun quatre fois. Les surfaces sont exprimées en mm^2.

TABLEAU I.

x_1 Doses mises benzoate de benzyle (BB)	x_3 Surface en mm^2	Doses calculées par : éq. 1	éq. 2	x_2 Doses mises cinnamate de benzyle (CB)	x_4 Surface en mm^2	Doses calculées par : éq. 3	éq. 4
20,5 mg	284,–	20,65	20,47	7,- mg	99,9	7,09	7,13
	281,5	20,47	20,28		98,5	6,97	7,03
	287,–	20,80	20,68		97,3	6,92	6,94
	282,–	20,48	20,32		97,37	6,91	6,95
30,- mg	422,–	30,18	30,46	10,- mg	137,7	9,95	9,87
	425,–	30,39	30,68		141,–	10,18	10,11
	416,5	29,93	30,06		143,–	10,29	10,26
	404,–	29,02	29,16		137,–	9,29	9,82
40,- mg	560,–	40,39	40,45	15,- mg	211,6	15,25	15,23
	552,7	39,87	39,92		209,–	15,06	15,05
	557,6	40,14	40,28		206,7	14,91	14,88
	553,–	39,80	39,95		204,–	14,72	14,68
45,- mg	621,6	45,13	44,91	18,- mg	255,–	18,32	18,38
	615,–	44,59	44,43		248,5	17,86	17,91
	620,–	44,89	44,80		247,7	17,82	17,85
	625,–	45,23	45,16		248,4	17,88	17,90

éq. 1 ; BB $= 0,0641 \, x_3 + 0,0186 \, x_4 + 0,57$ éq. 3 ; CB $= 0,0023 \, x_3 + 0,0676 \, x_4 + 0,32$
éq. 2 ; BB $= 0,0724 \, x_3 \qquad\qquad - 0,0997$ éq. 4 ; CB $= \qquad\quad 0,0726 \, x_4 - 0,1189$

A. Etalonnage par corrélation multiple.

Nous nous sommes demandé si le résultat du dosage de chaque constituant (benzoate de benzyle ou cinnamate de benzyle) n'était pas influencé par la présence de l'autre constituant.

Pour étudier cet effet, nous avons calculé la corrélation multiple entre ce constituant et les surfaces appariées.

Dans ce cas, l'équation de régression pour trois variables est :

$$x_i = b_{ij \cdot k} \, x_j + b_{ik \cdot j} \, x_k$$

dans laquelle les coefficients de régression b sont inconnus et les x donnés par les n expériences ont leurs moyennes nulles et leurs variances

$$s^2 = \frac{1}{n} \, \Sigma \, x^2$$

La matrice symétrique de corrélation pour les trois variables est :

$$\left\|\begin{array}{ccc} 1 & r_{ij} & r_{ik} \\ r_{ji} & 1 & r_{jk} \\ r_{ki} & r_{kj} & 1 \end{array}\right\|$$

A chaque corrélation correspond un cofacteur C indexé de la même façon ; le déterminant est noté $|C|$

Le coefficient de corrélation multiple entre les x_i d'une part et les x_j, r_k d'autre part, est par définition la racine positive de $R^2_{i(jk)}$ avec $o \leqslant R^2 \leqslant 1$.

Nous avons : $\quad 1 - R^2_{i(jk)} = s^2_{i \cdot jk} / s^2_i \quad (1)$
ou $1 - R^2_{i(jk)} = |C| / C_{ii} \quad (2)$

La comparaison entre (1) et (2), nous donne :

$$s^2_{i \cdot JK} = s \, \frac{|C|}{C_{ii}}$$

Il est évident que $R^2_{i(j)} = r^2_{ij}$

Les coefficients de régression sont estimés par les formules :

$$b_{ij \cdot k} = - \, \frac{s_i}{s_j} \, . \, \frac{C_{ij}}{C_{ii}} \quad \text{et} \quad b_{ik \cdot j} = - \, \frac{s_i}{s_k} \, . \, \frac{C_{ik}}{C_{ii}}$$

L'analyse de la variance peut dès lors s'établir comme suit :

Source de variation	Somme des carrés	Degrés Liberté	Carrés moyens
Régression : n var. E $(x_1/x_j, x_k)$ Résidu : n var. $(x_i - E(x_i x_j, x_k))$ ou variance erreur	$ns^2_i \, R^2_{i(jk)}$ $ns^2_i (1 - R^2_{i(jk)})$	p—1 n—p	$ns^2_{1 \cdot jk}/(n-p) = s^2_0$
Variation totale : n var. x_1	ns^2_i	n—1	

p est le nombre de variables et s_0, l'erreur standard.

Le t-test pour les coefficients de régression (pour p=3) se réduit à :
1) pour $b_{ij.k}$, $t = b_{ij.k} \, s_{j.k} / s_0$ où $s^2_{j.k} = s^2_j \, C_{ii}$
2) pour $b_{ik.j}$, $t = b_{ik.j} \, s_{k.j} / s_0$ où $s^2_{k.j} = s^2_k \, C_{ii}$

Dans l'étude présente, nous avons attribué aux indices les significations ci-après :

i $= 1$, concerne les concentrations en benzoate de benzyle
j $= 2$, concerne les concentrations en cinnamate de benzyle
k $= 3$, concerne les surfaces correspondant aux concentrations en benzoate de benzyle.
m $= 4$, concerne les surfaces correspondant aux concentrations en cinnamate de benzyle.

Les x_1, x_2, x_3, x_4 sont consignés dans le tableau I.

D'autre part,

$$s^2_i = \frac{1}{n} \left(\Sigma \, x^2_i - \bar{x}_i \, \Sigma \, x_i \right) \text{ et } \bar{x}_i = \frac{1}{n} \, \Sigma \, x_i$$

Donc,

$s^2_1 = 1{,}420{,}75 / 16$; $s^2_2 = 292 / 16$; $s^2_3 = 270.532{,}38 / 16$; $s^2_4 = 55.369{,}04/16$

$\bar{x}_1 = 33{,}875$; $\bar{x}_2 = 12{,}5$; $\bar{x}_3 = 469{,}18$; $\bar{x}_4 = 173{,}92$.

TABLEAU II

Etalonnages de :	benzoate de benzyle $E(x_1|x_3,x_4)$	cinnamate de benzyle $E(x_2|x_3,x_4)$
Régression	$x_1 - \bar{x}_1 = b_{13.4} (x_3 - \bar{x}_3)$ $+ b_{14.3} (x_4 - \bar{x}_4)$	$x_2 - \bar{x}_2 = b_{23.4} (x_3 - \bar{x}_3)$ $+ b_{24.3} (x_4 - \bar{x}_4)$
Matrice de corrélations	1 0,99923 0,99105 0,99923 1 0,98935 0,99105 0,98935 1	1 0,98994 0,99913 0,98994 1 0,98935 0,99913 0,98935 1
R^2	$R^2_{1(34)} = 0{,}99875$	$R^2_{2(34)} = 0{,}99836$
Coefficients b	$b_{13.4} = 0{,}0641$ $b_{14.3} = 0{,}0186$	$b_{23.4} = 0{,}0023$ $b_{24.3} = 0{,}0676$
analyse de la variance : régression : résidu s_0^2 : Total :	$16 \, s^2_1 R^2_{1(34)}/2 = 709{,}4870$ $16 \, s^2_{1.34}/13 \quad = \quad 0{,}1366$ $16 \, s^2_1 \quad /15 \quad = \quad 94{,}7167$	$16 \, s^2_1 R^2_{2(34)}/2 = 145{,}7605$ $16 \, s^2_{2.34}/13 \quad = \quad 0{,}0368$ $16 \, s^2_2 \quad /15 \quad = \quad 19{,}4667$
t-test des coefficients	$b_{13.4}$; $t = 13{,}12 >_{13}t_{0,05} = 2{,}16$ $b_{14.3}$; $t = 1{,}72 <_{13}t_{0,05} = 2{,}16$	$b_{23.4}$; $t = 0{,}88 <_{13}t_{0,05} = 2{,}16$ $b_{24.3}$; $t = 12{,}06 >_{13}t_{0,05} = 2{,}16$

Les limites fiduciaires d'un résultat au seuil 0,05 sont
pour le benzoate de benzyle : $x_1 \pm 0{,}798$
pour le cinnamate de benzyle : $x_2 \pm 0{,}415$

Les équations de régression du tableau II, peuvent s'écrire :
pour le benzoate de benzyle :

$$x_1 = 0,0641\, x_3 + 0,0186\, x_4 + (\bar{x}_1 - 0,0641\, \bar{x}_3 - 0,0186\, \bar{x}_4)$$

ou $x_1 = 0,0641\, x_3 + 0,0186\, x_4 + 0,57,$

et pour le cinnamate de benzyle :

$$x_2 = 0,0023\, x_3 + 0,0676\, x_4 + (\bar{x}_2 - 0,0023\, \bar{x}_3 - 0,0676\, \bar{x}_4)$$

ou $x_2 = 0,0023\, x_3 + 0,0676\, x_4 + 0,32.$

Sur la base de ces essais, nous n'avons pu démontrer la signification des paramètres d'affinement liés aux résultats du second corps en présence dans le mélange. Il s'agit en l'occurence, des coefficients de régression $b_{14.3}$ d'une part, et $b_{23.4}$ d'autre part.

***

B. Etalonnage par régression individuelle.

Habituellement, cette hypothèse se traduit par l'équation de la forme :

$$x = b\,L + c$$

où x est la concentration du corps et L la lecture à un appareil de mesure.

Cependant, dans le cadre de cet exposé, il nous a semblé préférable de considérer ce genre de régression comme un cas particulier de l'étalonnage par corrélation multiple.

TABLEAU III.

Etalonnages de :	benzoate de benzyle : $\dot{E}(x_1\, x_3)$	cinnamate de benzyle : $E(x_2\, x_4)$
Régression	$x_1 - \bar{x}_1 = b_{13}.\,(x_3 - \bar{x}_3)$	$x_2 - \bar{x}_2 \times b_{24}.\,(x_4 - \bar{x}_4)$
Matrice de corrélations	1 0,99923 0,99923 1	1 0,99913 0,99913 1
R^2	$R^2_{1(3)} = r^2_{13} = 0,99923$	$R^2_{2(4)} = r^2_{24} = 0,99913$
Coefficients b	$b_{13}. = 0,0724$	$b_{24}. = 0,0726$
analyses de la variance : Régression Résidu s^2_0 Total	$16\, s^2_1 R^2_{1(3)}/1 = 1.418,5680$ $16\, s^2_{1.3}/14 = 0,15585$ $16\, s^2_1/15 = 94,7167$	$16\, s^2_2 R^2_{2(4)}/1 = 291,49$ $16\, s^2_{2.4}/14 = 0,036295$ $16\, s^2_2/15 = 19,4667$

Les limites fiduciaires d'un résultat au seuil 0,05 sont :
pour le benzoate de benzyle : $x_1 \pm 0,847$ et
pour le cinnamate de benzyle : $x_2 \pm 0,409$

Les équations de régression peuvent aussi s'écrire sous la forme :
pour le benzoate de benzyle :

$$x_1 = 0,0724\, x_3 + (\bar{x}_1 - 0,0724\, \bar{x}_3)$$

ou $x_1 = 0,0724\, x_3 - 0,0997$

et pour le cinnamate de benzyle :

$$x_2 = 0,0726\, x_4 + (\bar{x}_2 - 0,0726\, \bar{x}_4)$$

ou $x_2 = 0,0726\, x_4 - 0,1189$

C. Conclusion.

Dans le tableau IV, nous avons comparé, au seuil 0,05, les résidus des analyses de la variance découlant des deux systèmes de régression :

TABLEAU IV.

benzoate de benzyle	cinnamate de benzyle
$F = \dfrac{16\,s^2_{1\cdot3}\,/14}{16\,s^2_{1\cdot34}/13} = \dfrac{0,15585}{0,1366}$	$F = \dfrac{16\,s^2_{2\cdot34}/13}{16\,s^2_{2\cdot4}\,/14} = \dfrac{0,0368}{0,036295}$
$= 1,141\ \ F_{(14,13)}$	$= 1,015\ \ F_{(13,14)}$

Sur la base de ce F-test, on ne peut conclure à une différence significative entre l'un ou l'autre mode de calcul.

Cependant, nous estimons que pour l'estimation des résultats de l'analyse de mélanges par chromatographie gazeuse, notamment, une investigation ultérieure s'impose. La méthode utilisant la corrélation multiple entre résultats associés permettrait de déceler éventuellement une équation d'étalonnage qui réduirait le résidu de l'analyse de la variance.

Bibliographie.

1. D.C. GARRATT - *The quantitative analysis of Drugs,* p. 509.
2. A. DENOEL - *Matière médicale végétale* (Pharmacognosie) p. 645, Ed. 1958.
3. KARSTEN, WEBER et STAHL - *Lehrbuch der Pharmakognosie* (9. Auflage), p. 579.
4. H. JORK - *Chromatographie - Symposium* II, Bruxelles 1962, p. 213.
5. M.G. KENDALL et A. STUART : « *The Advanced Theory of Statistics* », vol. **2** : 334, 1961.

Nouvelle perspective d'identification des vins d'hybrides blancs par chromatographie sur couches minces

par

Michel BOURZEIX et Marine MARINOV *
avec la collaboration technique de Nicolas HEREDIA
Station Centrale de Technologie des Produits Végétaux
Institut National de la Recherche Agronomique
Narbonne - Aude - France

Nous désignons par « *hybrides* » les cépages issus du croisement de deux *espèces* du genre Vitis, la dénomination « *métis* » étant réservée aux cépages qui, à l'intérieur d'une même espèce proviennent du croisement de deux *variétés*.

Les cépages qui existaient en Europe avant la dévastation du vignoble par le *phylloxera,* au XIXe siècle, appartenaient uniquement à l'espèce de Vitis Vinifera ; c'est d'ailleurs pour cela qu'on désigne parfois les cépages Vitis Vinifera, par l'expression « cépages européens ».

Aujourd'hui les cépages européens constituent la grande majorité du vignoble actuel. Les cépages hybrides qui avaient été créés à cause de leur résistance au phylloxera ont été rejetés presque partout par suite du goût désagréable de leurs raisins. Cependant, à cause de leur rusticité certains de ces hybrides ont subsisté.

Dans plusieurs pays, la *législation* interdit la commercialisation des vins d'hybrides ou leur retire tout droit à une appellation d'origine. Les règles relatives à l'exportation des vins sont plus draconiennes encore.

Le problème de la différenciation entre *vins rouges* issus de cépages européens (Vitis Vinifera) et ceux qui proviennent, même partiellement, de cépages hybrides est pratiquement résolu. En effet, la plupart des vins d'hybrides rouges contiennent des quantités importantes d'anthocyanes diglucosides, colorants rouges qui sont, soit absents, soit seulement à l'état de traces dans les vins de Vitis Vinifera.

---

* Stagiaire de l'Institut Supérieur des Industries Alimentaires, PLODIV — Bulgarie.

Cette différenciation entre hybrides et Vitis Vinifera s'est avérée beaucoup plus difficile dans le cas des *vins blancs* qui, naturellement, ne contiennent pas d'anthocyanes. L'utilisation de techniques modernes comme la *chromatographie en phase gazeuse* n'a pas permis de déceler la ou les substances responsables de l'arôme particulier aux vins d'hybrides blancs. Quant à l'analyse sensorielle de cet arôme elle n'a pas dégagé de caractéristiques exploitables ; ni d'ailleurs l'étude des pigments.

A l'heure actuelle, les oenologues n'ont toujours pas à leur disposition de *méthode pratique* d'identification des vins d'hybrides blancs dont ils ont pourtant un grand besoin.

En définitive le problème est de savoir s'il existe des constituants à la fois *spécifique* des hybrides blancs et faciles à mettre en évidence.

Nos recherches dans cette direction ont porté sur huit vins d'hybrides blancs et dix vins de Vitis Vinifera blancs.

Les résultats qui vont être exposés sont relatifs uniquement à ces dix-huit échantillons. Par ailleurs c'est la première fois que nous en faisons état. Ils feront l'objet d'un mémoire plus détaillé que cet exposé, qui paraîtra vraisemblablement en France dans les Annales de Technologie Agricole de l'Institut National de la Recherche Agronomique.

Sur les *huit vins d'hybrides,* six nous avaient été adressés par le laboratoire départemental d'analyses et de recherches de Tours. Il s'agit des vins des cépages suivants :

   — 1 — Ravat 6 (1967)
   — 2 — Seibel 11803 (1967)
   — 3 — Joannes Seyve 32.55 (1967)
   — 4 — Seyve-Villard 52.76 (1967)
   — 5 — Seibel 11803 (1967)
   — 6 — Ravat 6 — (1962)

Les trois premiers échantillons provenaient de vignes situées en Touraine, sur des terrains argilo sablonneux ; les trois derniers de vignes cultivées sur des terrains sablonneux du département de la Vienne.

Les deux autres vins d'hybrides avaient été envoyés par la Station Expérimentale I.N.R.A. de Cours-les-Cosnes, dans la Nièvre. Il s'agit des vins de :

   — Seibel 49.86 (1966)
   — Ravat 6 (1966)

Cinq des *dix vins* de Vitis Vinifera blancs expérimentés provenaient de la collection de l'Ecole Nationale Supérieure d'Agronomie de Montpellier. Il s'agit des vins de :

   — 1 — Clairette blanche (1967)
   — 2 — Chardonnay (1967)
   — 3 — Grenache blanc (1967)
   — 4 — Terret blanc (1967)
   — 5 — Ugni blanc (1967)

Les cinq autres avaient été mis à notre disposition par le Conseil Interprofessionnel des vins des Corbières et du Minervois à Lezignan, Aude. Il s'agit des vins de :

— 1 — Macabeu (1967)
— 2 — Marsanne (1967)
— 3 — Picpoul (1967)
— 4 — Sauvignon (1967)
— 5 — Picpoul (50 %) et Sauvignon (50 %) (1967)

Il s'agit donc de vins de *cépages sûrs,* cultivés dans des régions et des terrains différents.

Tous ces échantillons d'hybrides et de Vitis Vinifera blancs, étaient issus d'une *vinification « en blanc »,* c'est-à-dire sans macération des parties solides de la grappe comme cela se fait presque toujours dans le cas des vins blancs.

Pour tenter de différencier ces vins, nous avons utilisé la *chromatographie ascendante bidimensionnelle* sur couches minces de cellulose. De cette façon nous avons mis en évidence un composé présent dans les huit vins d'hybrides blancs étudiés, mais absent dans les dix vins de Vitis Vinifera blancs examinés parallèlement.

Nous avons employé des *plaques* carrées de 20 cm de côté, recouvertes d'une couche de poudre de cellulose de marque Macherey und Nagel M.N. 300 de 0,5 mm d'épaisseur.

Selon le mode opératoire établi, on *dépose* 200 microlitres de vin tel quel dans le coin du bas à gauche, de telle manière que le sens du premier développement soit le sens de l'étalement de la couche (sens de machine).

La *tâche* de départ peut atteindre 1 à 1,5 cm de rayon sans inconvénient pour la séparation ultérieure ; de sorte qu'on peut déposer le vin en deux affusions continues de 100 microlitres et sécher chacune d'elles par un courant d'air à température ambiante.

Avant le développement dans la première direction, on *humidifie* en plaçant dans l'enceinte chromatographique deux capsules à forme basse, contenant chacune quelques dizaines de millilitres d'eau distillée à 70° C. Les plaques sont disposées de façon à ce que les couches minces de cellulose soient du côté des capsules.

Précisons que les *enceintes* utilisées ont pour dimensions intérieures : 21 cm x 10,5 cm pour la base et 21 cm de hauteur.

Après au moins une demi-heure d'humidification préalable, on verse dans l'enceinte 150 millilitres d'une *solution décinormale de soude.* La migration jusqu'au bord supérieur de la plaque demande environ deux heures.

On *sèche* alors la couche à l'aide d'un courant d'air à température ambiante durant une heure au moins.

Le développement dans la *direction perpendiculaire* n'exige pas d'humidification préalable, de sorte qu'on peut mettre quatre plaques par enceinte, au lieu de deux pour le premier développement.

Pour ce second développement, on utilise 150 ml du mélange miscible, fraîchement préparé, d'alcool amylique primaire, d'acide acétique, et d'eau, dans les proportions 2 volumes, 1 volume, 1 volume.

La migration est arrêtée lorsque le front du solvant arrive à 2 cm du bord supérieur de la plaque.

Elle demande 4 heures et demie environ.

Le *séchage* s'effectue comme précédemment.

On peut laisser les plaques à température ambiante pendant la nuit, avant de révéler.

On *révèle* en pulvérisant sur chaque plaque :

— d'abord 5 ml d'une solution de 120 g de chlorure d'aluminium (Cl3 Al, 6 H_2O) par litre de méthanol.

— Puis, aussitôt après, 5 ml d'une solution saturée de carbonate de soude dans l'eau.

L'*examen des plaques* à la lumière de Wood (Lampe Mazda) peut se faire aussitôt après la révélation sans avoir besoin de sécher.

Résultats

Plusieurs constituants du vin apparaissent sous forme de *taches fluorescentes* jaunes, bleues, vertes, etc. Mais la comparaison des plaques montre qu'il existe dans les vins d'hybrides blancs un constituant responsable d'une *fluorescence bleue,* assez intense, qui est absent dans les dix vins de Vitis Vinifera blancs étudiés parallèlement.

Cette fluorescence bleue est bien rassemblée en forme de V dont l'axe est parallèle au sens du premier développement. Pour les huit vins d'hybrides blancs, elle se situe à l'intérieur d'une circonférence de 1,5 cm de rayon dont le centre a pour coordonnées : x = 4,2 cm, y = 9,4 cm, dans le système d'axes rectangulaires dont l'origine est le centre de la tache de départ.

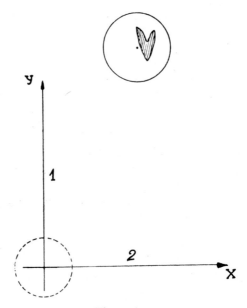

Figure 1.

On *détecte* plus facilement cette fluorescence en recouvrant la plaque d'un carton noir mat de 20 cm de côté, percé d'une fenêtre délimitée par cette circonférence.

La *nature du cépage* hybride n'a pas d'influence sensible sur l'intensité de cette fluorescence bleue.

Par ailleurs le vin de Ravat 6 de *1962* donne une fluorescence bleue un peu moins intense que celle du vin de Ravat 6 de *1967* mais toujours facilement décelable.

Si on *mélange* un vin d'hybride *blanc* à un vin de Vitis Vinifera *blanc* on détecte la fluorescence bleue si l'hybride représente au moins 15 % du coupage.

Dans le cas d'un *mélange* de vin d'hybride *blanc* avec des vins *rouges* de cépages de Vitis Vinifera parmi les plus répandus dans le Midi de la France, tels que ceux d'Aramon, de Carignan noir, de Cinsault, de Grenache noir, d'Alicante-Bouschet, on est gêné par la présence dans ces vins rouges d'un constituant qui donne une fluorescence d'un bleu un peu différent, mais située dans le voisinage immédiat de la fluorescence bleue caractéristique des huit vins d'hybrides blancs étudiés.

Actuellement nous cherchons à résoudre ce problème de la détection des vins d'hybrides blancs mélangés aux vins rouges de cépages européens en modifiant le système de solvants chromatographiques.

En conclusion

La chromatographie en couches minces de cellulose nous a permis de mettre en évidence une tache *fluorescente bleue* qui existe seulement dans *les huit vins d'hybrides blancs* étudiés et qui est due à un constituant encore non identifié.

Le *mode opératoire* utilisé est simple et ne demande aucun matériel onéreux. Il est donc susceptible de constituer une *méthode pratique* de détection des vins d'hybrides blancs, purs ou en mélange.

Mais il convient à présent d'examiner les vins des *autres cépages blancs,* hybrides et de Vitis Vinifera ; c'est pourquoi nous avons tenu à faire part de nos résultats, à tous les laboratoires intéressés, avant la période des vendanges. Nous tenons à remercier la Société Belge des sciences pharmaceutiques de nous avoir permis de venir faire cet exposé.

La composition de la partie glucidique de la Sénégine, saponoside de Polygala Senega

par

W. VAN DEN BOSSCHE et le Prof. Dr R. RUYSSEN
Université de Gand, Institut de Pharmacie.
Laboratoire de Chimie Médicale et Chimie Physique Biologique

Les données sur les propriétés des saponines sont souvent incertaines car la plupart de ces produits naturels ont été obtenus sous forme impure ou mal définie.

Nous avons réalisé l'extraction et la purification totale de la sénégine, saponine de la racine de Polygala Senega, dont la génine est un dérivé triterpénoïdique pentacyclique avec deux fonctions carboxyliques (1).

Nous en avons étudié la composition chimique et les propriétés physico-chimiques qui en solution aqueuse se rattachent à celles des tensides (1).

La sénégine est extraite de la poudre de racine de Polygala Senega par le méthanol et précipité par l'éther. Le traitement par charbon actif et l'adsorption sélective des flavonoïdes sur l'oxyde d'alumine ne donnent pas un saponoside pur. La séparation complète des impuretés non hémolytiques est réalisée par chromatographie sur colonne de gel de silice à

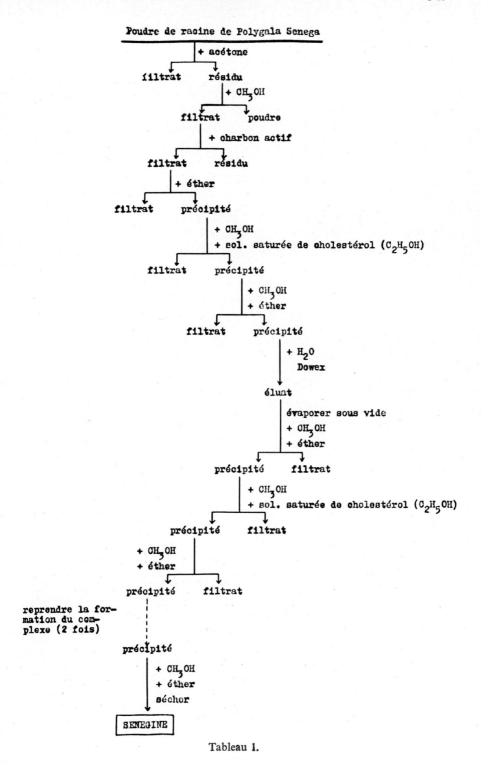

Tableau 1.

partir d'un mélange de benzène, isopentanol, méthanol et eau (8-8-11-2) et élution par méthanol pur. La solution aqueuse est déminéralisée par filtration sur colonne échangeur d'ions Dowex (50-W-X-2).

Toutes ces méthodes de purification ont été nettement simplifiées et rendues plus efficaces par la précipitation de la sénégine à l'état de complexe moléculaire avec le cholestérol.

La solubilité du saponoside limite la zone du milieu de précipitation, qui, après de nombreux essais à blanc a été réalisée à partir de deux volumes de solution saturée de cholestérol dans l'éthanol et d'un volume d'une solution de 7 % de sénégine dans le méthanol.

Le complexe est peu soluble dans l'alcool méthylique (2,5 g par 100 ml). Un traitement à l'éther permet la séparation quantitative de la saponine après élimination de l'excès de cholestérol par lavage. Le tableau I indique les opérations successives de purification.

Les saponines possèdent une structure amphipathique. La partie hydrophile ou polysaccharidique de la molécule se compose de monosaccharides dont le nombre et la nature sont très différents.

Les données concernant la composition de la partie glucidique de la sénégine sont assez confuses. On fait mention de l'identification de glucose, arabinose, une méthylpentose non spécifiée ou bien de glucose, xylose, rhamnose et fucose (2) (3). Les méthodes de séparation dont on disposait, n'ont pas toujours permis de résoudre ce problème. La purification préalable et la séparation complète de toute autre substance végétale est absolument nécessaire.

L'hydrolyse

Les recherches de la composition de la partie glucidique nous imposent le choix d'une méthode d'hydrolyse efficace. On emploie de préférence un acide minéral comme l'acide chlorhydrique, l'acide sulfurique, ou l'acide perchlorique. Les différentes méthodes d'hydrolyse ont été comparées. Une solution aqueuse du saponoside, acidifiée par l'acide chlorhydrique est chauffée pendant quelques heures sur bain marie. La sénégénine se précipite.

On filtre et le filtrat est neutralisé en ajoutant un excès de carbonate d'argent. On filtre de nouveau et le filtrat est saturé de H_2S, pour éliminer les ions d'argent. Cette méthode ne peut pas être recommandée pour les saponines car le sulfure d'argent ne précipite pas facilement à cause d'une diminution de la tension superficielle par la sénégine et la sénégénine (4). Dans le cas ou l'on emploie l'acide perchlorique, la neutralisation se fait par une solution d'hydroxyde potassique (5).

Concernant l'hydrolyse, nous préférons l'emploi de l'autoclave à 120° C en milieu d'acide sulfurique, car elle permet une concentration en acide plus faible et aboutit à une sénégénine, ainsi qu'un liquide surnageant, beaucoup moins colorés. La sénégine a été hydrolysée en vase clos dans l'autoclave à 120° C, respectivement dans 1 % et 2 % d'acide sulfurique. Le filtrat est neutralisé par le carbonate de baryum. Cette solution filtrée sert à l'examen chromatographique.

Chromatographie sur papier

Après de multiples essais d'orientation, nous avons choisi la méthode chromatographique descendante sur papier. Pour tous les essais, nous avons utilisé des feuilles de papier Whatman n° 4 (4) (5) (6) (7) (8).

Parmi les différents systèmes solvants essayés, nous avons retenu les deux phases suivantes :

$$
A : \begin{cases} \text{butanol-1 :} & 4 \\ CH_3COOH : & 1 \\ H_2O : & 5 \end{cases}
\quad \text{et} \quad
B : \begin{cases} \text{Acetate d'Ethyle :} & 3 \\ CH_3COOH : & 1 \\ H_2O : & 3 \end{cases}
$$

La révélation est pratiquée par pulvérisation d'une solution à base de benzidine, suivie d'un chauffage de 5 minutes à l'étuve à 100° C.

$$
\begin{cases} \text{Benzidine :} & 1 \text{ g} \\ CH_3COOH : & 10 \text{ ml} \\ C_2H_5OH : & 160 \text{ ml} \\ \text{Solution de } CCl_3COOH \text{ à } 40\% = 20 \text{ ml.} \end{cases}
$$

Les phases essayées sont mentionnées dans le tableau II avec les Rf auxquels elles ont donné lieu.

TABLEAU II

	Rf Phase A	Rf Phase B
Galactose	0,30	0,25
Glucose	0,33	0,27
Arabinose	0,40	0,44
Xylose	0,42	0,53
Ribose	0,47	0,70
Rhamnose	0,55	0,79

Le chromatogramme, développé dans la phase A (butanol-l-acide acétique glacial-eau : 4-1-5) présente quatre taches différentes (fig. 1).

La substance 1 correspond au glucose, galactose ou un mélange des deux monoses ; la substance 2 au xylose ou arabinose, et la substance 4 au rhamnose. Quant à la substance 3, il s'agit probablement d'un sucre, qui n'a pas été employé comme substance de référence.

La phase B (acétate d'éthyle-acide acétique glacial-eau : 3-1-3), retenue pour la chromatographie, nous donne un peu plus d'informations. La séparation est plus nette. On trouve de nouveau les quatre substances, dont la première et la troisième ne permettent aucune conclusion. Les taches 2 et 4 correspondent respectivement au xylose et rhamnose.

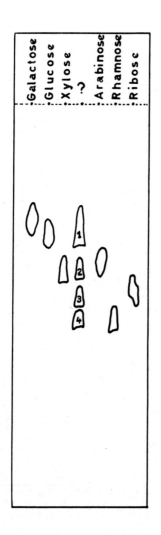

Phase A.

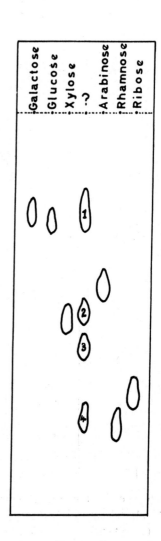

Phase B.

Aucun des systèmes chromatographiques ne permet une conclusion. Si nous chromatographions l'hydrolysat dans des autres phases, comme une solution aqueuse de collidine, nous obtenons la même image chromatographique. La séparation entre le glucose et le galactose n'est pas assez nette.

Chromatographie en couche mince

L'emploi de la chromatographie en couche mince peut résoudre ce problème.

Cette chromatographie, décrite par Waldi (9), est réalisée sur des chromatoplaques, préparées avec du « Kieselguhr-Merck » imprégné d'un tampon pH 5 (phosphate).

On utilise la phase : butanol-1 — acétone — tampon pH 5 (phosphate) : 4-5-1.

La révélation se fait par pulvérisation d'une solution d'acide sulfurique concentré, d'acide acétique cristallisable et d'anisaldehyde : (2-97-1), suivie d'un chauffage de 5 minutes à l'étuve à 120° C.

Nous avons identifié la présence de cinq monoses : glucose — galactose — rhamnose — xylose et fucose.

La séparation nette entre le glucose et le galactose, a démontré que la première substance de la chromatographie sur papier correspondait à un mélange des deux monoses, tandis que la substance non identifiée (substance 3), était le fucose.

En variant le temps d'hydrolyse de 30 minutes à 4 heures dans un milieu d'acide sulfurique de 1 et 2 %, on obtient toujours la même image chromatographique.

Le Professeur R. Ruyssen a bien voulu assumer la direction de ces travaux.

Bibliographie

(1) VAN DEN BOSSCHE W. ; Thèse pour l'obtention du grade de Docteur en Sciences Pharmaceutiques - Gand, 1968.
(2) KLEIN G. ; Handbuch der Pflanzenanalyse - Dritter Band, Zweiter Teil, Zweite Hälfte, Spezielle Analyse II. Wien, Verlag von Julius Springer, 1932.
(3) MOERMAN E. *Pharm. Tijdschr. Belgïë,* **39** ; 143, 1962.
(4) CZYSZEWSKA S. *Biuletyn Instytutu Roslin Leczniczych,* **9** ; 12, 1963.
(5) MOËS A. *J. Pharm. Belg.* **48** ; 347, 1966.
(6) SANDBERG F., AHLENIUS B. and THORSEN R. *Svensk. Farm. Tidskr.* **62** ; 541, 1958.
(7) CHANLEY J.D., LEDEEN R., WAX J., NIGRELLE R.F. and SOBOTKA H. *J. Am. Chem. Soc.* **81** ; 5180, 1959.
(8) HOTON-DORGE M. *J. Pharm. Belg.* **45** ; 3, 1963.
(9) WALDI D. *J. Chromatog.,* **18** ; 417, 1965.

Quelques applications de l'analyse immunoélectrophorétique à des médicaments d'origine animale

J. DONY, B. BEYS, A. RAPPE, H. MUYLDERMANS

INTRODUCTION

Il est actuellement acquis que les méthodes immunoélectrophorétiques ont apporté et apportent encore une importante contribution à l'analyse des protéines.

Initialement elles ont surtout été largement utilisées en analyse médicale pour l'essai des constituants des liquides biologiques et tissulaires mais, depuis quelques années, leur intérêt pour l'analyse des médicaments d'origine protéique s'est progressivement affirmé.

Nous avons déjà rapporté les résultats obtenus en nos laboratoires avec des préparations opothérapiques complexes et des enzymes incomplètement purifiés (1, 2, 3).

Nous exposerons ici succinctement * à titre d'exemple, les résultats de l'analyse de trois types de préparations d'origine animale à l'usage thérapeutique, purifiées à des degrés divers :

1. - une poudre de pancréas (pancréatine).
2. - un extrait d'estomac de porc (préparation de facteur intrinsèque).
3. - des préparations hormonales (lobe postérieur d'hypophyse et gonadotrophines).

La signification de ces essais sera discutée sous le double aspect de l'identification et de la pureté dans le cadre du contrôle des médicaments.

II. — MATERIEL ET METHODES.

B.A. — MATERIEL.

1. - Antigènes

— *Poudre de pancréas (pancréatine)* préparée suivant les normes du Codex français 7e éd. 1949.

---

* Pour répondre au désir des organisateurs nous nous limiterons à donner dans chacun des trois cas l'essentiel des résultats. Ils seront complétés par les auteurs dans des publications concernant spécifiquement chacun des trois sujets envisagés.

— *Poudre d'estomac (Facteurs intrinsèques)*
1. préparation N.F. : fournie par les services du National Formulary U.S.A. ;
2. une préparation commerciale (2727).

— *Préparations hormonales*
1. Poudre de lobe postérieur d'hypophyse de porc.
   a) Extraits aqueux ;
   b) Extrait acétique préparé suivant les normes de la BP 1963, p. 1132 ;
   c) une préparation commerciale.
2. Gonadotrophines chorionique et sérique ; pour chacune d'elles : un échantillon provenant de l'O.M.S. et une préparation commerciale.

2. - Immunserums :

Les immunsérums ont été préparés chez le lapin par injections sous-cutanées répétées de chacun des antigènes correspondants.

Nous disposions des immunsérums suivants :

immunsérum anti-pancréatine
immunsérum anti-trypsine
immunsérum anti-chymotrypsine
immunsérum anti-facteur intrinsèque N.F.
immunsérum anti-facteur intrinsèque — préparation commerciale 2727
immunsérum anti-lobe postérieur d'hypophyse de porc
    (extrait acétique)
immunsérum anti-gonadotrophine chorionique
immunsérum anti-gonadotrophine sérique
immunsérum anti-serum de porc
immunsérum anti-serum humain
immunsérum anti-serum de cheval.

Nous avons aussi utilisé les mêmes immunsérums après absorption par le sérum et dans le cas de préparations d'origine tissulaire par des extraits d'organes autres que celui ayant servi à la préparation de l'immunsérum considéré (extrait mixte d'organe).

B. — METHODES.

Pour l'analyse électrophorétique et immunoélectrophorétique nous nous sommes référés aux techniques décrites par Scheidegger (6). Sur les lames est coulé un gel d'agar ou d'agarose à 1 % en tampon véronal à pH 8,2 et de force ionique 0,05.

Les antigènes placés sur la couche de gélose de 1 à 2 mm d'épaisseur sont soumis à une différence de potentiel de 3 ou 6 volts/cm pendant une heure.

Pour les caractérisations chimiques et enzymatiques nous nous sommes référés aux techniques décrites par Grabar, Burtin (4) Uriel (5) (6) (7) et Wieme (8).

Lors de l'analyse immunoélectrophorétique, les plaques sont laissées 48 H. à température ordinaire, puis lavées pendant 3 jours avec du sérum

physiologique. Ce lavage terminé, les plaques sont soumises à la recherche
des activités enzymatiques et traitées au noir amide pour colorer les protéines.

I. — PANCREATINE :

1. - Electrophorèse :

Dans nos conditions opératoires, l'électrophorèse de la pancréatine
permet de séparer 7 sommets protéiniques et de mettre en évidence diverses
activités enzymatiques comme l'activité protéolytique (caséine), estérasique
(acétate de β naphtyl), amylolytique (amidon), lipolytique (huile d'olive) et
trypsique. (Benzoylarginine naphtylamide.)

Comme le montre la première figure, les activités protéolytique et
estérasique sont très intenses et liées à plusieurs constituants.

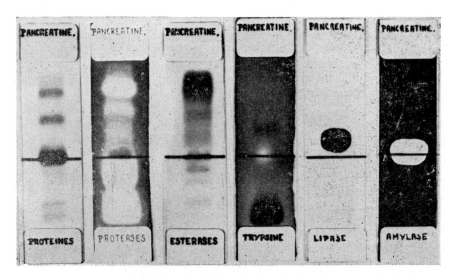

Fig. 1. — Images électrophorétiques de la *pancréatine* :

    — Protéines.
    — Activité protéolytique sur la caséine.
    — Activité estérasique sur l'acétate de β-naphtyl.
    — Activité trypsique.
    — Activité lipolytique.
    — Activité amylolytique.

        Electrophoretic separation of *pancreatin* :

    — Proteins
    — Proteolytic activity on casein
    — Esterasic activity on the acetate of β-naphtyl
    — Trypsic activity
    — Lipolytic activity
    — Amylolytic activity

L'activité trypsique est liée principalement à un sommet situé dans la zone cathodique, cependant une faible activité apparaît également dans la zone anodique lorsque nous travaillons à une concentration élevée (75 mg/ml).

Les autres activités sont liées à un seul sommet ; dans le cas de l'activité amylolytique il est situé sur le réservoir même tandis que l'activité lipolytique se marque du côté anodique.

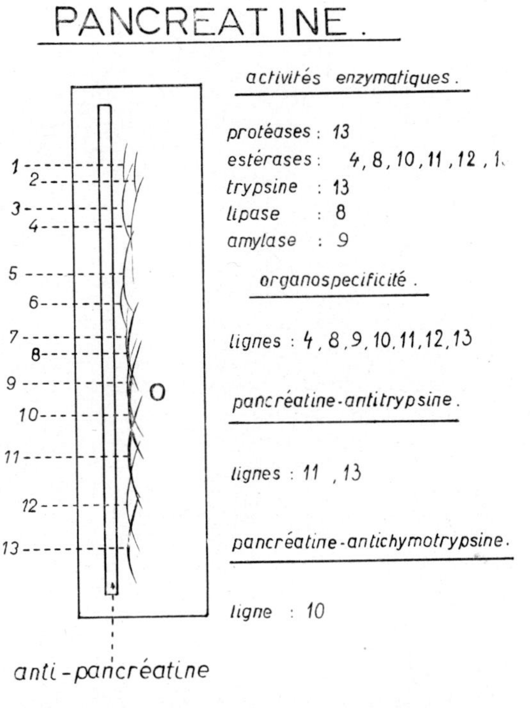

Fig. 2. — Image immunoélectrophorétique de la *pancréatine*.

Immunoelectrophoresis of *pancreatin*.

2. - **Analyse immunoélectrophorétique :** (voir fig. nº 2)

L'analyse immunoélectrophorétique de la pancréatine (à 25 mg/ml) à l'aide de son immunsérum révèle au moins 13 lignes de précipitation dont plusieurs possèdent une activité enzymatique. Ainsi 6 de ces arcs présentent une activité estérasique, 1 arc présente une activité amylolytique, un autre présente à la fois l'activité trypsique et protéolytique.

Il est intéressant de noter que, comme on pouvait s'y attendre, la pancréatine réagit également avec les immunsérums antitrypsine et antichymotrypsine. Ainsi, le premier donne 2 arcs dans la zone anodique, dont un présente une activité trypsique tandis que le second immunsérum fournit un arc près du réservoir avec activité chymotrypsique.

Lorsque l'analyse immunoélectrophorétique est réalisée avec l'immunsérum antipancréatine absorbé par du sérum de porc et un extrait mixte d'organes, nous obtenons encore 7 lignes de précipitation ; 2 de ces lignes correspondent aux arcs obtenus avec l'immunsérum antitrypsine et une autre à l'arc obtenu avec l'immunsérum antichymotrypsine.

En conclusion, l'électrophorèse de l'échantillon de pancréatine analysé permet la mise en évidence de différents types d'activités enzymatiques, caractéristiques de cette poudre.

L'analyse immunoélectrophorétique permet de préciser l'espèce animale d'où la poudre provient et montre la présence de plusieurs antigènes organospécifiques parmi lesquels il faut notamment citer ceux correspondant à la trypsine, la chymotrypsine et l'amylase.

II. — FACTEUR INTRINSEQUE :

1. - Electrophorèse :

Dans nos conditions opératoires, l'électrophorèse a séparé les constituants principaux des facteurs intrinsèques N.F. et 2727 en solution (à 25 mg/ml) en plusieurs zones distinctes qui peuvent être caractérisées par la vitesse de migration électrophorétique et par leur nature chimique.

Les limites sont fixées conventionnellement comme suit :
zone I (cathodique) qui donne la réaction des glucoprotéines.
zone II (centrale) où peut être localisée la réaction estérasique.
zone III (anodique) qui présente la réaction des protéines et des glucoprotéines.

La recherche des activités lacticodéshydrogénasique, catalasique et protéolytique s'est révélée négative.

Lorsqu'on soumet à l'électrophorèse les préparations de facteur intrinsèque N.F. et 2727 en présence de vitamine B12 on constate :
1. que la vitamine B12 migre vers l'anode contrairement à la vitamine B12 d'une lame témoin sans facteur intrinsèque qui migre vers la cathode.
2. que la vitamine B12 liée peut être localisée dans la zone de migration des β globulines.

2. - Analyse immunoélectrophorétique :

L'analyse immunoélectrophorétique a été réalisée sur le facteur intrinsèque N.F. (à 25 mg/ml) à l'aide de l'immunsérum anti N.F. de l'immunsérum anti N.F. épuisé par le sérum de porc et de l'immunsérum anti N.F. épuisé par le sérum de porc et un extrait mixte d'organes de porc. Nous avons analysé de la même manière les facteurs intrinsèques N.F. et 2727 à l'aide des immunsérums anti 2727, tels quels et épuisés.

Les immunsérums non épuisés ont révélé dans les facteurs intrinsèques N.F. et 2727, 2 ou 3 arcs de nature protéinique, dont le plus intense apparaît dans la zone des β globulines. (Ex. voir fig. 3.)

En présence des immunsérums épuisés par le sérum de porc, subsiste l'arc protéinique de la zone des β globulines. (Ex. voir fig. 3.)

Avec les immunsérums épuisés par le sérum de porc et un extrait mixte d'organes, l'activité estérasique a disparu mais il subsiste dans la zone des β globulines un arc de précipitation colorable par le noir amide. (Ex. voir fig. 3.)

L'analyse immunoélectrophorétique à l'aide des immunsérums épuisés a permis de déceler dans les préparations de facteur intrinsèque (N.F. et 2727) un antigène qui n'est présent ni dans le sérum de porc ni dans la plupart des organes de porc à l'exclusion de l'estomac et de l'intestin ; cet antigène organospécifique est présent dans la zone de migration électrophorétique où se retrouve la vitamine B12 liée après électrophorèse.

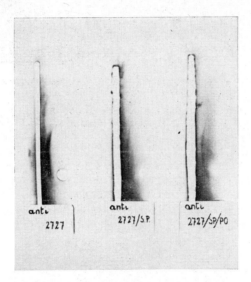

Fig. 3. — Analyse immunoélectrophorétique du facteur intrinsèque 2727 en présence de son immunsérum tel quel (anti 2727) et épuisé (anti 2727/SP et anti 2727/SP/Po.)

Immunoelectrophoretic analysis of the intrinsic factor 2727 in the presence of its « natural » immuneserum (anti 2727) and « inactive » (anti 2727/SP and anti 2727/SP/Po.)

III. — PREPARATIONS HORMONALES :

A. - Lobe postérieur d'hypophyse :

Analyse électrophorétique :

Si nous comparons les extraits acétiques de poudres de lobe postérieur d'hypophyse (de porc) aux extraits aqueux de ces mêmes poudres, nous constatons la présence de protéines et de glucoprotéines dans les deux types d'extraits.

La migration électrophorétique révèle la présence de plusieurs taches distinctes situées, pour la plupart, entre le réservoir et l'anode. — La préparation acétique du commerce, apparemment plus purifiée, n'a toutefois donné qu'une seule zone. (Fig. n° 4.)

Extrait acétique   Extrait aqueux
commerc.   pdre 1   pdre 1   pdre 2

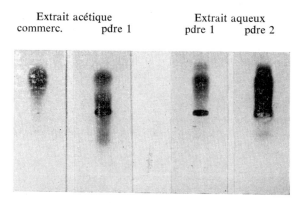

Fig. 4. — Images électrophorétiques de préparations de *lobe postérieur d'hypophyse.* — révélation par le noir amide.

Electrophoresis of preparations of *posterior lobe of hypophysis.* — revelation by amido-black staining.

Aucune activité enzymatique du type « estérase » ou « lacticodéshydrogénase » n'a été décelée dans les extraits acétiques, tandis que les deux types d'activité se retrouvent dans les extraits aqueux de poudres de lobe postérieur d'hypophyse, les zones actives se situant entre le réservoir et l'anode.

Notons que l'extrait aqueux d'une poudre de lobe postérieur d'hypophyse de *bœuf* s'est également montré actif du point de vue « estérases » et « lacticodeshydrogenases » mais en donnant une image électrophorétique légèrement différente de celles données par les extraits d'organes de porc : 2 zones anodiques « lacticodeshydrogenases » au lieu de 3 et une zone supplémentaire vers la cathode aussi bien pour l'activité estérasique que pour l'activité lacticodéshydrogénasique.

Aucun extrait n'a manifesté d'activité du type « phosphatase ».

Nous constatons ainsi, que si les préparations type BP 63 (extraits acétiques) de lobe postérieur d'hypophyse présentent une nature protéinique analogue à celle de la poudre d'origine, elles en ont toutefois perdu les propriétés enzymatiques.

Analyse immunoélectrophorétique (voir fig. n° 5)

Nous n'observons, dans les extraits acétiques que peu ou pas (préparation commercialisée) de substances précipitant avec l'immunsérum préparé vis-à-vis du sérum de porc (espèce animale d'origine). Ces substances sont, par contre plus nombreuses dans les extraits aqueux.

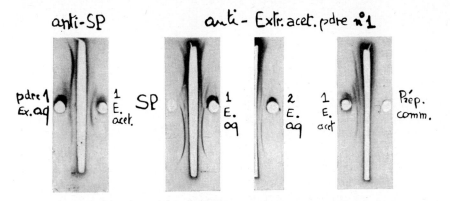

Fig. 5. — Images immunoélectrophorétiques de préparations de *lobe postérieur d'hypophyse* (extraits aqueux et acétiques) - révélation par l'immunsérum anti-sérum de porc (anti-S.P.) et l'immunsérum anti-extrait acétique de la poudre n° (1) (anti-Extr. acet. pdre n° 1)

Immunoelectrophoresis of preparations of *posterior lobe of hypophysis* (aqueous and acetic acid extracts) precipitation with anti-pig immuneserum (anti-SP) and anti-acetic acid extract of the powder (1) (anti-Extr. acet. pdre n° (1).

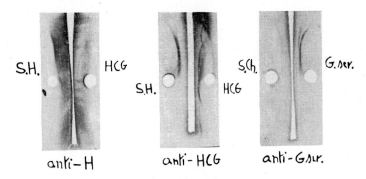

Fig. 6. — Images immunoélectrophorétiques de *gonadotrophines*, sérum humain et sérum de cheval.

HCG :	gonadotrophine chorionique
G. sér. :	gonadotrophine sérique
S.H. :	sérum humain
S.Ch. :	sérum de cheval
anti-H:	immunsérum anti-sérum humain
anti-HCG :	immunsérum anti-gonadotrophine chorionique
anti-G. sér. :	immunsérum anti-gonadotrophine sérique.

Immunoelectrophoresis of *gonadotropins*, human serum and horse serum.

HCG :	chorionic gonadotropin
G. sér. :	seric gonadotropin
SH :	human serum
S.Ch. :	horse serum
anti-H :	anti-human immuneserum
anti-HCG :	anti-chorionic gonadotropin immuneserum
anti-G. sér. :	anti-seric gonadotropin immuneserum

Notre immunsérum anti-extrait acétique de lobe postérieur hypophysaire a donné vis-à-vis des différents extraits 4 à 5 arcs de précipitations répartis tout au long de la lame. Notons que l'extrait acétique ayant servi à préparer le lapin ne donne pas l'arc extrême situé vers la cathode. Notons aussi que la préparation commercialisée n'a pas précipité avec cet immunsérum.

Jusqu'à présent, nous n'avons pas réussi à obtenir un immunsérum suffisamment riche que pour révéler des antigènes organospécifiques après épuisement par le sérum de l'espèce.

B. - **Gonadotrophines :**

L'analyse électrophorétique (réalisée sur des solutions à 5000 U/ml) confirme qu'il s'agit de préparations très purifiées.

Dans nos conditions opératoires, nous n'avons en effet décelé dans les gonadotrophines qu'un nombre réduit de constituants protéiniques et aucune activité des types « estérase » « phosphatase » ou « lacticodeshydrogénase », activités qui auraient pu constituer des impuretés provenant du matériel de départ.

Les colorations des polysaccharides et des glucoprotéines ont donné de faibles taches colorées dans le cas de la gonadotrophine sérique et des traces de coloration dans le cas de la gonadotrophine chorionique.

La coloration spécifique des protéines donne également de très faibles taches dans les deux cas, mais la réaction la plus faible était ici celle de la gonadotrophine sérique. Cette dernière semble donc encore plus purifiée que la gonadotrophine chorionique.

Analyse immunoélectrophorétique :

1) *Recherche des antigènes d'espèce :*

La gonadotrophine chorionique, d'origine humaine, a donné un arc de précipitation vis-à-vis de l'immunsérum anti-sérum humain (fig. n° 6).

La gonadotrophine sérique, extraite du sérum de cheval, n'a donné, par contre, aucune réaction vis-à-vis de l'immunsérum anti-sérum de cheval alors que nous décelions parfois de faibles réactions par la technique d'immunodiffusion simple. Il s'agit en effet d'un produit dont la purification vise à écarter les antigènes d'espèce animale, ceux-ci pouvant entraîner des accidents graves chez l'homme lors de l'utilisation thérapeutique répétée.

Quoique en faible proportion, ces antigènes d'espèce se sont révélés capables d'induire chez le lapin, la production d'anticorps précipitant les antigènes sériques.

En effet, l'immunsérum anti-gonadotrophine chorionique donne vis-à-vis du sérum humain un arc de précipitation, souvent double dans la zone des α globulines, et l'immunsérum anti-gonadotrophine sérique donne vis-à-vis du sérum de cheval un arc de précipitation situé entre le réservoir et l'anode (figure n° 6).

2) *Recherche des antigènes spécifiques des gonadotrophines* (voir fig. n° 7)

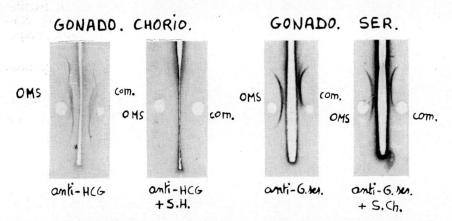

Fig. 7. — Images immunoélectrophorétiques de *gonadotrophines* O.M.S. et commer-cialisée (com.) - révélation par les immunsérums anti-gonadotrophine chorionique « tel quel » (anti-HCG) ou « épuisé » (anti-HCG + S.H.) et les immunsérums anti-gonado-trophine sérique « tel quel » (anti-G. sér.) ou « épuisé » (anti-G. sér. + S.Ch.).

Immunoelectrophoresis of World Health Organization standard (O.M.S.) and com-mercial (com.) *gonadotropins* - precipitation with anti-chorionic gonadotropin immune-serums, pure (anti-HCG) or mixed with human serum (anti-HCG + SH) and anti-seric gonadotropin immuneserums, pure (anti-G. sér.) or mixed with horse serum (anti-G. sér. + S.Ch.).

Dans le cas de la *gonadotrophine chorionique,* nous observons les résultats suivants :

a) *vis-à-vis de l'immunsérum anti-gonadotrophine chorionique* « tel quel » :

3 ou 4 arcs de précipitation répartis dans les zones α et β globulines du sérum humain.

b) *vis-à-vis de l'immunsérum anti-gonadotrophine chorionique préala-blement épuisé avec du sérum humain :*

1, 2 ou même 3 arcs de précipitation, suivant les gonadotrophines et suivant les immunsérums. L'arc situé vers le pôle positif et proche du réservoir est le plus intense et se retrouve dans tous les cas.

Un deuxième arc plus proche de l'anode est assez fréquent également, tandis que le troisième arc, situé dans la zone des β globulines, est plus rare et plus atténué.

Notons que le standard O.M.S. de référence a parfois donné ces 3 arcs mais que, le plus souvent, l'image ne comportait que les deux pre-miers arcs de précipitation.

Nous voyons donc (et le phénomène a été signalé par plusieurs au-teurs) que même vis-à-vis de l'immunsérum épuisé, les préparations de gonadotrophine chorionique et notamment le standard O.M.S. donnent plusieurs arcs de précipitation, ce qui témoigne de la complexité de ce produit. (11, 12, 13, 14, 15, 16, 17.)

De toute manière, l'essai immunoélectrophorétique, conduit parallèlement sur une préparation inconnue et sur le standard O.M.S. de référence, apportera des indications précises quant à l'identité des deux préparations et quant au degré de complexité de leur composition.

Dans le cas de la *gonadotrophine sérique,* nous observons les résultats suivants :

a) *vis-à-vis de l'immunsérum anti-gonadotrophine sérique « tel quel ».*

De 1 à 5 arcs de précipitation suivant les gonadotrophines et suivant les immunsérums.

Sauf un, ces arcs sont tous situés entre le réservoir et l'anode.

b) *vis-à-vis de l'immunsérum anti-gonadotrophine sérique préalablement épuisé avec du sérum de cheval :*

1 ou 2 arcs de précipitation dont l'un, dans certains cas, est formé de 2 légères courbures.

Ces arcs sont situés vers l'anode, le plus marqué étant celui qui touche le réservoir.

Remarquons que dans ce cas, la gonadotrophine standard O.M.S. ne donne le plus souvent qu'un seul arc de précipitation, tant vis-à-vis de l'immunsérum « tel quel » que vis-à-vis de l'immunsérum épuisé. Ce standard de gonadotrophine sérique serait donc plus purifié que le standard de gonadotrophine chorionique. On peut se demander si on ne serait pas en droit d'exiger le même degré de pureté pour les gonadotrophines sériques à usage thérapeutique, compte tenu de leur origine animale, ces antigènes d'espèce étrangère pouvant être la cause d'accidents anaphylactiques chez l'homme lors d'injections répétées.

Notons que pour pouvoir juger plus exactement du degré de pureté de préparations injectables d'origine tissulaire, il faudrait disposer d'immunsérums préparés au moyen d'extraits bruts qui renfermeraient, à un taux suffisant, des anticorps permettant de déceler les différents constituants de ces extraits autres que l'hormone elle-même. Ainsi que l'ont signalé différents auteurs, un tel essai de pureté s'avère extrêmement utile pour suivre les étapes d'une purification progressive et garantir une composition constante du produit final en substances protéiques ou plus généralement en substances sérologiquement identifiables.

De surcroît, nous avons pû vérifier qu'aucune réaction « croisée » ne s'observe entre un type donné de gonadotrophine et l'antisérum préparé vis-à-vis de l'autre type.

DISCUSSION ET CONCLUSIONS

Il ressort de l'ensemble des résultats qui précèdent que l'analyse électrophorétique et plus encore l'analyse immunoélectrophorétique, présentent un intérêt certain tant au point de vue de l'identification de préparations protéiniques à usage thérapeutique que de la détection d'impuretés.

L'électrophorèse simple permet déjà la séparation des principaux constituants protéiniques du matériel examiné qui peuvent être caractérisés par leur vitesse de migration et leur nature chimique ou enzymatique. L'ensemble des éléments recueillis permet d'établir ce qu'on pourrait appeler une « fiche signalétique globale » des matériaux examinés. Cette fiche signalétique comparée à celle de préparations témoins autorise déjà une première caractérisation des préparations expertisées. Bien entendu, cette identification ne se substitue pas dans le cas de la pancréatine aux mesures quantitatives prescrites dans les textes officiels mais les complète ; elle apporte surtout des renseignements utiles au point de vue qualitatif.

Au point de vue de la pureté, ces techniques peuvent déceler la présence dans les préparations expertisées de constituants protéiniques étrangers, et éventuellement de ceux présentant une activité enzymatique différente de celle justifiant l'utilisation thérapeutique de la préparation examinée.

Notons que ce type d'essai de pureté s'avère particulièrement utile dans le cas d'enzymes purifiés tels par exemple : la trypsine ou la chymotrypsine (18 - 19).

L'analyse immunoélectrophorétique complète l'électrophorèse. Sur le plan de l'identification, elle permet tout d'abord de mieux préciser la mobilité des différents constituants.

De plus, par l'utilisation d'immunsérums spécifiques, il est possible, dans certains cas, de déterminer l'origine animale et même l'origine tissulaire du matériel analysé et de voir s'il est conforme aux prescriptions des pharmacopées.

Ainsi l'analyse immunologique permet, dans le cas de la pancréatine, de rechercher l'origine porc ou bœuf de la préparation (exigence de la PB de rechercher l'origine porc ou bœuf de la préparation (exigence de la pB IV) et de vérifier qu'il s'agit bien de poudre de pancréas ; de même, dans le cas du facteur intrinsèque, elle permet de vérifier qu'il s'agit d'estomac ou d'intestin (de porc) (exigence du N.F.).

Dans le cas des gonadotrophines, il est possible de distinguer sans équivoque entre la gonadotrophine chorionique extraite de l'urine de femme enceinte et la gonadotrophine sérique préparée à partir du sérum de jument gravide.

En tant qu'épreuves de pureté les méthodes immunoélectrophorétiques s'avèrent particulièrement intéressantes dans le cas de préparations injectables d'origine non humaine dans lesquelles la persistance de protéines animales encore marquées de la spécificité d'espèce et de surcroît dénuées de tout intérêt thérapeutique est à redouter en raison principalement de leur caractère immunogène.

*_**

Nous tenons à remercier MM. S. FEJES, C. HOREMANS et G. MAUQUOY pour leur collaboration technique.

Bibliographie

(1) J. DONY. *J. Pharm. Belgique,* nᵒˢ 1 - 2 - 3, 1967.

(2) J. DONY, H. MUYLDERMANS. *J. Pharm. Belgique,* nᵒˢ 9 - 10, 1968.

(3) J. DONY, R. BONTINCK, H. MUYLDERMANS. Symposium IV : Chromatographie et Electrophorèse, Bruxelles, 1966.

(4) P. GRABAR et P. BURTIN. Analyse Immuno-Electrophorétique, Paris, Masson et Cie, 1960.

(5) J. URIEL. *Ann. Inst. Pasteur,* **101** : 104, 1961.

(6) J. URIEL. *Nature,* **188** : 853, 1960.

(7) J. URIEL. *Ann. N.Y. Acad. Sc.* **103** : 956, 1963.

(8) WIEME. « Agar Gel Electrophoresis », Amsterdam. Elsevier, 1965.

(9) J.J. SCHEIDEGGER. *Intern. Arch. Allergy Appl. Immunol.* **7** : 103, 1955.

(10) A. RAPPE. *Journ. Pharm. Belg.* nᵒˢ 7 - 8 : 343-380.

(11) S.S. RAO, S.K. SHAHANI. *Immunology,* **4** : 1, 1961.

(12) S. BRODY, G. CARLSTROM. *Acta Endocrin,* **42** : 485, 1963.

(13) GONADOTROPINS : Physicochemical and Immunological Properties, Ciba Foundation, 1965.

(14) H. VAN HELL, B.C. GOVERDE, A.H. SCHUURS, E. DE JAGER, R. MATTHYSEN, J.D. HOMAN. *Nature,* **261**, 1966.

(15) B.B. SAXENA, P. RATHMAN. *J. Biol. Chem.* **242** : 3769, 1967.

(16) J. HEUSE-HENRY, Cl. ROBYN, J.M. LIMBOSCH, P.O. HUBINONT. *Bull. Soc. Royale Belge Gynécol. Obstétrique,* **32** : 515, 1962.

(17) C. ROBYN. *Rev. Belge Pathol. Méd. Expérim.* **31** : 5 - 6, 334, 1965.

(18) J. DONY, H. MUYLDERMANS. Proc. of XI Intern. Congress of Microbiol. Standardization. Milan, 1968.

(19) 4th Report of Int. Com. for the Stand. of Pharm. Enz. J.M. Pharm. (à paraître).

(20) J. URIEL, S. AVRAMEAS. *Ann. Int. Pasteur.* **106** : 396, 1964.

Detection and determination of naphthalenesulfonic acid mixtures by thin-layer and paper chromatography

Dr. Carlo PRANDI

A.C.N.A., Centro Ricerche Materie Coloranti, Rico, Cesano Maderno, Milano.

Summary

Naphthalenesulfonic acids have a paramount importance in dye industry and their analysis by traditional methods is particularly time-consuming.

Thin layer and paper chromatography enables a rapid separation and determination with a good reproducibility of mixtures of mono-, di- and tri-sulfonic acids.

The primary factors influencing the chromatography separation are discussed.

I. — INTRODUCTION.

Naphthalenesulfonic acids have a paramount importance, in general, as intermediates in the aromatic compound chemistry and, in particular, in the dye and dyeing-assistant chemistry.

Their synthesis occurs through the sulfonation of naphthalene, mono- and polysulfonated isomer mixtures being obtained, the composition of which is dependent upon the variables involved in the preparation.

It would be particularly useful, therefore, that enough fast and precise detection and determination methods, be provided to enable :
— to study the influence of each variable ;
— to establish the optimum conditions for obtaining a required type and degree of sulfonation ;
— to control the course and the reproducibility of the manufacturing process, once it has been set up.

Since the sulfonation product analyses, by any eventual chemical method are particularly painstaking and timeconsuming, it has been thought to have resort to chromatographic analysis.

Techniques suitable only for the detection are reported (1) (4) (5) (6). This study enables :

— to improve the previous detection techniques by introducing the use of thin cellulose layers, so as to afford a more rapid, selective and sensitive analysis. Thus, rapid qualitative controls of the sulfonation masses and semi-quantitative evaluations of the impurities, present in reasonable amounts, are made possible ;

— to set up quantitative analysis techniques, by using chromatography along with spectrophotometry, which allow to carry out the determination of the main products and of the sulfonated by-products present in a remarkable amount.

The application of these techniques has enabled to face, by means of new and more precise investigation methods, a great many studies on the naphthalene sulfonation, thus, making easier the solution of the above problems.

This investigation represents only an example of the use of paper and thin layer chromatography that the Author makes in his laboratory, for the detection and determination of complex mixtures, difficult to analyse by chemical route.

II. — EXPERIMENTAL.

2.1.) Adsorbents.

— Cellulose MN 300 (Macherey Nagel), in the qualitative analysis ;
— Whatman 3 MM paper, in the quantitative analysis.

2.2) Eluents.

Aqueous Mg $SO_4.7H_2O$ R.P. (Carlo Erba) solutions of increasing strength.

2.3) Reagents.

Pinakryptol Yellow (photography grade) (Fluka A.G. Buchs S.G.).

2.4) Equipment.

— 10 μl graduated micropipette for qualitative and semi-quantitative analyses ;
— microsyringe for the quantitative analysis ;
— U.V. light lamp for the detection of chromatograms ;
— U.V. spectrophotometer for the quantitative analysis.

2.5) Qualitative analysis.

Thin layers of Cellulose MN 300, prepared according to the Macherey-Nagel's instructions, are used. The amount of tested product to be spotted on the chromatogram could range from 40 to 100 μg.

The elution is carried out in thin layer chromatography chambers by means of :

$$MgSO_4.7H_2O — water (55\text{-}100) \text{ wt/wt}$$

Once the elution is completed and the chromatograms oven dried at 70° C the bands are detected by the following techniques, previously reported in the literature :

— fluorescence and phosphorescence by direct exposition to U.V. light (λ 254 mμ) ;

— treatment wit Pinakryptol yellow followed by exposition to U.V. light (λ 254 and 350 mμ).

It is in our opinion that the latter technique is to be preferred in that it enables to obtain differently colored bands ; this makes easier the identification of impurities present in the tested samples.

In the chromatography of mixtures containing a preponderant amount of one product, the detection limit is 0.4 μg.

Table I shows an example of the separation of naphthalene- mono-, di- and tri-sulfonic acids.

This type of analysis allows a rapid identification of the composition of the sulfonation mass as such without previous treatment, and a qualitative evaluation of the purity of a finished product to be performed.

TABLE I

Rf	Chromato-gram sketch	Coloration under U.V. light λ 254 mμ.	Coloration after treatment with Pinakryptol Yellow	
			λ 254 mμ.	λ 350 mμ.
0.9		Yellow and		
0.8	1.3.5 NTS	brown fluo-	violet red	dark brown yellow
0.7	1.3.6 NTS	rescent	brown yellow	dark brown yellow
0.6	1.5 NDS	bands and	light yellow	light yellow
0.5	1.6 NDS	yellow and green phosphorescent	violet blue	brown violet
0.4	2.6 + 2.7 NDS	bands	ligt yellow	light yellow
0.3	Ac. 1 NS		pink	orange
0.2	Ac. 2 NS		orange yellow	orange yellow
0.1				

NS = Naphthalene-sulfonic acid.
NDS = Naphthalene-disulfonic acid.
NTS = Naphthalene-trisulfonic acid.

2.6) Semi-quantitative analysis.

The procedure is the same as for qualitative analysis ; and it is applicable to the identified impurities present in strengths ranging from 0.5 to 5 %.

The semi-quantitative evaluation is based on an approximate evaluation of the size and color intensity of the tested impurity spots, in comparison with standard-solution spots, on the same chromatogram.

2.7) Quantitative analysis.

The various steps of the quantitative analysis are the following :

2.7.1) Quantitative separation of the various components of the tested sample by paper chromatography.

This operation is carried out by spotting on Whatman Paper 3 MM a known amount of the tested product ($\sim$ 2 mg/200 ml) as a streak 11-12 cm long and by eluting with :

— $MgSO_4.7H_2O$: H_2O (30 : 100) wt/wt in the monosulfonic acid strength determination ;
— $MgSO_4.7H_2O$: H_2O (50 : 100) wt/wt, in the disulfonic and trisulfonic acids strength determination.

2.7.2) Extraction of the various components from the chromatogram.

The position of the various bands is established by irradiating the chromatogram itself by U.V. light (λ 254 mμ), the band contour being marked.

The zones corresponding to the various bands are cut out, chopped up and extracted with hot water, followed by filtration, collection of the filtrate in volumetric flakes and dilution to the mark.

2.7.3) Spectrophotometric determination.

The acids in the chromatogram extracts are determined by means of U.V. absorption spectrophotometry.

The wavelength at which readings were made for the various sulfonic acids are shown in table 2.

TABLE II

Product	Wavelength
1. naphthalenesulfonic acid	222.5 mμ
2. naphthalenesulfonic acid	226,5 mμ
1.6. naphthalenedisulfonic acid	231 mμ
1.5. naphthalenedisulfonic acid	225.5 mμ
1.3.6. naphthalenetrisulfonic acid	231.5 mμ
1.3.5. naphthalenetrisulfonic acid	235 mμ

This technique enables to determine the strength of the main product and of a high impurity content (from 5 to 20 %).

2.7.4) **The measurement of precision.**

Table 3 shows the variances and standard deviations obtained from the whole of tests carried out in the investigation.

These data are reported as two different strength levels and precisely :
        80-100 % (main product)
          5- 20 % (impurity).

The same table shows the 95 % confidence limits, calculated on the assumption that the analysis will be effected thrice.

TABLE III

Naphthalene sulfonic acids	80 - 100 % strength levels			
	No of tests	Variance	Standard deviation	Confidence limits of the analysis
1	16	0.500	± 0.707	± 0,90
2	20	0.482	± 0.694	± 0.85
1.5	21	0.303	± 0.550	± 0.65
1.3.6	19	0.422	± 0.649	± 0.80
	5 - 20 % strength levels			
	No of tests	Variance	Standard deviation	Confidence limits of the analysis
1	25	0.205	± 0.452	± 0.55
2	28	0.0826	± 0.287	± 0.35
1.3.5	22	0.0537	± 0.231	± 0.30

III. — DISCUSSION.

It is believed that the described chromatographic technique couldn't be strictly included in any of the three chromatographic classes ; i.e. adsorption, partition or ion-exchange chromatography.

In this case, the chromatographic behaviour of the tested products could be easily explained by assuming a particular type of partition chromatography, where the Rf does not depend on a solubility ratio, but probably on the solubility of the Mg salt of the sulfonated compound in the eluent, consisting in an aqueous magnesium sulfate solution.

In its turn, the solubility of the magnesium salt of the sulfonated product would depent upon :

— the salt concentration of the eluent ;
— the degree of sulfonation of naphthalene ;
— the position of the sulfonic acid groups.

3.1) Influence of the salt concentration of the eluent.

The solubility of the naphthalenesulfonic acids and, therefore, also their Rfs are remarkably affected by the salt concentration of the eluent.

The curves, shown in fig. 1 and obtained by following the conditions described in 2.5), point out clearly that the Rfs of the said products (1-Naphthalene-sulfonic, 1.6 naphthalenedisulfonic and 1.3.5-naphthalenetrisulfonic acids) decrease as the salt concentration of the eluent increases.

Further, the increase in the magnesium sulfate concentration was noticed to determine an increase in the time necessary to the front (Fs) to flow through a certain distance.

Finally, one can see that by using distilled water as eluent, the Rfs of the tested products are hardly different from each other and are very close to 1.

This fact leads to the exclusion of a sensible presence of adsorption and ion-exchange phenomena and corroborates, as a fundamental factor for the separation, the hypothesis previous ly put forward about the solubility of the magnesium salt of the tested product in the eluent.

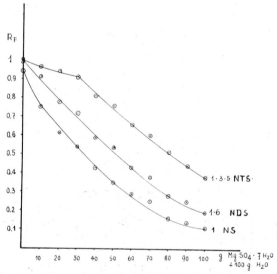

Fig. 1. — Dependence of the Rfs on the eluent strength.

3.2) Influence of the degree of sulfonation of naphthalene.

By using salt solutions of the same strength, the solubility of the tested sample increases as the number of sulfonic acid groups present in the naphthalene nucleus increases.

The Rfs of monosulfonic acids, therefore, will be smaller than those of disulfonic acids, and these, in their turn, smaller than those of trisulfonic acids, as it happens in effects.

3.3) Influence of the position occupied by the sulfonic acid group.

With the eluents, β-sulfonic acids give magnesium salt which are less soluble than those of the corresponding α-sulfonic acids.

This is partially corroborated also by some solubility values given in the literature and shown in table 4.

TABLE IV

Magnesium salt of the sulfonated product	Solubility of the dry product (g/100 ml H₂O)	
2	0.223	(2)
1	6.9	(2)
1.3.6	39.1	(3)
1.3.5	46.5	(3)

Consequently, it will happen that, with the same eluent strength and sulfonation degree, the Rf will decrease with the increase in number of sulfonic acid groups present in Beta-positions in the molecule.

3.4) Conclusions.

The hypotheses put forward in 3), 3.2) and 3.3), about the factors affecting the chromatographic separation, are experimentally corroborated by the example of separation shown in table 1.

In the latter one can see that :

— the Rf increases with the increase in sulfonation degree ;
— among naphthalene monosulfonic acids, the 2-sulfonic acid shows a smaller Rf than of 1-sulfonic acid ;
— among naphthalenedisulfonic acids, the 2.6- and 2.7-derivatives (2 sulfonic acid groups in Beta-position) show Rfs nearly equal to each other and smaller than that of the 1.6-derivative (1 sulfonic acid group in Beta-position) ; the Rf of the latter being smaller than that of the 1.3.5-derivative (1 sulfonic acid group in Beta-position) ;
— among naphthalenetrisulfonic acids, the 1.3.6-derivative (2 sulfonic acid groups in Beta-position) shows an Rf smaller than that of the 1.3.5-derivative (sulfonic acid group in Beta-position) ;

Acknowledgment.

The Author thanks the Aziende Coloranti Nazionali Affini — A.C.N.A., S.p.A., which has permitted the publication of this paper.

References

(1) G.V. HORNUFF, E. WIENMAUS: *Journal of Chromatography,* **8**: 90-95, 1962.

(2) EPHRAIM, SEGER: *Helvetica Chimica Acta,* **8**: 724 - 739, 1925.

(3) A.A.S. PRYSKOV, Trudy IVANOVSK: *Khim. Tekhnol. Inst.,* **8**: 12 - 13, 1958: C.A. **54**: 21939/d, 1960.

(4) J. LATINAK: *Collection Czechslov. Chem. Communs.,* **25**: 1649 - 1655, 1960.

(5) J. LATINAK: *Mikrochim. Acta;* 350, 1966.

(6) J. BORRY: *Journal of Chromatography* **2**: 612 - 614, 1959.

Identification des spans® et tweens® dans des poudres pour pâtisserie par chromatographie sur gel de silice silanisé

par

J. GOSSELE, S. SREBRNIK et C. CHARON

Institut d'Hygiène et d'Epidémiologie, Bruxelles.

I. INTRODUCTION

Dans le cadre de nos analyses de routine, nous avons reçu à analyser des poudres pour pâtisserie importées d'Amérique. Leur composition était la suivante : Sucre, farine blanchie, matières grasses avec des mono- et diglycérides et des agents humectants, sucre candi, cacao traité aux alcalis, levure, poudre de lait écrémé, sel, amidon modifié, glucose, monoesters de propylène glycol, cellulose, protéine de soya, monostéarate de sorbitane (= SPAN®), gomme cellulosique, gomme végétale, polysorbate 60 (= TWEEN®60), carragénine et des aromatisants artificiels.

Le Span et le Tween sont autorisés en Amérique, mais non en Belgique.

Le Span® est le nom commercial des esters d'acides gras et d'anhydrosorbitol. (2)

Span 20 = monolaurate de sorbitane
Span 40 = monopalmitate de sorbitane
Span 60 = monostéarate de sorbitane
Span 65 = tristéarate de sorbitane
Span 80 = monooléate de sorbitane
Span 85 = trioléate de sorbitane

En fait, ces produits ne représentent pas à 100 % l'acide gras qui est repris dans le nom. Selon des analyses de J. Cerdas et collaborateurs (4), le Span 20 contient 42 % d'acide laurique, 28 % d'acide myristique, 20 % d'acide palmitique, 9 % d'acide oléique, et 1 % d'acide caprique.

Le Span 40 contient 3 % d'acide myristique, 96 % d'acide palmitique et 1 % d'acide stéarique. Le Span 60 respectivement 2, 58 et 40 % des mêmes acides.

Les Spans forment des émulsions du type Eau dans l'Huile avec des H.L.B. (Balance hydrophile et lypophile) de 3 à 8.

Sorbitol

CH$_2$OH
H- C-OH
HO-C-H
H- C-OH
H- C-OH
CH$_2$OH

$-$ H$_2$O

1,5 anhydrosorbitol

+ HOOC-R → $-$ H$_2$O

1,4 anhydrosorbitol

+ HOOC-R → $-$ H$_2$O

$-$H$_2$O

1,4,3,6 dianhydro-sorbitol

+ HOOC-R → $-$ H$_2$O

SPANS

En introduisant des chaînes polyoxyéthylène, on obtient un autre groupe d'émulsifs du type Huile dans l'eau, avec les H.L.B. de 8 à 16. Ces produits sont connus sous le nom de TWEEN®.

Nous avons essayé la méthode de Wetterau (9) mise au point pour le dosage de monostéarate de sorbitane dans des poudres pour pâtisserie. Après extraction du Span par l'éthanol, on remplace l'éthanol par de l'heptane par une distillation azéotropique. L'extrait est ensuite passé sur colonne de gel de silice pour éliminer les triglycérides qui n'étaient pas encore précipités par l'éthanol, ainsi que d'autres impuretés. Après saponification de l'éluat, les acides gras sont éliminés sur échangeurs d'ions. La solution désionisée contient des polyols. Cette solution est analysée par chromatographie sur papier ou en phase gazeuse.

M. J. Hall (5) a identifié et dosé le Tween 80 dans de la pâtisserie par gravimétrie. Le tween 80 est saponifié, la partie glycol est extraite par du chloroforme, les acides gras sont éliminés à l'éther de pétrole. Ensuite le glycol est précipité par l'acide phosphomolybdique.

P. Barcklow (1) a identifié et dosé le Tween 80 dans des « pickles ». Il précipite les polyols par l'acide silicotungstique et complète son identification par spectrométrie I.R. du complexe.

La longueur de ces méthodes nous a incité à essayer une méthode de purification et de détection des deux additifs par chromatographie en couche mince.

E. Kröller (7) décrit une méthode élégante pour détecter les Spans dans des produits de confiserie : fondant, nougat, etc...

L'extrait benzénique des produits de confiserie est déposé sur une plaque de gel de silice G. La plaque est divisée en bandes en forme de triangle. Après développement de la plaque dans un mélange benzène-méthanol (80-20), les taches de Span sont localisées par pulvérisation d'une solution de tétraacétate de plomb. On obtient des taches blanches sur fond brun.

E. Kröller (6) décrit aussi une méthode pour l'isolement et la détection du Tween dans les limonades et des produits de base pour limonades. Il préfère la chromatographie sur papier en phase inversée : papier Schl. & Sch. 20436 Mgl imprégné de diméthylformamide. L'extrait est chromatographié avec un mélange de chloroforme, acétate d'éthyle et heptane (60-20-20). Après séchage pendant 2 à 3 heures à 100°C, le chromatogramme est pulvérisé avec le réactif modifié de Dragendorff. On obtient des taches rouge-orangé sur fond jaune.

R. Suffis et collaborateurs (8) ont analysé plusieurs surfactifs, après préparation de leurs dérivés T.M.S., par chromatographie en phase gazeuse, entre autres l'Arlacel, qui est un ester du sorbitane d'une qualité non alimentaire.

II. ETUDE PRELIMINAIRE

Examen chromatographique

a) *Chromatographie sur papier*

La chromatographie en phase inversée reste difficile à cause de la variation du taux d'imprégnation du papier.

b) *Chromatographie sur couche mince*

Nous avons essayé de séparer sur une même plaque de gel de silice G le Span et le Tween, avec le solvant benzène-méthanol (80-20).

Dans le cas du Span, on obtient quatre bandes, dont deux correspondent au sorbitol (spectre I.R.), et les deux autres donnent un spectre I.R. du Span.

Comme Kröller (6) le mentionne, nous obtenons pour le Tween une série de taches sur des plaques de gel de silice.

P. Breitburd et collaborateurs (3) utilisent un mélange acétone-ammoniaque à 20 % (85-15) pour l'identification des polyoxyéthylèneglycol. Nous avons repris ce solvant dans les mêmes proportions, mais avec de l'ammoniaque à 28 %. Ce solvant donne quatre taches pour le Tween.

Ce solvant est intéressant pour différencier le BRIJ, le MYR et le Tween (photo n° 1), mais aucun de ces deux solvants n'est valable pour séparer le Span et le Tween.

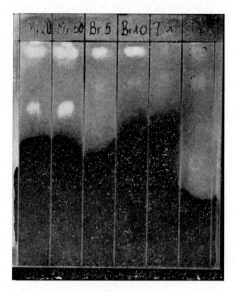

Photo 1.

Dans la pratique, l'extraction des Spans et des Tweens s'effectue à l'acétate d'éthyle au Soxhlet, ce qui entraîne des impuretés, entre autres des mono-, di- et triglycérides, et du propylèneglycolmonostéarate.

Si on dépose l'extrait sur une plaque de gel de silice, nous pouvons éliminer les mono- et diglycérides par deux développements à l'éther. En dessous des graisses, on trace un nouveau front de solvant. Deux autres développements à l'acétone, et deux à l'acétate d'éthyle font migrer le span au front de solvant, tandis que le Tween reste à la ligne de dépôt, le propylèneglycolstéarate à un Rf de 0,40.

Des spectres I.R. ont démontré qu'à la hauteur du Span, il n'y a que du Span, et qu'au point de départ on ne trouve que du Tween. Pour éviter des erreurs de détection du Tween, dues à la présence d'autres impuretés éventuelles, nous avons essayé de faire migrer le Tween afin d'obtenir un Rf caractéristique.

Si nous prenons au lieu de gel de silice ordinaire du gel de silice silanisé, avec le solvant acétone-ammoniaque 28 % (85-15) nous n'obtenons qu'une tache pour le Tween, à un Rf de 0,60. Une plaque à gradient gel de silice G-gel de silice silanisé HF254 démontre l'effet de la silanisation de l'adsorbant sur la séparation du Tween (photo n° 2).

Une plaque à gradient gel de silice G - Kieselgur (photo n° 3) nous donne à peu près la même image, seulement sur Kieselgur le Tween est au front du solvant.

En changeant les proportions acétone-ammoniaque, nous avons eu des traînées sur Kieselgur.

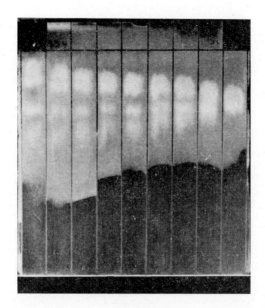

Photo 2.

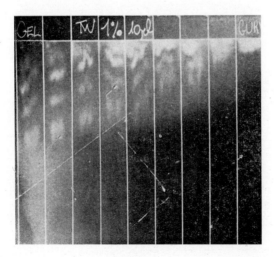

Photo 3.

C'est ainsi que nous avons continué nos essais, aussi bien pour le Span que pour le Tween, sur gel de silice silanisé.

c) *chromatographie gazeuse*

Cette technique est intéressante pour le Span, qui peut être chromatographié après préparation d'un dérivé silanisé. Le Tween donne, après hydrolyse, les acides gras qui ne nous apportent, comme l'a démontré J. Cerdas (4), que très peu d'information.

d) *Spectrométrie infrarouge*

La spectrométrie I.R. est une technique valable pour donner des informations complémentaires dans le domaine de l'analyse des émulsifs.

Nous avons utilisé le chloroforme comme désorbant d'une plaque silanisée, et l'alcool méthylique comme désorbant d'une plaque non silanisée. Après développement dans l'acétone-ammoniaque 28 % (85-15), le spectre du Tween isolé de la plaque est légèrement modifié.

La photo n° 4 montre les spectres du Span après désorption d'une plaque de gel de silice et de gel de silice silanisé, en comparaison avec un Span témoin.

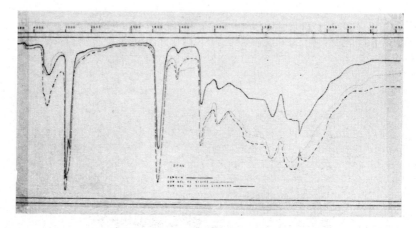

Photo 4.

La photo n° 5 montre les spectres du Tween déposé sur gel de silice, sans et après chromatographie avec le solvant acétone-ammoniaque 28 % (85-15) en comparaison avec un Tween témoin.

La photo n° 6 montre les spectres du Tween déposé sur une plaque de gel de silice silanisé, sans et après chromatographie en comparaison avec un Tween témoin.

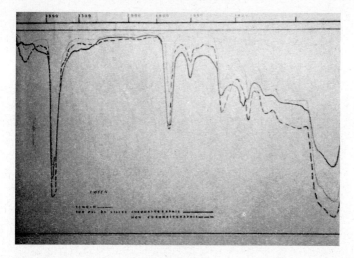

Photo 5.

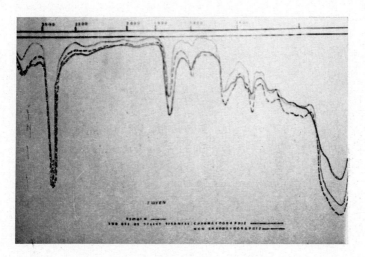

Photo 6.

III. MODE OPERATOIRE

a) *Réactifs*

    — acétate d'éthyle p.a.
    — chloroforme p.a.
    — acétone p.a.
    — éther p.a.
    — ammoniaque 28 % p.a.
    — solutions standards à 1 % dans le chloroforme de Span et de Tween.

— Plaques de gel de silice HF$_{254}$ silanisé (Merck n° 7750) agiter vigoureusement 30 g de gel de silice HF$_{254}$ silanisé avec 60 ml d'un mélange d'eau et de méthanol (2 + 1), jusqu'à ce que la suspension soit homogène. Ajouter à cette suspension 5 g de gel de silice HF$_{254}$ silanisé et agiter de nouveau. Cette quantité est suffisante à la préparation de 5 plaques. Etaler la suspension sur une épaisseur de 0,25 mm. Laisser sécher à l'air et 30 minutes à 105°C.

— Solvant : acétone-ammoniaque 28 % (85-15).

— Révélateurs.

Pour le Span

Solution mère : 5 g de tétraacétate de plomb sont mis en solution dans 100 ml d'acide acétique glacial.

Solution diluée : au moment de l'emploi, dans le rapport 1-2 en volume.

Pour le Tween (6)

Réactif de Dragendorff modifié selon Kröller.

b) *isolement*

Peser 50 g de poudre et transférer dans une cartouche d'extraction pour Soxhlet. Extraire à l'acétate d'éthyle pendant 6 heures. Dans ces conditions, on obtient la plus grande partie des Spans et Tweens. Une extraction plus longue augmente surtout la quantité d'impuretés. Evaporer la solution d'acétate d'éthyle à sec dans un évaporateur rotatif sous vide partiel à 80°C. Reprendre le résidu dans 5 ml de chloroforme.

Filtrer sur filtre sec.

c) *chromatographie*

Purification et identification

— Déposer ± 20 µl de la solution chloroformique sur une plaque de gel de silice silanisé.

— Développer la plaque deux fois dans l'éther sur une distance de 16 cm.

— Tracer une ligne en dessous des graisses qui sont visibles sous lumière U.V. (254 mµ).

— A ce moment, déposer 5 µl de la solution standard de Tween, et 20 µl de la solution de Span.

— Développer la plaque deux fois dans l'acétone jusqu'au nouveau front de solvant.

— Développer la plaque deux fois dans l'acétate d'éthyle.

Le Span se trouve au front du solvant, et le Tween reste à la ligne de dépôt.

— Ensuite, la plaque est développée dans l'acétone-ammoniaque à 28 % (85-15). Nous obtenons un Rf = 0,60 pour le Tween.

Remarque : Si l'extrait contient en plus du Span et du Tween, du stéarate de propylèneglycol, faire un essai comparatif sur plaque de gel de silice G.

d) *Spectres I.R.*

La désorption est réalisée par 10 ml de chloroforme.

La solution après filtration est évaporée à sec, et le résidu est repris par 0,25 ml de chloroforme pour I.R.

Les spectres sont enregistrés de 2 à 12 μ, sous une épaisseur de 0,49 mm. Comparaison avec une solution témoin à 1 %.

Références

(1) P. BARCKLOW, *J. of A.O.A.C.*, **50** : p. 1265-1268, n° 6, 1967.

(2) P. BERNDT, *Die Pharmacie, Heft* **6** : 359-364, 1965.

(3) P. BREITBURD, *Annales pharmaceutiques françaises,* **24** : 191-199, n° 3, 1967.

(4) J. CERDAS, *Annales pharmaceutiques françaises,* **25** : 553-559, n° 7-8, 1967.

(5) M. J. HALL, *J. of A.O.A.C.,* **47** : 685-688, n° 4, 1964.

(6) E. KROLLER, *Fette-Seifen-Anstrichmittel,* **66** : 583-586, 1964.

(7) E. KROLLER, *Fette-Seifen-Anstrichmittel,* **70** : 119-121, 1968.

(8) R. SUFFIS, *J. Soc. Cosmetic Chemists,* **16** : 783-794, 1965.

(9) F. P. WETTERAU, *J. Of Amer. Oil Chem. Soc.,* **41** : 791-795, n° 12, 1964.

Table des matières - Inhoudsopgave
Table of contents

584

586

Index des auteurs - Author index

SUBJECT INDEX *

* This index has been prepared from the key-words which appear in the texts in order to facilitate reference by the readers. This index does, however, not claim to be exhaustive.

590

Imprimé en Belgique — Printed in Belgium

Impr. E. HEYVAERT & Fils
(s.p.r.l.)
Rue de la Victoire, 102
BRUXELLES 6